International Conference on AI and Big Data Application (AIBDA 2019)

IOP Conference Series: Materials Science and Engineering
Volume 806

Guangzhou, China
20 - 22 December 2019

ISBN: 978-1-7138-3510-3
ISSN: 1757-8981

Printed from e-media with permission by:

Curran Associates, Inc.
57 Morehouse Lane
Red Hook, NY 12571

Some format issues inherent in the e-media version may also appear in this print version.

Institute of Physics
Dirac House, Temple Back
Bristol BS1 6BE UK

Phone: 44 1 17 929 7481
Fax: 44 1 17 920 0979

techtracking@iop.org

Additional copies of this publication are available from:

Curran Associates, Inc.
57 Morehouse Lane
Red Hook, NY 12571 USA
Phone: 845-758-0400
Fax: 845-758-2633
Email: curran@proceedings.com
Web: www.proceedings.com

TABLE OF CONTENTS

IOP Conf. Series: Materials Science and Engineering **806** (2020) 011001 doi:10.1088/1757-899X/806/1/011001

Preface

The IAECST-International Conference on AI and Big Data Application (AIBDA 2019), organized by Asia-Pacific Institute of Innovation and Economics, was held in Guangzhou, China on December 20-22, 2019. AIBDA 2019 linked experts and scholars in the areas of AI and Big Data Application around the world together to share the latest research achievements, confront the existing problems and challenges, and explore the advanced technologies. Everyone enjoyed three productive days.

Mr. Zhijian Li, the chairman of Asia-pacific Innovation Economic Research Institute of Guangdong, was invited to chair AIBDA 2019. And we are honored to have five sophisticated and distinguished experts to serve as our keynote speakers. They are Mr. Huang Yuezhen, President of GRG Banking, Prof. Wei Sun from Sun Yat-sen University, Mr. Wang Hongling, Vice President of CAICT in Guangzhou, Mr. Hongbo Xu, Chairman of Innochain, and Prof. Shengjie Zhao from Tongji University. Thanks to their insightful talks, which formed a lively and heated discussion on site.

Thanks to the great success of AIBDA 2019, we are happy to receive a lot of excellent paper featured in the following topics: Information Recognition, Artificial Intelligence, Big Data Management, Big Data Application, Cloud Computing Theory, Cloud Computing Application and other related areas. We have undergone rigid review process and selected and complied the proceedings of AIBDA 2019 to meet the requirements of international publication. We hope this proceeding can share some valuable research to all the scholars in the field of I and Big Data Application around the world.

We sincerely want to thank our conference chairman, the distinguished keynote speakers, and all attendants, without whom, this conference would not hold successfully. We also want to thank all sponsors for their great supports, and all authors for submitting their research results. We hope this conference can be held better in the future and we would like to see you next year.

The Committee of AIBDA 2019

IOP Conf. Series: Materials Science and Engineering **806** (2020) 011001 doi:10.1088/1757-899X/806/1/011001

Committee member

Chairman of the Seminar
Zhijian Li, Asia-pacific Innovation Economic Research Institute of Guangdong, China

Speaker at the seminar
Yuezhen Huang, Chairman of GRG Banking board, China
Wei Sun, School of electronic and Information Engineering, Sun Yat-sen University, China
Hongling Wang, Guangzhou Branch of China Academy of Information and Communication Technology (CAICT), China
Hongbo Xu, InnoHub Capital, China
Shengjie Zhao, School of Software Engineering, Tongji University, China

Academic Committee
Chengzhi Yuan, University of Rhode Island, USA
Ke-Lin Du, Concordia University, Canada
Juan Fang, Beijing University of Technology, China
Ke Gu, Beijing University of Technology, China
Jiaxin Han, Xi'an Shiyou University, China
Badrul Hisham bin Ahmad, University Teknikal Malaysia Melaka, Malaysia
Yigang He, Wuhan University, China
Omid Mahdi Ebadati E., Kharazmi University, Iran
Zhihong Qian, Jilin University, China
Rahul Saha, Lovely Professional University, India
KING-NING Tu, University of California, Los Angeles, USA
Jim Torresen, University of Oslo, Norway
Kuisheng Wang, Xi'an Shiyou University, China

Peer review statement

All papers published in this volume of *IOP Conference Series: Materials Science and Engineering* have been peer reviewed through processes administered by the proceedings Editors. Reviews were conducted by expert referees to the professional and scientific standards expected of a proceedings journal published by IOP Publishing.

IOP Conf. Series: Materials Science and Engineering **806** (2020) 012001 doi:10.1088/1757-899X/806/1/012001

Research on the application of big data technology in government management

Chen Mingtsung [1*], **He Jiaqi**[2]

[1] Department of Public Administration, Nanfang College of Sun Yat-sen University, Guangzhou, Guangdong, 510000,China

[2] Department of Public Administration, Nanfang College of Sun Yat-sen University, Guangzhou, Guangdong, 510000,China

*Corresponding author's e-mail:306470714@qq.com

Abstract. With the advent of the new technology times, big data technology has been widely used as an important strategy of comprehensive social governance. The development of big data not only represents huge amount of data, advanced information technology and huge business opportunities, but also is an effective method to solve problems, transform the world and explore new ideas and scientific applications in the future. However, big data in economic development does not mean that it can replace all the rational thinking of social problems. The logic of scientific development cannot be annihilated in the mass data. In this era of rapid development of intelligent hardware, an important problem for application developers is how to find the delicate balance between power, coverage, transmission rate and cost. Enterprise organizations can use relevant data and analysis to help them reduce costs, improve efficiency, develop new products and make more informed business decisions.

1. Preface

The launch of "big data research and development plan" by the US government in 2012 is regarded as the most indicative event in entering the era of big data. EU countries have successively opened up government information. In 2012, the UN released the big data report, setting off a global research and application boom. As early as before IBM officially coined the term "big data" in 2010, similar big data and cloud service cases have emerged in succession. With the advent of the society we live in, not only the technology is developed, the information is circulated, the communication between people is more and more close, and the life is more and more convenient. Big data is the product of this high-tech era, which has attracted more and more attention. In recent years, the development of big data has not only entered into commercial trade, but also affected many industries. It is also regarded as a key technology to enhance industrial competition and government efficiency. International research and application of big data have been actively invested. However, the purpose of enterprise management is to pursue profit, which is different from the government's low-cost management efficiency. As Harvard Business Review said, big data applications span financial management, retail channels, disease prevention, health care, energy conservation, transportation, human education and other technologies, and can even extend to social aspects such as disaster relief and humanitarian assistance. The following is a typical case of integrating big data in various fields of social management application, so as to explore the development of big data.

2. Application of big data in social management

2.1. Disease prevention

When it comes to the practice of big data in disease prevention, influenced by big data - the traditional way of tracking the spread of the epidemic in the United States is to collect doctors' visiting data to estimate. Because the general public feel sick, they usually go to see a doctor a few days later, and the reporting process also takes time, so the speed of grasping the epidemic is still slow about 1 to 2 weeks; Google has published another tracking method to pick out the most important one in the United States The 50 million commonly used search words were compared with the epidemic situation of influenza in 2003-2008 of the CDC. It was found that 45 search words (such as cough relieving and fever reducing) could be used in combination with a numerical model to successfully predict the spread of influenza in winter 2009, even in specific regions and states. If we make good use of this software, we can save the trouble of data collection and body detection. Only by using a large number of daily search records, we can accurately and real-time grasp the spread and intervention of the epidemic, such as early warning, isolation and dispatching of medical resources.

Table1. influenza epidemic in the United States - estimated number of cases by Google

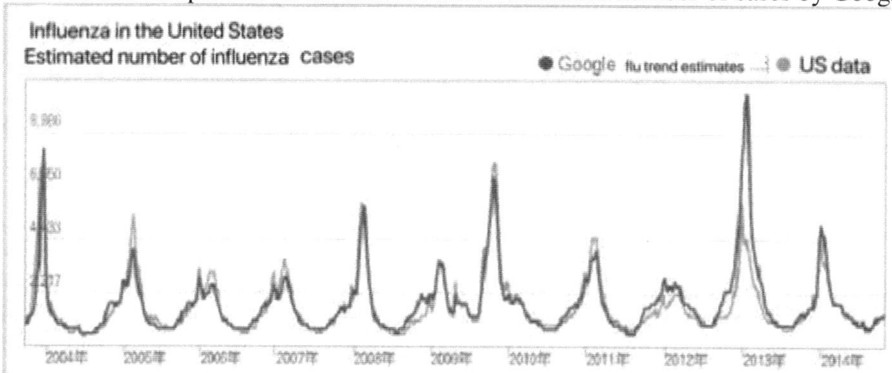

2.2. Health care

Big data has made great progress in clinical diagnosis, remote monitoring and drug research and development. Taking the other side as an example, at present, more than ten cities have launched digital medical treatment. Medical records, images, telemedicine, etc. all produce a large number of data, and form electronic medical records and health management files. Based on these massive data, the hospital can accurately analyze the disease, treatment cost and efficacy data, avoid excessive and high side-effect treatment, and can also be used for remote monitoring and chronic disease management. On June 21, 2014, Professor Nakagawa Kai, Affiliated Hospital of Toyama University, said at the 17th spring academic conference of all Japan medical information society that the Affiliated Hospital of Toyama University applied 17 million cases, 143 million prescriptions and 3 million diseases accumulated in the past nine years to assist doctors in analyzing the best prescriptions.

2.3. Transportation

Big data has been widely used to solve the complex traffic environment, because vehicle lbs positioning can accurately analyze the volume and speed of vehicle traffic at a specific time, provide guidance, scheduling and control prediction for the competent department, and remind the users to avoid the peak and blocked road sections and improve the transportation environment. TomTom, the leading brand of satellite navigation in the United States, uses 80 million mobile phones monitored, one million TomTom live satellite navigation machines to move at the road speed, with RDS-TMC road information system to build a real-time database. Through GPRS real-time information, such as the average speed of a road section, the switching frequency of red and green lights, the average speed of construction and accidents in different periods of the road section, every two minutes the frequency of one time can push and broadcast the satellite navigation machine in time. As long as the customer uses the TomTom navigation app on apple, iPhone or Android phone, it can provide the driver with an optimized path and save time. Even if they are trapped in the traffic array, they can accurately predict the delay time. According to the statistics system, the average driving time can be saved by 15%.

Table2. real time traffic data in Manhattan, New York. Real time traffic data system helps drivers avoid congestion or construction sections. (source: from TomTom live traffic)

2.4. Supervise governance

South Florida sun sentinel, in order to investigate the overspeed problem of police cars, through the analysis of big data system, found that 3900 police cars had more than 5100 speeding records in 13 months, and further screening analysis found that most of the speeding time actually occurred in the commuting time, so it was not due to official business. After the news was published, it caused public concern and nearly 800 police officers from 12 departments involved in the case were corrected or punished. This case shows that the application of big data can be very simple, not as complex as imagined. Digital data plus simple formula can discover potential problems, only need correct data and accurate calculation.

2.5. Humanitarian assistance

About 1300 people were killed and 350000 left their homes in a nationwide uprising over the 2007 election dispute of the leaders of the East African Republic of Kenya. At that time, a group of programmers and network groups jointly launched the "Ushahidi" plan. 1. Users can upload images in real time through SMS (short message service) or website, and then use Google map to demarcate the location, timely disclose the actual situation of the region and seek international assistance. After 2008, "Ushahidi" plan is often used for similar international humanitarian assistance, providing multi-national platforms for major event notification and crowd mapping, such as the tracking markers of the 2010 Republic of Haiti earthquake and 2011 northeast japan earthquake, setting a record of 600000 family name information markers, which can be regarded as a successful case of short-term humanitarian assistance information collection and integration.

3. Big data development in the future of government governance

Big data is a set of systematic, standardized and scientific complex concepts, including business and management majors, which need to have the byte concept of computer storage, as well as network application, system management and another professional knowledge. In order to turn complex concepts into systematic and simple explanations, the following formula is used to explain the essence of big data meaning:

Big data = transaction + interactions + observations

With the White House's "2014 big data white paper" published on May 1, 2014, which was included in the national strategic policy, advanced countries have caught up with each other and put forward national development policies. In December 2014, apple and IBM officially announced the launch of the enterprise version of mobile first app, which provides IBM's big data enterprise users with access to IBM's cloud big data analysis tools and business applications using iPhone and iPad, so that the future of big data industry will bloom. Looking forward to the future development of big data, it will present the following eight trends:

1. Data resource will become the most valuable asset.
2. The implementation of big data management in more traditional industries.
3. With the integration of big data and traditional business intelligence, customized solutions will emerge in the industry.
4. Data will be more and more open, and data sharing alliance will appear.
5. Big data security is getting more and more attention, and big data security market will become more and more important.
6. Big data promotes the development of intelligent city and is the engine of intelligent city.
7. Big data will give birth to a batch of new jobs and corresponding specialties.
8. Big data improves our life in many aspects.

4. Conclusion

Based on the traditional random sampling research mode, which has been gradually replaced by "big data convergence", the thinking decision-making mode of the past research methods faces the severe challenge of objective and accurate analysis of big data. When we are being penetrated rapidly by this invisible force, we should adjust our response in a timely manner in the face of the advent of the era of big data, based on the research position of crime prevention and control, and with brand-new thinking. The global revolution of big data has quietly taken shape, which is a huge change that is related to the country, society, people's livelihood and science and technology. With the gradual maturity of the development of network information, big data application technology is gradually subverting traditional thinking, changing our lives, influencing all kinds of specialties, all over the world. Accelerating the application of big data is absolutely an important strategy for the development of science and technology in China at this stage. It is hoped that the government can accurately calculate the needs of the public and look forward to governance through the application of big data in social management, and improve administrative efficiency. However, the significance of big data development does not mean that it can replace all rational thinking on social issues, and the logic of scientific development cannot be annihilated in massive data. It's true that we must be careful when developing big data technology.

References

[1] TOMTOM,ANNUALREPORT.(2014), Create designed corporate reports in-house, on the cloud, and in record time, http://annualreport2014.tomtom.com/management-board-report/business-and-financial-review- by-business-unit.
[2] What are the main values of big data? (2017), http://www.vwell.cn/news/82.html,2019
[3] 5 data breaches: From embarrassing to deadly,(2010), http://money.cnn.com/galleries/2010/technology/1012/gallery.5_data_breaches/
[4] Baidu, (2016), big data, https://baike.baidu.com/item/%E5%A4%A7%E6%95%B0%E6%8D%AE/1356941?fr=aladdin
[5] Wang,W.Q.(2014) application of big data processing technology in petrochemical enterprises, computer optical disk software and application, No. 2, Pp. 82.
[6] Li,J.J, (2015) on big data challenges and responses in Criminal Investigation, national lawyer, p. 64-66.
[7] Yan,T.(2019) on the opportunities and challenges brought by the big data era to the investigation of job-related crimes, Legal Expo, No. 22, Pp. 154-155.

[8] Zhang,G.M, Huang,W.Z,.(2019) big data in Criminal Investigation, Journal of Vocational School, No. 1, Pp. 84-89.
[9] Guo,Z. (2019) epistemology of investigating and handling job-related crimes in the era of big data, political and legal theory series, NO.3, Pp. 42-54.

IOP Conf. Series: Materials Science and Engineering **806** (2020) 012002 doi:10.1088/1757-899X/806/1/012002

Analysis of microblog public opinion based on big data technology

Cai Baojing [1]*, Li Wensi [2]

[1]Department of Public Administration, Nanfang College of Sun Yat-sen University, Guangzhou, Guangdong, 510000,China

[2]Department of Public Administration, Nanfang College of Sun Yat-sen University, Guangzhou, Guangdong, 510000,China

*Corresponding author's e-mail:35247838@qq.com

Abstract. With the advent of new technology era, big data technology has been widely used in the public communication of mainstream media and become an important strategy. With the continuous progress of the times, big data has entered thousands of households. The arrival of the mainstream media in the new era has changed the traditional public opinion environment, and also brought new challenges to the mainstream media how to deal with the sudden public opinion correctly. Sudden public events are easy to cause public opinion, which is easy to spread, ferment, and even lead to the aggravation of social contradictions. In the field of public communication vision, this paper analyzes the characteristics of public communication of new mainstream media, points out that the new mainstream media face the challenge of news release and public opinion guidance in the current major emergencies, focuses on the analysis of the public opinion guidance effect of people's daily micro blog in the communication process of Changchun long life vaccine event, and expounds the public communication of the current official public opinion field in new media What should be done in broadcasting. Under the influence of public opinion crisis, how the new mainstream media to use the communication strategy to guide the citizens' network participation scientifically and correctly, and then solve the social contradictions efficiently has become an important topic.

1. Preface

In recent years, people carry out various activities through the network, such as browsing the web, e-mail, communication, video and audio browsing, etc. up to now, the data transmitted by the network has grown rapidly, and the concept of big data has emerged. In the era of big data, the wide application of data indicates a new era of a new wave of production and consumption. In this context, the source and bearer of data are very extensive, including the Internet of things, mobile phones, the Internet of vehicles, cloud computing, tablet computers and a variety of sensors around the world. People can get information from many aspects. At present, the mainstream media in our country is mainly led by the official media operated by the government, most of which operate according to the standards set by the government, which leads to the limitations of public communication. One is the lag. Because of its administrative structure, the communication path of public communication shows the characteristics of one-way transmission layer by layer. If the information communication between departments is not smooth, it will lead to the lag of information transmission. The other is the incompleteness. When the government transmits information to the public, it is not completely open, but selectively disseminates information. When the audience understands the information incompletely, it will It will affect the trust degree of the mainstream media in the public heart if there is speculation and doubt and it develops for a long time. In the first part, this paper discusses the connotation, new characteristics, mechanism and problems of public communication in the new era, and tries to make a more accurate definition for the current stage of public communication. In the second part, by analyzing the specific content of the official microblog of people's daily in public events, this paper attempts to study the audience's attitude in two events in a qualitative way, and this paper gives some consideration criteria. The third part of the article analyzes the new changes in the field of communication brought about by Internet technology, and puts forward that the mainstream media should continue the professional quality of traditional media, maintain the sensitivity of news, tap the value of news, make a rational voice, and make correct guidance.

2. New ideas for the development of public communication in the era of big data

2.1. Public communication in the new era

According to the 42nd statistical report on the development of China's Internet issued by CNNIC, by June 30, 2018, the number of Internet users in China has reached 802 million, with a penetration rate of 57.7%; the number of mobile Internet users has reached 788 million, with 98.3% of the Internet users using mobile phones. According to the Weibo financial report data of sina Weibo data center in 2017, as of September 2017, there were 376 million Weibo monthly active users and 1.65 million daily active users. It can be seen that the network has become an important factor affecting social development, and the use of microblog and other new media to disseminate information has become a new trend of society. Among the numerous official news media microblogs, people's daily, CCTV news and Xinhua view point belong to the level of central level media microblogs, among which the number of users who pay attention to people's daily microblogs is the largest. People's daily micro blog was opened in 2009, and the number of fans has been increasing since its opening. So far, it has more than 77.9 million fans. The number of official micro blogs has reached 92576, and the original rate of micro blogs is as high as 95%. The average number of daily micro blogs is more than 40. At the time of major social and public events, people's daily micro blog has become a distribution center for netizens to express their opinions, suggestions and find official information.

The new media technology creates and designs a network platform for the public to discuss and exchange views, which to some extent dispels the discourse dominance of traditional media. In the past, TV, newspaper and radio were mostly used as the media to report events related to public interests. The public, as the receiver of information, could only passively understand the development process of events, and knew little or even no results of major events.The scope of news information received by the public could only be set according to the openness of traditional media. Nowadays, new Internet thinking influences and changes people's life style in different dimensions, and affects people's cognition and practice of the world in all aspects. In the field of public communication, new media has broken the previous one-way communication mode, giving power and opportunity to appeal to users of new media platform, especially some vulnerable groups in society 。 In the new media environment, in the process of information dissemination from the audience to users, the release of information gradually presents the trend of decentralization. The rise of social media ushers in new opportunities for public communication in the new era, which also leads to some unstable factors and risks in the process of information transmission. When Hu Baijing and Yang Yi talked about the basic issues of public communication research, they analyzed the characteristics of public communication from six aspects: the main body of communication, value norms and practice criteria, content, field, means, objectives and effects. This theoretical paradigm provides a new development idea for the research of public communication in the new era.

2.2. Subject, channel and content of public communication

With the rapid development of Web2.0, 4G and 5g mobile communication, the network has profoundly changed the communication and communication between individuals, between individuals and organizations, between organizations and organizations. New media has opened up new channels for citizens' network participation, activated citizens' enthusiasm for network participation, and improved citizens' network media literacy and cultural level. With its high speed and convenience, the Internet breaks the space-time limit of interpersonal communication, making people more convenient to communicate anytime and anywhere. When a large number of user resources are accumulated and gathered in social networking, social media emerges. Unlike the mass media, which are subject to certain systems, disciplines, norms and constraints of political, economic and cultural factors within the social system, such as television, newspapers and radio, social media is more accessible to individuals, more convenient to use, and more efficient.Therefore, social media has made rapid development in the Internet era.

In the new era, the main body, channel, content and form of public communication are more diversified. The diversity of communication subjects and channels benefits from the empowerment of the Internet, so that everyone can make a voice through the network platform and express their views and attitudes towards social events. As long as the masses of all social strata have certain ability to use text symbols and the funds to purchase mobile Internet equipment, they can express their opinions or suggestions through various network platforms; The content of public communication is diversified. First, it is based on the Internet's huge capacity of information storage space. Second, because the Internet has broken the "vision restriction effect" of traditional media, the public can understand all aspects of the world more closely and from multiple angles. People's visual latitude and value dimension are no longer single, and their understanding and thinking of things are more With personalized color, the objective world is more abundant and three-dimensional in the public's vision, which provides a practical basis for people to absorb and output diversified information content in the process of public communication; the form of public communication is diversified, because the progress of network technology makes more and more applications pregnant, they enter the public's vision through various ways, and with Innovation is constantly needed by the public. In the new era, the public communication has replaced the rigid, plane and one-way communication mode with its new paradigm of diversity, three-dimensional and interaction. The public participation in public events and social life has been greatly improved.

3. Case introduction and analysis

3.1. Changchun long life vaccine event

In the Changchun Changsheng vaccine case, the homepage of people's Daily's microblog was opened through microblog search, and the keyword "Changchun Changsheng" was entered into all the microblogs published to retrieve 29 microblogs related to the 2018 Changchun Changsheng vaccine case. Through the screening of 29 microblogs published by people's daily in Changchun Changsheng vaccine incident, the initial microblog selected was "black heart" published on July 15, 2018! A pharmaceutical company in Jilin Province made rabies vaccine production records and falsified the State Food and Drug Administration: the involved vaccine has not been delivered, the mid-term micro blog was released on August 16, 2018, about the "accountability list of Jilin Changchun Changsheng company's problem vaccine cases", and the final micro blog was released on October 16, 2018 by the official micro blog of the people's daily, "Changchun Changsheng problem vaccine compensation program announcement: leading to death See.

Table1.below for the specific parameter value of one-time compensation of 650000 yuan / person

	Number of points	Forwarding quantity	Comment quantity	Overall attitude
Initial micro-blog	1414	1708	1861	Support
Medium term micro-blog	181252	77886	52816	Support
Late micro-blog	7490	2516	6637	Neutral

It can be seen from the above table that most of the netizens have a positive attitude towards the contents of the people's daily microblog published in Changchun Changsheng vaccine event. Although there are some voices of doubt among the netizens in different periods, most of them express their hope to increase the punishment to the relevant personnel of the enterprises involved, but they do not show strong anti reaction to the event handling results. In the process of public transmission of this incident, people ' And so on. The people's Daily has published relevant micro blog contents in a rigorous and detailed manner, answered all kinds of problems related to vaccine safety to the masses in time, and promptly published remedial and compensation measures, which shows the people's Daily's humanistic concern. Among them, the most praised microblog was a combination of text and text entitled "Jilin Provincial Commission for Discipline Inspection and Supervision Commission launched investigation and accountability on corruption of Changchun Changsheng biological vaccine case" released on August 16, 2018. In the follow-up report of the incident, the title of people's daily is more prudent and objective. From the disclosure of the incident to the follow-up report of the relevant review results of the State Food and drug administration, public security organs, Jilin Provincial Commission for Discipline Inspection, and the investigation team of the State Council, the people's daily timely and correctly released the handling and compensation measures for the illegal production of Changchun Changsheng company to the public, as well as the relevant leaders The accountability list of the staff, and answered the public's questions about vaccine replacement and renewal in the follow-up Weibo content. It can be seen that in this incident, people's Daily's Micro blog publishing content closely linked with the most concerned issues of the masses, did not evade and hide the accountability of leaders, spread the authoritative voice of relevant institutions at the first time, effectively conducted public opinion guidance, and effectively solved social conflicts. At the same time, we can see that it is necessary for the mainstream media to correctly play the role of public opinion guidance in the process of public communication, timely establish communication between the government and the masses, closely contact the masses, understand the voice of the masses, answer the questions of the masses, and safeguard the interests of the masses.

3.2. Comprehensive analysis

The purpose of public communication is to safeguard social public interests. The main channel of public communication is new media platform. In this paper, Sina Weibo is taken as an example to explore the mechanism of new media platform in public communication. At present, the main contradiction of our society has been transformed into the contradiction between the people's growing needs for a better life and the unbalanced and inadequate development. Under the impact of today's globalization, the world is changing rapidly and rapidly. China is also experiencing the agitation and ups and downs of social conflicts from all over the world. Public communication has a new mission in the new era of social development. Wu Fei mentioned in the analysis of the social value and academic significance of public communication research that in the rapid development of social changes and globalization, Chinese communication research is faced with such contradictions as the contradiction between the public's free appeal and authoritative government governance and management, and the conflict between the government and society The problem of diversity in the form of outburst and confrontation. New media has built an equal and open public space for human society with the help of technological advantages. Sina Weibo is the representative product of this new media technology. A large number of Internet users are concentrated on the same open platform, and express their personal views, opinions and emotions directly or indirectly through microblog publishing, praising, forwarding, commenting and sharing When social events involve public interests, these concerns, forwarding and comments have strong power of public opinion supervision and agenda setting. Their onlooking and debate on events can often make an event enter the vision of public opinion.

On the microblog platform, there are three mechanisms of public communication. First, through the information released by netizens, the public can know the information that has not been reported by the government in advance or supplement the details that have not been reported by the government. In recent years, many public events are the first to enter the public's field of vision through microblog. For example, after the 812 Tianjin explosion, the first way for the outside world to know the event is through the live video released by netizens on microblog at the first time. A large number of subsequent microblogs on the scene supplement the reports of the government and the media. Second, netizens or opinion leaders first released an event and expressed a demand, which attracted the attention of the government and the media. For example, child abuse in Ctrip kindergarten, red, yellow and blue kindergarten, Changchun Changsheng vaccine, Hongmao liquor, etc. Third, netizens directly express their approval or query on social events and government decisions involving public interests. For example, the discussion on the "limited time dog walking" provision in Wenshan, Yunnan Province, the support for strengthening the supervision of didi drivers, and the initiative to set up a safe isolation area for bus drivers, etc. Generally speaking, netizens participate in the discussion of public events democratically through microblog and the government and media in the process of active communication and interactive consultation, which enhances the dialogue between the masses and the government and media, and can effectively alleviate and solve various contradictions, so as to make the development of public broadcasting peaceful and stable.

4. conclusion

With the rise of new technologies, big data has brought great changes to people's lives. Its development has broken the traditional one-way mass communication form. The interactivity and openness of cyberspace make many information become a "public resource" of network users. In this new media era of "everyone has a microphone", the release of information is no longer the "special" of traditional media Every network participant can make their own voice, participate in topic discussion, promote the power of event tracking process, and make public communication more open, inclusive and equal. The public discourse space it built provides a new platform for people to communicate with the government, and constantly enables the new mainstream media in China to present new features such as centralization, socialization and diversification in the field of public communication, making interactive discussion and consultation communication a new bridge for the construction of the new mainstream media discourse, which greatly improves the national quality and people's political participation Real meaning. In addition, as the mainstream media of the official public opinion field, we should improve our media quality, strengthen our communication ability, demand ourselves with a high sense of social responsibility, and take the public interest as the common goal, so as to promote the equal interaction of public communication among various groups, so as to better serve the social life of the people and the construction of various socialist undertakings in China.

References

[1] China net，（2018）42nd statistical report on the development of China's Internet，http://www.cac.gov.cn/2018-08/20/c_1123296882.htm.
[2] Sina Weibo data center,(2017),2017 Weibo user development report, https://www.useit.com.cn/thread-17562-1-1.html.
[3] Yin FJ, Wang L,(2018) research on the public opinion orientation and communication power of the official microblog of the people's daily,2:95-96.
[4] Liu WY,(2016) The limitation and Prospect of new media empowerment in microblog application,2:23-24.
[5] Hu BJ, Yang Y, (2016) Basic issues of public communication research and innovation of communication paradigm, international press,31:61-80.
[6] Wang Z, Han GF,(2019) In the new media era, analysis of the guiding principle of citizen network participation, administrative forum,6:129-132.
[7] Zhao QW, Liu XL,(2017) Social media Micro Communication from the perspective of public communication, young journalists,33:38-39.
[8] Wu F,(2012) Social value and academic significance of public communication research, Nanjing Social Sciences,5:102-111.
[9] Song W, Chang WF,(2019) The construction of public discourse space, news communication in the new media era,8:34-36.

IOP Conf. Series: Materials Science and Engineering **806** (2020) 012003 doi:10.1088/1757-899X/806/1/012003

Research on Personal Data Protection of EU General Data Protection Regulation

Hu pengnian[1], Qian Wei[2]

[1] Department of Public Administration, Nanfang College of Sun Yat-sen University, Guangzhou, Guangdong, 510000,China

[2] Department of Public Administration, Nanfang College of Sun Yat-sen University, Guangzhou, Guangdong, 510000,China

[*]Corresponding author's e-mail: 1492689709@qq.com

Abstract. The EU's General Data Protection Regulation (hereinafter referred to as GDPR), which came into effect on May 25, 2018, is the most concerned personal data protection legal document in the world, and it will inevitably have a profound impact on the construction of personal data protection systems in other countries and regions. It is generally believed that the GDPR is a further implementation and extension of the EU's consistent position to strengthen the protection of personal data as a basic right.

1. Foreword ntroduction

The General Data Protection Regulations ("GDPR") passed by the European Parliament on April 14, 2016, will come into effect on May 25, 2018 in EU member states. The scope of the regulation is extremely broad, and any organization that collects, transmits, retains or processes personal information in all EU member states is subject to the regulation. The GDPR has a total of 99 articles, with a complete structure and strict logic. Among them, the "principle" clause in Chapter II is responsible for the outline and provides the basis for the legal processing of personal data. The relevant spirit runs through the whole text.

Figure 1 General Data Protection Regulation

2. Personal data processing principles

Article 5 of the "Principles" section of the GDPR sets out seven personal data processing principles: legality, fairness and transparency, purpose restrictions, data minimization, accuracy, storage restrictions, integrity and confidentiality, and the principle of consistency of rights and responsibilities. Provisions are provided in the first paragraph and the principle of consistency of rights and responsibilities is provided in the second paragraph.

Compared to the predecessor of the 1995 EU Data Protection Directive (hereinafter referred to as Directive 95), GDPR adds two principles: the principle of transparency and the principle of consistency of rights and responsibilities. The principle of increased transparency reflects the EU's consideration and requirements for the exercise of data subjects' rights and the feasibility of supervision. According to the principle of consistency of rights and responsibilities, data controllers need to prove their compliance with other principles, that is, they must bear the burden of proof.

"Legal, fair and transparent" is the most principled requirement throughout the GDPR data collection, processing and utilization process. Among them, the requirement of legality needs to be comprehensively understood with Article 6.

"Restrictions on purpose" and "data minimization" require that the scope of data collection and processing should be limited to the necessary and minimum scope. On the one hand, it stipulates that it is limited to "specific, clear, and legitimate purposes; on the other hand, even based on the foregoing purposes, there must be sufficiency and relevance requirements to achieve the goal of data minimization." "The principle requires that from the dimension of storage time, after the limited purpose is reached, the data controller must not continue to store identifiable data.

The "restrictions on purpose" and "restrictions on storage" both provide for exceptions, that is, archiving based on public interest, scientific or historical research purposes, and statistical purposes. For this exceptional case, Article 89 of the GDPR provides more specific provisions. On the one hand, allowing member states to stipulate restrictions on the rights of data subjects based on these situations; on the other hand, even based on these exceptions, data controllers also need to take adequate security measures to avoid damage to data subjects, including necessary Technical measures and management measures, such as pseudonymization, and to ensure that data processing is minimized.

"Accuracy" and "integrity" point to quality requirements in data storage and processing. The implementation mechanism of these requirements is reflected in the relevant rights of the data subject stipulated in Chapter III[1], such as the data subject's right of access, correction, deletion (ie right to be forgotten), right to restrict processing, etc.

"Confidentiality" emphasizes the security protection of data, especially for unauthorized or illegal access and processing. This principle of data security and prevention of leakage has reached consensus in major countries and regions, and often attracts high public attention.

3. Legitimacy basis

The first principle to be observed in data processing is legality. Article 6 of the GDPR provides six types of legality basis, and basically follows the relevant provisions of Directive 95. Specifically include: the consent of the data subject, the performance of the contract, the performance of legal obligations, the protection of important interests, the protection of public interests, and the protection of the controller's priority interests.

In practice, the prevailing privacy policies, user agreements, and mandatory disclosure systems are the embodiment and implementation of the consent principle. It can be considered that user consent is the most important basis for the legality of personal data processing. On the one hand, the principle of consent is mainly reflected in the initial collection of personal data, and the data must be collected

---

[1] European Commission, Directive 95/46/EC of the European Parliament and of the Council of 24 October 1995 on the protection of individuals with regard to the processing of personal data and on the free movement of such data, EURPEAN COMMISSION, http://eur-lex.europa.eu/legal-content/en/TXT/?uri=CELEX%3A31995L0046(last visited May. 31, 2018).

with the consent of the user; on the other hand, other rights of the user are actually derived rights of the consent right, and the right to know is the effective exercise of the right to consent The basis of access, right to correction, right to delete (forgotten right), right to restrict processing, right to object, etc. are actually the concrete manifestation of the right to consent in different links and scenarios of data processing. The legislation, policies and judicial practices of various countries, and the construction and concreteness of a large number of rules, also focus on whether the data processor has truly guaranteed the user's right to know and consent. There are several places in the GDPR that provide specific safeguards for implementing the principle of consent. For example, Article 7 specifically specifies the requirements for consent. Regarding the specific form of consent, there are also important guidelines in the preamble.

Article 12 of Chapter 3 of the “Rights of Data Subjects” of the GDPR focuses on the requirement of transparency to ensure the right of the data subject to know. However, whether or not the principle of consent can bear the heavy burden of supporting the foundation of the entire personal data protection system at the present and future of the rapid development of data processing models has faced many questions and challenges.

The other five legality foundations do not require consent as a prerequisite, but all need to follow strict conditions. Among them, "contract performance" considers the agreement between the user and the data processor, including the necessary preparations for the contract, which is still essentially the consent of the data subject. Several other legitimacy foundations reflect the legislator's search for a balance between multiple conflicts of interest. For example, the two foundations of "fulfilling statutory obligations" and "protecting the public interest" indicate public interest considerations and statutory exceptions to the protection of personal data, and provide that each member state determines the specific circumstances. For these exceptions, Article 6 also provides strict restrictions to ensure that the principles set out in Article 5 can be effectively implemented.

Article 6, paragraph 4, provides for subsequent processing of data. Subsequent processing refers to processing that is not permitted by the data subject and is not permitted by EU law or Member State legislation. In this case, it is the responsibility of the controller to confirm whether the purpose of this new follow-up process is compatible with the purpose for which the data was originally collected and can therefore be allowed. This requirement is likely to add significant costs to data flow and sharing.

For data controllers and processors, in order to prove that their data processing behavior is based on the last five legal basis without obtaining consent, the wording of the law is very difficult, because most of these situations rely on individual judgment and With strict restrictions, it is safer to obtain user consent.

IOP Conf. Series: Materials Science and Engineering **806** (2020) 012003 doi:10.1088/1757-899X/806/1/012003

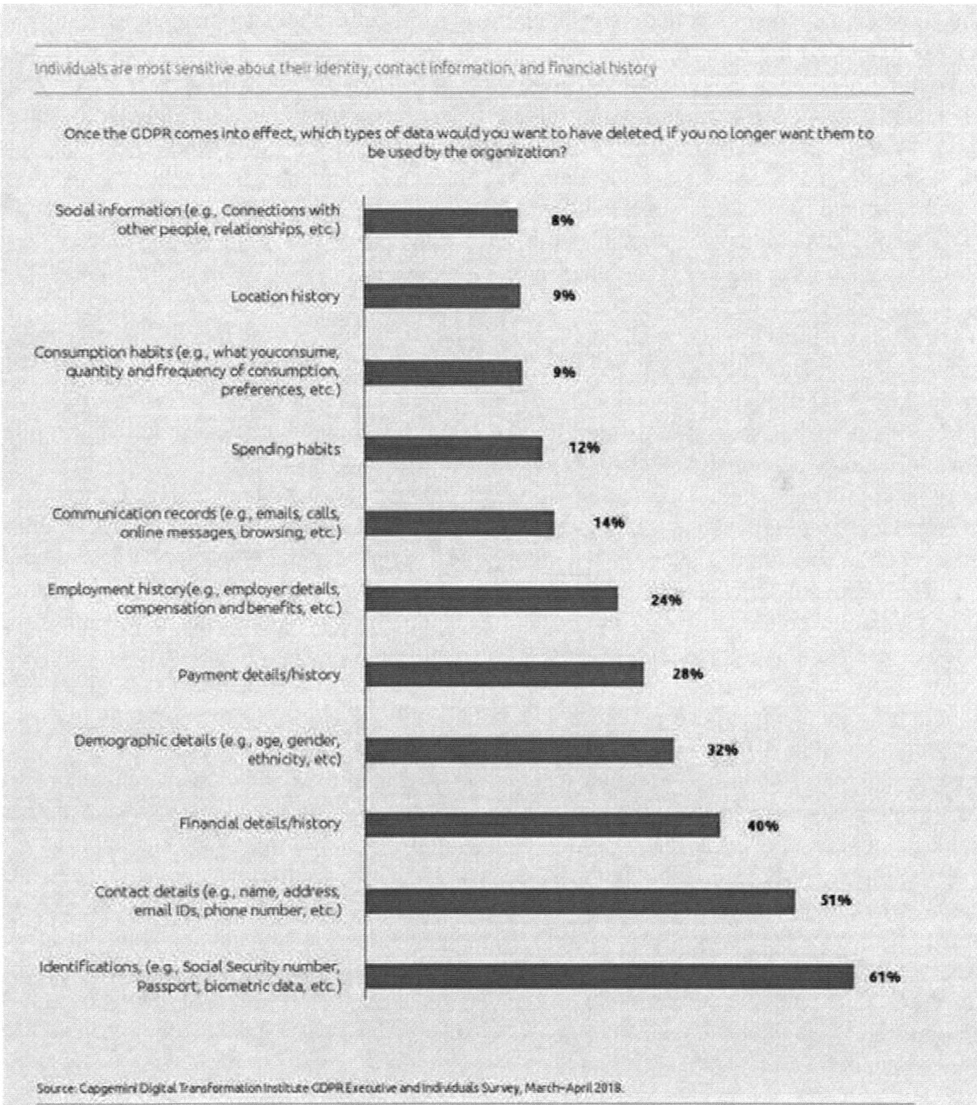

Figure 2 GDPR executive

4. The conditions of "Consent"

Personal consent in GDPR refers to the data subject's consent to the processing of their personal data through a written statement or through a clear affirmative action. The expression of will should be freely given, specific, specific, informed and clear. Silence, default checked dialogs, or inactions do not constitute consent.

Article 7 sets out the conditions for consent. Specifically, the article places a responsibility on the controller to prove that the data subject has agreed to the processing of his or her personal data. The second paragraph provides for other matters (commonly including comprehensive user agreement, etc.) in the written statement of obtaining consent. In this case, "agreement should be presented in a form that is significantly different from other matters, easy to understand and accessible, and in clear and common language." Paragraph 3 provides for the right of the data subject to withdraw consent. Paragraph 4 stipulates that if consent to data processing is a prerequisite for contract signing, and such data processing actually exceeds the scope necessary for the provision of services, it will violate the provisions on "consent should be made freely".

It is worth noting that the European Union used the concept of "explicit consent" when it proposed the draft GDPR, and did not add the "explicit" limit in the end. However, as stated earlier, "silence, default checked dialog, or inaction" does not constitute consent. Although there is room to determine consent by explaining certain behaviors of users in specific situations, given the overall strict protection tendency of GDPR, it is foreseeable that the interpretation space of "non-explicit" consent will not expand much.

Figure 3 personal data

5. Scope of application

The geographic scope of European privacy laws is expanding through GDPR. Because of this expansion of the scope of application, many organizations located outside the EU and not subject to current European privacy laws will have to apply EU privacy laws with the implementation of the GDPR. The GDPR requires these organizations to respond promptly to GDPR's "compliance" requirements.

First, GDPR applies to organizations with establishments in the EU, as long as these organizations process personal data in the activities of the business in the EU (regardless of whether such processing actually takes place in the EU).

The interpretation of the concept of "location" must be broad and flexible. The requirement is for effective, authentic activities (small activities are sufficient) through sustainable venues. The legal form (such as a branch or subsidiary with legal personality) is not a determining factor. Therefore, having a sole representative in the European Union is sufficient to qualify.

It is not required that personal data processing must be performed by the business organization itself. However, it is required that such processing activities occur in the activities of the business organization (usually having both of the above characteristics, but the conditions described in this sentence must always be met). If the activities of the business organization are inseparable from the activities of the parent company (for example, the existence of the business organization is for the economic benefit of the parent company), the relevant personal data processing behavior should be regulated by GDPR.

Therefore, if the activities of business entities (such as branch offices or liaison agents) are inseparable from the processing of personal data by organizations located outside the EU, GDPR should apply.

Second, if an organization does not have a business establishment in the European Union, it does process personal data of individuals in the European Union and such processing is related to the provision of goods or services to individuals in the European Union, regardless of whether such goods or services are charged, GDPR should also apply.

If a non-EU organization intends to provide goods or services to individuals in the EU, it will be considered to provide goods or services in the EU. Simply being able to access the website or its contact details from within the European Union or using the language commonly used in the country in which the organization is established is not in principle sufficient to be considered as having the above intention. Use the language and / or currency of one or more Member States, referrals from customers in the European Union (for example on a website), use of ads in search engines targeting one or more Member States and / or use of top-level domains (such as .eu or .nl) may result in the goods or services being deemed to be provided in the European Union.

Third, GDPR applies to the processing of personal data of individuals within the EU by non-EU organizations, as long as such processing involves monitoring the actions of those individuals and the processing occurs in the EU.

If, in particular, these individuals are tracked on the Internet in order to make decisions about them or to analyze or predict their personal preferences, behaviors, and attitudes, and processing techniques are used in the process to form portraits, etc., it constitutes Monitor behavior.

It is unclear to what extent monitoring will apply to GDPR. In principle, websites that use so-called "tracking cookies" and monitor used applications are covered by the GDPR (as long as those sites process personal data). The European Court of Justice recently ruled that in some cases, dynamic IP addresses can also be considered personal data. Therefore, owners of websites that store dynamic IP address log data may also be subject to GDPR regulations.

6. Conclusion

The entry into force of the European Union's General Data Protection Regulation (hereinafter referred to as the "Regulations ") has opened the curtain on the reform of personal information protection legislation in various countries. . The Regulations have made significant changes to the protection of personal information in many aspects, given individuals more information rights and restricted information transaction rules for information processors and controllers.The most fundamental purpose of these changes is to strengthen the information subject. The right to self-determination of information, thereby safeguarding the sovereignty of information among countries; on the other hand, it is also to meet the needs of the development of the information technology era and the construction of a unified digital market.

References

[1] Wang, H.C. (2019) Practical Evaluation and Enlightenment of the General Data Protection Regulation (GDPR). China's prices: 79–81.

[2] Zhang, J.P. (2019) The evolution of EU personal data rights and its implications. Studies in law and business:182–192.

[3] Wu, Y. (2019) On the Enlightenment of EU Data Protection Law Reform to China. Legal and Economic:99–102

[4] Liu, X.C. (2019) Analysis of the EU General Data Protection Regulation. China Market Regulation News: Page 2.

[5] Li Yuan. Research on personal information protection in the era of big data [D]. Southwest University of political science and law, 2016

[6] Nie Shengkui. Research on legal issues of personal information protection [D]. China University of political science and law, 2009

[7] Hong Hailin. Research on civil law protection of personal information [D]. Southwest University of political science and law, 2007

[8] Zhao Lijun. On the right to be forgotten [D]. Jiangxi University of technology, 2019

[9] Ren Xiaoyu. Research on personal information protection of mobile terminal users in the era of big data [D]. Guangzhou University. 2019

[10] Leonid Titkov,Stefan Poslad,Juan Jim Tan. Applied Artificial Intelligence. 2006 (2-4)

IOP Conf. Series: Materials Science and Engineering **806** (2020) 012004 doi:10.1088/1757-899X/806/1/012004

The development of artificial intelligence and the response of labor policy

Pan Xiaodie[1], Zhong Hongsen [2]

[1] Department of Public Administration, Nanfang College of Sun Yat-sen University, Guangzhou, Guangdong, 510000,China

[2] Department of Public Administration, Nanfang College of Sun Yat-sen University, Guangzhou, Guangdong, 510000,China

[*]Corresponding author's e-mail: 562849196@qq.com

Abstract. With the emergence of big data, information technology, digital technology and artificial intelligence, our society is making great strides towards an artificial intelligence society, and the foundation of economic society has undergone profound changes. Today's labor policy must not only adapt to the needs of marketization, but also need to respond legally due to the development of science and technology. It should reasonably anticipate the changes in labor relations caused by artificial intelligence, including the norms of digital labor and the status of labor policy of intelligent robots. And the protection of social rights after the replacement of traditional labor. In the era of high-tech being widely used, the rapid development of modern artificial intelligence is gradually spreading from the field of science and technology to the life of the general public. This will have a subversive impact on labor relations. This is also the future of labor The amendment of the Law is an important factor that cannot be ignored.

1. Labor policy in the age of artificial intelligence

In the field of labor relations involving artificial intelligence, the foundation of traditional labor policy has changed fundamentally. On the one hand, how will the replaced labor rights and interests be protected? Although traditional labor obtains greater freedom, its basic right to survival and development is facing a crisis; on the other hand, artificial intelligence involves in the traditional labor field, forming a new labor benchmark relationship, and the law should How to regulate, for example, whether labor protection is needed, whether labor remuneration should be paid, whether the eight hour working system is applicable, are all problems to be explored. It can be seen from this that AI is not only a pure technical problem, but also a close attention to the possible impact of AI development on the political, economic, social, cultural and legal systems. This is not only an impact on the existing ethical framework and social and economic systems, but also a challenge to the value and function of labor policy. How to reduce the impact of artificial intelligence on the labor market, how to effectively play the adjustment function of labor policy, and how to balance the demand of scientific and Technological Development and social security are the important issues of labor policy research in the era of artificial intelligence.

2. The influence of the application of artificial intelligence on labor relations

As mentioned before, before the era of artificial intelligence, the labor policy of various countries were mainly based on the labor relations in the industrial era, especially the manual labor in the background of factory machine production. After the rise of artificial intelligence, the pattern of labor has undergone a subversive change, such as the rise of intelligent cars, digital labor and intelligent robots. Such changes are reflected in the evolution of labor content. These changes are deeply changing the social and economic basis of labor relations, and labor policy faces new challenges.

2.1. Smart car

In practical application, intelligent vehicle is the representative of artificial intelligence. In the United States, Google has developed a driverless car in 2016. Although there have been traffic accidents in road tests and a slight collision with a bus, its technology has been quite mature. In China, Baidu and BOSCH also developed an autopilot vehicle, and carried out Road trials and related technologies. In December 2, 2017, 4 buses of the Shenzhen bus group, the Alpha Ba intelligent driving public transport system, were trial run in Futian Bonded Area. This is the first intelligent driving bus trial run on the open road in the world. The United States and Germany began to legislate to prevent the legal risks brought by driverless intelligent vehicles, and established some basic principles: "road safety is better than travel convenience, personal protection is better than other utilitarian considerations, the legal regulation of technology is to achieve a balance between personal freedom and the freedom and safety of others, and the protection of personal rights and interests must take precedence over the protection of personal rights and interests Protection of animal or property rights. "From the case of intelligent cars, it shows that the artificial intelligence in driving field has the attribute of" human like ", and this phenomenon is also in many industries such as news writing, big data analysis, etc. Whether it is intelligent automobile or other artificial intelligence applications, in essence, they all realize the replacement of traditional human labor, especially the replacement of intellectual labor, which will have a huge impact on labor relations.

2.2. Digital labor

Based on the development of information technology, especially the rise of social media, such as Google, Facebook, twitter, Amazon and other network enterprises, Christian Fuchs put forward the concept of "digital labor", which refers to the intangible assets composed of intellectual achievements relying on data and information, supported by big data, digital technology and the Internet, including industry, agriculture, economy Knowledge, information, there is a certain space, consumption of people's time in the form of information and network work. The difference between traditional labor and traditional labor is the intangible labor of culture, knowledge, information production and consumption. The labor tools of digital labor are mainly the user's brain and the network platform itself. Users spend a lot of time working in the network or being enslaved by the network reality environment. This kind of digital labor is also attributed to "net slaves", which is a free and exploited productive activity, and generally exists in the contemporary network society. Although digital labor seems to be voluntary and free labor, it implies the unequal relationship of social power under the control of technology, which is exactly what the law needs to pay attention to. Although digital labor is still a theoretical concept, with the development of artificial intelligence, under the condition of unequal technology and capital, people's voluntary and free non-material labor is still expanding its scope. It almost subverts the traditional concept of labor, quietly changes the foundation of labor relations, and lurks the erosion of labor rights.

2.3. Intelligent robot

The essence of labor in traditional labor policy is natural person. The development of artificial intelligence makes the connotation of labor expand, and even machines with more "human like" attribute appear. The biggest change it brings is that all kinds of intelligent robots (hereinafter referred to as robots) will become industrial employees. They are no longer machines and equipment that can only carry out simple and repetitive labor in traditional sense, but have learning ability Force, thinking ability, even simulation of emotion, and human communication robot. In recent years, based on the development of brain science, cognitive science and information science, the research of artificial intelligence has made great progress. For example, to prove the "Four Color Theorem" that human beings have never proved, the artificial neural network system can quickly find out the optimal path solution of urban travel agents. In addition, the diagnosis and treatment system of blood diseases beyond the level of human medicine is not the same as that of human experts geological prospecting system, unparalleled mathematical theorem proving system and meteorological forecasting system in speed, etc. Since 2017, with "alpha go" easily defeating humans in go, there are almost no profound problems in the world that cannot be solved. It can be seen that human intelligence can not only be surpassed, but also be replaced by robots, which means that in the future, there will be more jobs that need to be done by human intelligence, which may be replaced by robots, and robots will become the future The main body of labor.

3. The challenge of artificial intelligence development to labor policy

3.1. Personality guarantee of robot

When artificial intelligence is taken into consideration in labor policy, the first thing to face is the problem of "personality" of robot. The Latin meaning of personality is mask or role. Roman law recognizes that all kinds of unnatural people have rights and obligations. Smart cars, for example, can be regarded as "humanoid" and given legal personality, which can tie the ownership of cars with liability insurance. This arrangement will allow other innovative sanctions to be considered for car accidents. In 2016, the Legal Affairs Committee of the European Commission submitted a motion to position the most advanced automatic robot as an electronic person and prepare relevant legislation to confirm the legal status of the robot. In addition to giving it specific rights and obligations, it also proposed to register the intelligent automatic robot for tax payment, payment and claim The account number of the pension fund. If the legal motion is passed, it will undoubtedly shake the traditional civil subject system, which also shows that it is not only possible, but also necessary for robots to obtain the civil subject status. Of course, from the existing theory and legislative practice, some regard robots as "animals", some define them as legal status of "similar to legal person", some think robots are more realistic than corporate legal person, but due to the value of human subjectivity, jurists are more cautious about robots as "people", and robots cannot become law because of legal imitation The main body of law. Even so, with the development of artificial intelligence in more and more fields of labor, exploring the legal rights and obligations of robots is bound to become an important issue that can not be ignored in the legal field.

3.2. Robot ethics charter

At present, artificial intelligence people are in the development stage, and their types are quite complex. We can divide them into weak artificial intelligence (ANI) and strong artificial intelligence (AGI). Although the former provides many auxiliary functions, some of which even surpass the ability of human beings, in essence, it still belongs to the category of "tool", which belongs to the consideration of technology promotion and cost within the enterprise. Its work and regulation can be regarded as the internal operation standard of the factory, not included in the category of labor policy. The latter can focus on completing a specific task, and has the function of deep learning. Some scientists even think that as long as we have the detailed data of the real brain, we can simulate the brain in the biological sense, which has the ability of deep self-learning, adaptability in new fields and perfect perception technology, coupled with the continuous development of biotechnology, to create a high degree of simulation Human robots are the reality that is happening. If so, how to solve the problem of human like artificial intelligence's rights in labor policy can't be ignored any more. For example, labor remuneration may not only be paid to its owners.

Table1.The Three Laws of Robotics

THE THREE LAWS OF ROBOTICS

1. A ROBOT MAY NOT INJURE A HUMAN BEING, OR THROUGH INACTION, ALLOW A HUMAN BEING TO COME TO HARM.
2. A ROBOT MUST OBEY THE ORDERS GIVEN IT BY THE ONE POSSESSING CONTROL OR AUTHORITY EXCEPT WHERE SUCH ORDERS WOULD CONFLICT WITH THE FIRST LAW.
3. A ROBOT MUST PROTECT ITS OWN EXISTENCE AS LONG AS SUCH PROTECTION DOES NOT CONFLICT WITH THE FIRST OR SECOND LAW.

If robots have the status of labor, it's hard to say that robots won't get independent income. With the strengthening of robot's subject consciousness, they can even break away from human beings and open up Create your own civilization. As far as labor policy is concerned, whether artificial intelligence should be paid attention to and whether robot labor rights should be given to protect human self-interest at the same time is a problem that needs to be solved soon. Therefore, the law needs to respond to the essence of robot, especially its personality problem. The European Law Commission's motion claims that robots have "worker status" and should be given specific rights and obligations to workers. In addition, South Korea has also begun to draft the "robot ethics charter", which prohibits the illegal use of robots and prevents human abuse of robots. These legislative explorations show that although robots are called machines, they may even have the status of "human" in the future. They have reason to be the adjustment object of labor policy, and their own legal rights should be taken seriously.

3.3. Labor policy after robot replacement

Under the legal system with fairness as the basic value orientation, how to better protect the "vulnerable groups" in future labor is a problem that the labor policy has to face. With the rise of artificial intelligence, the serious crisis is that a lot of traditional work will be replaced by robots, and this phenomenon has appeared, and will happen in the near future. It can be seen that in the era of artificial intelligence, the first thing that labor policy should protect is the basic rights of traditional labor, that is, "human". Because the development of science and technology is to benefit human beings, rather than create human difficulties. What the future labor policy needs to do is how to protect the basic survival rights of the workers who have been replaced by robots, and how to deal with the social unrest caused by the stagnation of employment. In the face of this crisis, in the future, the government must launch relevant contingency mechanisms, such as the implementation of the basic income guarantee for the whole people, that is, everyone, rich or poor, will be paid a salary, so that they will not fall into the life of basic dignity. Therefore, whether it is the protection of traditional labor or the reconstruction of social security, appropriate arrangements must be made through the legal system. Therefore, the labor policy in the era of artificial intelligence is no longer just a labor policy focusing on labor relations, nor a retirement security based on a certain number of years of labor, but a social security law of the whole people must be considered, which comes from the social security with the goal of maintaining basic survival under the humanitarianism. ections should be numbered with a dot following the number and then separated by a single space:

4. Conclusion

There are many possibilities for the development of artificial intelligence. The coming of artificial intelligence has become an irresistible trend of modern science and technology, especially the impact of artificial intelligence on labor policy has gradually emerged. The thinking of labor policy has gradually evolved from industrial society to artificial intelligence society. As a labor policy involving the vital interests of labor, it must be mentioned that Make arrangement and response before. The main purpose of the development of artificial intelligence must be to serve human beings and improve their living standards. The response of labor policy also needs to focus on this purpose, especially to fully consider the application of artificial intelligence technology, establish a people-oriented concept,

IOP Conf. Series: Materials Science and Engineering **806** (2020) 012004 doi:10.1088/1757-899X/806/1/012004

protect the basic rights of labor, and build a social security system, which is the challenge of the development of artificial intelligence to labor policy.

References

[1] Weng, Y.L.(2019) Function Adjustment of Labor Law in the Age of Artificial Intelligence. Journal of Xi'an Jiaotong University (Social Science Edition) Number 1: 145-152.

[2] He, S.S. (2017) The United States and Germany, two kinds of auto-driving legislative exploration. http://www.sohu.com/a/196925591_114778

[3] Wu, H.D.(2017) The institutional arrangement and legal regulation in the age of artificial intelligence, https://www.iolaw.org.cn/showNews.aspx?id=61404

[4] Kurzwell，R.（2016）How to Create a Mind: The Secret of Human Thought Revealed. Zhejiang People's Publishing House. Hangzhou.

[5] Fuchs，C.（2017）Social Media: A Critical Introduction. Wunan Publishing House. Taiwan.

[6] Li，H.M.（2011）Laborers in Labor Law. Tsinghua Law Journal：115-129.

[7] Han，W.（2018）Future challenges of artificial intelligence and labor law. Heilongjiang Social Sciences Number 5：40-42.

[8] Yu，Y.C.（2017）The age of artificial intelligence is starting. Guangming Daily.

[9] AidanMiller，H.（2017）The rise of robots and human law. Law and modernization Number 4：62-71.

[10] Si，X.（2017）On the Civil Liability of Artificial Intelligence: Starting with Self-driving Cars and Intelligent Robots. Science of Law Number 5：166-173.

IOP Conf. Series: Materials Science and Engineering **806** (2020) 012005 doi:10.1088/1757-899X/806/1/012005

Fraud Analysis Using Machine Learning Algorithms

Molei Qin

Peiyang Park Campus, Tianjin University, No.135 Yaguan Road, Haihe Education Park, Tianjin, China

Abstract. In the previous studies done by others, the methods of picking abnormal numbers and data have been developed. This paper will focus on using machine-learning method to find abnormal value. Machine learning has been applied in all kinds of fields, including picture identifying and regression which is widely used in finding anomaly. The author finds some of the anomalies during the investigation and the fraud detection rate is higher than 80 percent. This paper not only aims at using traditional way of finding abnormal value, but also aims at using the modern computer science to deal with huge amount of data. Traditionally, when it comes to the topic of finding anomalies, people tend to think about t-checking or f-checking[1,2], which shares the same limitation that the target has to be fixed when being measured. However, a more realistic problem is that the target and the measuring environment is always changing. To fix the problem, this paper uses machine learning to do regressions of the environment and the target. Not only can it help us find the relationship between our target and 'environment', but also it can be written into computer elegantly so that it could be able to deal with huge mount of data.

1. Introduction

Economic behaviors, such as paying taxes for a particular item or a credit card transaction, are playing a more and more important role in people's daily life, which might cause frauds such as telling a lie when asked about the price of a poverty, or stealing people's credit card to buy a lot of things before the owner of the card knowing about it. Actually, these kinds of frauds are more prevalent than ever in that people can even make a great life by doing this kind of dirty work. Therefore, a method must be established to prevent this kind of things from happening. It is not easy, though, as fraud is only a small part of the huge amount of data, which makes check each one of them by human hand impossible. Computers, however, can do vastly considerable mount of calculation in a rather short time, making it a perfect tool for catching fraud. All kinds of frauds are for sure anomaly compared with regular data, which means that the fraud record may disobey some certain rules that other regular data obeys. Catching fraud, in this case, is like catching anomaly in a huge amount of data.

2. Methods and Analysis

2.1. Collecting data and data cleaning

First, enough mount of data are needed and the author should make a data quality report to make sure that there are few mistakes in our data set. Next step is to clean the data, which means that the author should fill in the blank with certain number to make every records usable in the following data analysis. The principal here is to fill in the blanks with normal number, making sure that the record will not appear strange when it is completed.

IOP Conf. Series: Materials Science and Engineering **806** (2020) 012005 doi:10.1088/1757-899X/806/1/012005

2.2. *Variables-creating*

The following part is creating variables, which is an essential part before building the model. Whether an algorithm is good or not can be largely dependent on the variables involved. As many variables as possible should be created in case that some important variables which could make the model work better might be left. Then the number of the variables should increase from tens to hundreds. In this process, the columns filled with text are deleted and the author only uses numerical variables to describe the records.

2.3. *Supervised algorithms*

Supervised model means that whether a record in a learning group is normal or not is known, as long as examples of frauds are given, what the author has to do is to develop a algorithm that can help separate the normal from abnormal, so this model can be used to predict whether it is fraud or not.

In the supervised model, the number of the dimensions should be reduced because it will take an absolutely long time to run the whole model with so many variables, and when the number of the dimensions is too high, it will be much harder for the models to find the nonlinear relationships among so many variables. Thus it is necessary to reduce dimensions. What has to be done here is to pick those variables that can make a difference to the result. There are three general methods to complete this process, including filter, wrapper and embedded. Filter is basically a method for picking certain variables, which can separate the fraud record from the regular ones when the author is dealing with binary classification issue. Fraud detection rate and the KS distance between the distributions of anomaly and the normal ones are great criterion in this process.

$$KS = \max_{x} \int_{x_{min}}^{x} [P_i(x') - P_j(x')]\, dx' \tag{1}$$

This method can work extremely fast but it does not work very well because the ability of variables to separate fraud from the regular data varies little, so in this paper the author only uses this kind of method to get away the first half of the variables. Wrapper is a process using simple models such as logistic regression or linear regression which involves different subsets of variables to get rid of the variables containing the models which work badly. It can reduce one variable at a time and can repeat the process again and again until getting a proper number of variables which is based on how many records the author has in all. Then embedded in the algorithms can be used and will be discussed in detail in the algorithms building part.

The supervised model means that the author has the examples of fraud, and there are two general algorithms that can be used, including neural network and decision tree.

The first one is neural network and the author uses the label which can help identify whether it is a fraud or not as the target output. And it is easy to see that the output is binary. So naturally the logistic function is used as the transfer function because the result of it is between 0 and 1. However, it is so delicate of the neural network that it can be easy to be overfitting. The real process of learning can be very meandering thus the machine may stop training when it achieves a local optimum. Fortunately, there are some details that make the neural network more robust. The first thing is that the learning rate can be changed while the model is training. Learning rate is basically the step length for each learning opportunity, which is known as the chance to change the value of the weight matrix. The author can use comparatively large number for learning rate at first and decrease it gradually. This process is like finding the lowest spot in a meandering surface. If you take a ball and let it slide naturally, it will probably find itself in a local optimum. However, if you shake the table, it will be more possible to find the lowest point. The second thing is that the author can optimize the way of changing the weight matrix. The gradient is normally used as the indicator of adjusting the weight matrix. However, there are more things that can be done. For example, it is acceptable to use the gradient as the only part of the indicator and add some adjustment term such as a fixed number to make the model more stable and make the model with only one node and one layer as well as gradually increase the complexity of the model. When the overfitting occurs, it is able to make best decision concerning super coefficients.

The second method for supervised model is the decision tree. The author simply divides the whole data set and gives the record in the same area with the same label. It is a delicate thing as a computer can have a simple solution as it can simply divide every record. However, it will not work for the testing. So there are some other algorithms based on decision tree performing better due to the specific structure they have. Two of them will be introduced. One is boosting[6] tree and the other is random forest.

Random forest is a combination of hundreds of single decision trees. However, the structure of each tree varies a lot. They use different subset of variables as the decision indicator. If you draw a picture, it means that each decision tree is using different axes. The author combines all the decision tree and finally gets a result. The computer will automatically decide the position of the boundary to minimize the error which is similar to the neural network model's loss.

Boosting is very similar to the random forest by its definition and they are all combinations of a large number of decision trees. What can separate boosting tree from random forest is that each decision tree has the same structure with the boosting tree. The author produces such series by adding an adjustment term for mistakes. It is more similar to a Taylor series. For instance, the author makes a simple division of records and then makes mistakes. Next, the author adds a decision tree to correct the mistakes and makes fewer mistakes. The process has been repeated again and again before final model become stable.

2.4. Unsupervised algorithms

Unsupervised model means that the author does not have any examples of frauds, so here is how the author deals with the problem. The author first scales the variables so they are basically in the same distribution where z-scale is a great example. Z-scale is standardizing the variables by using the subtraction of the real number and the mean divided by the standard deviation, making the mean of every variable 0 and making the standard deviation of every variable 1. And then the author can do PCA, which stands for principal component analysis. During the process, the author could not only remove the linear relationship among variables, but also select variables that have fluctuation. In data analysis, more fluctuations mean more information, so the author naturally selects variables that contain the most information. In this way, the author can reduce the dimensions of the unsupervised.

Now, what the author has for supervised model is records with selected features and what the author has for unsupervised model is records with tens of principle components. The author is going to build a criterion that can tell which record is regular and which is not.

For unsupervised model, the first thing the author needs to do is to z-scale the principal components because they certainly have different distributions and z-scale can make them equally important. Then there are two ways of analyzing the problem. The first way is simple and the author can view the record as a vector in a high-dimension space. Each of the variable can be viewed as a dimension or an axis in the high-dimension space. The author can simply use the distance between the record vector and the origin as the criterion because the author has already z-scaled the record and it should form a spherical cloud around the origin with the radius of one. If it is an anomaly, it should be far away from the origin. The second method is using a machine-learning algorithm which is called autoencoder. Autoencoder[5] is a specific kind of neural network.

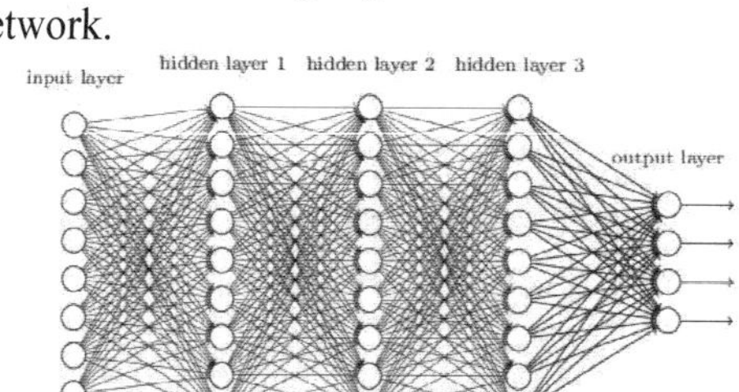

Fig. 1. Structure of a neural net work

Neural network is basically a model which can automatically evolve, and here is how it works. The structure of a neural net work is shown in figure 1. It consists of three parts. The first one is the input layer, where records are put. The second part is the hidden layer, where there are several layers hidden in this part. It is very similar to the way a brain works. There is a matrix in the interval of layers and the output vector multiply this matrix to form the input of the next layer. Through a transfer function, which can provide some nonlinear terms in this algorithm, the input of the layer transform into the output of the layer. The third part is the output layer, where the output is produced. Then the output and the real result of the input will be compared together and produce a value measuring loss through a loss function. The whole system is self-evolving because the loss is the controller of the the matrix and this process will automatically reduce the loss. The number of ways to control the matrix is more than considered, and here is a basic method about how to realize the control. Since the transfer function, the input and the output are all fixed and the loss function as a function only concerning the matrix can be seen. That is to say, there is a fixed functional relationship between the matrix and the loss. Surely loss can be represented by the matrix and the author can take the partial derivative for the loss by one variable in the matrix. When the result is below 0, the author can make the variable in the matrix a little higher to reduce the loss and make the variable a little bit lower when the partial derivatives is above 0 to reduce the loss.

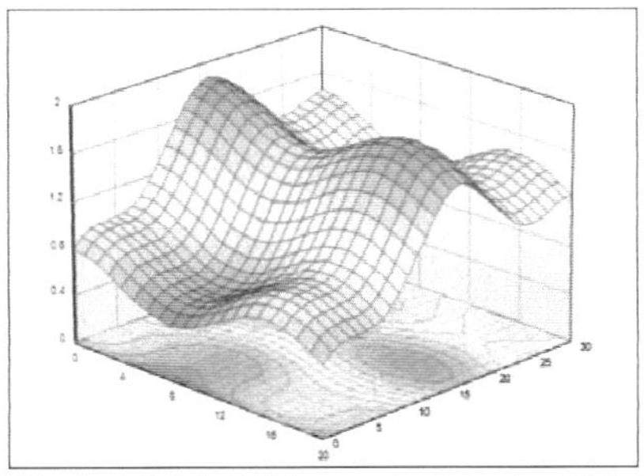

Fig. 2. Think this way for model building process, algorithm design and selection

Additionally, autoencoder is basically a neural network whose input is exactly the same as the target output. It is primarily used in the dimensional reduction, because when the functional relationship is written down, there is a hidden functional relationship among the variables. Also, some changes to the equation can be done, and finally the one variable can be represented by other variables. Here, the author does not use it to reduce the number of dimensions but uses it to find the hidden functional relationship among variables. When the record is normal, the variables in this record will certainly obey the hidden relationship. As a result, the record can be reproduced well. However, when the record is abnormal, the variables tend not to obey the hidden relationship so the record is reproduced horribly. The autoencoder does not need to be perfect because the error part is needed only. The bigger the error for a record, the more abnormal it tends to be.

The next step is to combine the two results which can be concluded from methods above for unsupervised model. One way is ranking. Due to the varieties of the definitions of the distance and the specific structure of the autoencoder, the absolute value of each record in these two methods means nothing. However, the author can combine the ranks for each record together and make a whole ranking order for them. And now, it is known which record is the most-likely anomaly while others tend not to be.

2.5. Result analysis

For unsupervised model, only ranked suspicious list is provided. The author can simplify and investigate it one by one, from high possibility of fraud to low possibility of fraud. If enough numbers of fraud are found, the algorithms will work well. For supervised model, what the author has is records with the possibility of being a fraud. The author can simply apply the result to a new data set and see how it works. Fraud detection rate is always a great criterion in this part.

3. Conclusion

Given different kind of initial conditions, the problems could be divided into supervised regression and unsupervised problems. After cleaning the data and make some feature selection process, we use machine learning to fix the regression problem and it works well. The machine-learning solution to the regression problem could be a great way of solving other business problems such as prediction[3] and detection[4].

References

[1] Comparison between linear model and machine learning model in prediction of yak weight, *Modern Agricultural Sciences and Technology*, 2019. [Online] Available: http://kns.cnki.net/kcms/detail/34.1278.S.20191209.1349.242.html [Accessed on Oct. 15, 2019]

[2] P. C. Yu and L. W. Zhong, Prediction of product testing time based on linear regression, *China Water Transport*, vol. 18, pp.79-80,83, 2018.

[3] J. P. Pu, The application of machine learning in the prediction of commodity sales, *Practical Electronics*, vol. 22, pp.81-82,70, 2018.

[4] M. J. Zhang, W. J. Yu ,W. B. Li, X. D. Zhu, Design of license plate recognition system based on machine learning [J/OL]. *Computer technology and development*, 2020 (04): 1-9, [Online] Available: Http://kns.cnki.net/kcms/detail/61.1450.tp.20191218.1113.046.html [Accessed on Dec. 19, 2019]

[5] C. H. Zhang, X. T. Zhou, Y. A. Zhang, J. M. Yao, T. L. Guo, Q. Yan, Research on Application of deep self encoder in data anomaly detection [J/OL], *Computer engineering and application*: 1-9 [Online] Available: Http://kns.cnki.net/kcms/detail/11.2127.tp.20191213.1448.010.html[Accessed on Dec. 15, 2019]

[6] Rui Sun, Guanyu Wang, Wenyu Zhang, Li-Ta Hsu, Washington Y. Ochieng, A gradient boosting decision tree based GPS signal reception classification algorithm[J], *Applied Soft Computing Journal*, 2019.

IOP Conf. Series: Materials Science and Engineering **806** (2020) 012006 doi:10.1088/1757-899X/806/1/012006

Simulation Research on Influence of Neural Networks Scale on its Performance

Xilei Wu

College of Computer Science and Electronic Engineering, Hunan University, Tianma Student Apartment, Juzizhou Street, Yuelu District, Changsha, Hunan.

Email: angela@cas-harbour.org

Abstract: When using deep learning methods to deal with specific problems, the structure and configuration complexity of neural networks model have a great impact on the performance of the system. This paper constructed a multi-layer perceptual neural networks model basing on classification problem of film and television reviews. The paper also analyzed the forward and back propagation, processing algorithm and key parameter optimization method, and built the simulation environment. By adjusting the network depth and the number of neurons in each layer, the paper analyzed the influence of the complexity of the neural networks model on the network performance. Then the paper gave the analysis results under the condition of different model scale. Finally the paper purported conclusion in the last section that the performance of the neural networks model changes greatly with number of neurons in each layer changed, but the performance changes little with network depth changed. It can provide an important reference for two classification problems in different fields.

1. Introduction

Nowadays, the deep learning method based on artificial neural networks develops rapidly and is widely applied[1]. When using deep learning methods to solve binary classification problems such as comments classification of film, television program, literature as well as art, and recognition of emotions and images[2]. The structure and configuration complexity of neural networks model have a great impact on the performance of the system. Many articles focus on the improvement of neural networks forward and back propagation algorithm on the basis of deep learning[3], or focus on application studies in specific fields such as sea load data processing[4] and etc. However, there is little quantitative analysis of the impact of network complexity on system performance. Considering the influence of model structure and scale on model performance, it is generally believed that multi-layer perception can have better performance with the same input [5]. In this paper, through the construction of multi-layer perceptual neural networks model and the establishment of simulation environment, by adjusting the depth of the networks and the number of neurons in each layer, we quantitatively analyze the impact of neural networks complexity on system performance. The models, simulation analysis methods and conclusions in this paper bring convenience to those who seek to find proper neural networks scale for their dichotomous or multiple classification applications and more importantly have a wide range of reference and application value in solving various binary classification and even multi-classification problems.

2. Principle Analysis of Model

Artificial Neural Networks (ANNs) is also called connection model. It is an algorithm mathematical model of distributed parallel information processing [6]. A typical neural networks consists of several neurons distributed in several layers and connections with different weights between adjacent layers. Figure 1 shows a multi-layer perception neural networks model which is also the theoretical analysis model of our research target.

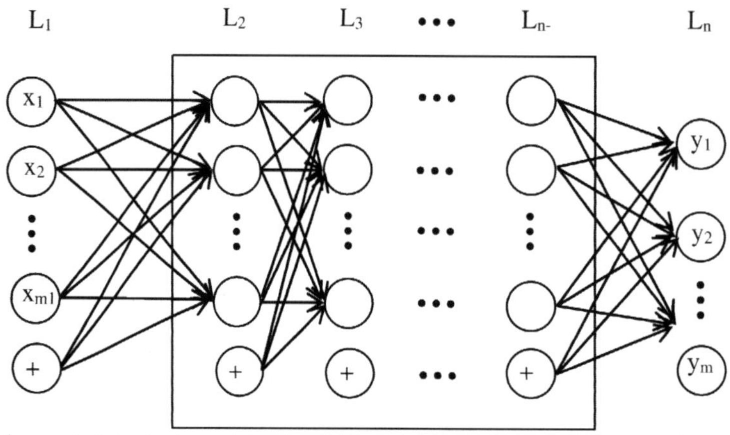

Figure 1.A typical model of multi-layer perception neural networks.

The model consists of n layers in total, including an input layer, an output layer and n-2 middle layers which are also called hidden layers. m_p (p = 1, 2, 3, ... , n) represents number of neurons in the pth layer L_p. Vector x = (x_1, x_2, x_3, ... , x_{m1}) is the input of the neural networks, and vector y = (y_1, y_2, y_3, ... , y_{mn}) is the output of the neural networks.

The process of neural network optimization is divided into two stages: forward propagation and back propagation. In the first stage, the predicted value is calculated by the forward propagation algorithm, and the error between the predicted value and the real value is obtained. In the second stage, the cost function and the gradient of each parameter that needs to be adjusted (such as weight w and bias b) are calculated by back propagation. And then the weight and bias of each connection are updated according to the gradient and learning rate by using the gradient descent algorithm.

Figure 2 represents the model of transmission and processing procedure of a certain neuron in forward propagation, which reflects the relationship among neuron input, weight connection, processing and neuron output.

In the model shown in figure 2, all the weights $w_{ik}^{(p)}$ (k = 1, 2, 3, ... , m_{p-1}) are randomly distributed. With input vector $a^{(p-1)} = [a_1^{(p-1)}, a_2^{(p-1)}, \ldots, a_{n(p-1)}^{(p-1)}]^T$, the status $z_i(p)$ and output $a_i(p)$ of the ith neuron in the pth layer are respectively represented in formula (1) and (2).

$$z_i^{(p)} = \sum_{j=1}^{m_{(p-1)}} w_{ij}^{(p)} \bullet a_j^{(p-1)} + b_i^{(p)} \tag{1}$$

$$a_i^{(p)} = f(z_i^{(p)}) = f\left[\sum_{j=1}^{m_{(p-1)}} w_{ij}^{(p)} \bullet a_j^{(p-1)} + b_i^{(p)}\right] \tag{2}$$

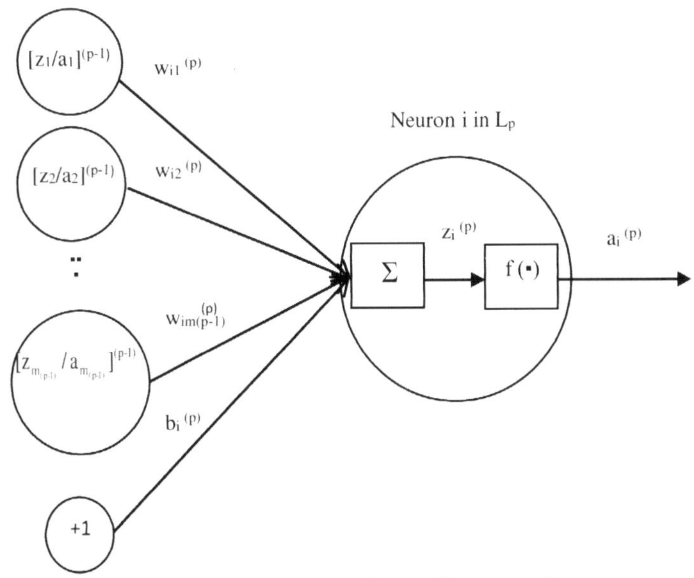

Figure 2.A neuron in the forward propagation.

In formula (1), $w_{ij}^{(p)}$ represents the weight of connection between the jth neuron in the $(p-1)^{th}$ layer and the ith neuron in the pth layer; $b_i^{(p)}$ represents the input bias from the ith neuron in the $(p-1)^{th}$ layer to the ith neuron in the pth layer; $z_i^{(p)}$ represents the status of the ith neuron in the pth layer; $a_i^{(p)}$ represents the input (or say activation) of the ith neuron in the pth layer; f(•) is the activation function as which we often use sigmoid , ReLU and so on. In figure 2, when L_{p-1} is the first layer (input layer), input vector is $a^{(1)} = x = (x_1, x_2, \ldots, x_{m1})^T$.

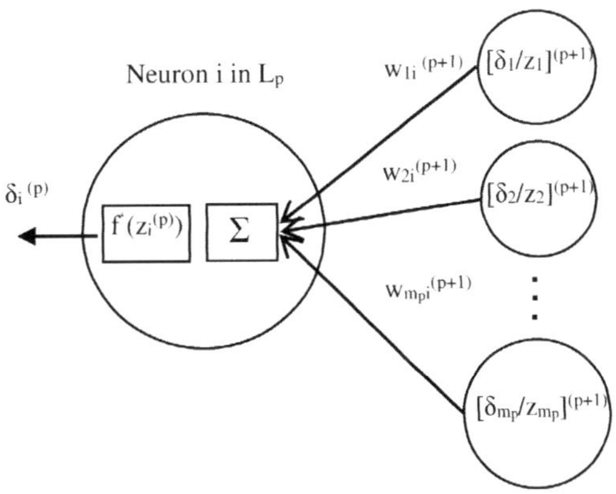

Figure 3.A neuron in the back propagation.

Figure 3 shows the model of transmission and processing procedure of a certain neuron in the error back propagation. In figure 3, the output $\delta_i^{(p)}$ can be expressed as formula (3).

$$\delta_i^{(p)} = \sum_{j=1}^{m_{(p+1)}} \delta_j^{(p+1)} w_{ji}^{(p+1)} \bullet f'(z_i^{(p)}) \tag{3}$$

In the formula (3), we use $\delta^{(p+1)}$ in layer L_{p+1} to calculate $\delta^{(p)}$ in layer L_p. Then, through training

IOP Conf. Series: Materials Science and Engineering **806** (2020) 012006 doi:10.1088/1757-899X/806/1/012006

from input layer to output layer and iteration of error back propagation, the parameters of neural networks are finally optimized. In back propagation, cost function reflects the difference between the expected value and actual value. And in order to optimize the performance of the neural networks model, making the actual value close to expected value, we should keep the cost function as small as possible.

The total average cost function of the network is expressed as formula (4)

$$E = \frac{1}{N}\sum_{k=1}^{N} E(k) \tag{4}$$

In the formula (4), $E(k) = \frac{1}{2}\sum_{i=1}^{m_n}[y_i(k) - a_i^{(n)}]^2$ is the cost function of the network under the kth training dataset.

Then according to gradient descent algorithm, calculation and optimization of weight w and bias b are expressed as formula (5) and (6).

$$\frac{\partial E}{\partial w_{ij}^{(p)}} = \delta_i^{(p)} a_j^{(p-1)} \tag{5}$$

$$\frac{\partial E}{\partial b_i^{(p)}} = \delta_i^{(p)} \tag{6}$$

3. Simulation and Analysis

3.1. Simulation Process and Related Parameters

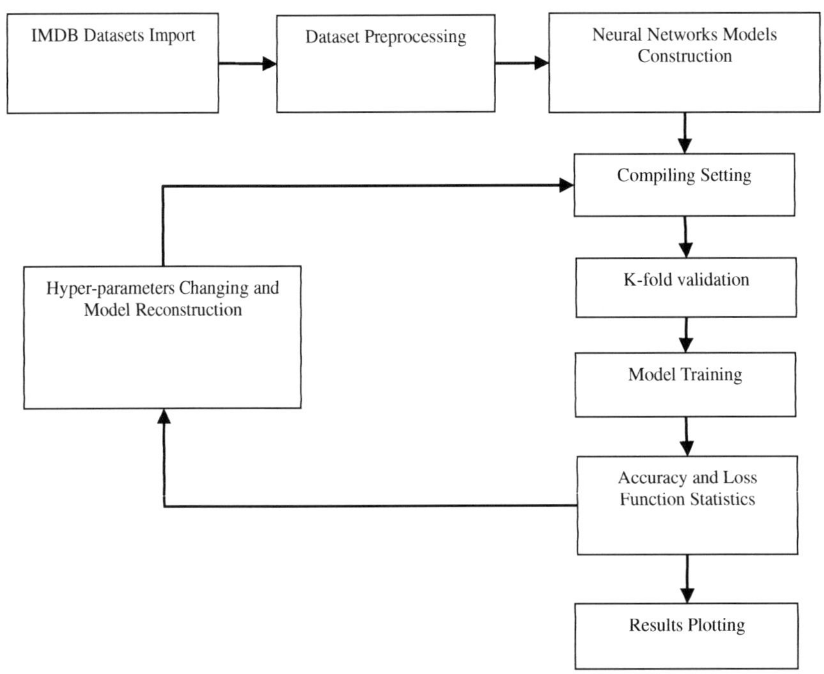

Figure 4.The working process of the neural networks model.

In this section, the paper describes the simulation preparation process and how the model of neural networks is constructed. A multi-layer perceptual neural network model is constructed based on the classification problem of film and television reviews, and relative model parameters are given. sentiment text classification of film comments is a dichotomous problem with only positive and negative target categories [7]. The working process of the neural network is given in figure 4.

In figure 4, IMDB (Internet Movie Database) contains several film comments data in the Keras library. In the model, select 10000 high-frequency English words for simplifying data processing, and vectorize the data of film comments to fit the requirement of neural networks input. Then use K-fold validation to set part of the data to validation data. The dataset have enough training data (25000 film comments), in order that simply dividing the data into training data and validation data. What to do next is build a model of neural networks for training data. Keras has highly modular and packaged model function, so it can just import model from Keras and add layers to it. Relative parameters of the model are presented in table 1 below.

Table 1.Simulation environment and relative parameters.

Dataset	Activator of hidden layers	Activator of output layer	Optimizer	Loss function	Metrics (parameters to monitor)	Number of training epochs
IMDB dataset in Keras Library	ReLU (Rectified Linear Units)	Sigmoid	RMSprop (Root Mean Square Prop)	Binary crossentropy	Accuracy & Loss	20

In the model, the number of neurons in each layer is changed firstly and then the number of layers in the model. Then the author builds 3 models with same parameters in the table 1 and with number of neurons or number of layers changed. Table 2 and Table 3 show experimental variables settings in part of the evaluations.

Table 2.Variable settings with numbers of neurons changed.

Model	Number of hidden layers	First evaluation		Second evaluation		Third evaluation	
1st model (model 0)	2	Number of neurons in each hidden layer	4	Number of neurons in each hidden layer	4	Number of neurons in each hidden layer	4
2nd model (model 1)	2	Number of neurons in each hidden layer	8	Number of neurons in each hidden layer	16	Number of neurons in each hidden layer	16
3rd model (model 2)	2	Number of neurons in each hidden layer	256	Number of neurons in each hidden layer	1024	Number of neurons in each hidden layer	4096

Table 3.Variable settings with numbers of layers changed.

Model	Number of neurons in each layer	First evaluation		Second evaluation		Third evaluation	
1st model (model 0)	16	Number of layers	1	Number of layers	2	Number of layers	2
2nd model (model 1)	16	Number of layers	2	Number of layers	3	Number of layers	3
3rd model (model 2)	16	Number of layers	3	Number of layers	5	Number of layers	9

Finally, results of accuracy and loss which are the metrics to evaluate performance of the model are plot in curve graphs. In order to have a better view of the comparison of the results, in one evaluation the accuracy and loss of these 3 models are plot in the same curve graph simultaneously, which will be presented in the next section. The specific outcomes of sentiment judgment of each film comment will not be shown in the results displayed in the next section because they are not in the scope of this paper.

IOP Conf. Series: Materials Science and Engineering **806** (2020) 012006 doi:10.1088/1757-899X/806/1/012006

3.2. Results Analysis

In this section, the paper gives part of the experiment results and results analysis. First, the number of neurons in each layer is changed and the number of layers remains the same. Figure 5 - 7 respectively show the accuracy and loss results of the 1st, 2nd and 3rd evaluation.

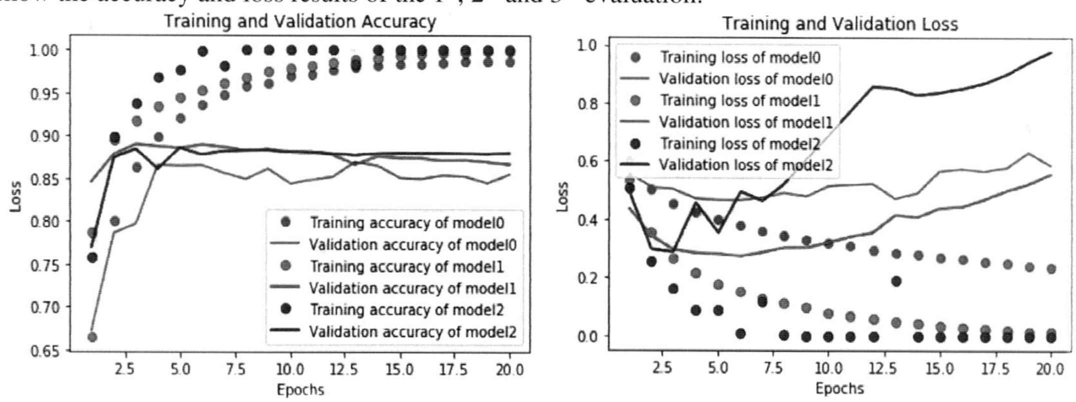

Figure 5.Result of the 1st evaluation with 4, 8, 256 neurons in each layer of model 0, 1, 2 respectively.

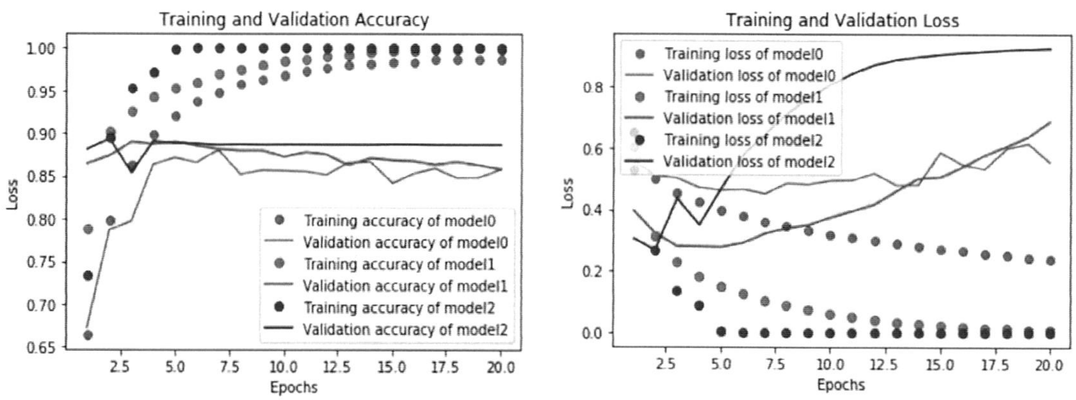

Figure 6.Result of the 2nd evaluation with 4, 16, 1024 neurons in each layer of model 0, 1, 2 respectively.

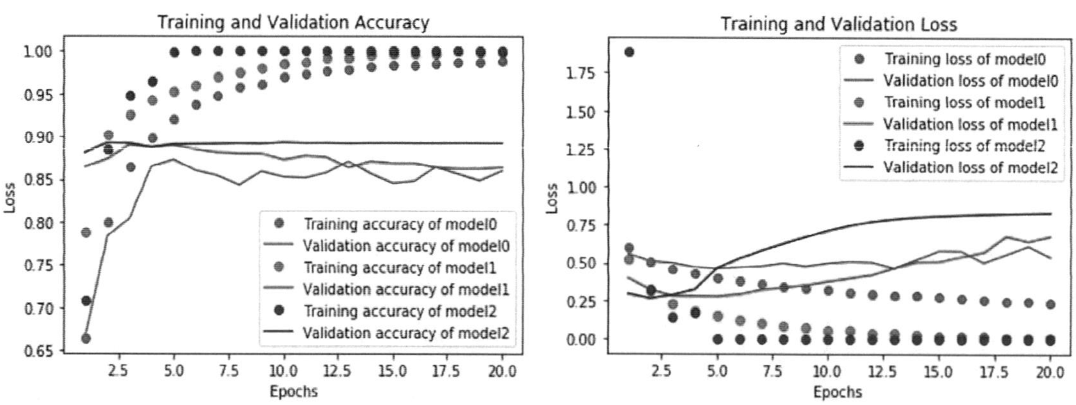

Figure 7.Result of the 3rd evaluation with 4, 16, 4096 neurons in each layer of model 0, 1, 2 respectively.

As shown in figure 5 - 7, a clear comparison of accuracy and loss of each model is displayed. Generally speaking, when dealing with a dichotomous problem, first, the more neurons there are, the larger the capacity of the neural networks will be, and the loss function of training data rapidly reaches the minimum value with the accuracy rate of training data quickly going to the top. When there are relatively less neurons in each layer, the increasing rate of accuracy and the declining rate of loss function get smaller, requiring more epochs of training to have an ideal training performance. Second, in terms of validation data, when there are more neurons in each layer, the loss function soon drop to the minimum and over-fit occurs. After over-fit, loss function rise sharply and the gradually flatten. Accuracy of validation data can just be improve to a limited extent (about 87%) and it raises fast to the maximum and then remains nearly invariable with bigger capacity of neural networks. Besides, even though not presented in the paper, when adding excessive neurons in each layer, such as 8096 neurons, the model loses the ability to learn from the training data. It also occurs when adding too little neurons to each layer.

Then, the number of layers in each model is changed while number of neurons remains the same in each layer. Accuracy and loss results of the 1st, 2nd and 3rd evaluation are presented in figure 8 - 10 respectively.

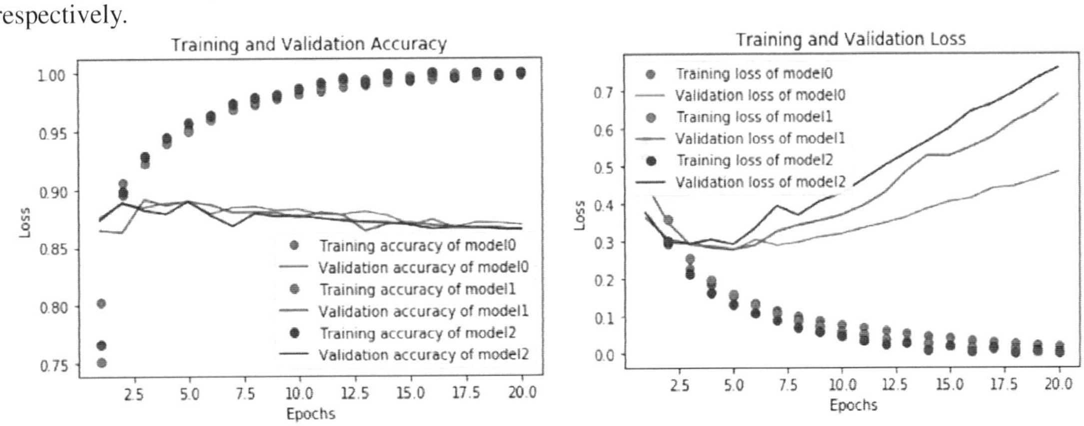

Figure 8.Result of the 1st evaluation with 1, 2, 3 layers in model 0, 1, 2 respectively.

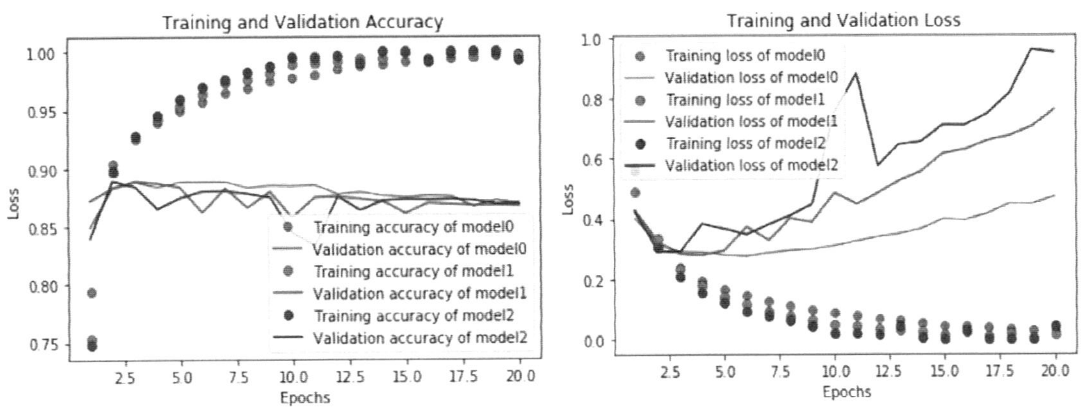

Figure 9.Result of the 2nd evaluation with 2, 3, 5 layers in model 0, 1, 2 respectively.

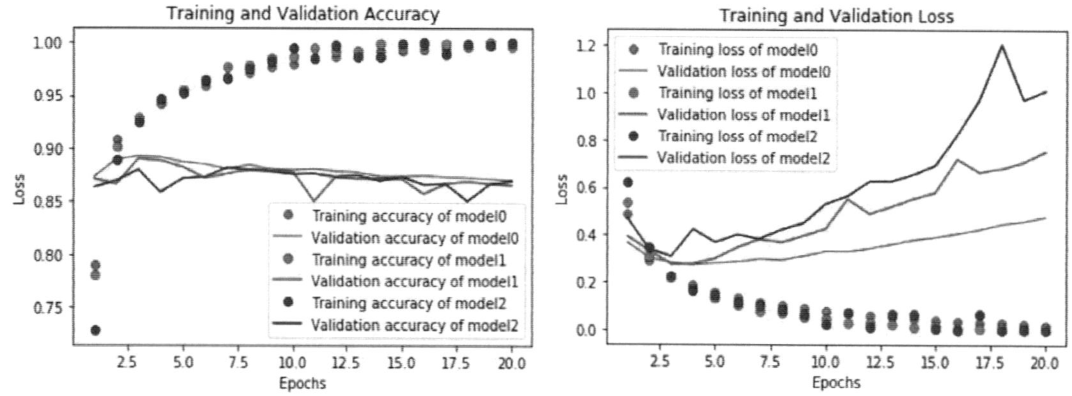

Figure 10.Result of the 3rd evaluation with 2,3,9 layers in model 0, 1, 2 respectively.

When number of layer increases, training accuracy and loss function curve are nearly the same, even if number of neurons is set to a bigger value like 256 or 512. In terms of the validation data, then loss function deteriorate sooner with more layers, while accuracy change little. Compared to the change of number of neurons in each layer, the change of number of layers has less effect on validation performance. What's more, similarly, model is unable to be trained if number of layers is excessive, such as 30 layers in a model.

4. Conclusions

The paper assesses the performance of neural networks which deals with a dichotomous problem by modeling training and validation process, and part of the evaluation results are presented. According to the results above, with bigger number of neurons in each layer, the validation loss function deteriorates more quickly and seriously. However, what is noticeable is that the general validation accuracy in each experiment set is higher than the other two models of smaller scale. As a consequence, the author thinks that the common statement that over-fit will leads to bad performance of neural networks is not precise. Moreover, the same experiment on another multi-classification problem (handwritten digits recognition) proves that it is very important to select the appropriate size of neural networks in accordance with different data scale to improve model performance. The above process has shown that changing the number of neurons and changing the number of neural networks layers have different effect on model performance, which may provide a reference for selecting different neural network structures for different data. Besides, more research on other kind of problems have to be carried out to further validate the conclusion that over-fitting does not necessarily lead to performance degradation.

Reference

[1] Ma,S.L., Li,X.P., (2016) Overview of big data and deep learning [J]. Journal of intelligent systems, 11(6):728-741.

[2] He,K., Zhang,X., Ren,S., etal. (2016)Identity Mappings in Deep Residual Networks[C]. European Conference on Computer Vision. Springer,Cham., 630-645.

[3] Kobayashi,N., Inui,K., Matsumoto,Y., etal. (2004) Collecting Evaluative Expressions for Opinion Extraction[J].Journal of Natural Language Processing, 12: 596-605.

[4] Liu, Y., Xu,L.X., (2018) High-performance back-propagation neural network algorithm for classification of sea load data [J]. Automation of power system,42(21):96-103.

[5] Shan,S.G., Meina,K., Liu,X., (2016) Deep learning: Renaissance and reform of multi-layer neural network [J]. Science and technology herald, 34 (14):60-70.

[6] Zhang,L., Zhang,Y., (2016)Bigdata analysis by infinite deep neural networks [J].Journal of Computer Research and Development,,53 (1):68-79.

[7] Zan,H.Y., Guo,M., Chai,Y.M.,etal. (2010) An investigation on the emotional orientation of news reports [J]. Computer engineering,36(15):20-22.

Application of Markov Chains

Qianfan Luo

International Curriculum Center, Shenzhen Foreign Languages School, Guangdong Province, 518000, China

Corresponding author's e-mail:angela@cas-harbour.org

Abstract. This paper introduces the theory of Markov Chains and illustrates the deduction of its transition probability matrix. A mathematical model for the growth of a bacteria colony and a mathematical model of the artificial intelligence in the venous transfusion are built. They are based on Markov chains and accord with the properties of transition matrix. The limitation of the model is discussed. The material is cited from authoritative websites, research done by academic institutions, and scientific essays.

1. Introduction

Markov chains are widely used in mathematics, especially in the field of statistics and information theory. They are widely used in economics, game theory, communication, genetics and finance. [4] When it comes to real-world problems, they are used to hypothesize solutions to study cruise-control systems in motor vehicles, queues or queues of customers arriving at airports, exchange rates and so on [2]. However, there are still some technical problems. For example, venous transfusion is a common treatment in people's daily lives. However, irregular use of venous transfusion can cause some dangers. Therefore, it's necessary to design a cheap and efficient intelligent intravenous infusion management system to prevent medical accidents caused by improper venous transfusion. By using the transition matrix theory, this paper draws a picture of the application of Markov Chain in this field and gives a proof of the nth matrix equation.

2. Analysis

2.1. Application of Markov chains in the prediction of the growth of bacteria

2.1.1. Assumption

Firstly, only one bacteria cell can duplicate in a time interval. The remaining bacteria are assumed not changed during that time. Secondly, during the duplication process, the probability for the bacteria to replicate or die is 50%. Thirdly, the population of the bacteria is limited to N and whether the bacteria can survive is 50%. Fourthly, the bacteria can only decay for limited generations. Under these circumstances, assume that the total number of the bacteria is N=4, and the probability of the bacteria to decay is P (decay) and the probability of the bacteria to die is P (die), suppose that the initial state of the bacteria is i and the final state of the bacteria is j, We find that the bacteria decay from state i to state j in a particular k generations, where k is an integral. Assume that the maximal capacity of the bacteria is 4, ie k=1, 2, 3, or 4, and the problem is how to find the possibility of passing from a specific state to another state within particular generations.

IOP Conf. Series: Materials Science and Engineering **806** (2020) 012007 doi:10.1088/1757-899X/806/1/012007

2.1.2. Formalism

To display the information in a more compact way, we can put the possible states into a matrix. Because there is only one possibility in the initial state, P(00)=1. Similarly, we can depict a diagram of the bacterial decay where the numbers represent the total number of bacteria and the rows refer to the generation of bacteria.

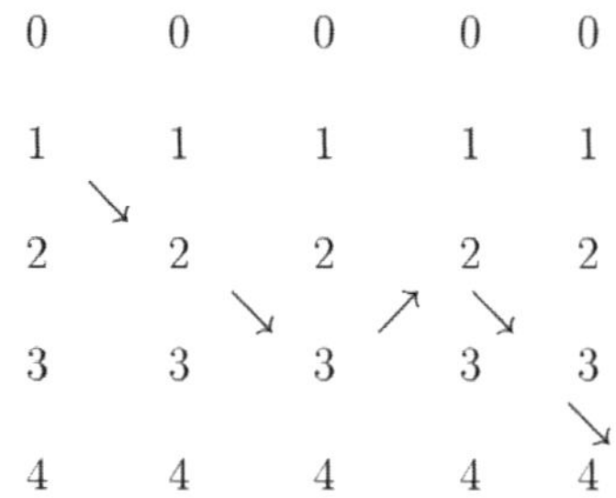

Each number stands for the state of the bacteria, while the arrows represent the transition of the state. Therefore, we can get a sequence of probabilities:

$$\{\ P00, P01, P02, \cdots, P44\ \}$$

Because when a bacterium decays, it only has one possibility to decay into two bacteria, so the matrix which is given from state i to state j by one step can be shown as:

$$\begin{pmatrix} P00 & P01 & P02 & P03 & P04 \\ P10 & P11 & P12 & P13 & P14 \\ P20 & P21 & P22 & P23 & P24 \\ P30 & P31 & P32 & P33 & P34 \\ P40 & P41 & P42 & P43 & P44 \end{pmatrix}$$

where the number serve as the probability of bacteria in each state.

2.1.3. Solutions

According to the second assumption, we can set up a model where P (decay) = P (die) = 1/2. Based on the assumption above, we can conclude that when bacteria decay, it has two options, decay or die, and the probability of each option is equal to 50%. In our assumption, the maximal generation is equal to 4. We have already known that from state i to state j, the probability of each state is given in the matrix above. Therefore, according to the Multiplication Principle, we can infer that in order to calculate the probability of bacteria decay in each state in two steps, we need to calculate the square of the matrix, that is, multiply the matrix by itself.

$$\begin{pmatrix} 1 & 0 & 0 & 0 & 0 \\ \frac{1}{2} & 0 & \frac{1}{2} & 0 & 0 \\ 0 & \frac{1}{2} & 0 & \frac{1}{2} & 0 \\ 0 & 0 & \frac{1}{2} & 0 & \frac{1}{2} \\ 0 & 0 & 0 & 0 & 1 \end{pmatrix}$$

According to the Matrix Multiplication Rule we have discussed in the third section, we can get the results of the probability of each state where bacteria decay:

$$\begin{pmatrix} 1 & 0 & 0 & 0 & 0 \\ \frac{1}{2} & \frac{1}{4} & 0 & \frac{1}{4} & 0 \\ \frac{1}{4} & 0 & \frac{1}{2} & 0 & \frac{1}{4} \\ 0 & \frac{1}{4} & 0 & \frac{1}{4} & \frac{1}{2} \\ 0 & 0 & 0 & 0 & 1 \end{pmatrix}$$

Similarly, when k=3, which means that we have to go three steps to get the result from state i to state j, the result is the cube of the original matrix, or we can calculate the product of the square sum of the matrices:

$$\begin{pmatrix} 1 & 0 & 0 & 0 & 0 \\ \frac{5}{8} & 0 & \frac{1}{4} & 0 & \frac{1}{8} \\ \frac{1}{4} & \frac{1}{4} & 0 & \frac{1}{4} & \frac{1}{4} \\ \frac{1}{8} & 0 & \frac{1}{4} & 0 & \frac{5}{8} \\ 0 & 0 & 0 & 0 & 1 \end{pmatrix}$$

In the same way, we can calculate the result when k is equal to 4, that is, we have to go four steps to get from state i to state j by identifying the biquadratic of the matrix. By multiplying the cube of the matrix, which we have already calculated above, with the original matrix, we can get the following results:

$$\begin{pmatrix} 1 & 0 & 0 & 0 & 0 \\ \frac{5}{8} & 0 & \frac{1}{4} & 0 & \frac{1}{8} \\ \frac{1}{4} & \frac{1}{4} & 0 & \frac{1}{4} & \frac{1}{4} \\ \frac{1}{8} & 0 & \frac{1}{4} & 0 & \frac{5}{8} \\ 0 & 0 & 0 & 0 & 1 \end{pmatrix}$$

We can use the same method to calculate the rest of the results.

Thus, we can figure out the general term formula when there are n steps. To make the calculation easier, we can find the relationship between two numbers in the same position in the two matrices and conclude that when k=n, the result will be

$$P^{(n)} = \begin{pmatrix} 1 & 0 & 0 & 0 & 0 \\ \frac{3}{4}+\alpha_n & \beta_n & \gamma_n & \epsilon_n & \frac{1}{4}+\alpha_n \\ \frac{1}{2}+\alpha_{n-1} & \beta_{n-1} & \gamma_{n-1} & \epsilon_{n-1} & \frac{1}{2}+\alpha_{n-1} \\ \frac{3}{4}+\alpha_n & \beta_n & \gamma_n & \epsilon_{n-1} & \frac{3}{4}+\alpha_n \\ 0 & 0 & 0 & 0 & 1 \end{pmatrix}$$

Define that:

$\alpha_n = (\frac{\sqrt{2}}{2})^n[-\frac{1}{4} - \frac{\sqrt{2}}{8} + (-1)^n(-\frac{1}{4} + \frac{\sqrt{2}}{8})]$

$\alpha_{n-1} = (\frac{\sqrt{2}}{2})^{n-1}[-\frac{1}{4} - \frac{\sqrt{2}}{8} + (-1)^{n-1}(-\frac{1}{4} + \frac{\sqrt{2}}{8})]$

$\beta_n = \frac{1}{4}[(\frac{-\sqrt{2}}{2})^n + (\frac{\sqrt{2}}{2})^n$

$\beta_{n-1} = \frac{1}{4}[(\frac{-\sqrt{2}}{2})^{n-1} + (\frac{\sqrt{2}}{2})^{n-1}$

$\gamma_n = \frac{\sqrt{2}}{4}[(\frac{\sqrt{2}}{2})^n - (\frac{\sqrt{2}}{2})^n]$

$\gamma_{n-1} = \frac{\sqrt{2}}{4}[(\frac{\sqrt{2}}{2})^{n-1} - (\frac{\sqrt{2}}{2})^{n-1}]$

$\epsilon_n = \frac{1}{4}[(\frac{-\sqrt{2}}{2})^n + (\frac{\sqrt{2}}{2})^n]$

$\epsilon_{n-1} = \frac{1}{4}[(\frac{-\sqrt{2}}{2})^n + (\frac{\sqrt{2}}{2})^n]$

What's more, to explain the concepts in a more specific way in the result matrix, $P(12)=\frac{1}{2}\cdot 0+0\cdot 0+\frac{1}{2}\cdot\frac{1}{2}+0\cdot 0+0\cdot 0$ due to the Matrix Multiplication Rule.

We assume that each decay is only influenced by the decay before it, which can explain the theory of Markov Chains. As a result, when there are two steps for the bacteria to decay into another state, we have to identify the probability of each step, and according to the Multiplication Principle, the two possibilities should be multiplied. Similarly, in the case of n steps, we find that the result of the probability will be the product of P and the power of n. So, when we go n steps to get from state i to state j, the probability is P^n.

2.2. Application of Markov chains in artificial intelligence using in the venous transfusion

2.2.1. Situation

We use a trial module to simulate the real situation of an intravenous injection (programmed with Arduino electronic blocks) that is supposed to be connected to an operator controlled by a nurse. The nurse need to operate the computer to adjust Arduino electronic blocks that are connected with Bluetooth, and to regulate the speed of the intravenous drip bottle. We also designed a smart bracelet to monitor the heart rate of the patient. However, we cannot avoid accidents. For example, the sudden breakdown of the speed controller, network signal problems, and information transmission errors.

2.2.2. Formalism

Under this condition, we can only consider the error probability of the speed controlling machine, the smart bracelet as well as the internet connection. Assume that P(x) is the probability of getting wrong information of each state, P(m) is the probability of the transfusion speed controlling machine making mistakes, P(b) is the error probability of smart bracelet, and k is the number of times of information transmitting. Consequently, for each time the information transmitted, the previous non-mistaken message will not influence whether the information will be transmitted successfully at this time, that is,

$$P(X_{t+1}|..., X_{t-2}, X_{t-1}, X_t) = P(X_{t+1}|X_t)$$

At the same time, any event is independent of the other events, so the only thing that can affect an event is the probability of the previous event. Consequently, we can apply Markov Chains to this system. The process of transmitting information starts from the smart bracelet detecting the patient's heart rate. When it finds something unusual, it will send a message to the computer which is monitored by a nurse. Then the nurse can manipulate the computer to control the speed of transfusion. When the Arduino electronic block receives the message, it will adjust the infusion speed artificially. Therefore, information needs to be transmitted four times by the machines in the system. In our experiment, we found out that the Arduino electronic block failed every twenty times we test it, that is, the error probability is roughly equal to 5%. Approximately, we can ignore the errors when we render the error probability from 0 to 10%.

According to the test on the smart bracelet, the probability of the system's stability is 93.27%, that is, 6.73% of the data detected by the smart bracelet is inaccurate. The computer, which is operated by the nurse, only has 10.6% probability to get wrong information or miss the signal from the Bluetooth. As above, we consider the data of which error being less than 5% is a normal experimental data.

2.2.3. Solutions

Above all these assumptions, each information transfer choice has a Markov chain when Xi is the number of the exchange in the ith state. Assume that at first there is only one piece of information, which means the initial state is 1, then the possible routes through different states are as follows:

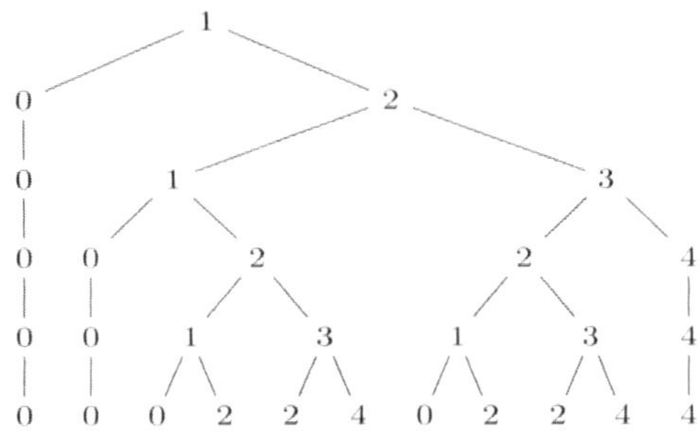

The information can only be transferred four times, so k can only be equal to 0, 1, 2, 3, or 4. Under the conditions described above, we first need to determine the expected value of the error probability of the whole system to better clarify the expected values of our experiment.

$$EV=\frac{1}{4}\cdot 5\%+\frac{1}{4}\cdot 6.73\%+\frac{1}{2}\cdot 10.6\%=8.2325\%$$

If we want to figure out the probability of state i to state, we can first draw a matrix which might be related to the bacterial culture problem:

$$\begin{pmatrix} P00 & P01 & P02 & P03 & P04 \\ P10 & P11 & P12 & P13 & P14 \\ P20 & P21 & P22 & P23 & P24 \\ P30 & P31 & P32 & P33 & P34 \\ P40 & P41 & P42 & P43 & P44 \end{pmatrix}$$

which is equivalent to:

$$\begin{pmatrix} 1 & 0 & 0 & 0 & 0 \\ 0.082325 & 0 & 0.082325 & 0 & 0 \\ 0 & 0.082325 & 0 & 0.082325 & 0 \\ 0 & 0 & 0.082325 & 0 & 0.082325 \\ 0 & 0 & 0 & 0 & 1 \end{pmatrix}$$

Naturally, if a matrix satisfies the two properties above and k is given as a constant, then it is a transition matrix. Because if we have such a matrix and know every element of it, we can construct Markov chains and let $\mathbb{P}_{ij}^{(k)} = a_{ij}^{(k)}$.

Thus, this matrix has to be a transition matrix.

Depending on the bacterial culture instance, we can easily find the solution to the venous transfusion problem. When k=2, the result will be the previous matrix multiplied by itself because of the Multiplication Principle, so we can get the result:

$$\begin{pmatrix} 1 & 0 & 0 & 0 & 0 \\ 0.082325 & 0.006777 & 0 & 0.006777 & 0 \\ 0.006777 & 0 & 0.013555 & 0 & 0.006777 \\ 0 & 0.006777 & 0 & 0.006777 & 0.082325 \\ 0 & 0 & 0 & 0 & 1 \end{pmatrix}$$

Similarly, we are able to calculate the error probability of state i to state j, where k equals to 3, or 4.
Therefore, the theory $P^{(n)}= P^{(n)}$ can be proved. [3]

Proof. We need to prove that each element in the two matrices is equal. Assume that $a_{ij}^{(n)}$ is the element of the ith row, jth column in $P^{(n)}$, and $b_{ij}^{(n)}$ is the element of the ith row, jth column in $P^{(n)}$. Now we assume that $P^{n-1} = P^{(n-1)}$ is true.

Obviously,

$$\sum_{k=1}^{M} a_{ik}a_{kj} = b_{ij}^{(2)}$$

To calculate $a_{ij}^{(n)}$, firstly, it has to travel (n-1) steps. In fact, we already knew the (n-1)-step transition matrix because we have assumed that $P^{n-1}=P^{(n-1)}$. According to the Markov property, we can assume that it will first travel to the k state in the (n−1)th step. For different k, these events are disjoint. So,

$$\sum_{k=1}^{M} \mathbb{P}_{ik}\mathbb{P}_{kj} = \mathbb{P}_{ij}^{(2)}$$

$$\sum_{k=1}^{M} a_{ik}a_{kj} = a_{ij}^{(2)}$$

Therefore, the two matrices are equal.

3. Comment and Evaluation

3.1. Limitations of Bacteria Decay

Firstly, there are about "1,000,000 bacteria per milliliter (1/5 of a teaspoon) of water in the coastal ocean" according to the *Cryomics Lab* [3]. Hence, the number of bacteria cannot be four. Secondly, bacteria can double in a short period of time. Therefore, bacteria do not decay at a constant speed. Thirdly, bacteria need specific environment to reproduce. Thus, the probability of the bacteria decay or die cannot always be equal to 1/2.

3.2. Limitations of factors

In our assumption, there are only four factors that influence the efficiency of messages transmission; however, it is obviously not true in real life. Factors that may affect the final result include external signal interference, signal connection, human interference and others.

3.3. Suggestions and Improvements

In order to make the experiment more precise, we need to expand our sample space. Additionally, we can consider some uncertainties in our data or set some random variables in the number, such as setting the probability of the bacteria decay from 40% to 60%, so that we are able to reduce the error.

For the venous transfusion part, we can simulate the real world by adding numerical constraints, but human behavior is the hardest to predict, so we can't make accurate predictions, only rough data values and a reasonable uncertainty range. As a result, we can only set the random variable of the machine or human performance possibility to the veracity of our data.

4. Conclusion

The conclusion leads to the basic application of Markov chain in daily life, and a derivation to its transition probability is maken by providing the detailed calculation process. However, there are still several problems that this paper does not solve. Firstly, there is no general way to calculate all the $P^{(n)}$ if we rarely know that $P^{(n)}$ is a transition matrix. The main reason is that not every transition matrix can be diagonalized. However, if we know that the state i can only move to state (i+1) or state (i-1) in one step and the number of states is limited, we can solve this problem as we have done before. Besides, if the probability for a bacterium to duplicate itself successfully is not 1/2 butp, the transition matrix can still be diagonalized. In this case, we can divide the whole fluctuation process into several short periods, during which we consider the transition probability as a constant value. Through the step-by-step calculation process, we can eventually apply the Markov chain to solve the problem.

Future papers would focus on the unsolved problems in this paper and try to improve the accuracy of data collection and calculate the probability of eventual outcomes.

Reference

[1] R Eric Collins (2009), How many bacteria are in the ocean? And how far is it between bacteria in the ocean? http://www.reric.org/wordpress/archives/648
[2] Paul A. Gagniuc (2017), Markov Chains: From Theory to Implementation and Experimentation (1st edition). Wiley publishing, New York City.
[3] Joseph K. Blitzstein (2008), Markov Chain, Introduction to Probability (2nd edition), Athena Scientific publishing, Belmont, MA. pp. 406-409
[4] Sejal Jaiswal (2018), Markov Chains in Python: Beginner Tutorial. https://www.datacamp.com/community/tutorials/markov-chains-python-tutorial

IOP Conf. Series: Materials Science and Engineering **806** (2020) 012008 doi:10.1088/1757-899X/806/1/012008

Study on Dissolved Oxygen, Ammonia-nitrogen and Permanganate Index in Water of Lake Dianchi

Lucen Li

College of Computer Science, Chong Qing University, Chongqing, 400044, China

angela@cas-harbour.org

Abstract. The water environment of Lake Dianchi in Yunnan province is threatened by eutrophication, which leads to the decrease of dissolved oxygen in water. Meanwhile, as the index reflecting the pollution of organic and inorganic oxidizable substances in water, permanganate index is also affected. China has been developing the automatic monitoring system for a series of water quality factors, such as dissolved oxygen and permanganate index. Additionally, the monitoring and prediction of ammonia-nitrogen, one of the water eutrophication sources, are also important. This study analyzed the water quality factor data in Lake Dianchi from 2015 to 2018 and implemented the prediction based on the Long Short-Term Memory neural network algorithm. It is found that the water quality of Lake Dianchi will be in a relatively ideal state in the future, but it cannot be improved further by itself. The Kunming government should implement more targeted improvement measures according to the characteristics of water quality factors, such as the seasonality of Lake Dianchi. Those measures will improve the ecological service quality of Lake Dianchi effectively, ensure the daily water safety of people in Kunming, and bring more economic values to the city.

1. Introduction

China is a country with many lakes. With stable conditions, lakes have the abilities to provide water for producing and living, rendering aquatic resources, regulating atmospheric and so on[1]. Water pollution is a crucial threat to maintaining the values of lakes. Lake Dianchi is the primary water source in Yunnan Province. With the rapid development of economy and society, the water quality in Lake Dianchi declines and pollution is increasingly severe. Since the 1990s, the overall water quality in Lake Dianchi becomes worse than type V[2].

Table 1 Basic Items of Environmental Quality Standards for Surface Water Unit: mg/L

Number	Standard value / Type / Item	I	II	III	IV	V
1	Temperature (° C)	Water temperature changed by human should be limited to: maximum rise of weekly mean temperature ≤ 1 maximum drop of weekly mean temperature ≤ 2				
2	PH Value	6 ~ 9				
3	Dissolved oxygen ≥	7.5	6	5	3	2
4	Permanganate index ≤	2	4	6	10	15
5	Chemical oxygen demand (COD) ≤	15	15	20	30	40
6	Five days' biochemical oxygen demand (BOD_5) ≤	3	3	4	6	10
7	Ammonia-nitrogen (NH_3-N) ≤	0.15	0.5	1.0	1.5	2.0
8	total phosphorus (TP) ≤	0.01	0.025	0.05	0.1	0.2
9	total phosphorus (TN) ≤	0.2	0.5	1.0	1.5	2.0

Water pollution has changed the value structure of Lake Dianchi. From 2006 to 2015, the average annual ecosystem service value of Lake Dianchi was 386.55 billion yuan and the water pollution in Lake Dianchi caused a loss of 1.523 billion yuan to the economic and social development in Yunnan Province. The principal economic losses mainly came from the cost of water recharge and cyanobacteria removal[1].

Previous researchers have found that Lake Dianchi was eutrophicated and polluted. In 2018, the water quality of Caohai of Lake Dianchi raised to class IV[3], which means that the treatment of Lake Dianchi has made some achievements in local prevention and treatment. However, due to the complexity and uncertainty of the ecological restoration process of lake water, there is no fundamental solution to the pollution control in Lake Dianchi. Therefore, this research aimed to analyze the feasibility of maintaining Permanganate index, NH_3-N and Dissolved oxygen of Lake Diachi that meet the requirement of type IV, and upgrading water quality to type III in the future.

Molecular oxygen in air dissolves in water becomes the dissolved oxygen, which is usually called DO. The amount of DO in water is an index to measure the self-purification ability of water. If DO in water is consumed and it takes a short time to restore to the initial state, it is indicated that the water has a strong self-purification ability or the water pollution is not serious. Permanganate index refers to the amount of consumed oxidant when treating water samples with potassium permanganate as oxidant under certain conditions. It represents the total amount of reducing substances (mainly organic pollutants) in the water sample that can be oxidized by potassium permanganate. The higher the permanganate index is, the more serious the water is polluted by organic matter[4]. NH_3-N is a nutrient in water, which can lead to water eutrophication. It is a major oxygen consuming pollution in water, which is toxic to fish and other aquatic organisms[5].

2. Literature review

Eutrophication is one of the significant challenges to the lake environment in China.

Eutrophication is a kind of water pollution caused by excessive nitrogen, phosphorus and other plant nutrients. It causes algae and other plankton proliferate rapidly, resulting in reduced dissolved oxygen in the water. Monitoring and controlling the content of NH_3-N, DO and Permanganate index is an important measure to protect the water quality of Lake Dianchi. At present, many scholars have made lots of researches on the water quality index that affects eutrophication and its modeling methods.

Yang Yinkang and Zou Xuexian[6] selected environmental quality assessment parameters, determined the assessment standards of various pollutants, calculated the comprehensIVe assessment map of environmental quality and obtained the pollution situation of each part of Lake Dianchi according to the map. Therefore, this study selected three parameters: Permanganate index, NH_3-N and DO, to determine the pollution condition of Lake Dianchi. Moreover, Chen Yuling and Zhang Shisen[7] selected four main limiting factors raised by Snincdder: the limit values of total phosphorus concentration, chlorophyll concentration, transparency and saturation of dissolved oxygen in the deep water layer as the discriminant of the eutrophication process of Lake Dianchi. Through the experiment, Chen Yuling established a phosphorus model to describe the total phosphorus concentration over time. The model was calibrated through the measured data of 1988, and verified with the measured data of 1989. It used the mean square root verification error to represent the simulation accuracy. Chen Canyun, Zhang Baoxu and Li Yuliang[8] established a model to determine apparent settling velocity S of total phosphorus on the basis of the measured date of the Miyun Reservoir from 1985 to 1995 and used the apparent settling velocity S to predict the change of TP concentration. Therefore, this study focused on the change trend of observed data of Permanganate index, NH_3-N and DO to predict these three parameters of Lake Dianchi. Liu Liping[9] adopted chlorophyll concentration, total phosphorus concentration and water transparency as the discriminant, used the modified Carlson index to evaluate the eutrophication of Lake Dianchi and analyzed the eutrophication level of Lake Dianchi and the possible causes from 1988 to 1999.

As the rise of more water quality control schemes in Lake Dianchi, more experts studied the water environment of Lake Dianchi. Liang Zhongyao and Liu Yong[10] took Waihai of Lake Dianchi as the analysis object, selected eight detection points to detect the water quality indicators and used STL and RSI method to decompose the water quality indicators to study and analyze the change trend. Therefore, this study identified the variation trend of water quality index in Lake Dianchi based on the long time series of water quality data. In addition, many scholars have predicted the trend of water quality according to different models. For example, Zou Zhihong and Wang Lejuan[11] applied the Gray Markov model to predict the eutrophication trend of Lake Taihu, Lake Dianchi and Lake Chaohu. Yang Zhiying[12] established a BP network model based on the observed data of six environmental quality assessment parameters and used the model to predict the pollution trend of Lake Honghu. Moreover, Zhang Hucai and Chang Fengqin[13] used the detected data of four points of Dianchi Lake from April to September in 2015 to analyze the transformation of each parameter from the dry season to the rainy season, as well as the characteristics and spatial changes of the rainy season. This study also studied water quality changes under different seasons and analyzed the fluctuation of water quality in Lake Dianchi. Chen Yingyi, Cheng Qianqian[14] used PCA to select the water quality parameters and established a prediction model for dissolved oxygen in water for aquaculture based on long short-term memory neural. Through the experiment, the PCA-LSTM network model was proved to meet the actual demand of predicting dissolved oxygen. Therefore, this study adopted the LSTM neural network to predict Permanganate index, NH_3-N and DO in Lake Dianchi.

This project focused on Permanganate index, NH_3-N and Dissolved oxygen (DO) in Lake Dianchi, used scientific computing tools to analyze the pollution condition of Lake Dianchi, and employed Deep learning prediction model (LSTM) to evaluate the feasibility of improving water quality of the whole Lake Dianchi to break from type IV in the future. If the water environment in Lake Dianchi is improved in a certain situation in the computer simulation, it means that it might be a feasible solution to the pollution treatment of Lake Dianchi. Meanwhile, it can provide the Kunming Government with more precise guidance in pollution treatment and strategic planning. Moreover, it is helpful for Kunming Government to reduce economic losses in water treatment and restore the ecosystem services of Lake Dianchi.

3. Research methods

3.1. Long Short-Term Memory neural network

LSTM was proposed in 1997 by Sepp Hochreiter and Jürgen Schmidhuber[15]. It is an special kind of

Recurrent Neural Networks (RNN). Standard RNN cannot deal with long minimal time lags between relevant signals. Unlike RNN, a common LSTM unit is composed of a cell, an input gate, an output gate and a forget gate. LSTM adds or removes information to the cell state through the gate structure and the cell can remember values over arbitrary time intervals. Therefore, LSTM has ability to deal with the issues that are highly related to time series.

3.2. LSTM water quality factor prediction model

3.2.1. The data sample

This project took the surface water quality monitoring data of Lake Dinachi, Guanyin Mountain site, in Kunming, Yunnan province as the research object, and used LSTM to construct the prediction model of water quality factors.

China environmental monitoring station uses automatic water quality monitoring system to detect water quality. At present, most of the monitoring instruments in China are mainly used to monitor the comprehensive pollution index, and there is a lack of instruments to monitor individual pollutants, such as the instruments to monitor the organic pollutants[16].

This project took Permanganate index, NH_3-N and DO value from 2015 to 2018 as the input parameters to the model and predicted their change trend. Partial data of selected data samples are shown in Table 2.

Table 2 Partial data of selected data samples

	time	name	PH value	DO(mg/L)	NH_3-N(mg/L)	Permanganate index(mg/L)
1	2017-08-01 00:00:00	Guanyin Mountain	8.36	5.56	0.52	12.10
2	2017-08-01 04:00:00	Guanyin Mountain	8.20	5.19	0.61	12.60
3	2017-08-01 08:00:00	Guanyin Mountain	7.98	4.77	0.65	11.08
4	2017-08-01 12:00:00	Guanyin Mountain	7.65	3.76	0.70	10.43

In order to ensure the stability of time series data and the high efficiency of the model, the data is first normalized and converted to [0, 1]. This project adopted the Min-Max scaling method. The formula (1) is shown as follows:

$$X_{norm} = \frac{X - X_{\min}}{X_{\max} - X_{\min}} \quad (1)$$

3.2.2. Test results

Using the Python 3.6.3 as programming language, this project was implemented under the development environment of Anaconda 4.3.30 and based on TensorFlow 1.13.1. Parameters related to the experiment were set as follows: LSTM time step 3, hidden_size of hidden neurons 32, batch_size 10, num_layers 1, training_steps 2000. This project predicted the water quality data of the next 12 months based on water quality data samples from 2015 to 2018.

In order to display the prediction results of the LSTM algorithm for various water quality factors visually, the real data and prediction data graphs of three factors are drawn respectively, as shown in Figure 1-3. The figures show the change trend of each factor in the next 12 months. Since the time step set is 3, the data of the first 3 months are omitted from the figure.

IOP Conf. Series: Materials Science and Engineering 806 (2020) 012008 doi:10.1088/1757-899X/806/1/012008

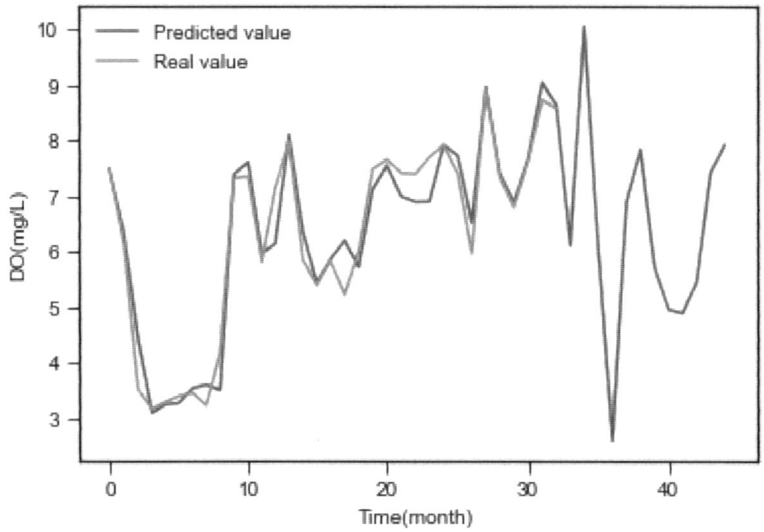

Fig. 1. DO prediction results

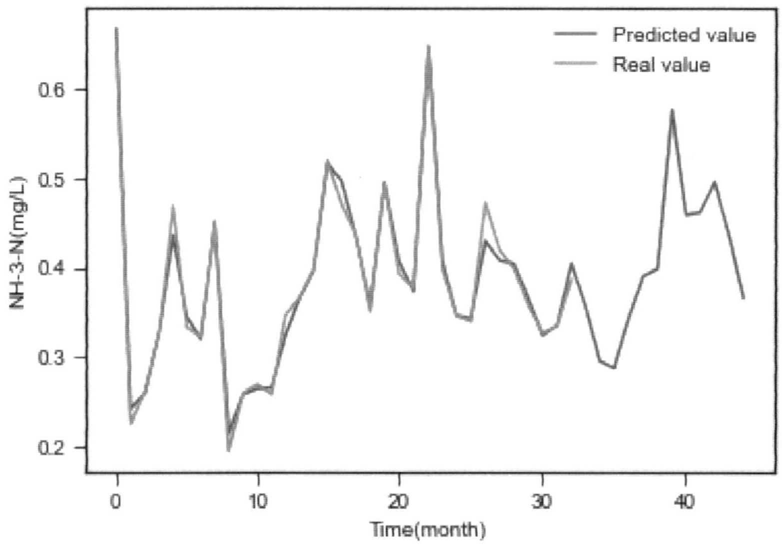

Fig. 2. NH_3-N prediction results

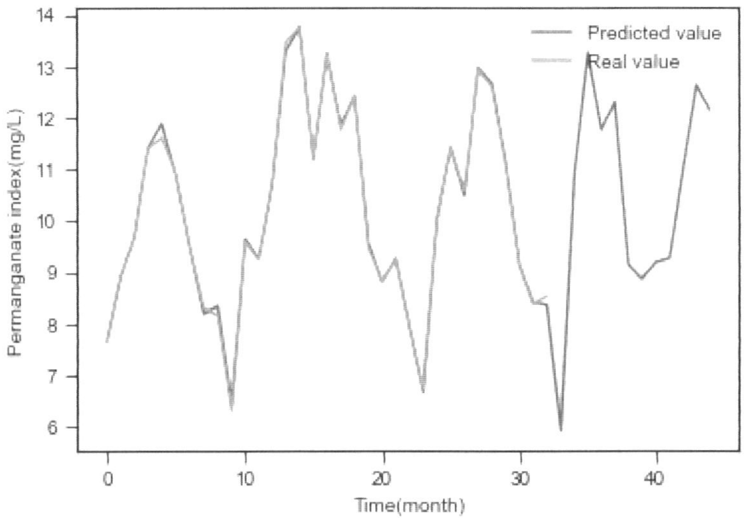

Fig. 3. Permanganate index prediction results

3.2.3. *The model evaluation indices*

This project adopted the unIVersal mean square error (MSE) to evaluate the regression prediction model. The calculation formula is shown in formula (2).

$$MSE(y, y') = \frac{\sum_{i=1}^{n} (y_i - y_i')^2}{n} \qquad (2)$$

The calculation results of MSE of each water quality factor are shown in Table 3. It can be seen from the table that the relatIVe error of the water quality prediction model based on LSTM is low.

Table 3 Mean square error of predictors (%)

predictors	DO	NH_3-N	Permanganate index
MSE	5.386	1.408	0.0423

4. Experimental summary

Water pollution has affected the ecosystem service of Lake Dianchi severely and caused serious economic losses. This project analyzed three main water quality factors in Lake Dianchi in the past three years and predicted the change trend of each factor in the next twelve months based on LSTM algorithm. Four results are summarized as followed:

1. Both actual and predicted values of NH_3-N meet the requirements of type III. Although it might fluctuate in the next 12 months, most of the time is still lower than 0.5 and has reached the class III water quality requirement. Therefore, the ammonia nitrogen is not the main pollutant in Lake Dianchi.

2. Only one of the predicted values of DO does not meet the requirements of type III. It suggests that Lake Dianchi still has the self-purification ability and DO is less likely to be reduced to type IV standard in the future.

3. Permanganate index is the main pollution factor. The measured data of the water quality of Lake Dianchi over thirty-six months are uniformly over 6, which does not meet the requirements of class III water quality. In addition, more than half of the time does not meet the requirements of class IV water quality. Moreover, it shows a certain seasonality, which is at a high level in the middle of each year. And this seasonality is also reflected in the predicted results. Therefore, the Kunming government should pay more attention to the seasonality of permanganate index, and adjust the preventive measures. The feasibility of breaking away from type IV and upgrading to type III is high, if the Kunming government attaches importance to the control of permanganate index.

4. The prediction model of water quality based on LSTM shows prediction results with small errors, which provides a guiding method for the development of automatic water quality monitoring system in China. By increasing the prediction function of water quality monitoring system, it can find and improve the deficiencies of equipment timely and provide better monitoring scheme.

5. Conclusion

This project used the data measured by the automatic surface water monitoring system from the Ministry of Ecology and Environmental of the People's Republic of China. The function of automatic water monitoring system is not perfect and the data defaults and exceptions are common, which increases the difficulty of data analysis. If water quality data are measured in a more regulated approach, the data will be more accurate, which will improve the accuracy of prediction and provide more specific guidance to water improvement and the development of automatic water monitoring.

References

[1] Wei, G. Du, Z.P., Yan, C.A., Chen,Y. (2019) Evaluating net ecosystem services value of a polluted lake: A case study of Lake Dianchi. Acta Ecologica Sinica, 39 (5).
[2] Feng, N., Mao, F., Li, X.Y., Zhang, A.D. (2010) Research on Ecological Security Assessment of Dianchi Lake. Environmental Science, 31 (2).
[3] Monthly report on water quality monitoring of lakes and plateaus. (2018) Yunnan Provincial Department of Ecological Environment.
[4] Lai, J.F., Wu, J.N., Liu, L.B. (2018) Study on the correlation of surface water dissolved oxygen with permanganate index and chemical oxygen demand. City and Town Water Supply.
[5] Jiang, R., Zeng, Y.H., Wang, Q. (2013) Research progress of ammonia nitrogen wastewater treatment technology Environmental Science and Management, 38 (06): 131-134.
[6] Yang, Y.K., Zou, X.X. (1982) Study on environmental quality evaluation of Lake Dianchi. Acta Academia Medicine Kunming.
[7] Chen, Y.L., Zhang, S.S. (1991) A Simulation Model for Eutrophication Prediction in Lake Dianchi. Journal of Chengdu University of Science and Technology, 6 (60).
[8] Chen, Y.C., Zhang, B.X., Li, Y.L.(1998) Nutrient analysis and prediction of Miyun reservoir. Shuili Xuebao.
[9] Liu, L.P. (2001)Analysis on Eutrophication of Dianchi Lake and its Control Measures. Yunan Environmental Science.
[10]Liang, Z.Y., Liu, Y., Sheng, H., *et al.* (2014) Identification of water quality trend in Lake Dianchi using an integrated STL and RSI approach Acta Science Circumstantiae, 34(3).
[11]Zou, Z.H., Wang, L.j. (2009) Gray Markov forecast of lake eutrophication. Acta Scientiae Circumstantiae, 29(2).
[12]Zhang, Q. (2016) Prediction of Water Quality Index of Honghu Lake based on Back Propagation Neural Network Model, 14(2).
[13]Zhang, H.C., Chang, F.Q., Duan, L.Z., *et al.* (2017) Water quality characteristics and variations of Lake Dianchi. Advanced in Earth Science, 32 (6).
[14]Chen, Y.Y., Cheng, Q.Q., *et al.* (2018) Principal component analysis and long short-term memory neural network for predicting dissolved oxygen in water for aquaculture. Transactions of the Chinese Society of Agricultural Engineering, 34(17).
[15]Hochreiter, S., Schmidhuber, J. (1997) Long Short-Term Memory. MIT Press.
[16]Yang, H.D. (2019) Interpretation of Application and Development Trend of Surface Water Environment Automatic Monitoring Technology. Energy Conservation & Environmental Protection.

IOP Conf. Series: Materials Science and Engineering **806** (2020) 012009 doi:10.1088/1757-899X/806/1/012009

Problems And Solutions Of China's Credit Rating Industry Under The Background Of Big Data

Liang Liu

School of computer science and technology, Harbin University of Science and Technology, Harbin, Heilongjiang, 150080, China

Corresponding author's e-mail: angela@cas-harbour.org

Abstract. Customer credit rating adopts scientific methods and standardized procedures to investigate, analyze, evaluate, measure and review the ability of rating objects. The objective of the system is to fulfill corresponding economic commitments. Today, big data enriches the data source of credit data that surpasses that of traditional financial data. This paper starts with the concept of enterprise credit rating system, through the understanding of credit rating system and the combination with big data, to understand the existing problems and the causes of the current credit rating system in China.

1. Introduction

For a long time, China's capital has been experiencing soaring corporate and personal debts based on an inefficient credit rating system. Presently, the nation divides the enterprise credit rating into industrial and commercial enterprises, financial as well as public institutions. Using the existing credit rating standards to evaluate creative enterprises, often cause unfair, objective, unscientific results that lack credibility. Therefore, the enterprise credit rating needs to consider the particularity of industry and enterprise. The development of China's credit rating industry is secondary to that of foreign countries in rating technology and market demand. Similarly, there lacks proper positioning of rating agencies and national credit system construction and management. This paper starts with the concept of enterprise credit rating system, through the understanding of credit rating system and the combination with big data, finds out the existing problems and the causes of the current credit rating system in China.

2. Literature review/ Related work

A majority of Chinese individuals lack faith in the integrity of the government due to years of environmental disaster, Ponzi schemes, and food scares[2]. Critics believe that the system has the ability to minimize the freedom of speech since the government will be able to control the credit score[2]. China lacks a common regulatory body for the credit rating system. There are more than 100 credit rating agencies in China, which are far higher than the average level of developed countries in Europe and the United States[4]. The high number increases competition in the industry, causing a lack of coordination. The Chinese researchers have always had an issue in finding independent and reliable data. The issue is due to the government's stance on researching sensitive topics, which is a punishable offense[3]. The systematic nature of the social credit system in China means the lack of infrastructure to ensure reliability of the data. The inadequate collection of data will cause an issue with scoring patterns and the role of the judiciary. Hence, the government requires adequate infrastructure to ensure proper collection of data to enable an operational and effective system.

3. Theoretical basis and research hypothesis

The theoretical basis focusses on the developments in the credit scoring ecosystem. The new forms of credit agencies are experiencing resistance from already established ones. Until 2011, more than 460 million Chinese citizens lacked a credit history or bank account an occurrence that prompted the creation of credit agencies[5]. Additionally, companies have been defaulting payments for the past decade, which negatively affects the Chinese economy. According to data by Bloomberg (2019), Chinese firms defaulted on over 39 billion yuan of domestic bonds in the initial four months of 2019, which is 3 times that of 2018 [1]. When compared to previous years, the rate of default is more than triple, which shows the need for a functional credit system to avoid higher figures in 2020. The statistic led to the development of agencies that did not just measure credit scores, but evaluated social credit scores.

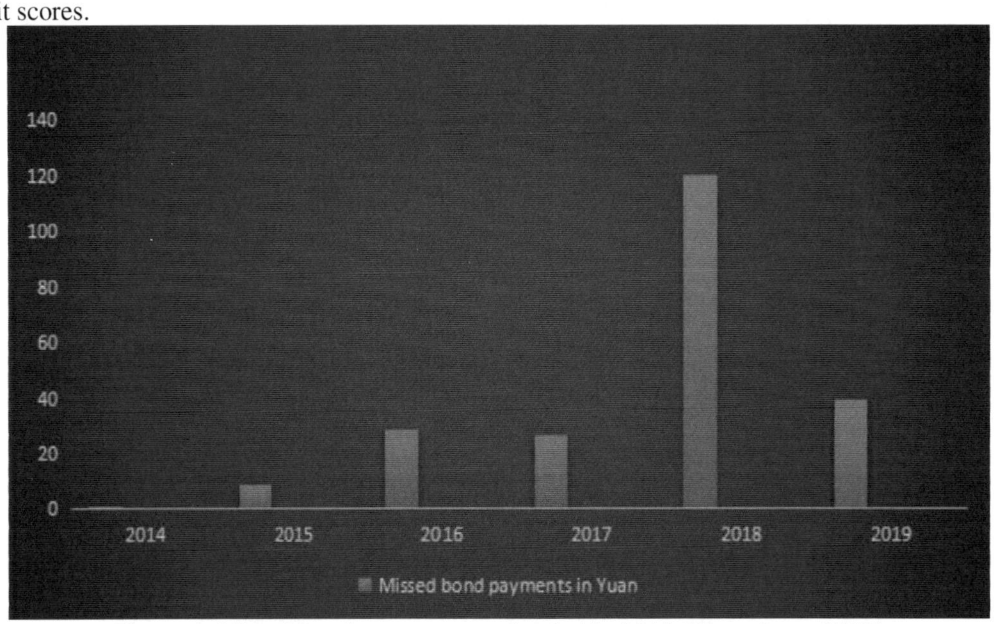

Figure 1. Missed bond payments quadrupled in 2018 (Bloomberg, 2019)

4. Credit evaluation model of Chinese companies

While most people believe the government controls the credit system, the current system involves several public and private agencies. The success of the new credit system compelled the government to include private agencies as a way of expanding nationwide scoring. However, the establishment of this environment is affected by competition. For instance, Alibaba and Tencent cannot disclose their operations under Baihang. When the Chinese government was recognizing eight private companies to initiate social credit operations in 2015, Alibaba's Ant Financial decided to exploit the license by establishing Zhima Credit (Sesame Credit)[5]. Sesame Credit has over 520 million users. Similarly, in 2017, Tencent established Tencent Credit to cater for an approximate 800 million monthly active WeChat users[5]. The emergence of these new agencies were met with the demand for increased safeguarding of user data and privacy for e-commerce. However, the inclusion of private and public agencies created conflict based on company policies. Tencent and Alibaba are hesitant to share their data with the government agencies because of their consumer privacy concerns and fear of providing the government with control over valuable consumer information [6]. The issue associated with data is due to the different operations of the new forms of financial credit systems function.

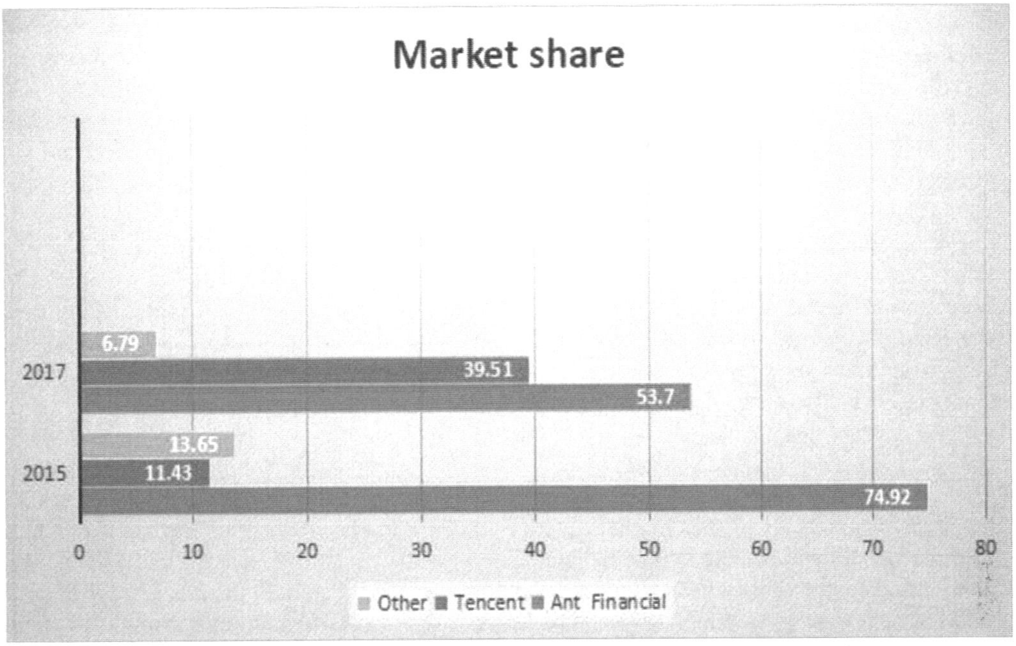

Figure 2. Market share for China's mobile payments (Seo, 2017)

4.1. Sesame credit

From the perspective of the credit system constructed by Sesame credit, the credit score of the company is obtained after processing, sorting and calculating. The standard score range is 350 to 950 and represents the credit level, whereby a good one is a result of a higher score. The dimensions that determine the operations at Sesame credit include credit history that involves past credit account repayment records and credit account history, behavior preference that focuses on the preference and stability in shopping, payment, transfer, financial management to identify the job and social role of the individual, performance ability that includes enjoying all kinds of credit services and ensuring timely performance, such as whether car rental is returned on time, or the timely payment of water, electricity and gas, Identity characteristics that includes sufficient and reliable basic personal information left during the use of relevant services and relationship that includes the identity characteristics of friends and the degree of interaction with friends. Transfer relationship and alumni relationship are regarded as one of the bases to judging personal credit.

4.2. 51 credit card

51 credit card is mainly based on the historical analysis of users' credit card e-bills, strong cross validation of e-commerce and social relations. Cross validation of multi-dimensional information is used to determine the user's risk level and whether to loan to the user. The data is obtained according to the user's credit card data, the purchase behavior corresponding to the e-commerce data open to the platform, the phone operator's call situation, registration information, among others. 51 credit card risk level consists of five dimensions: bill management time, bill performance, time limit for mobile access, operator to judge the user's own reliability through the effective call records in recent 4 months and whether there are negative contacts in the address book and Taobao to investigates whether the commonly used name and telephone number are consistent with the reserved number of the applicant.

5. Problems in the Development of China's Credit Rating Industry

5.1. Lack of coordination in credit rating industry

Presently, various departments offer some requirements for credit rating from the perspective of their business supervision. However, they do not adhere to any laws and regulations set for the credit industry. the lack of coordination exists between institutions like the CIRC, the CSRC, and the People's Bank of China. For instance, while the CIRC accredits five rating agencies, the People's Bank of China identifies nine rating agencies that are eligible for bond rating. Therefore, the current system lacks a competent authority for credit rating agencies; hence, an ineffective regulatory system.

5.2. Lack of regulation

The system lacks relevant business norms, rating procedures and methods of rating agencies. Rating results are very different, which affects the standard criteria for determining scores. There is also a lack of unified review and supervision agencies for the rating results nationwide, which makes many credit rating agencies strive for the rating business at the expense of the evaluation quality. The occurrence seriously affects the healthy development of the credit rating industry.

5.3. Lack of basic market access restrictions

In the management of market access, each department only accepts the rating results of several rating agencies. This approach establishes market access only from the perspective of the target rated, rather than from the overall perspective of rating agencies to conduct a comprehensive analysis. Currently, there is no competent authority to conduct a more comprehensive industry access review and annual business activity review for rating agencies. The Chinese onshore bond market is completely different from that of developed markets. The difference affects investors that desire to access sections of the onshore markets past the ones included in the extensive global benchmarks. In China, some credit rating agencies rated AAA or AA are not recognized outside China, but issue bonds at high percentages [7].

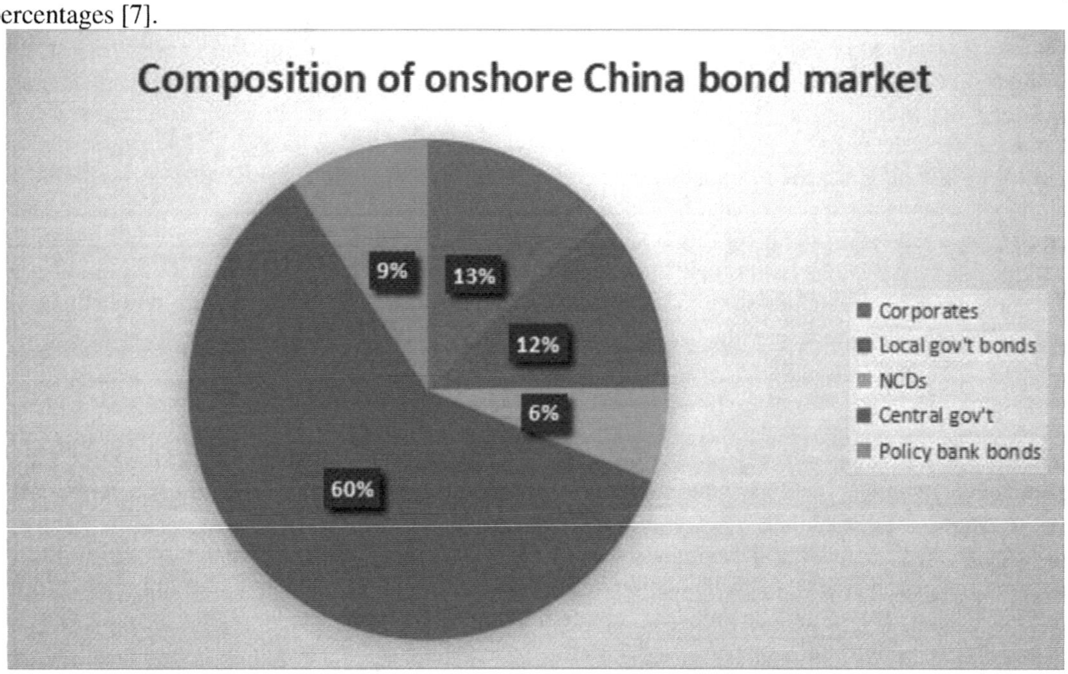

Figure 3. Composition of Onshore China Bond Market

5.4. Unhealthy competition

The underdeveloped bond market in China means the issuing interest rate of corporate bonds is not linked to the credit rating of bonds. The stipulation cannot exceed 40% of the fixed deposit interest rate of banks in the same period. The importance of credit rating is not significant. This makes the overall rating market demand too small, narrow business scope. As the market demand is too small, dozens of rating agencies, large and small, compete for market share and launch vicious competition. The direct consequence of vicious competition is that it seriously damages the image of the rating industry and affects the authoritativeness of the rating results.

6. Solutions of China's Credit Rating Industry

6.1. Establish a rating licensing system

The credit rating industry is still new in China and lacks any form of regulatory system that can assist in creating order. Credit rating agencies are competing unfairly and the system does not adhere to global standards when compared with the west. Therefore, it is necessary to implement a government-led licensing or accreditation system of credit rating business in the initial stage of the development of the credit rating industry to ensure the quality of credit rating. There are two ways: first, the relevant departments issue licenses to rating agencies for certain rating business; second, the relevant regulatory authorities, as users of credit rating results, approve the rating results of certain rating agencies.

6.2. Establish a "delisting" system for rating agencies

In order to effectively supervise the credit rating agencies, authorities must establish a system of reviewing and evaluating the rating results, and formulate a corresponding "delisting" system. By investigating the default rate of rating results and other indicators, on the one hand, the rating agencies with high default rate of rating results and irregular operation ought to be disqualified from their rating business. On the other hand, the evaluation companies with high rating quality must be encouraged to become large rating companies nationwide as soon as possible, so as to ensure that the relative concentration of the rating market is based on high-quality rating results.

6.3. Establish daily management systems

The government should draft and formulate basic regulations on the management of the rating market and clarify the conditions for the establishment of credit rating agencies. Furthermore, the administration should approve procedures, business rules, prohibitions and penalties.

6.4. Coordinated action and unified supervision of all departments

Considering that the Ministry of Finance participates in the construction and development of various fields of finance as a macroeconomic department, there is no conflict of interest in the field of rating business. Additionally, the rating work is closely related to the accounting, auditing, asset evaluation and other business under the supervision of the Ministry of Finance. Therefore, the Ministry of Finance can facilitate coordination by establishing a small joint work with the People's Bank of China, the Development and Reform Commission, the Banking Regulatory Commission, the Securities Regulatory Commission and the Insurance Regulatory Commission. Group B shall establish a coordination mechanism to unify the management of the rating industry in all aspects. Moreover, it can jointly determine the norms in the field of rating business and the qualification certification of the agencies, and jointly evaluate the rating work of the rating agencies. The Joint Working Group confirmed that member departments could issue licenses to credit rating agencies for certain types of rating business, or users as credit rating results could endorse the rating results of certain rating agencies. On the basis of sufficient coordination, the member departments of the joint group can also formulate their respective management measures, rules and regulations. The department can also constantly improve the regulatory system of China's rating market, so as to make the operation of credit rating agencies legal and rule-based.

7. Conclusions and future work

China's recent development of its credit rating industry has been exposed to diverse obstacles associated with both public and private institutions. Before the restructuring of the credit system,

China had a large number of defaulting companies and individuals. However, the advent of technology and popularity of big data compelled the Chinese government to identify ways of improving credit scores. However, the rating system experiences immense internal competition, lack of infrastructure for research, lack of regulation, and lack of basic market access restrictions. The government should vigorously promote the development of China's credit rating industry by establishing a rating licensing system, a rating agency delisting system and a daily management system. Credit investigation is the cornerstone of the society, it also needs infrastructure, such as a sound legal system and clear regulatory policies. The development also requires proper coordination between the government and the private sector. Private companies need to liaise with public ones to provide consumer data that will assist in calculating credit scores. It is hoped that all parties concerned will pay close attention to the problems existing in China's credit rating industry and formulate countermeasures to improve the nation's credit rating system.

Reference

[1] Bloomberg. (2019, May 8). China Defaults Hit Record in 2018. 2019 Pace Is Triple That. Retrieved from https://www.bloomberg.com/news/articles/2019-05-07/china-defaults-hit-record-in-2018-the-2019-pace-is-triple-that

[2] Chorzempa, M., Triolo, P., & Sacks, S. (2018). *China's Social Credit System: A Mark of Progress or a Threat to Privacy?* Peterson Institute.

[3] Kostka, G. (2018). China's Social Credit Systems and Public Opinion: Explaining High Levels of Approval. *SSRN Electronic Journal*. doi:10.2139/ssrn.3215138

[4] Livingston, M., Poon, W. P., & Zhou, L. (2017). Are Chinese Credit Ratings Relevant? A Study of the Chinese Bond Market and Credit Rating Industry. *SSRN Electronic Journal*. doi:10.2139/ssrn.2893776

[5] Mishra, V. (2019, October 3). Credit Scores to Social Credit Scores: Developments in China's Social Credit Scoring Ecosystem. Retrieved from https://gomedici.com/credit-scores-social-credit-scores-developments-china-social-credit-scoring-ecosystem

[6] Seo, B. K. (2017). Globalization of Chinese Credit Rating Corporations and Implication to Korean Credit Rating Industry. *The Journal of Modern China Studies*, *19*(2), 165-202. doi:10.35820/jmcs.19.2.5

[7] Smith, C. (2019). The evolving Chinese onshore bond market: Be wary of additional risks. Retrieved from https://www.livewiremarkets.com/wires/the-evolving-chinese-onshore-bond-market-be-wary-of-additional-risks

IOP Conf. Series: Materials Science and Engineering **806** (2020) 012010 doi:10.1088/1757-899X/806/1/012010

Discussion of the Risks and Supervision of Financial Big Data Applications

Ma Liyun

Northwestern Polytechnical University Mingde College 710000

Corresponding author's e-mail:aierlankafei666@163.com

Abstract: With the innovation of big data finance, the existing financial industry and financial institution structure have been affected to a certain extent, and the risk of financial market has undergone a profound change.In the development of Internet finance, big data technology has been recognized and widely used in various industries, financial big data and Internet finance are two important players to transform the competitiveness of the financial industry for Chinese financial ecology. In order to identify the risks in financial markets in the application of financial big data, we need to make clear the characteristics of financial big data. The essence of big data finance is financial innovation.In the process of financial innovation, it not only increases the volatility of the financial system, but also causes the homogenization of financial market. The purpose of this paper is to analyze the risks of big data in finance, to clarify the important matters of financial big data supervision, and to improve the relevant laws and regulations, strengthen the information sharing mechanism and other big data financial supervision measures in view of the existing data monopoly, personal privacy leakage and many other issues,and to provide theoretical reference for the development of big data finance and the ecological stability of financial market in China.

1. Introduction

Big data financial risk is a kind of resource, optimise allocation of risk can effectively constrain potential big data risk crises through big data financial regulation, good risk allocation promotes maximizing the benefits of financial innovation, therefore, the research of big data financial risk and the optimization of big data financial risk supervision are mainly aimed at promoting the further development of the financial advantages of big data. At present, the literature research boom in big data finance aspect is aimed at financial innovation and financial supervision two aspects in China. Therefore, this paper takes big data as the main research background, and it will ensure the stability of the financial market industry as the research significance.This paper begins with an analysis from two angles: the first is a brief explanation of the development trend of big data and the characteristics of big data, and the second is a brief analysis of Internet finance. Then the domestic financial industry is the main research object in order to analyse the impact of big data on Chinese financial industry field. Finally, this paper expands a comprehensive discussion on the risks and regulatory measures faced by financial big data applications.

2. the Trend and Characteristics of Big Data

2.1. the Trend of Big Data

Related findings show that global big data is rising at a geometric rate , IBM predicts that the total amount of data and information in the world will reach 35ZB by 2020 under the application and development of the Internet. If analyse the relations of ZB and GB in detail, we can know that a trillion GB is equal to a unit of ZB, so wo can know the condition of the global data information quantity by 2020. The total amount of global data by 2020 is 8 times that of the current global data through summing the amount of data in the world today, but the total amount of global data by 2020 will be 50 times that of the sum of data generated over all years before the global twentieth century. The trend of big data is a sharp increase in data volume through survey and comparison of data , which is reaching a trend that people could not have imagined before.

2.2. Analysis of the Characteristics of Big Data

2.2.1. Online

The generation, invocation, and delivery of big data need to be done online. In fact, a lot of big data in the process of collection has been handled through the subjective consciousness of the human brain. Applications of big data in various industry sectors store and transfer the data generated by market research in order to achieve the corresponding objectives. People don't "deliberately" deal with big data in the process of data storage, production, or invocation. New data information processing methods such as cloud computing can realize real-time online, real-time storage, real-time calls and real-time analysis of big data.

2.2.2. Mix

Traditional data age sampling and interception are the most common ways to obtain data, the way of obtaining the data has direct effect on the results of data analysis. The data analyzed by the limited data analysis means are generally refined and quantified. But in the age of big data, because of the all-in-one and online characteristics of big data, if sum up the above three significant characteristics , the characteristics of big data are mixed.

3. the Summarize of Big Data Finance

3.1. the Development of Big Data Finance

The first stage: the intelligent stage of service systems and the industrial phase of traditional financial institutions from the 1990s to 2005. Traditional financial institutions use the Internet economy platform to realize the construction of online service system based on the internet, big data technology, such as online investment, online insurance, etc.

The second stage: the stage of increasing the trend of big data finance and economic earnings is from 2005 to 2011.Third-party payments in this phase gradually reached into the hearts of the public, but did not get the full public approval.

The third stage: it is a stage of rapid development and innovation in the financial content of big data from 2011 to 2015, it can also be considered that this stage is the peak period of the development of big data finance.

The fourth stage: since 2015, big data finance undergoes regulation, change and regulation. In 2015, China updated the related regulations on financial supervision of big data, what is mapped is that big data financial industry states and service systems will be fully regulated and regulated.

3.2. The Main Body of Big Data Finance

In the process of transitioning to big data finance, Internet enterprises are mainly targeting the needs and service systems in the financial industry and develop the corresponding Internet economy platform. On the one hand, it can meet the needs of social financial market small and small enterprises, individual financing, on the other hand, it can expand and extend the Internet industry field.

4. Risk Analysis of Financial Big Data Applications

4.1. Fintech Giants Could Create Data Monopolies
Fintech giants can access and master large amounts of data by taking advantage of the inherent advantages of Internet platforms, computer technology, and big data technology and analyse the phenomenon objectively. There is a probability of data oligarchy, which leads to the problem of data monopoly in the financial industry. For example, some financial institutions have mastered the core credit data information, e-commerce transaction data information, offline transaction data information and so on. In essence, data monopoly is that the sharing mechanism has not been fully implemented, the sharing incentive mechanism is not in place, the concept of knowledge data information sharing conflicts with credit.

4.2. Data Silos and Data Fusion Issues
The government and enterprises are facing the danger and dilemma of data silos under the development of big data finance and the application of big data. In the era of big data, there is bound to be some confidential types of data information in various industry and government-generated data. So data information has become an important resource. Companies are reluctant to share data information in order to protect trade secrets, In the process of big data application, the government pays attention to application light disclosure, which leads to the existence and malignant development of data island phenomenon.

4.3. Data Security and Personal Privacy Issues
Various financial institutions can realize the access to user data information through various means such as accumulation and cooperative transactions on their own platforms in the applications of big data financial. Although the access to customer information data has brought convenience to the corresponding financial industry institutions, but from the customer's point of view, it involves the invasion problems of customer privacy. If from the point of view of data information security, once there is illegal data transaction or data information theft, then the development of the domestic financial big data industry will be more seriously affected.

The figure below shows the impact of the asymmetry of big data financial information on financial enterprises, as shown in figure 1:

Figure 1: Big Data Financial Information Chart

5. Supervision of Big Data in Finance

5.1. Strengthen the Improvement of Laws and Regulations on the Protection of Personal Information
The perfection of policies and regulations should also conform to the development of society, In the

context of the development of big data finance, in order to ensure information security, data security and do not infringe on the privacy of users, the most important measure is to improve the corresponding laws and regulations and make the regulatory system as the main guide to regulate the financial big data industry in all regions of china. The following points are specified, first, the current laws and regulations can be broadened to a certain extent through combining the development of the domestic overall big data finance; Second, strict and clear definition of the scope of responsibility of big data financial regulators and ensure that regulatory authority held by regulators can adapt to the benign control of process of current big data financial development; Third, protect the body of big data, that is, the rights of the corresponding users and customers, such as the right to privacy, the right to know, etc.

5.2. Strengthening the Construction of Information Sharing Mechanism

The construction of information sharing mechanism is the most important means to solve the problem of data silos. First of all, the corresponding government should carry out information data disclosure activities and break up the data bastion drives information sharing across industry, realize the full play of big data and drive rapid transformation and upgrading across industries in regional cities .

6. Conclusion

Combined with the development trend of computer technology in our country, we know that the Internet has become an important mode of social and economic development in our country. Under the background of the application of computer technology, people can achieve the corresponding goals through network consumption, network profitmaking and so on. At present, China has continuously optimized the economic operation organization. Internet finance has been listed as an important member of the current organization to optimize the operation of the economy. So for big data finance and Internet finance, in the process of specific optimization, the main aspect is focusing on the activation of idle social resources, and realizes the double-sided interaction between demand and supply, which can enhance the creativity of individual domestic economies and help further optimize the allocation of market resources.

Reference

[1]Chen Shuyi,Risk and Supervision of Financial Big Data Applications [J]Modern Marketing (Business Edition), 2019 (11): 92

[2]Tang Tan.,On Internet Financial Risk Analysis and Regulatory Recommendations [J] Modern Economic Information, 2019(17):308.

[3]Xie Li, Internet Financial Risk Analysis and Regulatory Recommendations [J] Management and Technology of Medium-Sized and Small Enterprises (the Book in the First Ten-Day Period of a Month), 2018 (02): 64-65.

[4]Sun Guofeng, Risks and Regulation of Financial Big Data Applications [J]Finance Comment from Tsinghua,2017(10):93-96.

[5]Ke Jiayi, Analysis of Risks and Challenges of Big Data Finance [J]Today's Wealth, 2017(13):42-43.

IOP Conf. Series: Materials Science and Engineering **806** (2020) 012011 doi:10.1088/1757-899X/806/1/012011

Mapping Big Data and Artificial Intelligence in arts and humanities across time: an exploratory scientometric analysis based on the WoS database

Zhichao Xu[1], Yuan Feng[2], Han-Teng Liao[1*] and Xiaolin Jiang[1]

[1] New Media Research Center, Sun Yat-sen University Nanfang College, Guangzhou, Guangdong, 510970, China

[2] School of Communication and Design, Sun Yat-sen University, Guangzhou, Guangdong, 510275, China

*Corresponding author's e-mail: h.liao@oxon.org

Abstract. The ways in which artificial intelligence (AI) and Big Data shape the work and the role of artists and humanities scholars have been at the forefront discussions among advanced universities and institutions, including examples such as computer-generated artwork, ethics, and philosophy, etc. With the aim to systematically review the related literature in arts and humanities, it is important to examine the discipline dynamics across time, thereby exploring both historical and current status. Based on the bibliographic data of 2,707 articles collected from Web of Science from 1956 to 2020, this paper has identified main disciplines and publications across three time periods, using various clustering and visualization methods. Based on the analytical and visualization findings, this article provides both a roadmap and data processes for artists, researchers, and policy-makers to understand how AI and Big Data have been discussed in arts and humanities disciplines.

1. Introduction

In recent years, the impact of Big Data and artificial intelligence (AI) applications on humanities, art, and philosophy have been discussed, especially about the promises and potential dangers of such applications. For instance, to ensure the development of responsible artificial intelligence, how to develop AI responsibly for human beings is a pressing issue for researchers and policy makers[1,2]. As another example, as forms of human expressions and societies are fundamental to the practices and theories of arts and humanities scholars, Yang Qian, the President of the 2019 International Joint Conference on Artificial Intelligence (IJCAI 2019), stated that "the second phase of AI will truly reshape human society, giving it its future form"[3]. Thus, the ways in which AI, along with the Big Data of human expressions and activities, reshape human and social forms of expressions become a central question facing arts and humanities scholars and practitioners. Advanced universities such as Stanford[4] and Oxford [5] have begun to develop humanities-related institutions surrounding AI, echoing the need to address various ethical implications and exploring the ways in which AI technology benefit humanity more broadly [1,3,6].

The discussions and applications of Big Data and AI in arts and humanities appears to be growing. For instance, philosophy of information has become fertile grounds for research and policies[7]. Digital humanities also play an important role[8]. For another, data visualization can provide new knowledge for understanding human experience, social relationships and network relevance[9]. AI also has begun to reshape music expressions and compositions[10,11]. Nonetheless, few literature review papers have been published regarding the overall picture of the status of Big Data and AI in arts and humanities. With the purpose to fill such a gap, this paper aims to provide a systematic account of current status.

2. Data and methods

Following the conventions of scientometric analysis, the section describes the query design and methods.

2.1. Data sources

Often used for bibliometric analysis[12], the ISI Web of Science (WoS) is used to gather related literatures. In order to cover comprehensively arts and humanities research in relation to the topic of AI and Big Data (coded by the WoS field as TS), we have executed the Advanced Search query as below, covering 15 WoS research areas (coded by the WoS field tag as SU):

(TS = (AI OR "Artificial Intelligence" OR "machine learning" OR "Big Data")) AND (SU=("Arts & Humanities" OR "Architecture" OR "Art" OR "Arts & Humanities Other Topics" OR "Asian Studies" OR "Classics" OR "Dance" OR "Film, Radio & Television" OR "History" OR "History & Philosophy of Science" OR "Literature" OR "Music" OR "Philosophy" OR "Religion" OR "Theater"))

The bibliometric data was collected on 2019.11.04 with 2,707 articles.

2.2. Methods and Tools

To assist our analysis of the data, we first used VOSviewer to explore and analyse the network maps using journal publication sources as the main unit of analysis.

3. Research mapping results

This section presents the findings using a historical perspective.

3.1. Annual trends

Specific time periods have been identified based on the linear regression modelling of the number of articles (in log values) across time. As shown in Figure 1(a)-(d), the number of publications has continued to grow exponentially, except for the time period between 1986 and 2003.

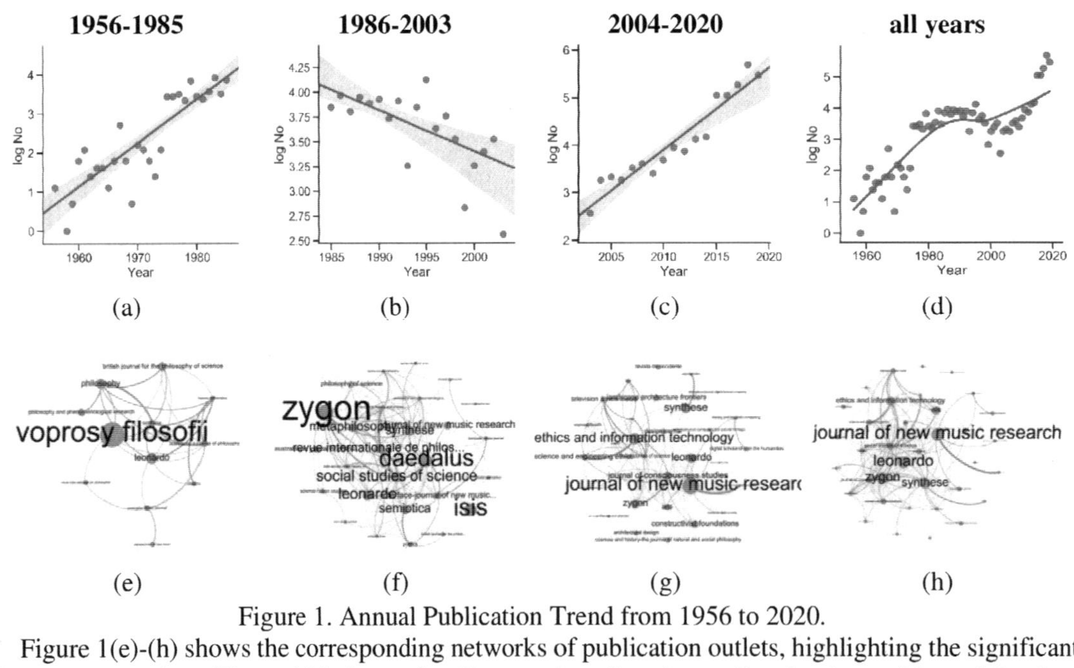

Figure 1. Annual Publication Trend from 1956 to 2020.

Figure 1(e)-(h) shows the corresponding networks of publication outlets, highlighting the significant changes across time. Figure 1(e) shows that *Voprosy Istorii* as the top Russian humanities publication.

Figure 1(f) shows *Zygon*, a.k.a. *the Journal of Religion and Science* as the top. Figure 1(g)-(h) shows the *Journal of New Music Research*, which is a peer-reviewed academic journal covering research on musicology (including music theory), philosophy, psychology, acoustics, computer science, etc.

3.2. Top publication outlets

Table 1 lists the top publication outlets across three time periods, indicating significant shifts in main publication outlets and the dynamics of interdisciplinary collaboration.

Table 1. Top publication outlets

1956-1985	1986-2003	2004-2020
• Voprosy Istorii • American Historical Review • Voprosy Filosofii • Istoriya Sssr • Deutsche Zeitschrift Fur Philosophie	• Voprosy Istorii • Arbor-Ciencia Pensamiento Y Cultura • Daedalus • Isis • American Historical Review	• Journal of New Music Research • Ethics and Information Technology • Leonardo • Synthese • Journal of Consciousness Studies

The publication outlets during the time period of 2004-2020 clearly differ from the other two.

For instance, dedicating to the advancement between moral philosophy and information and communication technology (ICT), the journal *Ethics and Information Technology* has become the one of the main publication outlets after 2003. Also, covering the application of contemporary science and technology to the arts and music, *Journal of New Music Research,* and *Leonardo* have become important publication outlets after 2004, indicating strong research interest in the integration of science, technology, art, and humanities. Note that during the first time periods before 2003, there were several European continental publication outlets.

3.3. Publication clusters

Figure 2 shows the clustering outcomes based on Figure 1(g), revealing important outlets for each cluster.

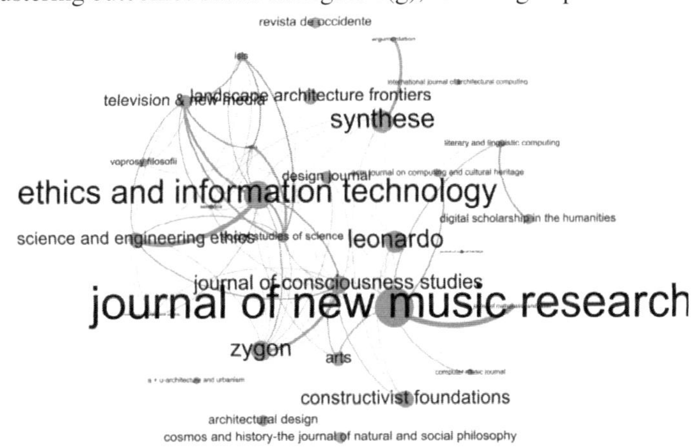

Figure 2. Clustering outcomes of publication outlets from 2004-2020.

Four clusters have been identified and visualized: the ethics/philosophy cluster (in blue color, with *Synthese* at its core position), the arts/consciousness cluster (in orange), the music/technology cluster (in purple color, with *Journal of New Music Research*), and the ethics/technology cluster (in green color, with *Ethics and Information Technology* at its core position).

4. Conclusion

This literature review contributes to a better understanding of how humanities literature related to AI and Big Data change across time. By presenting findings across three time periods, the paper shows the mainstreaming of Big Data and AI applications, especially expressions of music. We also have found that, in the humanities field, cross-disciplinary collaboration is evident especially in research that intersects with science and technologies.

Bibliometric methods and regression modelling have helped to identify major time periods and publication outlets. Focusing on the most recent period of outlets from 2004-2020, based on the network analysis of the publication outlets, four clusters have been identified: the ethics/philosophy cluster, the arts/consciousness cluster, the music/technology cluster, and the ethics/technology cluster. Although we may miss the related research published in technology journals such as computer-generated graphics, the paper nonetheless provides important findings and methods for researchers, funding agencies, policymakers, and industry professionals to understand the dynamics in this area. Future research can build upon our research to advance such understanding.

Acknowledgments

The research is funded by the Application-oriented Curriculum Development Projects of "API, Machine Learning and Artificial Intelligence" (NFU 02-40248) and "Big Data" (NFU 02-40249), under the Guangdong Province Department of Education 2018 "Innovation-Strengthening Higher Education Program" Grants, and partly by a project of HCI for Education Development (2018WQNCX283), under the 2018 Major Projects of Guangdong Province Department of Education "Young Innovative Talents" Grants. Zhichao Xu, Yuan Feng and Han-Teng Liao have made equal contributions to the paper, including the data collection and analysis work and have approved the current version for publication.

References

[1] UNESCO (2018) Towards a global code of ethics for artificial intelligence research. UNESCO 2018: 3.
[2] Dignum, V. (2019) Responsible Artificial Intelligence: How to Develop and Use AI in a Responsible Way. (Springer International Publishing)
[3] Yang, Q., Wang, C. (2018) The second phase of AI will truly reshape human society, giving it its future form. UNESCO
[4] Stanford University (2019) Stewarding AI in the Age of Big Tech. Stanf. Inst. Hum.-Centered Artif. Intell.
[5] Choudhury, N. (2019) £150 million funding for AI research at University of Oxford. Open Access Gov.
[6] Boddington, P. (2017) Towards a Code of Ethics for Artificial Intelligence. (Springer International Publishing)
[7] Floridi, L. (2002) What is the Philosophy of Information? Metaphilosophy 33: 123–145.
[8] Sculley, D., Pasanek, B.M. (2008) Meaning and mining: the impact of implicit assumptions in data mining for the humanities. Lit Linguist Comput 23.0: 409.
[9] McCosker, A., Wilken, R. (2014) Rethinking 'big data'as visual knowledge: the sublime and the diagrammatic in data visualisation. Vis. Stud. 29: 155–164.
[10] Widmer, G. (2002) Machine discoveries: A few simple, robust local expression principles. J. New Music Res. 31: 37–50.
[11] Moroni, A., Manzolli, J., Zuben, F.V., Gudwin, R. (2000) Vox populi: An interactive evolutionary system for algorithmic music composition. Leonardo Music J. 49–54.
[12] Cui, T., Zhang, J. (2018) Bibliometric and review of the research on circular economy through the evolution of Chinese public policy. Scientometrics. 1013–37.

IOP Conf. Series: Materials Science and Engineering **806** (2020) 012012 doi:10.1088/1757-899X/806/1/012012

A Non-line-of-Sight Propagation Identification Algorithm in Wireless Sensor Network

H Chu*, P Ji, S Lian, H Yang and R Wu

Faculty of Robot Science and Engineering, Northeastern University, Shenyang, Liaoning, 110000, China.

*Corresponding author's e-mail: chuhao_neurobot@163.com

Abstract. With the rapid developing of communication technology, Wireless Sensor Network(WSN) has been applied in many fields. The accuracy of localization is the vital key to all these applications. In order to mitigate the non-line of-sight(NLOS) influence, we propose a propagation identification algorithm based on Anderson-Darling(AD) test. Firstly, the status of signal is determined by the AD testing. Secondly, the particle swarm optimization algorithm is adopted to estimate the mobile node based on the identification results. Finally, the effectiveness of the algorithm is verified by simulation experiments.

1. Introduction

As a distributed network, Wireless Sensor Network(WSN) senses and monitors the environment. It has been widely applied in the fields of ecological monitoring, electrical automation and aerospace etc[1]. With rapid developing of communication technology, the accuracy of localization is demanding especially in the complex environment. However the non-line-of-sight(NLOS) propagation influences greatly on the localization results. Therefore, it is essential to mitigate NLOS error to improve the localization accuracy.

Many research works have been done to estimate the NLOS propagation status. A NLOS determination method based on a known propagation model is proposed to distinguish both signals in [2]. In [3], a method combining Hypothesis Test (HT) and Triangular Inequality Properties of Euclidean Space is introduced to detect whether ranging is affected by bias caused in NLOS conditions. A hybrid localization algorithm combining WTD is proposed to distinguish Line-of-sight(LOS) and NLOS paths in [4]. In [5], the Wylie algorithm is also used to identify the existence of NLOS. A Support Vector Machine (SVM) classifier is proposed to judge the signal propagation status in [6]. However, novel algorithms focusing on mobile node localization and complex environment are still in desperate need.

In this paper, we propose a NLOS propagation identification algorithm based on Anderson-Darling(AD) testing in the WSN. The proposed algorithm will detect the NLOS status with relatively low computation and the mobile node can be located with particle swarm algorithm. The rest of paper organizes as follows: in Section 2, we introduce the propagation state identification algorithm based on Anderson-Darling test; in Section 3 we perform simulation implementation and result analysis of the algorithm. And Section 4 concludes the paper.

2. Node localization algorithm based on Anderson-Darling test

2.1. Build the ranging model

There are one unknown node and N beacon nodes randomly deployed in a certain region. And the position of the unknown node is $\mathbf{U} = [x, y]$, the position of *ith* beacon node is $\mathbf{X}_i = [x_i, y_i]$. We select Time-of-arrival(TOA) as the ranging model. Therefore, the true distance between the *ith* beacon node and the unknown node is:

$$d_i = \|\mathbf{X}_i - \mathbf{U}\| \tag{2.1}$$

The distance measurement between unknown node and beacon node can be represented as:

$$\hat{d}_i = \begin{cases} d_i + n_i, LOS \\ d_i + n_i + n_{NLOS}, NLOS \end{cases} \tag{2.2}$$

where n_i is the measurement noise that follows Gaussian distribution $N(0, \sigma_i^2)$, n_{NLOS} is the NLOS error which is respective with measurement noise, n_{NLOS} follows Gaussian distribution $(N(\mu_{nlos}, \sigma_{nlos}^2))$, uniform distribution$(U(a, b))$ or exponential distribution$(E(1/\lambda))$ under different environments.

2.2. *Structure of the proposed Algorithm*

The flowchart of proposed algorithm is illustrated in Figure 1. When the measurement model is set up, Anderson-darling test is deployed for the NLOS propagation. If the status is LOS, the objective function is established according to the distance; else the measured value is taken as a constraint in the localization algorithm. Finally, the particle swarm optimization localization algorithm is used for positioning.

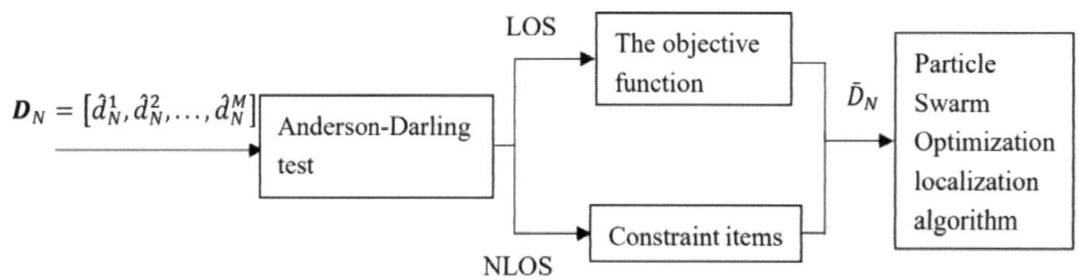

Figure 1. The flowchart of the proposed algorithm

2.3. *Propagation identification algorithm based on Anderson-darling test*

Assuming that the distance between the *ith* beacon node and the unknown node can measure M times in a short time. The measuring distance set is presented by $\mathbf{D}_i = [\hat{d}_i^1, \hat{d}_i^2, \ldots, \hat{d}_i^M]$. We obtain the following assumptions:

$$H_0: \hat{d}_i^j \sim N(0,1) \tag{2.3}$$

The mean and variance of the current measurement are as follows:

$$\hat{\mu}_i = \frac{\sum_{j=1}^{M} \hat{d}_i^j}{M} \tag{2.4}$$

$$\hat{\sigma}_i^2 = \frac{\sum_{j=1}^{M} (\hat{d}_i^j - \hat{\mu}_i)}{M} \tag{2.5}$$

The AD statistics are: $$AD = -\frac{1}{M} \sum_{j=1}^{M} (2j - 1)[\ln z_j + \ln(1 - z_{M+1-j})] - M \tag{2.6}$$

$$Z_j = \varphi(\frac{d_i^j - \hat{\mu}_i}{\hat{\sigma}_i^2}) \tag{2.7}$$

where $\varphi(d_i)$ represents the standard normal distribution function. Discriminant standard CV is:

$$CV = \frac{0.752}{1 + \frac{0.75}{M} + \frac{2.25}{M^2}} \tag{2.8}$$

If AD>CV, H_0 is rejected. It is considered as NLOS propagation. Otherwise, it is considered as LOS propagation.

2.4. *Particle swarm optimization localization algorithm*

Assuming that at least one of the measured values is line-of-sight, the mean value of the measurement in LOS environment is $\left[\bar{D}_1^{LOS}, \bar{D}_2^{LOS}, \ldots, \bar{D}_q^{LOS}\right]$, where $\bar{D}_i = \frac{1}{M}\sum_{j=1}^{M} \hat{d}_i^j$, The mean value of the measurement in NLOS environment is $\left[\bar{D}_1^{NLOS}, \bar{D}_2^{NLOS}, \ldots, \bar{D}_p^{NLOS}\right]$. The location objective function is as follows:

$$min \sum_{i=1}^{q} \left\| \bar{D}_i^{LOS} - \sqrt{(\hat{x} - x_i)^2 + (\hat{y} - y_i)^2} \right\|$$

$$s.t. \sqrt{\left(\hat{x} - x_j\right)^2 + \left(\hat{y} - y_j\right)^2} \le \bar{D}_j^{NLOS}, (j = 1, \ldots, p) \quad (2.9)$$

The estimated unknown node position can be obtained by solving the minimum value of the above equation. Because the formula is a nonlinear function with constraints, it is difficult to obtain the resolution directly. Therefore, particle swarm optimization localization algorithm is adopted to solve the above equation.

The parameters in the particle swarm optimization localization algorithm are defined below. The position of the particles is presented as S={P1,...PL}, where L is the number of particles. For the *ith* particle in step k, the updated equation for speed and position is as follows:

$$v_i(k) = \omega v_i(k-1) + c_1\xi\left(pbes_i - P_i(k-1)\right) + c_2\eta\left(gbest - P_i(k-1)\right) \quad (2.10)$$

$$P_i(k) = P_i(k-1) + v_i(k) \quad (2.11)$$

where $v_i(k)$ and $P_i(k)$ are the speed and position of the *ith* particle at the k step, $pbest_i$ represents the historical optimal position of the *ith* particle search, $gbest$ represents the historical optimal position of all particles search. c_1 and c_2 particle acceleration is constant. In this paper, we set $c_1 = c_2 = 2$. ξ and η are two random numbers in the range [0,1], and the maximum number of iterations is 60.

3. Simulation results

3.1. Simulation environment and experimental evaluation

This section mainly evaluates the success rate and localization accuracy of NLOS, The simulation environment is as follows: N beacon nodes are deployed randomly in a square area of 60m×60m, and the parameters of the node measurement model are shown in the Table 1.

Table 1. The default parameters

Parameter	Symbol	Default
Number of beacon nodes	N	7
Measurement noise standard deviation	σ_i	1
Exponential distribution NLOS error	$E(1/\lambda)$	$E(1/4)$
The number of measurements	M	500
The level of significance	α	0.05
The number of units	r	7
The number of particles	L	15

We compare the proposed algorithm with Maximum Likelihood(ML) and Residual Weighting Algorithm(Rwgh). And the average localization error is deployed to evaluate the accuracy of the localization algorithm:

$$AVE = \frac{1}{MC}\sum_{i=1}^{MC} \sqrt{\left\| \hat{U}_i - U_i \right\|^2} \quad (3.1)$$

where U_i is the real location of the unknown node in the *ith* experiment, and $\hat{U}_i$ is the estimated location of the unknown node in the *ith* experiment.

3.2. Simulation results and analysis

Figure 2 shows the relationship between the success rate of the NLOS identification algorithm and parameters λ when the NLOS error follows exponential distribution($n_{NLOS} \sim E(1/\lambda)$). The successful identification rate decreases as the parameter λ increases. When the NLOS error follows the exponential distribution, the success rate of the NLOS identification algorithm is above 93%. Moreover, it also indicates that the proposed algorithm can work robustly in complex environments.

Figure 3 shows the relationship between the measured noise standard deviation σ_i and the mean localization error when the NLOS error follows the exponential distribution E(1/4) and E(1/8). The experiments show that the proposed algorithm improves the localization accuracy by 83.89% and 63.62% compared with the ML and Rwgh algorithms.

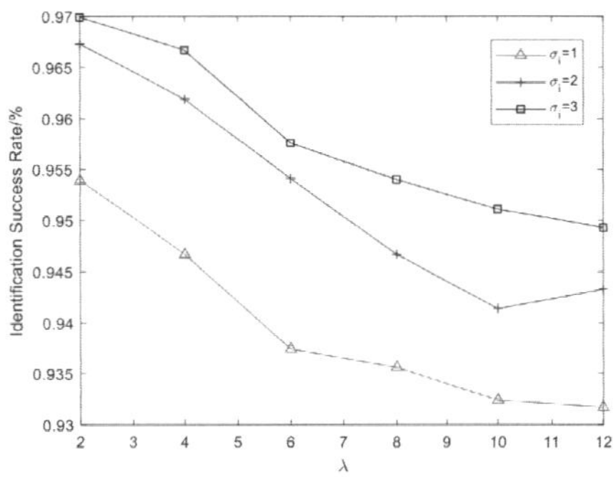

Figure 2. Relationship between parameter b and test success rate.

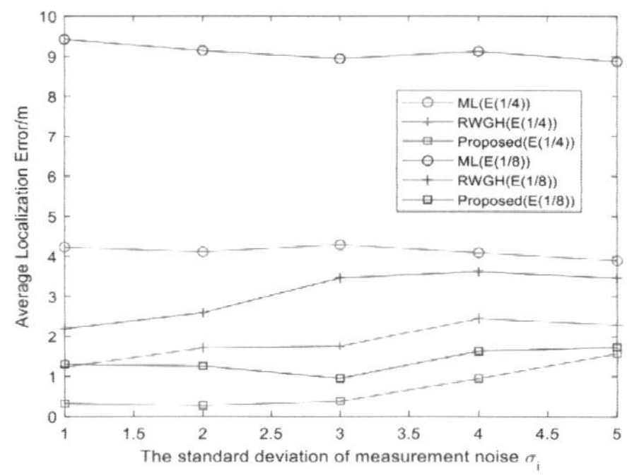

Figure 3. Relationship between measurement noise standard deviation and mean localization error

4. Conclusion

This paper proposes an NLOS propagation identification method based on Anderson-Darling test. Firstly, through multiple sets of distance measurement values, statistical analysis is used to determine the signal propagation state, and then based on the determination results, a particle swarm optimization algorithm is used for positioning. Finally, the proposed algorithm is simulated. Simulation results show that the proposed algorithm can effectively suppress NLOS errors of many different distributions.

Acknowledgments

This work was supported in part by the Fundamental Research Fund for the Central Universities of China N172604004, N182613001, N172604003 and N172603001; the National Natural Science Foundation of China under Grant nos. 61901098, 61701101, 61803077, 61603080, and the National Key Robot Project under Grant nos. 2017YFB1301103.

Reference

[1] Li, H. (2013) Study of wireless sensor network applications in network optimization. Sens Transducers, 157:180–189.
[2] Venkatesh, S., Buehrer, R. (2007) Non-line-of-sight identification in ultra-wideband systems based on received signal statistics. Microwaves, IET Microwaves, Antennas and Propagation, 1:1120-1130.
[3] Destino, G., Macagnano, D., de Abreu, G.T.F. (2008) Hypothesis testing and iterative WLS minimization for WSN localization under LOS/NLOS conditions. In: 41st Asilomar Conference on Signals, Systems and Computers. Pacific Grove, CA, USA. pp. 2150-2155.
[4] Zhang, L., Chen, F., Yu, Y. (2015) Research on hybrid location algorithm with high accuracy in indoor environment. In: 34th Chinese Control Conference (CCC). Hangzhou, China. pp. 8764-8767.
[5] Li, H., Deng, Z., Yu, Y. (2011) Investigation on a NLOS error mitigation algorithm for TDOA mobile location. In: IET International Conference on Communication Technology and Application. Beijing, China. pp. 839-843.
[6] Destino, G., Macagnano, D., de Abreu, G.T.F. (2008) Hypothesis testing and iterative WLS minimization for WSN localization under LOS/NLOS conditions. In: 41st Asilomar Conference on Signals, Systems and Computers. Pacific Grove, CA, USA. pp. 2150-2155.

IOP Conf. Series: Materials Science and Engineering **806** (2020) 012013 doi:10.1088/1757-899X/806/1/012013

Mathematical Model of Seat Arrangement in Large Gymnasium

Shu-e Sun

Xi'an Shiyou University, Xi'an, Shaanxi, 710065, China

e-mail: sse63@126. com

Abstract. In China, where the population ranks first in the world and the total economic output ranks second in the world, the construction of large-scale stadiums or Olympic sports centers has become a trend. When we arrange the seats on the viewing platform, we should not only consider the rationality, practicability and aesthetic value of the seat distribution, but also consider the safety of evacuation.Therefore, under the comprehensive consideration of the above factors, the mathematical model of periodic jump sequence is established, that is, an arithmetic sequence of periodical arithmetic jump, which provides an effective theoretical basis for engineering designers to use computers to process large data and arrange seats scientifically, and also provides an operable accurate model for later programming calculation and simulation technology research.

1. Introduction

The scale and level of urban public infrastructure construction are important indicators to measure a country's economic development, and also important aspects to reflect the comprehensive strength. In order to strengthen the consciousness of national culture and promote the harmonious development of its physical and mental health, the state should pay attention to the construction of infrastructure, because it has a profound impact on improving the overall quality of the nation. A country that is moving towards modern civilization should pursue the unity of science, practicality and beauty.

The world's largest stadium should be built in China, depending on China's population and economic trends, but a large stadium with a capacity of 100, 000 people has yet to be built. The largest Beijing national Stadium, built in 2008, can hold 91000 people. Other major cities have built or are in the process of building some large stadiums. As shown in figures 1 and 2. Here are the real pictures of the seating arrangement of Hebei Olympic Sports Center and Dalian Sports Center respectively. Therefore, with the vigorous development of national economy and sports, China needs to build more and larger sports centers to meet the needs of people's lives. However, the layout of this kind of stadium audience seats should be practical, beautiful and safe. In order to pursue the scientific, practical, safe and beautiful overall effect, the corresponding mathematical model is established.

Mathematical modeling has a long history like mathematics. Since the beginning of the 20th century, with the penetration of mathematics into all fields with unprecedented breadth and depth, and the emergence and rapid development of electronic computers, mathematical modeling has a great potential in the field of engineering technology, high-tech and non-physical interdisciplinary fields[1]. As Marx said, "A science only when the successful use of mathematics, can be considered to have reached the point of perfection".

In order to set up the viewing platform and seats reasonably in the large public places of the Olympic Stadium, I first need to establish the corresponding mathematical model, and then use the computer to process the big data and scientifically control it. In reality, the number of seats generally presents an arithmetic increase sequence from the first row to the back, and every few rows need to set up a corridor to facilitate the flow of people or safe evacuation, so this arithmetic increase sequence will jump periodically once. Therefore, the change of the number of seats actually corresponds to an arithmetic sequence of periodical jump. The following is a study of this sequence.

Figure 1. Hebei Olympic Sports Center

Figure 2. Dalian Olympic Sports Center

2. Study of the model of " sequence of periodic jump "

Defined 1 If the sequence { a_n } is an arithmetic sequence with a common difference d from item (kt + 1)th to item (kt + t)th(k = 0, 1, 2,..., t is a constant natural number), and $a_{(k+1)t+1} = a_{(k+1)t} + e$(e is any constant real number). Then the sequence { a_n } is called an arithmetic sequence of periodical arithmetic jump, where t is jump period, e is its jump common difference[2], and d is its intra-period common difference. Its general form is as follows (notate as a sequence (1)):

$$\begin{aligned} & a_1,\ a_1+d,\ a_1+2d,\ldots,\ a_1+(t-1)d;\ a_1+e+(t-1)d, \\ & a_1+e+td,\ a_1+e+(t+1)d,\ldots,a_1+e+(2t-2)d; \\ & a_1+2e+(2t-2)d,\ a_1+2e+(2t-1)d,\ a_1+2e+2td, \\ & \ldots,\ a_1+2e+(3t-3)d;\ldots \end{aligned} \tag{1}$$

In particular, when t = 1, the sequence { a_n } is an arithmetic sequence with a normal common difference e; When t = 2, the sequence { a_n } is a double arithmetic sequence [3].

Conclusion 1 The general term formula for arithmetic sequence of periodic arithmetic jump (1) is

$$a_n = a_1 + \left[\frac{n-1}{t}\right]e + \left(n-1-\left[\frac{n-1}{t}\right]\right)d \tag{2}$$

Analysis of proof thinking According to definition 1, the proof process of this conclusion 1 should be divided into two layers. The first level, In any period, the difference between the latter and its former is equal to d; At the second level, the difference between the first item of any subsequent period and the last item of its former period is equal to e.

Proof 1° When $n = kt+1,\ kt+2,\ldots,\ kt+t\ (k = 0,\ 1,\ 2,\ldots)$, $\left[\frac{n-1}{t}\right] = k$, the general formula of the (k +1)th period of the sequence (1) be obtained by the formula (2) is as follows

$$a_n = a_1 + ke + (n-1-k)\,d,$$

When $kt+1 \le n < kt+t$,

$$a_n+1-a_n = (a_1+ke+(n-k)d) - (a_1+ke+(n-1-k)d) = d.$$

That is, { a_n } is an arithmetic sequence with a common difference d in the $(k + 1)$th period of the sequence { a_n }. In particular, its last item is

$$a_{kt+t} = a_1 + ke + ((k+1)t - (k-1))d.$$

2°when $n = (k+1)t+1, \left\lceil \frac{n-1}{t} \right\rceil = k+1,$ at this time, a_n is the first item of the (k+2)th period in the sequence { a_n }, Obtain the following expression from (2)

$$a_n = a_{(k+1)t+1} = a_1 + (k+1)e + (k+1)t - (k+1))d.$$

So, the difference between the first item in the(k + 2)th period and the last item in the (k + 1)th period in the sequence { a_n } is $a_{(k+1)t+1} - a_{kt+t} = e.$

By synthesizing 1 °, 2 °, we know that formula(2) is a general item formula that satisfies definition 1.

Conclusion 2 The sequence formed by the sum of t items in each period of arithmetic sequence(1) of periodical arithmetic jump is a_n arithmetic sequence with the first item is $ta_1 + \frac{t(t-1)}{2}d$, and the common difference is $te + t(t-1)d$. Then the sequence { a_n } is called a sequence of within-cycle sum.

Proof If the sum of t items in the k-th period of the sequence(1) is denoted by $S^{(k)}, (k = 1, 2, ...)$

Obviously, there is $S^{(1)} = ta_1 + \frac{t(t-1)}{2}d$. And the first item in the k-th period of the sequence (1) is $a_{(k-1)t+1} = a_1 + (k-1)e + ((k-1)t - (k-1))d.$

$$S^{(k)} = ta_{(k-1)t+1} + \frac{t(t-1)}{2}d$$
$$= ta_1 + t(k-1)e + \frac{2k-1}{2}t(t-1)d$$

the first item in the (k+1)th period of the sequence{ a_n } is $a_{kt+1} = a_1 + ke + (kt - k)d,$

$$\therefore \quad S^{(k+1)} = ta_{kt+1} + \frac{t(t-1)}{2}d$$
$$= ta_1 + tke + \frac{2k+1}{2}t(t-1)d$$

$\therefore$ $S^{(k+1)} - S^{(k)} = te + t(t-1)d,$ therefore, conclusion 2 is established.

In the following, we'll find the formula for the sum S_n of the first n items of the sequence (1).

For the first n items of the sequence (1). It contains all the items in the first $\left\lfloor \frac{n}{t} \right\rfloor$ periods and the first $n - \left\lfloor \frac{n}{t} \right\rfloor t$ items in the $(\left\lfloor \frac{n}{t} \right\rfloor + 1)$th period. So we can divide the first n items of the sequence into the first $\left\lfloor \frac{n}{t} \right\rfloor$ items of the sequence of within-cycle sum and the first $n - \left\lfloor \frac{n}{t} \right\rfloor t$ items in the $(\left\lfloor \frac{n}{t} \right\rfloor + 1)$th period. From conclusion 2, it is known that the sum of the first $\left\lfloor \frac{n}{t} \right\rfloor$ items of the sequence of within-cycle sum is

$$(ta_1+\frac{t(t-1)}{2}d)\left[\frac{n}{t}\right]+\frac{\left[\frac{n}{t}\right](\left[\frac{n}{t}\right]-1)}{2}(te+t(t-1)d)$$

$$=\left[\frac{n}{t}\right]ta_1+\frac{\left[\frac{n}{t}\right](\left[\frac{n}{t}\right]-1)}{2}te+\frac{t(t-1)}{2}\left[\frac{n}{t}\right]^2 d.$$

And the first item of the $(\left[\frac{n}{t}\right]+1)$th period of the sequence (1) is

$$a_{\left[\frac{n}{t}\right]t+1}=a_1+\left[\frac{n}{t}\right]e+\left[\frac{n}{t}\right](t-1)d.$$

∴ The sum of the first $n-\left[\frac{n}{t}\right]t$ items of the $(\left[\frac{n}{t}\right]+1)$th period is

$$(a_1+\left[\frac{n}{t}\right]e+\left[\frac{n}{t}\right](t-1)d)(n-\left[\frac{n}{t}\right]t)+\frac{(n-\left[\frac{n}{t}\right]t)(n-\left[\frac{n}{t}\right]t-1)}{2}d.$$

The following conclusions can thus be obtained.

Conclusion 3 The sum S_n of the first n items of the arithmetic sequence of periodical arithmetic jump is

$$S_n=na_1+\frac{\left[\frac{n}{t}\right]\left(\left[\frac{n}{t}\right]-1\right)}{2}te+\frac{t(t-1)}{2}\left[\frac{n}{t}\right]^2 d$$
$$+\left(\left[\frac{n}{t}\right]e+\left[\frac{n}{t}\right](t-1)d\right)\left(n-\left[\frac{n}{t}\right]t\right)$$
$$+\frac{\left(n-\left[\frac{n}{t}\right]t\right)\left(n-\left[\frac{n}{t}\right]t-1\right)}{2}d.$$

In particular, when $t\mid n$ and $\frac{n}{t}=k$, $S_n=na_1+\frac{k(k-1)}{2}te+\frac{t(t-1)}{2}k^2a.$

3. Model application and its theory Promotion

Example There are 72 rows of seats in one stand of a large gymnasium. There are 180 seats in the first row, 26 more seats in each row than in the front row, one aisle every 13 rows, and 48 more seats in the row behind each aisle than in the front row.

solution From the meaning of the question, we know that the number of seats in each row here forms arithmetic sequence of periodical arithmetic jump. The first item of these is a_1= 180, jump period is t = 13, jump common difference is e = 48, intra-period common difference is d = 26, n = 72.

So there are $\left[\frac{n}{t}\right]=5$ complete periods in the sequence, and there are the first 7 items in the 6th period. We know from conclusion 3 that the total number of seats in this stand is

$$\begin{aligned} S &= \mathbf{72}\times \mathbf{180}+\frac{\mathbf{1}}{2}\times \mathbf{5}\times \mathbf{4}\times \mathbf{13}\times \mathbf{48}+\frac{\mathbf{1}}{2}\times \mathbf{13}\times \mathbf{12}\times \mathbf{52}\times \mathbf{26} \\ &+ (\mathbf{5}\times \mathbf{48}+\mathbf{5}\times \mathbf{12}\times \mathbf{26})\times(\mathbf{72}-\mathbf{5}\times\mathbf{13}) \\ &+\frac{\mathbf{1}}{2}\times(\mathbf{72}-\mathbf{5}\times\mathbf{13})\times(\mathbf{72}-\mathbf{5}\times\mathbf{13}-\mathbf{1})\times \mathbf{26} \\ &= \mathbf{71706}(per) \end{aligned}$$

By analogy, it is natural to deduce geometric sequence of periodical geometric jump sequence and its related conclusions from the arithmetic sequence of periodical arithmetic jump.

Defined 2 If the sequence from the item $(kt+1)$th to the item $(kt+t)$th $(k = 0, 1, 2, \ldots)$is a geometric sequence with common ratio q, and $a_{(k+1)t+1} = a_{(k+1)t}\,p(p \text{ is any constant})$, then the sequence is said to be a geometric sequence of periodical geometric jump, where t (t is a constant natural number)) is the jump period, p is the jump common ratio, and q is the common ratio in a period. The form is as follows (recorded as a sequence (3))

$$a_1, a_1q, a_1q^2, ?, a_1q^{t-1};\ a_1pq^{t-1}, a_1pq^t, a_1pq^{t+1},\ldots,$$
$$a_1pq^{2t-2};\ a_1p^2q^{2t-2}, a_1p^2q^{2t-1}, a_1p^2q^{2t}, ?, a_1p^2q^{3t-3};\ \ldots \quad (3)$$

In particular, when $t = 1$, the sequence $\{ a_n \}$ is an ordinary geometric sequence of periodic geometric jump with a common ratio p; When $t = 2$, the sequence $\{ a_n \}$ is a double geometric sequence.

Conclusion 4 The general term formula of the geometric sequence(3) of periodical geometric jump is $a_n = a_1 p^{\left[\frac{n-1}{t}\right]} q^{(n-1)-\left[\frac{n-1}{t}\right]}$.

Conclusion 5 The sequence formed by the sum of t items in each period of the geometric sequence(3) of periodical geometric jump is a geometric sequence with the first item $\frac{a_1(1-q^t)}{1-q}$ and the common ratio pq^{t-1}.

Conclusion 6 The sum S_n of the first n items of the geometric sequence(3) of periodical geometric jump is

$$S_n = \frac{\frac{a_1(1-q^t)}{1-q}\left(1-(pq^{t-1})^{\left[\frac{n}{t}\right]}\right)}{1-pq^{t-1}} + \frac{a_1p^{\left[\frac{n}{t}\right]}q^{\left[\frac{n}{t}\right](t-1)}\left(1-q^{n-\left[\frac{n}{t}\right]t}\right)}{1-q}$$

In particular, when $t\,|\,n$ and $\frac{n}{t}=k$, $S_n = \frac{a_1(1-q^t)(1-(pq^{t-1})^k)}{(1-q)(1-pq^{t-1})}$

The proof of the conclusion 4, 5, 6 is omitted.

4. Summary

The above theory is a scientific and effective mathematical model for engineering designers. They can determine the corresponding parameter values according to the actual needs of the specific site, and substitute them into the model for calculation.

Computers have brought great convenience to big data processing. People can achieve the final results by programming, just like dealing with problems in engineering, sports, economy and other fields, first establish mathematical models, and then use computer language programming or direct use of relevant software to achieve. In paper[4], the author has established the equation for calculating the equilibrium composition of the furnace gas accurately. When the atmosphere in the furnace changes very much, the data or the parameters of the mathematical model need to be adjusted according to the actual problems, which provides convenience for the computer to process the data.

In the later research, I will use computer simulation technology, not only to achieve the calculation of large data, but also to control the change of data, in order to correct and update the model in various situations, and promote its application in the solution of practical problems with similar characteristics. As mentioned in paper [5], the author can infer the nature and motion change of the flight control system of UAV through online interactive simulation technology, and update and correct the flight control system of UAV through simulation model and update and correct model, so as to ensure the accuracy and precision of the flight control system of UAV.

References

[1] Jiang Qiyuan, Xie Jinxing, Ye Jun. (2011)Mathematical Model. higher Education Press, Beijing.
[2] Hou Xinchang. (1997)sequenc of periodic jump. Mathematical bulletin, 3:15-17.
[3] Gao Hongbin. (1995)double arithmetic sequence. Mathematical bulletin, 7:14-16.
[4] Jiang Yan. (1994)The mathematical model of the calculation of the balance component of the furnace and the programming of the computer. Metal heat treatment, 6:36-37.
[5] Liu Hui Fang. (2017)Brief introduction of on-line interactive simulation technology of UAV flight control system. China Communications, 6: 128.

A Scientometric Analysis of Self-tracking in Relation to Artificial Intelligence and Big Data

Xue Wu[1], Man Zhao[1], Han-Teng Liao[1*] and Si-Pan Sun[1]

[1] New Media Research Centre, Sun Yat-sen University Nanfang College, Guangzhou, Guangdong, 510970, China

[*]Corresponding author's e-mail: h.liao@oxon.org

Abstract. As an increasing amount of personal data has been gathered by wearable and mobile devices, self-tracking, or the practice that people keep track, has become an important topic in Artificial Intelligence (AI) and big data applications. With the aim to provide a systematic review of the literature on self-tracking, this paper presents a scientometric analysis of 109 articles since 2000 collected from the Web of Science. Based on keyword co-occurrence network analysis, the paper has identified four major clusters: (1) wearables as quantified-self applications; (2) big data and critical theory; (3) data and privacy; (4) personal informatics. The further keywords-in-context (KWIC) analysis of the abstracts of the dataset clarifies the seemingly-interchangeable notions of "self-tracking" and "quantified-self": While "self-tracking" refers to more general activities, practices, technologies, and applications of keeping tracks, "quantified-self" refers to the more conscious efforts and meaning-making outcomes of the self-tracking activities. Such clarification, along with the keyword network analysis, suggests that self-tracking has become a specific and major type of datafication of human conditions or existence and that quantified-self is the construction of self through such datafication. A more integrated conceptual framework is needed for future research to better understand what amounts to meaningful datafication of human conditions and existence, thereby helping researchers and designers to discern the classic notions of health, wellness, and happiness for better research and design outcomes.

1. INTRODUCTION

As the practice of people keeping tracks has been recently amplified and enhanced by mobile and data technologies, a social and cultural phenomenon called "self-tracking" refers to the process of people turning their everyday experiences (especially wellness- and health-related) into data and data applications[1]. A major case of proactive self-tracking is the Quantified Self movement, originated from San Francisco in 2007 by Gary Wolf and Kevin Kelly from Wired magazine. The Quantified Self is defined as "an individual engaged in the self-tracking of any kind of biological, physical, behavioral, or environmental data"[2].

Self-tracking and the Quantified Self thus provide opportunities for meaningful collaboration across disciplines, especially social sciences and precision medicine[3]. The related work not only concerns health researchers and practitioners [4], but also may lead to new discoveries and innovations in health informatics [5]. Indeed, as increasing amount of personal data has been gathered by wearable and mobile devices [6], self-tracking has become an important topic in Big Data and Artificial Intelligence (AI) applications [7], with important discussions such as health-monitoring practices [8], open-access and privacy standards for personal data [9].

Several review articles have been conducted around the topic matter. One research has examined the notion of health self-tracking based on a 26 empirical studies [4]. Another has explored the socioeconomic and political implications of the risks associated with digital health technologies, which include self-tracking technologies [8]. Yet another has conducted a research on self-tracking in education settings using machine learning. However, such review work has not yet systematically examined the self-tracking research in intersections with Big Data and AI. This article therefore aims fill the gap by providing such a systematic review.

2. DATA AND METHODS

Following scientometric research conventions, the paper first describes the query design that covers both self-tracking and quantified self, in relation to AI and Big Data as follows:

("self-tracking" OR "quantified self" OR "self-quantification" OR "quantified-selfers" OR "personal quantification" OR "quantified academic selves" OR "autobiologies" OR "self-diagnosis “) AND ("Artificial Intelligence" OR "machine learning" OR "Big Data”)

About 109 articles are retrieved in November 2019 from the platform of Web of Science. Major research indexes such as SCI-EXPANDED, SSCI, A&HCI, and ESCI are included in the scope of this research. We used both VOSviewer and AntConc for an exploratory analysis of the dataset.

3. RESEARCH RESULTS

The following research findings first describe how literature are clustered based on keywords, and then discuss the semantic relationship among them, based on VOSviewer visualization outcomes. The findings also allow us to clarify and define the notions of self-tracking and quantified-self.

3.1. Keyword Map and Clusters

Using a constructed thesaurus that combine synonyms, we produced a keyword map (see Figure 1).

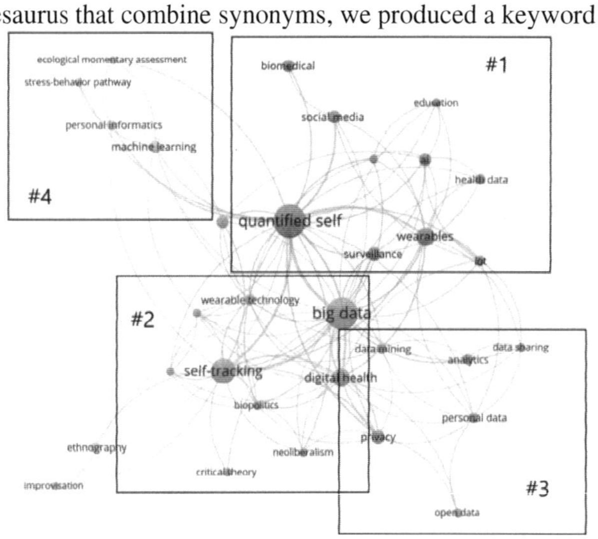

Figure 1. A keyword map based on co-occurrence data of literature: clustered outcomes.

Four main clusters have been identified based on the network of keyword co-occurrences. Each cluster reveals research focus in relation to other clusters/foci. The cluster #1 (wearables as quantified-self applications) includes AI, quantified self, biomedical, health data, wearables, IoT, social media, surveillance, and TORMES methodology. It contains research from benefits wearable devices and big data bring to AIED (Artificial Intelligence in Education) [10], to user motives of habitual integration of wearables in everyday [11]. Note that the topic “wearables” in the cluster #2 is different from that of “wearable technology” in the cluster #2. It appears to us that the term “wearable” refers to the general notion of wearable devices and artefacts, whereas the term “wearable technology” means technologies such as embedded sensors and algorithms that track, analysis and guide wearers' behavior. The cluster #2 (big data and critical theory) includes topics such as big data, self-tracking, self-care, wearable technology, datafication, digital health, critical theory, algorithms, biopolitics, neoliberalism. Based mostly on critical theories, these studies explore social, cultural and political background of self-tracking. For example, research on datafication of clinical and self-care practices[12], and the Taylorist influence on wearable technologies within neoliberal workplaces [13]. The cluster #3 (data and privacy) consists of topics such as personal data, open data, privacy, data mining and analytics. It includes work such as the barriers to the use of personal health data [14], self-tracking data as posterity [15], etc. The cluster #4 (personal informatics) contains topics such as machine learning, personal informatics, stress-behavior pathway, and ecological momentary assessment. For example, how do we capture stress-behavior pathway with mobile technology and machine learning analytics [16].

3.2. The Relation of Self-tracking, Quantified-self, AI and Big Data

As shown in Figure 1, quantified-self has been discussed along with AI, IoT (internet of things) technologies and applications[17]. Also, wearables are considered as a novel interaction technology from "IOme" (internet of me), a concept that highlights personalized self-tracking experience compared to IoT(internet of things) [10]. Both advocate that AI make quantified-self applications smarter.

Quantified-self wearables are often regarded as an outcomes of active and positive engagement with Big Data, which is different from data fetishists[18]. Also, self-tracking wearables can be digital compasses that help consumers navigate everyday choices[19]. The findings indicate that self-tracking constitutes the everyday life practices and applications to health and wellness for constructing quantified-self.

3.3. The Definition of Self-tracking and Quantified-self

However, what makes "quantified-self" differ from "self-tracking"? Table 1 present the findings based on the keyword in context (KWIC) analysis of all abstracts of the dataset, using AntConc.

Table 1. Keyword in Context Comparison: (a) self-tracking vs. (b) quantified-self

	Keyword in Context (KWIC)		source
	being restricted to an individualized form,	**self-tracking practices** are also becoming part	AjanaB_2017
	to the data they generate in	**self-tracking practices**. We argue that the	SharonTZand_2017
	intentionally, as in the case of	**self-tracking activities**. Self-tracking	Kneidinger-
	to follow from the availability of	**self-tracking technologies** and new volumes	McFallLMoor_2018
	as physical energy. In this context,	**self-tracking technologies** and related	TillC_2019
	corporate wellness programmes using	**self-tracking technologies** shows that their	TillC_2019
(a)	source on data about themselves,	**self-tracking technologies** reflect the	CrawfordKLi_2015
	and fitness companies. But can this	**self-tracking technology** sink its teeth into	BucherTAriy_2016
	for rich, personalized collections of digital	**self-tracking records.** Using qualitative	TraceCBZhan_2019
	which escape this data fetishist critique:	**self-tracking as a practice of mindfulness,**	SharonTZand_2017
	of the ideologies and rationalities	**self-tracking culture**. I argue that such	AjanaB_2017
	of sensor input and output when	**self-tracking wellness**. The conclusion is that	BomanMSanch_2015
	intensity levels in order to allow	**self-tracking while providing motivation**. In	GrillonAPer_2017
	users with higher mastery-orientation	**quantified-self design** more important. Users	HamariJHass_2018
	with positive evaluations of gamification	**quantified-self design** classes. Users with	HamariJHass_2018
	for users with high goal specificity	**quantified-self features** were important. The	HamariJHass_2018
	fall under the classes of (1) gamification,	**quantified-self** and **(3) social networking**	HamariJHass_2018
(b)	-Self information in the workplace. The	**Quantified-Self movement** has enabled	LavalliereM_2015
	apps are part of a wider "	**quantified self" movement** and many opt-in	WeberIAchan_2016
	era of new digital tools, the	**quantified self-movement, and precision**	TorousJSumm_2016
	accompanying mindset of 'datafication,'	**quantified self-movement,'** and the shift from	vandeVenI_2018
	by the do-it-yourself and	**quantified-self movements**, it is now possible	SantosOC_2016

examine the biopolitical aspects of the	**Quantified Self practices,** exploring some of	AjanaB_2017
light of big data and the	**quantified self (QS) movement.** Can we	KaranasiouAP_2016
The impact of the	**Quantified Self (QS) movement,** or the	EsmondeK_2019
, gut microbiome, clinical laboratory tests	**quantified self measures** from each individual	HoodLLovejo_2015

Table 1 (a) lists the terms often-associated with "self-tracking". As self-tracking technologies mean the knowledge-making processes between data and bodies[9], its primary concern becomes the capture and control of wellness rather than only physical health outcomes [20]. The datafication of users and their activities become the goal for a wide range of technologies, applications, activities, etc. to achieve.

Table 1 (b) shows those with "quantified-self". Based on the fact that it is associated with "Quantified Self movement"[21], we further define quantified-self as a result or phenomenon of datafication of users, including quantified-self movement, practices, measures, features, and so on.

Quantified-self means the outcomes of user datafication, while self-tracking refers to the means (often technical) of such datafication. Not all user datafication qualifies as quantified self, which may require additional meaning-making or group-identity efforts. One can use self-tracking technologies without becoming an active member of quantified-self movement.

The distinction between the two key concepts can be productive, suggesting the need to enrich the meaning of "quantified-self". For instance, one research argued that, based on the ontological premise with mind over the body, users should measure their health and well-being states[13]. Furthermore, another research proposed the concept of "holistic quantified self", with the aim to capture not only physical condition but also emotional status, social relationship status, digital consumption behavior status [22]. Different modes of user datafication using different sets of self-tracking technologies may lead to different construction of quantified-selves.

4. CONCLUSION

With the purpose to understand to what extent AI and Big Data constitute new understanding of self-tracking devices, technologies, services, and practices, the systematic review has presented crucial findings that reveal research opportunities and challenges regarding self-tracking and quantified-self.

Overall, the self-tracking research in relation to AI and Big Data has not only enriched the notion of self-tracking, but also opened up new conversations between self-tracking and quantified-self.

Self-tracking can be defined as datafication of human conditions or existence; quantified-self is the construction of self through such datafication. In other words, self-tracking practices and technologies may amount to new design, features, movements of Quantified-Self.

In anticipating such datafication processes of many aspects of our everyday life, we authors sense that the holistic approach [23] and the notion of "Holistic Quantified Self" [22] indicate an emerging need to provide a more holistic and integrated framework. Such a framework will be instrumental for theoretical and design development, so as to allow us to harness the potentials of the datafication of human activities for self-knowledge. To do so, further theoretical and technological advances must be developed not only to discern various aspects (e.g. health, wellness and happiness) of life itself for datafication but also to acknowledge both the potentials and dangers of datafication. As self-tracking becomes part of the wider datafication processes of human and social phenomenon, the future research needs to summarize and apply what has been discussed on datafication in the wider Big Data and AI literature to develop a conceptual framework of self-tracking as a major component of datafication of everyday life when more aspects of human activities, practices, records, etc. have be datafied.

ACKNOWLEDGMENTS

The research is funded by the Application-oriented Curriculum Development Project of "Information Visualization Design" (NFU 02-40250), under the Guangdong Province Department of Education 2018 "Innovation-Strengthening Higher Education Program" Grants, and partly by a project of HCI for Education Development (2018WQNCX283), under the 2018 Major Projects of Guangdong Province Department of Education "Young Innovative Talents" Grants.

REFERENCE

[1] Neff, G., Nafus, D. (2016) Self-tracking. (Cambridge, Massachusetts: The MIT Press)

[2] Swan, M. (2018) Quantified Self. Encyclopedia of Behavioral Medicine ed M Gellman (New York, NY: Springer) pp 1–4

[3] Tzovaras, B.G., Angrist, M., Arvai, K.J., Dulaney, M., Estrada-Galiñanes, V., Gunderson, B., Head, T., Lewis, D., Nov, O., Shaer, O., Tzovara, A., Bobe, J.R., Ball, M.P. (2019) Open Humans: A platform for participant-centered research and personal data exploration. GigaScience

[4] Almalki, M., Gray, K., Martin-Sanchez, F. (2016) Activity Theory as a Theoretical Framework for Health Self-Quantification: A Systematic Review of Empirical Studies. J. Med. Internet Res. 18:

[5] Saheb, T., Saheb, M. (2019) Analyzing and Visualizing Knowledge Structures of Health Informatics from 1974 to 2018: A Bibliometric and Social Network Analysis. Healthcare informatics research

[6] Rehman, M.H. ur, Liew, C.S., Wah, T.Y., Shuja, J., Daghighi, B. (2015) Mining Personal Data Using Smartphones and Wearable Devices: A Survey. Sensors 15: 4430.

[7] Swan, M. (2013) The Quantified Self: Fundamental Disruption in Big Data Science and Biological Discovery. Big Data 1: 85–99.

[8] del Río Carral, M., Roux, P., Bruchez, C., Santiago-Delefosse, M. (2017) Digital health: Promises, challenges, and fears. A literature review. Prat. Psychol. 23: 61–77.

[9] Crawford, K., Lingel, J., Karppi, T. (2015) Our metrics, ourselves: A hundred years of self-tracking from the weight scale to the wrist wearable device. Eur. J. Cult. Stud. 18: 479–96.

[10] Santos, O.C. (2016) Training the Body: The Potential of AIED to Support Personalized Motor Skills Learning. Int. J. Artif. Intell. Educ. 26: 730–55.

[11] Stragier, J., Vanden Abeele, M., Mechant, P., De Marez, L. (2016) Understanding persistence in the use of Online Fitness Communities: Comparing novice and experienced users. Comput Hum Behav 64.0: 34.

[12] Ruckenstein, M., Schüll, N.D. (2017) The Datafication of Health. Annu. Rev. Anthropol. 46: 261–78.

[13] Moore, P., Robinson, A. (2016) The quantified self: What counts in the neoliberal workplace. New Media Soc. 18: 2774–92.

[14] Bietz, M.J., Bloss, C.S., Calvert, S., Godino, J.G., Gregory, J., Claffey, M.P., Sheehan, J., Patrick, K. (2016) Opportunities and challenges in the use of personal health data for health research. J. Am. Med. Inform. Assoc. 23: e42–8.

[15] Trace, C.B., Zhang, Y. (2019) Extended Abstracts of the 2019 CHI Conference on Human Factors in Computing Systems. 2019 CHI Conference (Glasgow, Scotland Uk: ACM Press) pp 1–6

[16] Cheung, Y., Hsueh, P.-Y., Qian, M., Yoon, S., Meli, L., Diaz, K., Schwartz, J., Kronish, I., Davidson, K. (2017) Are Nomothetic or Ideographic Approaches Superior in Predicting Daily Exercise Behaviors: Analyzing N-of-1 mHealth Data. Methods Inf. Med. 56: 452–60.

[17] Gruyer, D., Magnier, V., Hamdi, K. (2017) Perception, information processing and modeling: Critical stages for autonomous driving applications. Annu. Rev. Control 44: 323–41.

[18] Sharon, T., Zandbergen, D. (2017) From data fetishism to quantifying selves: Self-tracking practices and the other values of data. New Media Soc. 19: 1695–709.

[19] Schüll, N.D. (2016) Data for life: Wearable technology and the design of self-care. BioSocieties 11: 317–33.

[20] Till, C. (2019) Creating 'automatic subjects': Corporate wellness and self-tracking. Health Interdiscip. J. Soc. Study Health Illn. Med. 23: 418–35.

[21] Nafus, D., Sherman, J. (2014) This One Does Not Go Up To 11: The Quantified Self Movement as an Alternative Big Data Practice. Int. J. Commun. 8: 11.

[22] Woo, W. (2018) Augmented Human: Augmented Reality and Beyond. Proceedings of the 3rd International Workshop on Multimedia Alternate Realities - AltMM'18 the 3rd International Workshop (Seoul, Republic of Korea: ACM Press) pp 1–1

[23] Banos, O., Bilal Amin, M., Ali Khan, W., Afzal, M., Hussain, M., Kang, B.H., Lee, S. (2016) The Mining Minds digital health and wellness framework. Biomed. Eng. OnLine 15: 76.

IOP Conf. Series: Materials Science and Engineering **806** (2020) 012015 doi:10.1088/1757-899X/806/1/012015

Based on PPI Network and Deep Learning Predicte Protein Function Module Algorithm

SicongHuo[1], TaoLu[2]

[1] Nanning University, Nanning, Guangxi, 530200,China

[2] Nanning University, Nanning, Guangxi, 530200,China

*Sicong Huo e-mail: 237290097@qq.com

Abstract. The existing Protein protein interaction(PPI) based prediction algorithm for protein functional modules has the disadvantages of low accuracy and less number of functional modules. This paper presents an algorithm based on PPI network and in-depth learning and prediction of protein function module. In this method, three important attributes (node location, PPI network structure and core nodes) are integrated to improve the prediction accuracy and increase the number of protein functional modules. In this paper, First use the Local Density-Based Methods (LDBM) improved density clustering algorithm for clustering analysis. After clustering, use the improved to isometric mapping dimension reduction algorithm for principal component analysis (PPIPCA); then use Multi Layer Perceptron (MLP) for training and function module selection. Last Based on PPI network and deep learning based prediction protein function module algorithm (LPMM) compared with other algorithms. On the basis of yeast dip data set, the accuracy, F value, dimensionality reduction rate and functional module of recognition protein of LPMM algorithm and acc-fmd, MCL and mcode algorithm based on PPI network were compared. Experimental results show that LPMM algorithm is superior to other algorithmsin accuracy, F value, dimensionality reduction rate, number of recognition modules and so on.

1. Introduction

In recent years, as the bioinformation technology develops, prediction of protein function module has become a research hotspot in the field of bioinformation. At present, many researchers predict protein function modules according to the characteristics of Protein protein interaction (PPI), and a large number of protein data are mined out. Prediction of protein function module involves many technologies. More and more researchers use the three-dimensional data structure of protein to predict protein function module, and successfully predict more and more new protein function modules. This kind of research direction has become a new hotspot for researchers to predict protein function module. At present, the intelligent algorithm plays an important role in the prediction of protein function module. This kind of algorithm has good nonlinear fitting ability. It can map the linear nodes of PPI network to any complex nonlinear relationship. The learning rules are simple, the programming implementation is easy to achieve, and it has strong robustness, memory ability and global approximation ability, which solves the problem The basic problems of protein function module prediction. However, the new algorithm still has some shortcomings, such as accuracy, low F value, and the number of protein functional modules. In view of these shortcomings, researchers use the characteristics of PPI network three-dimensional data structure to predict protein function module [1].

The reliability of protein function module based on PPI network is high. This paper proposes an algorithm based on PPI network and deep learning to predict protein function module, which integrates three important information of PPI network (protein core node, protein each node location, PPI network structure) as the important attribute of protein function module, proposes an Local Density-Based Methods (LDBM) for clustering analysis. After clustering,use the improved to isometric mapping dimension reduction algorithm for principal component analysis (PPIPCA), and then the Multi Layer Perceptron (MLP) is used to train the protein function module prediction method .Based on PPI network and deep learning based prediction protein function module algorithm (LPMM) has higher accuracy and F value than other algorithms --ACC-FMD, MCL,MCODE.

2. Algorithm of protein function module based on PPI network and deep learning

PPI network is usually represented as an undirected graph G (V, E) [1], where $V = \{V_1, V_2, V_3, \ldots\ldots V_N\}$, expressed as node set, $E = \{E_1, E_2, E_3, \ldots\ldots E_M\}$ is expressed as a set of connecting edges of each node. In this paper, a protein function prediction algorithm is based on LPMM and PPI network proposed. As shown in Figure 1, LPMM algorithm is divided into four parts: 1. Data input; 2. Training model; 3. Comparison with known protein function module; 4. Data output. First of all, the input data is the known protein function module interaction annotation term data and the main characteristics of PPI network (protein core node, protein each node location, PPI network structure); the input data is normalized; secondly, the multi-layer sensor (MLP) is used for training; then two classifiers are used to classify the predicted protein function module; finally, the function module is selected for output.

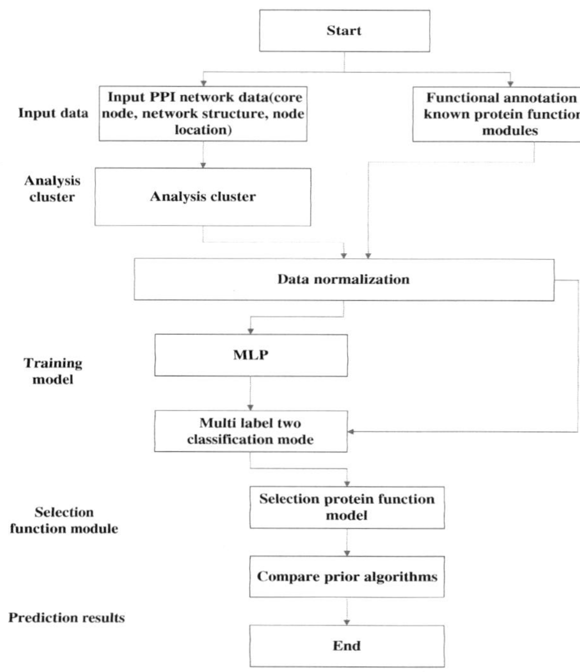

Fig 1 LPMM algorithm flowchar

2.1. Extract the main attributes of PPI network

The research shows that the closer the two core nodes in PPI network are, the greater the probability of predicting the protein function modules [2]. Using the characteristics of PPI network, we can first mine some known or unknown protein function modules by clustering, and then use the main characteristics of clustering to predict the protein function modules. The noise of this method is small and robust.

Density clustering algorithm is often used to predict the function modules of PPI network. The traditional density algorithm needs human intervention to determine the protein core nodes. In this paper, an improved density clustering algorithm is proposed to predict protein function modules. According to formula (1) and formula (2), the local density is calculated and the central node is determined. This is the result of the first clustering.

$$\rho_i = \sum_j x(d_{ij} - d_c) \tag{1}$$

$$\delta_i = \min_{j:\rho_j > \rho_i}(d_{ij} - d_c) \tag{2}$$

The remaining undivided node sets were calculated again to obtain the maximum local density and determine the next core node of PPI network, and then clustering analysis was conducted. Repeat the above process. Until all node data are clustered or sparse PPI network nodes are treated as noise, a threshold condition is finally set to determine whether the remaining data nodes are regarded as noise. The specific steps of the algorithm are shown in the figure below.

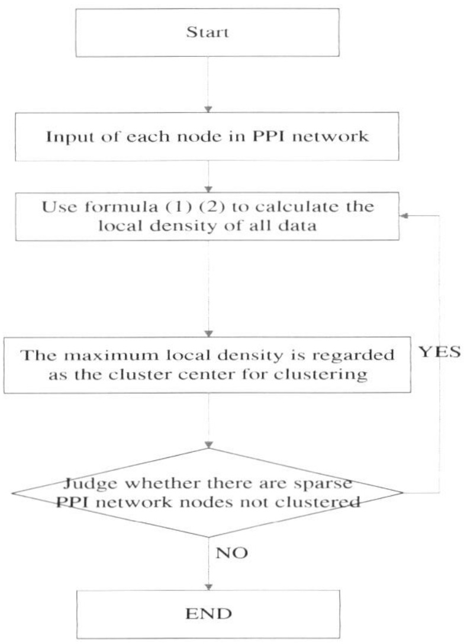

Fig 2 Flow chart of LDBM algorithm

2.2. Principal component analysis

As mentioned above, three main attributes (core node, core node location, PPI network structure) in PPI network play a key role in the prediction of protein function modules. These three main attributes are used to predict protein function modules to improve accuracy and reduce data noise. In this paper, MLP is used in training. If the clusters are trained, the high dimension of clusters will lead to high noise and affect the prediction results.[3] Considering the correlation of protein functional attributes, it is necessary to analyze the clusters after clustering. In order to maintain the consistency of data structure before and after dimensionality reduction, an improved isometric mapping dimensionality reduction algorithm is proposed. The basic idea is as follows: according to the principle of local linearity, the matrix is established by clustering clusters; then the minimum distance between any cluster nodes is calculated, which is expressed as the geodesic distance; finally, the low-dimensional representation of high-dimensional data is obtained by using multiple dimensional scaling (MDS) [8] algorithm.

The specific process of the improved isometric mapping dimensionality reduction algorithm is as follows: (1) construct the adjacency matrix. Firstly, the clustering cluster is regarded as the neighborhood point. The method of selecting neighborhood points: ε is the neighborhood radius, and when$|x_i\ x_j| < \varepsilon$, X_j is the neighborhood point of X_i. Establish weighted adjacency matrix:

$$w_{ij} = \begin{cases} 1: & x_i \ and \ x_j \ \text{mutual domain points} \\ 0: & \text{Other cases} \end{cases} \quad (3)$$

Calculate the shortest path. If Xi and XJ are domain points, the geodesic distance can be replaced by the Euclidean distance s (i, j) between them. Otherwise, the algorithm sets the starting geodesic distance Sc (i, j) = ∞, and constructs the shortest path as follows:

$$s_c(x_i, x_j) = \min_{l=1\cdots n}\{s_c(x_i, x_j), s_c(x_i, x_l) + s_c(x_l, x_j)\} \quad (4)$$

The above methods, the geodesic distance in the neighborhood and the geodesic distance outside the neighborhood are obtained respectively, and the shortest path moment representing the geodesic distance between sample points is obtained. $S_C = [S_C(i, j)]_{n \times m}$ shortest path distance converges to geodesic distance S_C, and then w matrix is applied to MDS algorithm for dimension reduction.

In order to improve the quality of dimensionality reduction, the geodesic is modified as follows and calculated by formula (5):

$$S_C = \left[s_C(x_i, x_j)\right]_{N \times N}, i, j = 1, 2, \cdots, N \quad (5)$$

Where $S_C(X_i, X_J)$ is determined by the following formula (6):

$$s_C(x_i, x_j) = \min\left\{\sum_{k=j}^{i} s(x_k, x_{k+1}), |\sum_{k=j}^{N=1} s(x_k, x_{k+1}) + s(x_N, x_1) + \sum_{k=1}^{i} d(x_k, x_{k+1})\right\} \quad (6)$$

The distance s shown in formula (7) is Euclidean distance, and the core matrix (kernelmatrix) k is calculated by geodesic distance:

$$K = HS_C^2H \quad (7)$$

Formula (7) matrix H is calculated by the following formula:

$$H = I_n - \frac{1}{N} e_N e_N^T \quad (8)$$

Where $e_n = [0\ 1]\ t \in w_n$. The matrix K is eigendecom posed to get the maxi n values $\lambda_1 \lambda_2 \lambda_3 \ldots \lambda_n$, and the corresponding eigenvector $\omega_1 \omega_2 \omega_3 \ldots \omega_n$ Matrix composed of ω_n, using matrix $\omega = \{\omega_1, \omega_2, \omega_3\ \omega_n\}$ $\omega \rightarrow x'_{ij}$ gives the corresponding low dimensional data X_{ij}:

$$X'_{ij} = \left\{x_{ij} \mid x_{ij} \in R^S, j = 1, 2, \cdots, N\right\} \quad (9)$$

2.3. *Multi sensor training and functional protein selection*

The core cluster of PPI network after clustering and dimensionality reduction reflects the importance of protein functional modules in PPI network. The higher the importance of core cluster nodes, the more functional properties of protein.[4] In this paper, the core cluster after dimensionality reduction is regarded as a feature matrix, which is combined with the matrix composed of the known main components of protein function module attributes, and a new feature matrix is obtained. In order to improve the quality of training, the characteristic matrix is standardized by min max to make each element data linear transformation.

In this paper, MLP with single hidden layer is used as classifier. Compared with the traditional classifier, MLP classifier is more flexible, which overcomes the weakness that perceptron can't recognize the linear non separable data. It uses PPI network to form the model of node number, activation function and loss function of cluster, and uses hidden layer to mine the complex relationship between the predicted protein function module and the known protein function module. In PPI network, the characteristic vectors of the data terms of the known protein function modules are used as the input, and the annotation of the known protein function modules is used as the output to establish the mapping relationship between the multi features and multi functions in the protein function modules.

3. Experimental data set and evaluation index

In order to verify the effectiveness of the LPMM algorithm proposed in this paper, PPI network interaction data is relatively complete and reliable as the experimental data. In the experiment, firstly, we download PPI network data from dip database, use uniprotkb / Swiss prot to perform ID conversion on proteins in PPI network, process and filter repeated interaction and untranslatable protein terms through serial affinity purification, and then obtain dip term number interpro term number according to uniprotkb / Swiss prot number of each protein this paper, precision, recall and F-measure and OS (a, b) are the indexes to measure the prediction effect of the algorithm. [9] the formula is as follows:

$$\text{Precision} = \frac{TP}{TP + FP} \tag{10}$$

$$\text{Recall} = \frac{TP}{TP + FN} \tag{11}$$

$$\text{F-Measure} = \frac{2 * \text{Precision*Recall}}{\text{Precision} + \text{Recall}} \tag{12}$$

$$OS(A, B) = \frac{|A \cap B|}{|A||B|} \tag{13}$$

3.1. Experimental data analysis

In order to analyze the protein function modules detected by LPMM algorithm on PPI network in detail, based on the PPI network data provided by dip database, LPMM algorithm is compared with ACC-FMD [5], MCL [6], MCODE [7] algorithm.

The number of predicted protein functional modules (PM) of LPMM algorithm compared with other three algorithms is shown in Table 1.

Table 1 Basic information of functional modules for various algorithms mining

Approachs	PM	TP	FULL
LPMM	392	253	19
ACC-FMD	374	101	9
MCL	126	52	5
MCODE	254	129	8

The number of significant protein function modules (SC) mined by LPMM algorithm and other three algorithms is shown in Table 2.

Table 2 Comparisons of the number of significant functional modules in each algorithm mining

Approachs	PM	SC	TP/PM
LPMM	392	153	39.03%
ACC-FMD	374	85	22.73%
MCL	126	36	28.57%
MCODE	254	98	38.58%

Table 3 shows the comparison of dimension reduction rate (1-N / M) between lpmm algorithm and other three algorithms.

Table 3 Comparison of dimension reduction rates of various algorithms

Approachs	M	N	1-N/M
LPMM	905	223	75.36%
ACC-FMD	823	488	40.70%
MCL	764	346	54.71%
MCODE	819	459	43.96%

The comparison between lpmm algorithm and other three algorithms, precision, recall, F-measure, shown in Table 4.

Table 4 Efficiency analysis of LPMM algorithm and three other algorithms

Approachs	Precision	Recall	F-Measure
LPMM	0.381	0.352	0.366
ACC-FMD	0.231	0.187	0.207
MCL	0.129	0.404	0.196
MCODE	0.328	0.268	0.295

The experimental results show that 253 of the protein function modules (PM) mined by LPMM algorithm are successfully matched by known protein function modules, which indicates that LPMM algorithm can recognize more protein function modules. The dimensionality reduction rate (1-N / M) is 75.36%, which is higher than other three algorithms. As shown in Table 4, the LPMM algorithm reflects the efficiency of the LPMM algorithm in the evaluation and prediction of the index accuracy of the protein function module (precision) 0.381, recall (recall) 0.352 and F-measure (F-measure) 0.366.

4. Conclusion

In the face of the continuous growth of PPI network data, how to efficiently predict protein functional modules is an important research issue in modern bioinformatics. It is a popular method to predict protein function module based on PPI network. This kind of method has low cost and high efficiency, but the data obtained is noisy. This paper proposes an LPMM algorithm for this problem. Experimental results show that LPMM algorithm can effectively predict protein function modules. In the next step, Researcher can study predict protein function modules from the following directions. How to predict the protein function module based on PPI network and swarm intelligence optimization algorithm, such as Bayesian intelligence algorithm, and neural network intelligence algorithm Prediction of protein functional modules.

Appendices

This study is supported by the National Natural Science Foundation of China under Grant Nos. 61741203,61866006; Guangxi Natural Science Foundation (2016GXNSFAA380243,); Guangxi Nanning Science and Technology Development Planning Project(20181015-5)

References

[1] Huo Sicong, Li Ying. A review of clustering methods for protein functional module detection [J]. Computer engineering and application, 2019,55 (08): 17-26.

[2] Dynamic weighted PPI network complex mining based on ant colony clustering [J / OL]. Computer application research: 1-10 [2019-06-13]. Https://doi.org/10.19734/j.issn.1001-3695.2018.07.0518.

[3] Chen B, Fan W, Liu J, et al. Identifying protein coMLPexes and functional modules--from static PPI networks to dynamic PPI networks.[J]. Briefings in Bioinformatics, 2014, 15(2):177.

[4] Liu Wei, Liu Shang, Zhou Xuan. Study on BP neural network sub batch learning method[J]. Journal of intelligent systems, 2016, 11 (2): 226-232.

[5] Liu Hongxin. Research on PPI network functional module detection algorithm based on ant colony clustering mechanism [D]. Beijing University of technology, 2013

[6] Hu Qingsheng, Lei Xiujuan. Improved Markov clustering algorithm for PPI network [J]. Computer science, 2015, 42 (7): 108-113

[7] Lei X, Li H, Zhang A, et al. iOPTICS-GSO for identifying protein coMLPexes from dynamic PPI networks[J]. Bmc Medical Genomics, 2017, 10(Suppl 5):80

[8] Zhang X F, Ouyang L, Zhu Y, et al. Determining minimum set of driver nodes in protein-protein interaction networks[J]. Bmc Bioinformatics, 2015, 16(1):146.

[9] Song J,Singh M.How and When should Interactome-derived Clusters be Used to PredictFunctional Modules and Protein Function[J].Bioinformatics,2009,25(29):3143-3150.

IOP Conf. Series: Materials Science and Engineering **806** (2020) 012016 doi:10.1088/1757-899X/806/1/012016

The Hotspots of Smart Education in China: Base on the Bibliometric Analysis and Knowledge Mapping

Wang HaiBin[1], Chen Qian[1], Hong Aoyuan[1], Wang Xiaoyan[1], Hou Ziyi[1*], Cheng Lin[2*]

[1] School of Educational Science, Huangshan University, Huangshan, 245041, China

[2] School of Tourism, Huangshan University, Huangshan, 245041, China

*Corresponding author. Hou Ziyi, zhou7@hsu.edu.cn or Cheng Lin, chenglin@hsu.edu.cn

Abstract. To explore the hotspots in the area of Smart education in China, 245 related articles form the China National Knowledge Infrastructure (CKNI) were analyzed by the visualization software of CiteSpace. The results show that: (1) the studies of smart education began in 2003 while there were sporadic papers published in the next ten years. There was an increasing trend year by year from 2013 to 2017 and the number of articles reached the historical peak in 2017 with 52 articles. Xianming Yang, Jiangsu Normal University and E-Education Research topped the lists of contributing authors, organizations and publications respectively. (2) "smart education", "educational informationization", "intelligent education" and "smart campus" were the top four largest clusters in the domain of smart education in China.

1. Introduction

With the development of educational informatization, smart education has received more and more attention, for example, the State Council issued the "New Generation Artificial Intelligence Development Plan" in 2017, proposing to improve the artificial intelligence education system and implement intelligent education [1-2]. The Education Informationization 2.0 Action Plan issued by the Ministry of Education in 2018 marked the beginning of a new era in smart education in China [3-4]. Smart education, that is, education informatization, refers to the process of comprehensively and deeply applying modern information technology in education (education management, education and education, and scientific research) to promote education reform and development [5]. The researchers from Malaysia, South Korea, Singapore, the United States, and the United Kingdom have paid more attention to the smart education in the 21st century, for example, IBM originally proposed the concept of Smart Planet, and then extended to smart education [6]. However, the research of smart education in China is still in its infancy when theoretical research still stays on a shallow level in general [6]. The earliest research on smart education started in 2003, which mainly discussed the experience of foreign smart education [7]. In the next six years, the research on smart education was less while there was an upward trend until 2013, and the research on smart education in China can be summarized in four aspects including theory research, environment research, technology-supported research and practice research of smart education [8]. However, the development of smart education and the specific content of the related research in China are still unclear. This study, therefore, set out to explore the status and hotspots in the domain of smart education in China with the help of the visualization software Citespace.

2. Methodology

2.1. data sources
245 articles were yielded by the search conducted in the China National Knowledge Infrastructure (CKNI) under the heading of "smart education", and the journal source categories were limited to Peking University core journals, CSSCI and CSCD. Among the 245 articles, there were 204 journal articles and 41 Dissertation.

2.2. data processing
The RIS format of the 245 articles was carried out by the bibliometric analysis and visual analysis by the visualization software CiteSpace.5.5. R2 [9].

3. Results

3.1. Bibliometric analysis

3.1.1. Publishing trend. Fig. 1 demonstrates the publishing trend of studies on smart education in China. The earliest research of smart education in China began in 2003 while there were sporadic papers published in the next ten years. After 2013, the research on smart education in China has shown an increasing trend year by year and appeared at a historical peak in 2017 with 52 articles.

Fig. 1 Publishing trend.

3.1.2. Author statistics. As shown in Tab.1, all of the top 10 contributing authors have published more than four articles and Xianming Yang topped the list with fifteen articles, followed by Lin Chen and Zhiting Zhu, each of them published more than ten articles.

Tab.1 Top 10 contributing authors

Rank	Authors	Frequency
1	Xianming Yang	15
2	Lin Chen	12
3	Zhiting Zhu	11
4	Fan Wang	9
5	Longju Tian	5
6	Qiujin Zhao	5
7	Hongchao Peng	4
8	Mengmeng Sun	4
9	Jing Chen	4
10	Shengquan Yu	4

3.1.3. Affiliation statistics. As seen in Tab.2, all of the top 10 organizations have contributed more than three articles on the studies of smart education in China. Jiangsu Normal University was the best performer among the list, followed by East China Normal University and Beijing Normal University, each of them contributed more than fifteen articles.

Tab.2 Top 10 contributing organizations

Rank	organizations	Frequency
1	Jiangsu Normal University	27
2	East China Normal University	20
3	Beijing Normal University	16
4	South China Normal University	4
5	Shaanxi Normal University	4
6	Tsinghua University	3
7	Zhejiang Normal University	3
8	Bohai University	3
9	Nanjing Normal University	3
10	Northeast Normal University	3

3.1.4. Source publication statistics. As seen in Tab.3, all of the top 10 source publications contributed more than four articles on the studies of smart education in China. E-Education Research topped the list with twenty-nine articles, followed by Modern Educational Technology and China Educational Technology, each of them published more than twenty articles.

Tab.3 Top 10 contributing source publication

Rank	Source publication	Frequency
1	e-Education Research	29
2	Modern Educational Technology	28
3	China Educational Technology	24
4	Journal of the Chinese Society of	14
5	Journal of Distance Education	6
6	Theory and Practice of Education	5
7	Modern Distance Education Research	4
8	People's Education	4
9	Open Education Research	4
10	Distance Education in China	4

3.2. Knowledge mapping results

3.2.1. High-frequency keywords. The top 10 high-frequency keywords in the domain of smart education in China, as well as their frequencies on the studies of smart education were shown in Tab.4 and Fig.2. "smart education" was the most frequent keywords, and "educational informationization", "artificial intelligence", "smart" and "smart campus" were positioned in the 2nd, 3rd 4th and 5th places on the list. It is indicated that educational informationization, artificial intelligence, smart and smart campus attracted researchers' attention.

Tab.4 Top 10 keywords

Rank	Keywords	Frequency
1	smart education	186
2	educational informationization	43
3	artificial intelligence	17
4	smart	15
5	smart campus	9
6	big data	7
7	smart learning	6
8	smart curriculum	5
9	smart classroom	5
10	intelligent education	4

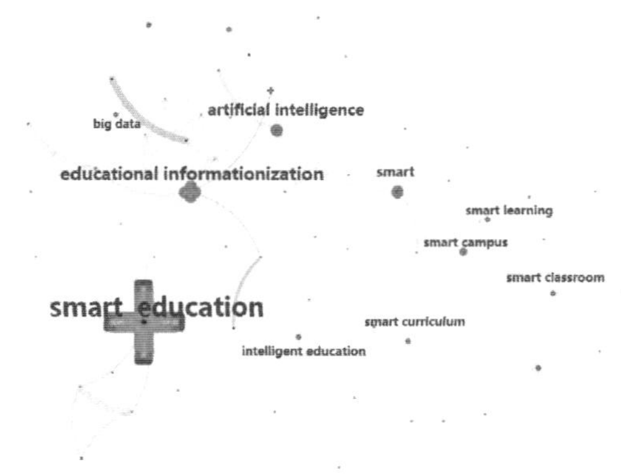

Fig. 2 Top 10 keywords

3.2.2. Popular research topics. Fig 3 outlines the top four largest clusters on the studies of smart education in China, which were "smart education", "educational informationization", "intelligent education" and "smart campus" and the Modularity Q value is 0.76, the Mean Silhouette value is 0.51. As shown, the largest cluster is "smart education (#0)", which contains 11 articles and its silhouette value is 0.85. The high-frequency keywords of this cluster contain "smart education", "smart education era" and "the integration of production and education". The second-largest cluster is "educational informationization (#1)", which contains 10 articles and its silhouette value is 0.97. The high-frequency keywords of this cluster contain "educational informationization", "educational technology", "big data" and "artificial intelligence". The third-largest cluster is "intelligent education" (#2), which contains 9 articles and its silhouette value is 1.00. The high-frequency keywords of this cluster contain "intelligent education", "cloud computing", "cloud platform" and "Education Information 2.0". The fourth-largest cluster is "smart campus" (#3), which contains 8 articles and its silhouette value is 0.87. The high-frequency keywords of this cluster contain "smart campus", "smart learning", "smart classroom" and "smart curriculum".

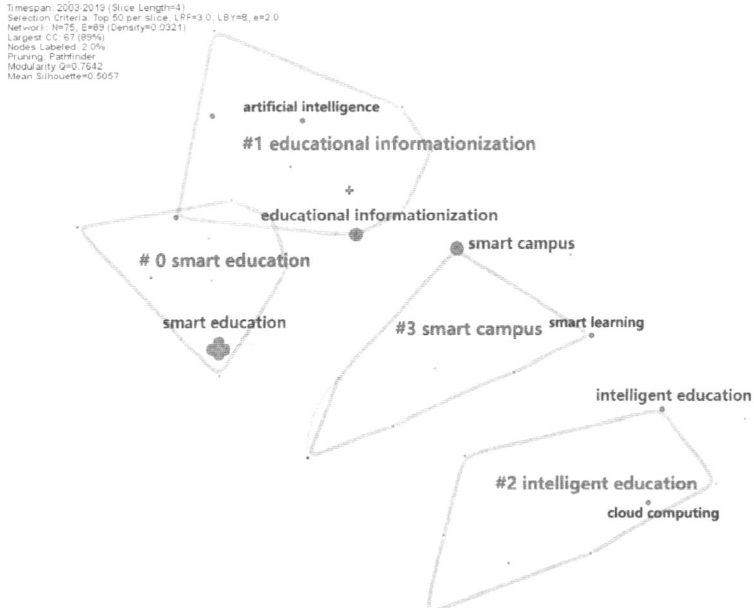

Fig. 3 Popular research topics

4. Conclusions and further research

To explore the status and hotspots in the domain of smart education in China with the help of the visualization software Citespace. The bibliometric analysis results showed that the studies of smart education in China while there were sporadic papers published in the next ten years. There was an increasing trend year by year from 2013 to 2017 and the number of articles reached the historical peak in 2017 with 52 articles. Xianming Yang topped the list of contributing authors with fifteen articles, followed by Lin Chen and Zhiting Zhu, each of them published more than ten articles. Jiangsu Normal University was the best performer among the list of the contributing organizations, followed by East China Normal University and Beijing Normal University, each of them contributed more than fifteen articles, indicating that the normal universities are the main research institution on the studies of smart education in China. E-education Research topped the list of the contributing source publication, followed by Modern Educational Technology and China Educational Technology, each of them published more than twenty articles, indicating that smart education particularly attracted the educational technology researchers' attention in China.

In the section of Knowledge mapping analysis, the results of high-frequency keywords showed that "smart education" was the most frequent keywords, and "educational informationization", "artificial intelligence", "smart" and "smart campus" were positioned in the 2nd, 3rd 4th and 5th places on the list of the high-frequency keywords, indicating that educational informationization, artificial intelligence, smart and smart campus attracted researchers' attention in China. The clusters analysis results showed that "smart education", "educational informationization", "intelligent education" and "smart campus" were the top four largest clusters on the area of smart education in China. Among them, "smart education (#0)" is the largest cluster and the high-frequency keywords of this cluster contain "smart education", "smart education era" and "the integration of production and education", indicating that smart education is the product and trend of deep integration of science, technology, and education in China.

Acknowledgement
This work was financially supported by the Later-stage projects funded by the National Social Science in China (Grant NO. 18FKS007); Key project for university outstanding young talents support program of Anhui province (Grant NO. gxyqZD2019068); Pilot project for smart classroom of Anhui province (Grant NO. 2017zhkt414) and Project for education science research of Anhui province (Grant NO. JK18071).

References

[1] Hongmei, L. (2004). Quality education, smart education and exam-oriented education. *Heilongjiang Higher Education Research* (01), 10-12.

[2] Xiaoxiang, L. (2004). Smart education: interpretation of whitehead's educational thoughts. *Education guide* (05), 50-52.

[3] Xiaozong, S. (2005). Going to smart education. *Educational Research and Experiment* (02), 24-26.

[4] Yuling, L., Xiaozong, S. (2005). The trend of modern education reform——smart education.*Educational theory and practice* (08), 4-6.

[5] Lin, C., Mengmeng, S., Xuefei, L. (2017). Study on the origin of wisdom education. *e-Education Research, 38*(02), 13-18.

[6] Peijie, C. (2018). Smart Education：The educational reform at the age of artificial intelligence. *Educational Research, 39*(08), 121-128.

[7] Jiaxian, Z. (2003). Towards smart education: design of exploratory courses - interview with dr. grant wiggins, american education specialist. *Global Education, 32*(06), 3-7.

[8] Xin, Y., Yueguang, X. (2019). Smart teaching ability: teacher ability dimension in the age of smart education. *Educational Research, 40*(08), 150-159.

[9] Chen, C. (2012). Predictive effects of structural variation on citation counts. *Journal of the Association for Information Science & Technology, 63*(3), 431-449.

IOP Conf. Series: Materials Science and Engineering **806** (2020) 012017 doi:10.1088/1757-899X/806/1/012017

Research on short-term traffic flow prediction model based on RNN-LSTM

Shuying Liu, Li Zhang, Hangfu Li

School of Information Engineering, Anhui Xinhua University, Hefei, Anhui,230088, China

29064581@qq.com

Abstract. With the rapid development of economy and the increasing number of cars, it is the first task to effectively predict the traffic flow of road traffic to alleviate road congestion and reduce traffic accidents. Aiming at the uncertainty and nonlinearity of traffic flow data, the traffic flow prediction model based on long-term and short-term memory is designed and compared with the traditional BP model and RNN model. The simulation results show that the LSTM model has a higher prediction accuracy and better prediction effect on traffic flow.

1. Introduction

With the rapid development of economy and the gradual improvement of people's living standards, t the demand for travel is also growing higher. Therefore, cars are more and more appearing in people's daily life. Although the increase of car ownership brings convenience to people's life, it also increases environmental pollution, road congestion and frequent traffic accidents. Therefore, the most important task of the road management center is to plan the road reasonably, predict the traffic flow accurately and efficiently, and induce the traffic flow in time. Under the current road conditions, intelligent transportation system (ITS)[1] is applied. It is an intelligent traffic management system based on traffic control ,guidance and traffic flow prediction algorithm. Therefore, it is of great significance to predict traffic flow in real time and accurately, because it provides timely road conditions for travelers and provides information guarantee for dynamic planning of travel routes.

2. Introduction to traffic flow prediction

According to different time span, traffic flow prediction is generally divided into long-term flow prediction and short-term flow prediction[2]. The long-term flow forecast is the traffic flow based on year, month and day, which mainly provides the basis for road planning and road design. The short-term flow forecast is generally no more than 15 minutes, or even less than 5 minutes, which provides the basis for the traffic control or guidance system to make the next decision.

2.1. Introduction to prediction model

Experts at home and abroad, after analyzing and studying the characteristics of traffic flow, such as strong uncertainty and weak regularity[3], put forward a variety of traffic flow prediction models. According to the different principles of prediction, these models are roughly divided into two categories. One is the data model represented by Kalman filter and ARIMA model. This kind of model is simple in structure and convenient in calculation, but it is greatly affected by the fluctuation of traffic flow, and it can not predict the non-linear and non-stationary data with strong randomness.

Therefore, it is difficult to meet the requirements of short-term traffic flow prediction; The other is the non mathematical model represented by support vector regression machine, non parametric regression model and neural network model, this kind of model has a strong dependence on historical data. As long as there is a large amount of historical data, there is no need to build a large redundant model. According to the nonlinear characteristics of short-term traffic flow, it increases the adaptability to nonlinear[4]. The prediction effect of short-term traffic flow is good.

2.2. RNN model introduction

The model of recurrent neural network (RNN) [5]for processing time series data, The biggest difference between it and the traditional neural network is that the traditional neural network layer is fully connected, and the nodes between each layer are unconnected, however, there are connections between the hidden layer nodes in RNN. In RNN, the hidden layer node not only receives the information transmitted by the input layer at the current time point, but also contains the information transmitted by the hidden layer at the previous time. Therefore, the network constructed by RNN retains the information of historical time points and applies the information left by memory to the output calculation of current neurons. It is a simple chain structure, that is, the same neural network is copied many times to enter the cycle structure.

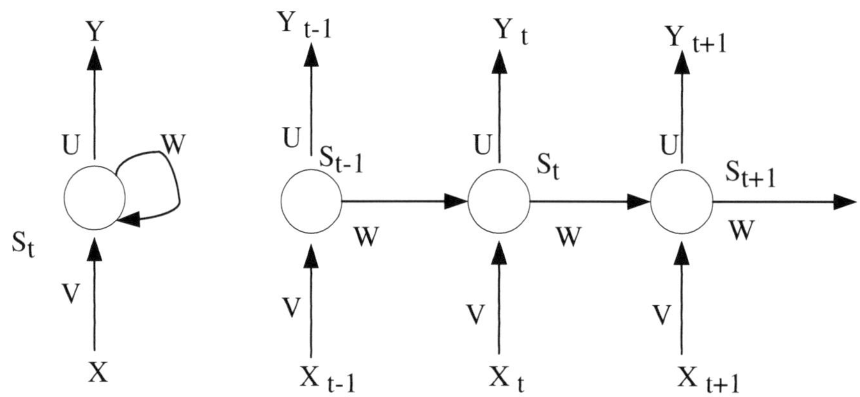

Figure 1. RNN unit expansion

A simple RNN structure is expanded as shown in Figure 1. Where st is the state of step t of the hidden layer, it is the memory unit of the network. It can be seen from the above figure that the hidden layer node at time t is not only affected by the current time XT, but also by the last time reserved ST-1, similarly, the state st retained at time t affects time t+1. Therefore, RNN is mostly used to forecast time series data, such as weather forecast, stock forecast and traffic flow forecast.

Although RNN has memory function for data processing, the time span of information that can be used in practice is limited, When the time span is small, RNN can effectively learn the information of historical time, but when the time span is large, RNN's learning ability will be reduced, That is, the problem of RNN gradient disappearance or gradient explosion. As shown in Figure 2. The memory of t1 is transferred to tn through long journey, RNN adopts the principle of error reverse transfer, error

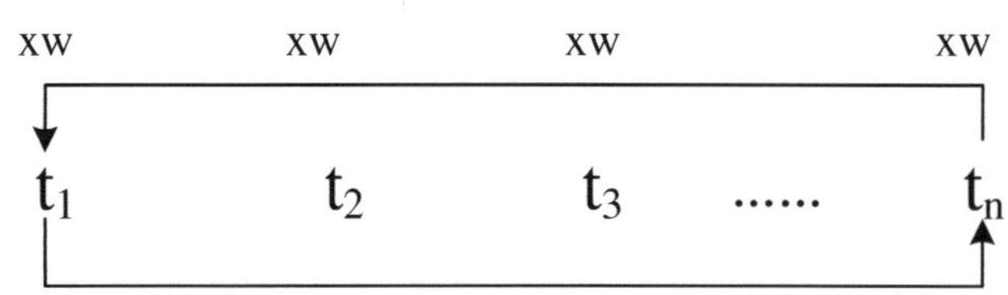

Figure 2. Gradient disappearance

x obtained at TN needs to be multiplied by the parameter W at each step, if W is less than 1, the error may be very small when it is transferred to T1, close to 0, so the error for the initial t1 disappears, the gradient disappears; On the contrary, if W is greater than 1, there will be gradient explosion.

To solve the problem of RNN gradient disappearance, experts and scholars have given a variety of solutions, among them, Hochreiter and schmidhuber's long-term and short-term memory algorithm proposed in 1997 is the most popular, it can complete the long-term dependence more accurately and has the ability to learn long-term information better[6].

2.3. LSTM model

LSTM long short term memory network is an improved cyclic neural network[7], it overcomes the problem of gradient disappearing in RNN and the problem of long-term dependence which can not be dealt with in RNN.

LSTM evolved on the basis of RNN[8]. In the standard RNN network, the structure of cyclic module is simple, with only one tanh layer, As shown in Figure 3, where a represents the relationship between the input layer and the hidden layer.

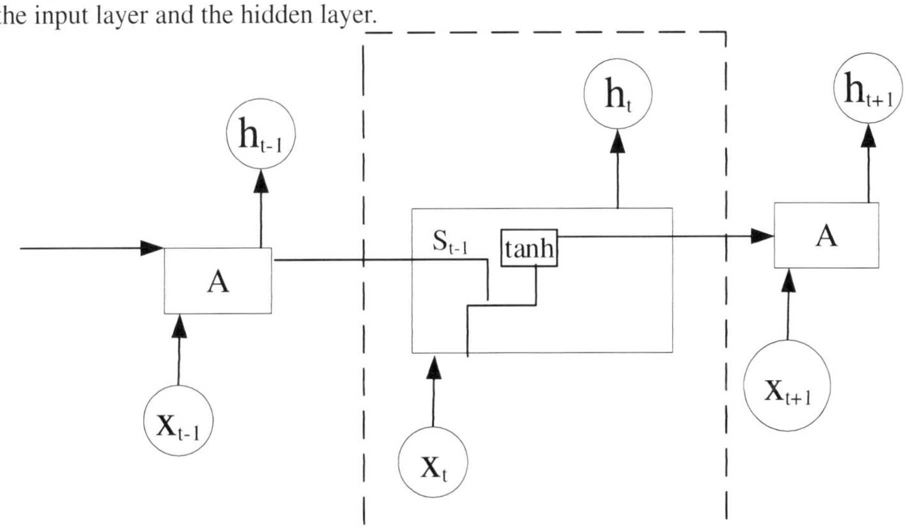

Figure 3. Repeated module chain in standard RNN

LSTM is a special RNN. Its memory unit is mainly composed of three gates: An input gate, an output gate and a forget gate[9]. The three gates symbolize the gate of information. They control the transmission of neuron information. The memory function of the network is realized by these gate nodes. When the gate is opened, the information retained by the front neuron is transmitted to the current neuron, while when the gate is closed, the information of the front neuron will not affect the calculation of the current neuron. Therefore, by adjusting the gate switch, we can realize the influence of historical time series on the later neurons, If the current time information is not expected to affect the calculation of the back neurons, just close the gate. As shown in Figure 4, the input at time t1 affects the output of t4.

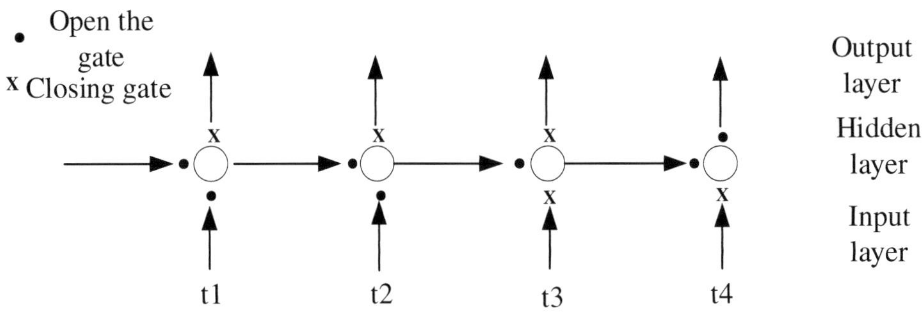

Figure 4. Display of LSTM gate

An important part of the LSTM is the door, which interacts with each other to control the operation of the module[10]. The opening and closing of the gate determines the retention or deletion of historical information. The basic working principle of the gate is shown in Figure 5. Under the action of activation function, three doors will generate a number between [0,1], which is used to determine the opening and closing of doors.

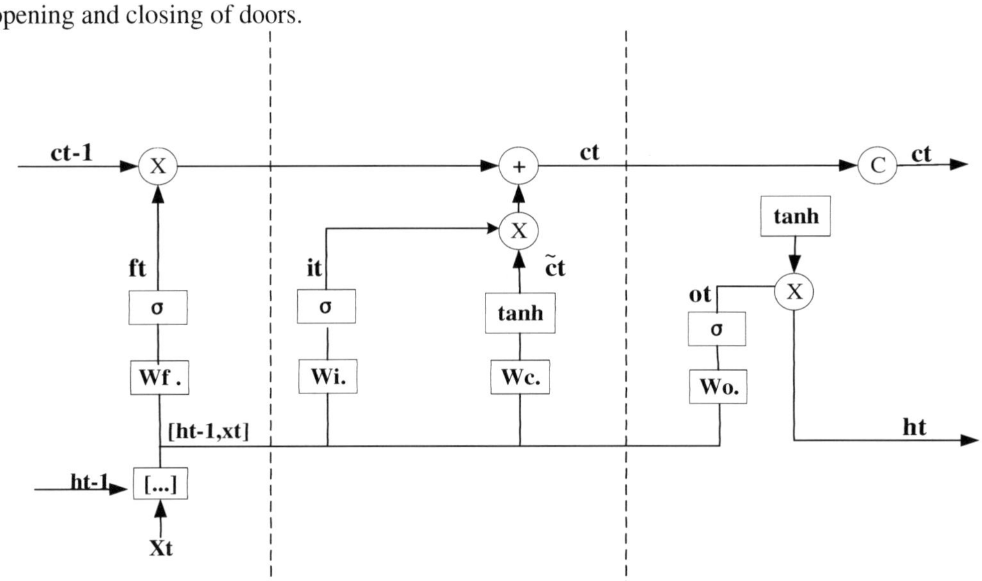

Figure 5. Basic working principle of LSTM

- the input gate determines how much information of the input XT of the network is retained to the unit state CT at the current time, so as to avoid the current irrelevant content entering the memory module.
- the forget gate determines how much of the cell state CT-1 of the previous time is reserved to the current time CT and what information is discarded.
- the output gate controls how much state CT of the unit outputs to the current output value ht of the LSTM, and controls the effect of long-term memory on the current input.

The calculation formula of forgetting gate is as follows:

$$Ft=\sigma(wf\cdot[Ht-1,Xt]+bf) \quad (1)$$

The calculation formula of the input door is:

$$It=\sigma(wi\cdot[Ht-1,Xt]+bi) \quad (2)$$

The calculation formula of current input cell state CT is:

$$Ct=Ft*Ct-1+It*\tilde{C}t \quad (3)$$

$$\widetilde{C}t = \tanh(wc\cdot[Ht-1,Xt]+bc) \quad (4)$$

$\widetilde{C}t$ is the current memory. The current memory $\widetilde{C}t$ is combined with the long-term memory Ct-1 to form a new unit state Ct, which is updated in the memory module without segments.

The calculation formula of the output gate is:

$$Ot=\sigma(wo\cdot[Ht-1,Xt]+bo) \quad (5)$$

$$Ht=Ot*\tanh(Ct) \quad (6)$$

In the above formula, FT is the forgetting gate vector at time t, It is the input gate vector at time t, OT is the output gate vector at time t, Ht is the output vector at time t, and Xt is the input vector at time t. w, bare weights and offsets of different gates.

Two activation functions are used in the formula: sigmoid function and tanh function. The sigmoid function is also known as the S function, it can introduce nonlinear factors into LSTM, When a real value is input to the sigmoid function, it can be compressed to [0,1] output, in which the extremely large positive number is mapped to 1, and the extremely small negative number is supplemented to 0. The formula is as follows:

$$S(x) = \frac{1}{1+e^{-x}} \quad (7)$$

Tanh function is the same as hyperbolic tangent function. It maps the real value of input to the range of [- 1,1]. When the input value is 0, the mapping output of tanh function is 0, which conforms to the form of activation function. Its formula is as follows:

$$\tanh(x) = \frac{\text{Sinhx}}{\text{Coshx}} = \frac{e^{x}-e^{-x}}{e^{x}+e^{-x}} \quad (8)$$

3. Verification and simulation

3.1. Data sources

In this paper, the data used in the verification experiment is the measured data of the traffic flow at an intersection in Hefei. Select 5min as the time interval for data collection. Select the data from September 24, 2017 to September 30, 2017 from 8:00 a.m. to 12:00 a.m. to generate Figure 6. From Figure 6, it can be seen that the traffic volume and time distribution in this data set are regular and can be used as the prediction data set of traffic flow.

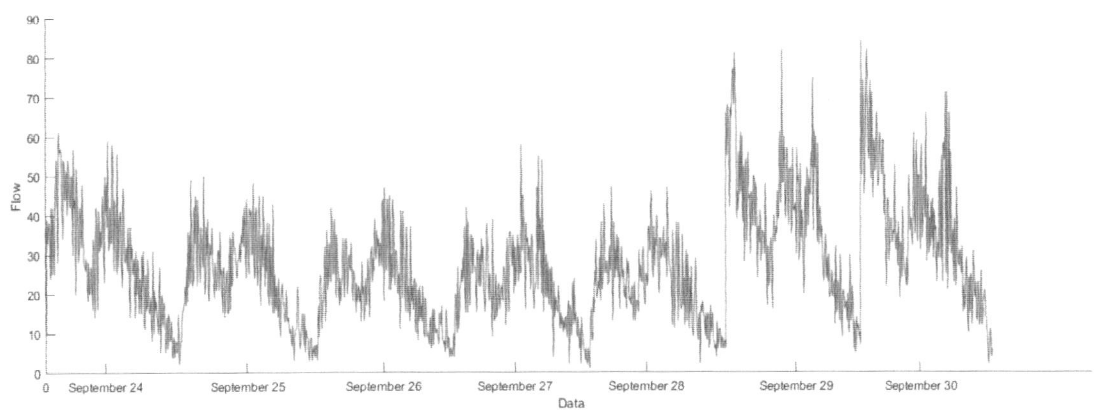

Figure 6. One week traffic flow chart

In the validation experiment, from September 18, 2017 to October 11, 2017 was selected as the training data set to train the model, and the data from October 12, 2017 to October 18, 2017 was used as the test set, to verify the validity of the model. For the data in the data set, abnormal data caused by equipment aging, instrument failure, storage failure and other reasons will affect the accuracy of the final prediction, so this experiment adopts the method of first deleting and then adding. For abnormal data, the mean value of the data in the first five days of the same period is used instead.

3.2. Evaluation index

The quality of a prediction model needs to be evaluated quantitatively. The evaluation indexes used in this simulation experiment are MAE, RMSE and MAPE.

(1) MAE——Mean Absolute Error The formula is:

$$\mathrm{MAE} = \frac{1}{n}\sum_{i=1}^{n}\left|y_p - y_r\right| \quad (9)$$

(2) RMSE——Root Mean Squae Error The formula is:

$$\mathrm{RMSE} = \sqrt{\frac{\sum_{i=1}^{n}(y_p - y_r)^2}{n}} \quad (10)$$

(3) MAPE——Mean Absolte Percentage Error The formula is:

$$\mathrm{MAPE} = \frac{1}{n}\sum_{i=1}^{n}\left|\frac{y_p - y_r}{y_r}\right| * 100\% \quad (11)$$

In the above formula, n is the number of samples, $y_p$is predict value, y_r is true value, The smaller the values of MAE, RMSE and MAPE are, the closer the predicted value is to the real value, the better the prediction effect is.

3.3. Verification results and analysis

In order to verify the effect of LSTM in traffic flow prediction, three models are constructed in this simulation: BP model, RNN model and LSTM model.

According to the verification results of reference 6, the BP model uses three hidden layers, and the number of hidden layer nodes is 11. The main parameter in the LSTM model is the number of cycles of the gates, which changes with the number of gates according to the loss function (MSE), as shown in Figure 7. As can be seen from the figure, when the number of cycles of the door is 400, the value of loss almost no longer changes with the increase of the number of cycles. Therefore, 400 gate control cycle units are selected in the hidden layer of the LSTM model simulation experiment.

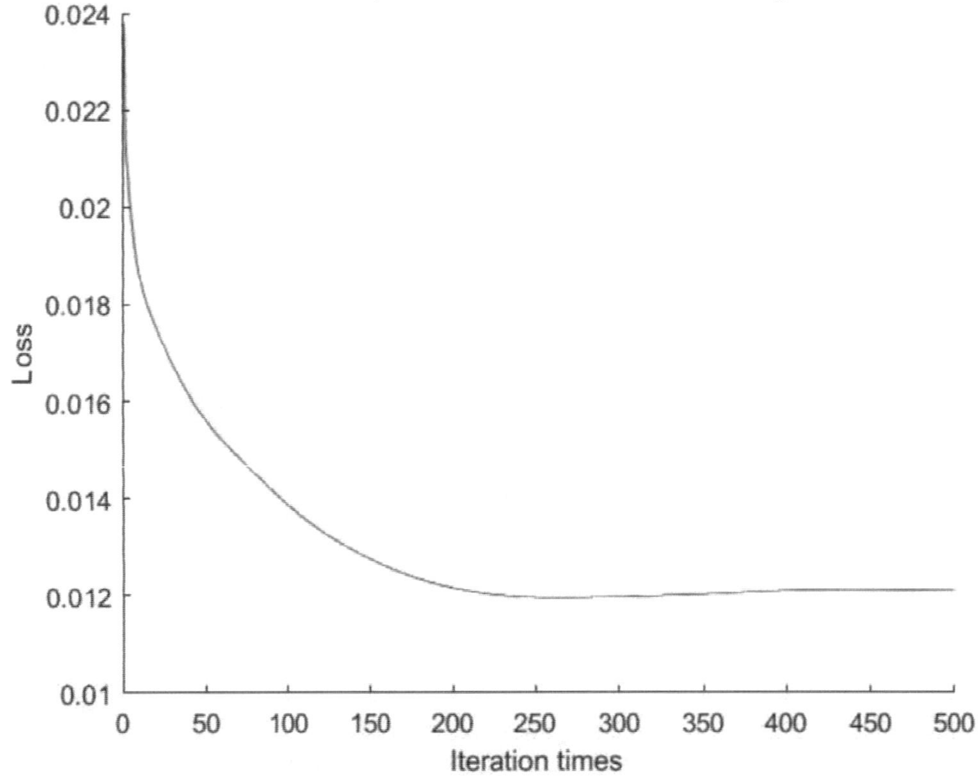

Figure 7. The change of loss function with the number of iterations

After the verification of the three models, the corresponding evaluation index values are obtained, as shown in Table 1.

Table 1. Comparison of indicators of different models

Model	MAE	RMSE	MAPE
BP	9.7372	12.4387	0.474
RNN	8.149	10.6517	0.3039
LSTM	7.2054	9.7062	0.2126

From the data in the table, it can be seen that among the three models, LSTM has the best verification effect, RNN takes the second place, and BP has the worst effect. It can be seen that LSTM model is suitable for traffic flow prediction. The real data comparison of the predicted data of the LSTM model is shown in Figure 8.

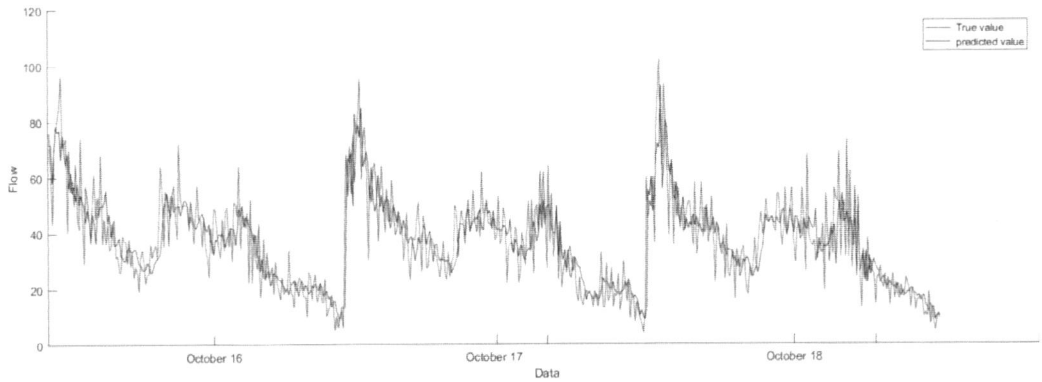

Figure 8. Comparison of LSTM prediction data and real data

4. Conclusion

In this paper, the working principle of LSTM model is introduced in detail, and three models of BP, RNN and LSTM are constructed for simulation verification. After comparison, LSTM model can predict traffic flow more effectively and can be used for traffic flow prediction of urban roads. At the same time, in this experiment, only historical traffic data are used to simulate and predict traffic flow, but the influence of weather, holidays and other factors on traffic flow is not considered. In the later stage, based on the further optimization of the model, more influence factors are added to improve the prediction accuracy.

Acknowledgments

This paper was financially supported by the quality Engineering Project of Anhui NO. 2018 mooc595 and the quality Project of Anhui Xinhua Universityunder Grant NO.2018 ppzyx03.

References

[1] Chuntao Man, Danqing Kang. (2019) Considering the short-term traffic flow prediction of the upstream and downstream LSTM. Journal of Harbin University of technology, 5: 101-107.

[2] Chenghong Fu, Shumin Yang. (2019) Short term traffic flow prediction by improved supp ort vector regression. Transportation system engineering and information, 11: 130-134 + 148.

[3] Xiaolei Ma, Zhimin Tao. (2015) Long short-term memory neural network for traffic speed prediction using remote microwave sensor data. Transportation Research, 3:158-162.

[4] Zhicong Han, Yanguo Fan, Huisheng Wu, Huiyan Liu. (2017) Research on short-term traff ic flow prediction method based on GA-SVR model. Highway traffic technology, 1: 96-103

[5] Yu Tao. (2018) Short-term traffic flow forecasting and Realization Based on SVM and BP neural network [D]. Nanjing University of Posts and Telecommunications,NanJin.

[6] Hongju Cheng, Zhe Xie.(2019) Data prediction model in wireless sensor networks based on bidirectional LSTM.EURASIP Journal on Wireless Communications and Networking. 1:1-12.
[7] Yi Lin,Jian-wei Zhang,Hong Liu. (2019) Deep learning based short-term air traffic flow pr ediction considering temporal–spatial correlation. Aerospace Science and Technology,1 1:93-97.
[8] Zhen Yan, Zhongzhong Yu, Lu Han, Weijun Su. (2019) Short term traffic flow prediction method based on CNN + LSTM. Computer engineering and design, 9: 2620-2624 + 2659.
[9] Yuelong Li, Dehua Tang, Guiyuan Jiang, Zhitao Xiao. (2019) Short term traffic flow pred iction of residual LSTM based on dimension weighting. Computer Engineering, 6: 1-5
[10] Xiangxue Wang, Lunhui Xu.(2018) Short-term traffic flow forecasting based on in-depth l earning. Transportation system engineering and information, 18: 81-88.

IOP Conf. Series: Materials Science and Engineering **806** (2020) 012018 doi:10.1088/1757-899X/806/1/012018

Accounting, Auditing, Accountability: Key Insights from a Multidisciplinary Review of AI and Big Data Literature

Chung-Lien Pan[1], Yu-Chun Pan[2*], Zhichao Xu[3], Zizhen Chen[1], Lin Yu[1], Jingjing Qiu[1], Zhuoshan Lin[1], and JiaLong Li[1]

[1] Accounting School, [3] New Media Research Center, Nanfang College of Sun Yat-sen University, Guangzhou, Guangdong, 510970, China

[2] Graduate Institute of Finance, National Taiwan University of Science and Technology, Taipei, 10607, Taiwan

[*]Corresponding author's e-mail: alice719tw@gmail.com

Abstract. Big Data and Artificial Intelligence (AI) projects have been rapidly applied by the diversity of applications, as well as effectively uncover the secret behind the data. In recent years, Big Data and AI have also had a huge impact on accounting and auditing, such as the launch of intelligent robots. This research puts forward a review of literature on Big Data, AI, and Machine Learning with Accounting, Auditing, and Accountability, based on 259 articles retrieved from Web of Science Database since 1989 to 2019. With the database, the author analyzed occurrences and produced the data contact network with the United States as the largest data resource subject and associated in three directions. Through the comparison of the three figures, it is not difficult to find that the countries with advantages in data resources have more influence in an organizational and sources cluster, and the United States is the most obvious example.

1. Introduction

The social, economic, and political implications of the advancement and adoption of Big Data and AI applications have been at the forefront of high-level dialogues in recent years, discussing its promises and potential dangers[1]. The Association of Chartered Certified Accountants(2013) published the report " Big data: it's power and perils", which proposed that big data could provide Accountants and accounting professionals with opportunities for radical innovation, as well as opportunities for more strategic and "future-oriented" positions in enterprises[2]. Deloitte (2019) released the Global AI industry white paper, which pointed out that today's AI is being commercialized in an all-round way, accelerating enterprise digitization, improving industrial chain structure and improving information utilization efficiency[3].

Previous research on the application of big data and AI, accounting, auditing and accountability do not provide a full picture. For example, Krahel & Titera (2015) proposed changing the accounting standard to focus on the process and analysis of data generation, which will add value and relevance to the accounting industry and improve the efficiency of the capital market[4]. Brown-Liburd et al.,(2015) discussed challenges encountered by auditors in integrating big data into audit analysis, as well as various analytical tools currently used by companies in big data analysis[5]. Griffin & Wright(2015) outlined the agenda of the relationship between social media, big data, and accounting functions. It is expected that accounting, accountant, social media, and big data will become part of future dialogue

[6]. Katyal(2019) discussed the conflict between civil rights protection and AI and helped solve the opacity and accountability of AI by using various tools[7].

To sum up, the application of AI and big data in accounting, auditing, and accountability system has become a hot topic discussed by all circles in recent years. In this context, we provide a general picture of the field, placing the latest research concerns and developments related to big data and AI in the context of accounting, auditing, and accountability. Such maps will help track and explore cross-disciplinary collaboration over the years, thus providing a foundation for the application of ai and big data in accounting-related fields.

2. Data and methods

The paper first describes the query design that defines literacy about big data, AI and machine learning, and then explains the mapping protocols and strategies as follows:

From the WoS database, we executed queries below to gather literature data on the various terms of literacy in relation to Big Data and AI on Nov 2019:

- TS = ("Artificial Intelligence" OR "machine learning" OR "Big Data") AND TS= ("accounting" OR "auditing" OR "accountability " OR "accountants") AND SU= ("Business & Economics "OR" Government & Law" OR "Social Sciences Other Topics").

Various terms of literacy were chosen from the Oxford bibliography [8]. As a result, 259 articles are retrieved on Nov 2019. VOSviewer and Python visualization packages were used for mapping. It reveals heightened interests and perhaps paper quality on the topic.

3. Research mapping results

3.1. Annual trends

The statistics show a small number of articles and references published between 1989 and 2010. However, the number began to spike in 2010 and peaked in 2018. This suggests that more and more people are likely to study the topic and improve the quality of the papers.

3.2. Top WoS categories and research areas

The classification of WoS is shown in table 1. According to the proportion of WoS, the categories of business and finance rank first, followed by management, social sciences and interdisciplinary, etc. The number of articles published in adjacent categories is similar, but the number of publications in the top three categories accounts for about 1/2 of the total publications. It indicates that these categories are clustered, that is, people are more inclined to study this field. Other top categories include law, economics, and education.

Table 1. Top WoS categories

WoS categories	No. of Articles	415 in total
Business, Finance	96	23.13%
Management	66	15.90%
Social Sciences, Interdisciplinary	46	11.08%
Business	40	9.64%
Law	36	8.67%
Economics	28	6.75%
Education & Educational Research	10	2.41%
Political Science	9	2.17%

Table 2 lists the top research fields with more than 7 articles in 9 fields. It can be found that business and economics[9] still hold the top position, accounting for about half of the total articles published, far ahead of social sciences [10], politics and law [11]. The first three fields are roughly the same as the WoS category, but the rankings vary, such as education and education research, which ranks fourth in the research field and seventh in the WoS category.

Table 2. Top research areas

Research areas	No. of Articles	%
Business & Economics	168	48%
Social Sciences - Other Topics	52	15%
Government & Law	42	12%
Education & Educational Research	10	3%
Arts & Humanities - Other Topics	8	2%
Operations Research & Management Science	8	2%
Computer Science	7	2%
Public Administration	7	2%
Information Science & Library Science	7	2%
Others	41	12%

Table 3. Top Countries

Countries	No. of Articles	%
USA	105	33%
PEOPLES R CHINA	55	17%
ENGLAND	28	9%
AUSTRALIA	27	8%
ITALY	11	3%
GERMANY	9	3%
ROMANIA	9	3%
CANADA	7	3%
FRANCE	6	2%
DENMARK	5	2%
NETHERLANDS	5	1%
NEW ZEALAND	5	1%
SCOTLAND	5	1%
Others	45	14%

3.3. Top organizations
From the institutional distribution can be seen in table 3, research institutions, led by the United States, published several accounts for one-third of the total, followed by China and the UK, the top three countries in a paper published several proportions is as high as 72.59%, indicates that these countries dominated in the field in this paper, and China is the only developing country.

3.4. Top publications
In the most frequently cited article, Cao et al. (2015) argued that big data analysis can improve the efficiency and effectiveness of financial statement auditing[12]. Other top 10 most cited articles also focus on the impact of big data or AI on corporate accounting or audit analysis and judgment (such as Vasarhelyi, MA; Kogan, A and Tuttle, BM[13]; Warren, JD; Moffitt, KC and burns, P[14]; Brown - Liburd, H; Issa, H and Lombardy, D[15]). It indicates that the influence of big data and AI on accounting and auditing is far-reaching and worth discussing.

3.5. Clusters
As shown in (a), (b) and (c) of Figure 1, three keywords related to technology are striking on the map. Among them, compared with "organization" and "source", the "country" has the widest contact area. As shown in Figure 1 (a), the relationship description of this figure covers 13 countries. At the same time, Figure 1 (a) shows us the data contact network with the United States as the largest data resource subjected and associated in three directions. The data branch headed by the United States is related to five major countries including the China, the Netherlands, Canada, France, and Romania. The data branch headed by China is related to three major countries including the United States, England and Germany. This branch of data, with the Australia, Italy and New Zealand as the main body, not only integrates with the data of the United States, but also connects the data systems of England and Scotland. Denmark and Germany, as the other kind of correlated data subjects shown in the figure, although the data proportion is small, they also have correlated communication with each country through the main branches of the first two types. In the group of "organizations", Figure 1 (b) clearly shows that the three major organizational plates, namely "Rutgers state univ", "Nyu" and "Stevens inst

techno", are closely related to each other. Obviously, among the 9 publishing channels in Figure 1 (c), "*Journal of Information Systems*", "*Journal of Emerging Technologies in Accounting*" and "*International Journal of Accounting Information Systems*" occupy the most important positions and are closely related to each other.

Through the comparison of the three clusters, it is not difficult to find that the countries with advantages in data resources have more influence in an organizational and sources cluster, and the United States is the most obvious example.

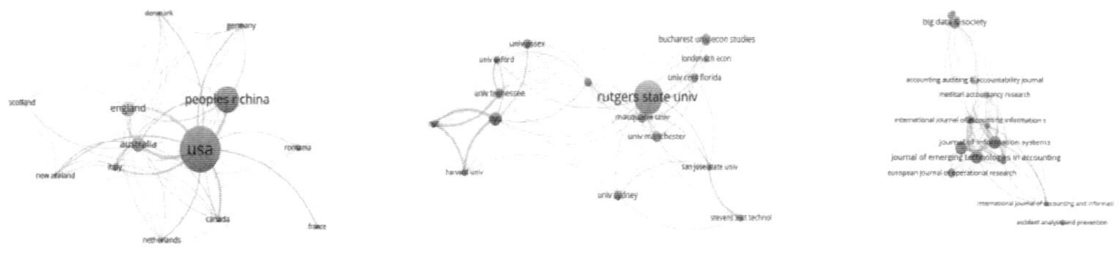

(a)Country cluster (b)Organization cluster (c)Sources cluster

Figure 1. A keyword map based on co-occurrence data

4. Conclusion

The paper intends to contribute to the body of knowledge towards the relationship between domains of accounting and technological keywords. To understand the links and gaps of literature related to accounting, auditing, accountability, big data and AI, this paper visualizes the keyword analysis networks of key publications from the WoS database. The data branch headed by United States is related to six major countries, including China, England, Netherlands, Canada, France and Romania. The data branch headed by China is related to three major countries including the United States, England and Germany.

The analytical results of this paper provide some gaps and potential to better understand the relationship between domains of accounting and technological keywords. Through the comparison of the three clusters, it can find that the United States with advantages in data resources have more influence in organizational cluster and sources cluster.

Acknowledgments

We thank our colleague Han-Teng Liao and Zhichao Xu who provided expertise and insight into the bibliometric data that greatly helped the research.

References

[1] UN News (2019) Artificial intelligence summit focuses on fighting hunger, climate crisis and transition to ‘smart sustainable cities.’ UN News

[2] Faye Chua， ACCA & IMA, Accountancy Futures Academy.(2013). Big data： it’s power a nd perils ,as shown(p.5). Retrieved from https://www.accaglobal.com/gb/en/technical-ac tivities/technical-resources-search/2013/december/big-data-its-power-and-perils.html

[3] Deloitte, (2019). Global artificial intelligence industry whitepaper [white paper]. Retrieved S eptember 19， 2019, from Deloitte official: https://www.deloitte.com/cn/en/pages/techn ology-media-and-telecommunications/articles/global-ai-development-white-paper.html

[4] Krahel, J. P., & Titera, W. R. (2015). Consequences of Big Data and formalization on accounting and auditing standards. Accounting Horizons, 29(2), 409-422.

[5] Brown-Liburd, H., Issa, H., & Lombardi, D. (2015). Behavioral implications of Big Data's impact on audit judgment and decision making and future research directions. Accounting Horizons, 29(2), 451-468.

[6] Griffin, P. A., & Wright, A. M. (2015). Commentaries on Big Data's importance for accounting and auditing. Accounting Horizons, 29(2), 377-379.

[7] Katyal, S. K. (2019). Private Accountability in the Age of Artificial Intelligence. UCLA L. Rev., 66, 54.

[8] Anon Digital Literacy - Communication - Oxford Bibliographies - obo.

[9] Bhimani, A., & Willcocks, L. (2014). Digitisation,'Big Data'and the transformation of accounting information. Accounting and Business Research, 44(4), 469-490.

[10] Van der Vlist, F. N. (2016). Accounting for the social: Investigating commensuration and Big Data practices at Facebook. Big Data & Society, 3(1), 2053951716631365.

[11] Fan, M. D. (2018). Body Cameras, Big Data, and Police Accountability. Law & Social Inquiry, 43(4), 1236-1256.

[12] Cao, M., Chychyla, R., & Stewart, T. (2015). Big Data analytics in financial statement audits. Accounting Horizons, 29(2), 423-429.

[13] VASARHELYI, Miklos A.; KOGAN, Alexander; TUTTLE, Brad M. Big Data in accounting: An overview. Accounting Horizons, 2015, 29.2: 381-396.

[14] Warren Jr, J. D., Moffitt, K. C., & Byrnes, P. (2015). How Big Data will change accounting. Accounting Horizons, 29(2), 397-407.

[15] Brown-Liburd, H., Issa, H., & Lombardi, D. (2015). Behavioral implications of Big Data's impact on audit judgment and decision making and future research directions. Accounting Horizons, 29(2), 451-468.

Exploring the cross-disciplinary collaboration: a scientometric analysis of social science research related to Artificial Intelligence and Big Data application

Han-Teng Liao[1], Zijia Wang[1*] and Yu Liu[1]

[1] New Media Research Center, Sun Yat-sen University Nanfang College, Guangzhou, Guangdong, 510970, China

[*]Corresponding author's e-mail: wang.zijia@outlook.com

Abstract. Given the significant impacts of Big Data and Artificial Intelligence (AI) on the planet and people, it is important for researchers from all disciplines to pay attention to the social, economic and policy needs and challenges with its applications. With the aim to provide a systematic review of social science literatures on the topic, this article has summarized the annual trends since 1962, main research disciplines and publication outlets based on a scientometric analysis of 10,229 articles collected from the WoS database. The research findings show that disciplines/areas such as management and psychology have been constantly engaging the Big Data and AI, whereas new emerging disciplines/areas such as interdisciplinary social sciences and geography have experienced exponential growth in recent years. Both the application and critiques of Big Data and AI have been published in various new journals such as *Big Data & Society*, *Value in Health*, etc. The article therefore provides essential groundwork for understanding and contextualizing the Big Data and AI applications, so that researchers, funding bodies, policy-makers and industry professionals can enhance cross-disciplinary collaboration.

1. Introduction

The social, economic, and political implications of the advancement and adoption of Big Data and Artificial Intelligence (AI) application have been in the forefront of high-level dialogues in recent years, discussing its promises and potential dangers[1]. AI will have "good, bad, transformative and plain weird effects" on societies; thus, major organizations such as Google, Microsoft, Facebook , UNICEF, etc., founded the Partnership on Artificial Intelligence to Benefit People and Society in 2016 [2,3]. For another, the ITU has established the "AI for Good" Global Summit with ACM and other UN organizations as the UN "platform for global and inclusive dialogue on AI"[4]. UN Chief Information Technology Officer also argued that we need to reflect on both positive and negative ramifications and we need to "transform ourselves as innovators"[5]. In addition to AI, various partnerships have been built as part of "Data Revolution", leveraging Big Data for sustainable development[6]. The United Nations Statistical Commission also created the Global Working Group (GWG) on Big Data for Official Statistics to investigate its benefits and challenges[7]. Acknowledging the significant impacts of Big Data and AI on the planet and people, these examples demonstrate the strong demand for responsible AI and Big Data for a positive future.

Thus, any Big Data and AI application must pay attention to the social, economic and policy needs and challenges. Some literature review efforts have been made to explore the application of Big Data or AI on specific areas of social sciences [e.g. 8–11], but none of them are comprehensive enough in terms of the scope of research areas and time periods to provide an overview picture. It is then vital to map out, in relation to AI and Big Data, how the social science literatures and their sources evolve. To provide an overview picture, the article aims to provide a systematic literature review to contextualize such latest research concerns and developments within social sciences in relation to the Big Data and AI. Such a map will be useful for tracing and exploring the cross-disciplinary collaboration across time, thereby providing groundwork for contextualizing and understanding the application of AI and Big Data in relation to social sciences.

2. Data and methods

To explore the dynamics of social sciences literatures in tackling Big Data and AI related issues and applications, this article starts with the query design that covers the social sciences domain.

The Thomson Reuters Web of Science (WoS) database has 25 WoS "research areas" under the main area of social sciences, ranging from archaeology to women's studies. Major research indexes such as SCI-EXPANDED, SSCI, A&HCI, and ESCI are included in the scope of this research.

The below "Advanced Search" query shows how our data query design incorporates "Artificial Intelligence", "machine learning" and Big Data" as topics (coded by the WoS Field Tag as TS) and "Social Sciences" as research areas (coded by the WoS Field Tag as SU).

- TS = ("Artificial Intelligence" OR "machine learning" OR "Big Data") AND SU="Social Sciences"

Together with other narrower 25 social science areas, total 26 queries executed in November 2019, resulted in 10,229 articles. VOSviewer and Python visualization packages were used for mapping.

3. Research mapping results

Given the scope and time span of collected data, it may be helpful to identify several time periods and then focus the historical, publication outlets, and disciplinary dynamics to frame the data sets.

3.1. Annual trends

In order to identify time periods, we conducted linear regressions between the variable of years and that of number of articles (in log values), and three time periods were found through several iterations.

As shown in Figure 1, the number of publications has continued to rise exponentially, except for the time period between 1995 and 2004. Also, Figure 1(c) shows that the growth rate after 2005 is also faster than the time period before 1995, based on the slope value of regression lines.

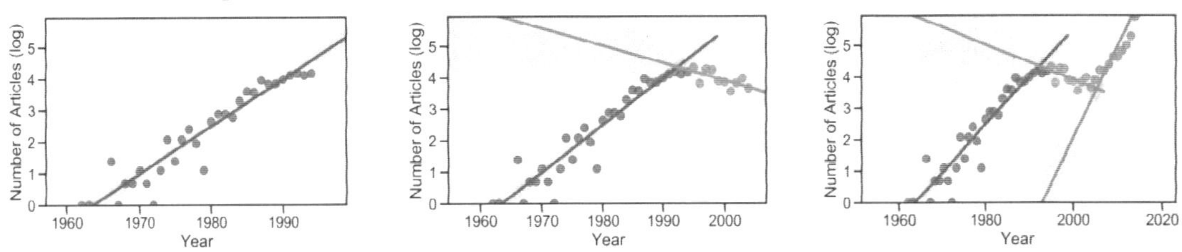

(a) 1962-1994 (b) 1962-2004 (c) all years since 1962

Figure 1. Annual Publications Trend

3.2. Top WoS categories and research areas

Table 1 lists the top WoS categories across three time periods. While management has been at the top category throughout, research categories such as economics, business, geography and communication have emerged after 2005, taking the place of those such as operations research and psychology. Offering slightly different disciplinary categorization, Table 2 indicates that the social sciences research around the world not only focus on the classic disciplines such as psychology, business, operation research and education, but also increasingly cover "Other Topics" of Social Sciences.

Table 1. Top WoS categories (WoS tag: WC)

1962-1994	1995-2004	2005-2020
• Management • Psychology, Multidisciplinary • Operations Research & Management Science • Psychology • Education & Educational Research	• Management • Operations Research & Management Science • Psychology, Multidisciplinary • Psychology, Experimental • Social Sciences, Interdisciplinary	• Management • Economics • Business • Geography, Physical • Communication

Table 2. Top research areas (WoS tag: SC)

1962-1994	1995-2004	2005-2020
• Psychology • Business & Economics • Operations Research & Management Science • Education & Educational Research • Computer Science	• Business & Economics • Psychology • Operations Research & Management Science • Computer Science • Social Sciences - Other Topics	• Business & Economics • Psychology • Social Sciences - Other Topics • Education & Educational Research • Physical Geography

3.3. Top publication outlets

While some publication outlets may be representatives of a discipline or research area's work, some may serve as sites of collaboration across disciplines. Table 3 lists the top publication outlets across different time periods, indicating similar (or corresponding) patterns found above: For instance, the journal *Big Data and Society* belongs to the WoS research area of "Interdisciplinary Social Sciences"; the journal *Value in Health* belongs to that of "Economics; Health Care Sciences & Services; Health Policy and Services". Both of which indicates significant presence of cross-disciplinary collaboration.

Table 3. Top publication outlets

1962-1994	1995-2004	2005-2020
•European Journal of Operational Research • Contemporary Psychology • Interfaces • Journal of the Operational Research Society • Behavioral and brain sciences	• European Journal of Operational Research • Journal of the Operational Research Society • International Journal of Human-Computer Studies • Computational Linguistics • Behavioral and Brain sciences	• Big Data & Society • ISPRS International Journal of Geo-information • European Journal of Operational Research • Value in Health • IEEE Journal of Selected Topics in Applied Earth Observations and Remote Sensing

3.4. Bibliographic coupling network and clusters

Since citation-based bibliographic coupling can be used to identify research fronts and intellectual bases [12], we conducted network mapping based on bibliographic coupling of publication outlets. As shown in Figure 2, four clusters have been identified and visualized for two time periods: (a) from 1962 to 1994 and (b) from 2005 to 2018. The operation/management cluster (in red color, with *European Journal of Operational Research* at its prominent core) and the psychology cluster (in green color) have been consistently prominent. In contrast, the interdisciplinary social sciences cluster (in blue color) and the geography cluster (in yellow color) have only become much more prominent in the last decade. Indeed, as also indicated by the top publication outlets in Table 3, the journals *Big Data and Society* and *ISPRS International Journal of Geo-information* correspond to these two emerging clusters of interdisciplinary social sciences and geography respectively.

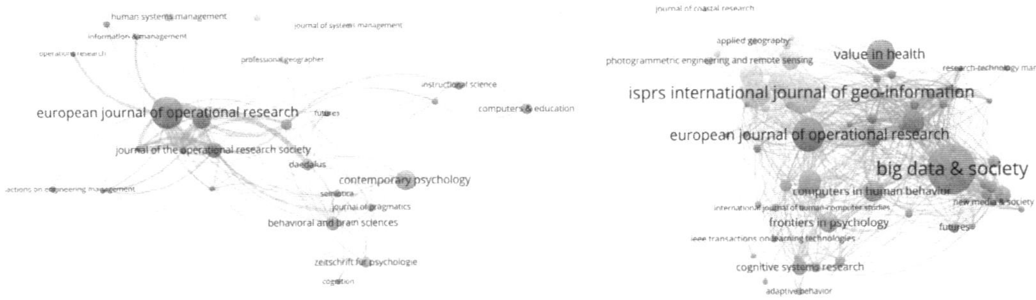

(a) From 1962 to 1994 (b) From 2005 to 2018

Figure 2. Intellectual base and research fronts of main social science publication outlets

Further calculation of betweenness centrality, a widely-used measure of network positional significance as "bridges", reveals several important publication outlets. *European Journal of Operational Research*, *International Journal of Forecasting*, *Technological Forecasting and Social Change*, *MIS quarterly*, and *Business Process Management Journal* are among the top ones in the operation/management cluster. *Behaviour research methods*, *Frontiers in Psychology*, *Frontiers in human neuroscience*, etc., are among the top ones in the psychology cluster. *Policy and Internet*, *Big Data & Society*, *Computers in Human Behavior*, *EPJ data science*, etc., are among the top ones in the interdisciplinary social sciences cluster. Finally, *Transactions in GIS*, *Computers Environment and Urban Systems*, *International Journal of Geographical Information Science*, etc., are among the top in the geography cluster.

3.5. Two types of cross-disciplinary collaboration

At least two types of cross-disciplinary collaboration are identified when looking from the top cited articles to the other. On one hand, by seeing Big Data and AI as new technologies, important questions have been raised from the classic research concerns of social sciences: How will it transform how we understand human communication and culture[13]? How do they drive the transformation of society and business, thereby changing our current practices and knowledge on employment, business models, etc.?[14] What are the key areas of such innovation in terms of the ethical risks [15]? It appears to us that the interdisciplinary social sciences cluster (in blue color) has been the fruitful research front for such research questions, with the journals such as *Computers in human behavior* and *EPJ Data Science* as the bridging or in-between nodes to other clusters.

On the other, research efforts have also been made on the adoption of Big Data and AI on the production of social science knowledge. For instance, machine learning methods such as support vector machines have been applied to the analysis of remote sensing data[8], predicting market outcomes [16], assessing text readability [17], etc. For another, the role of data in social science research in the age of Big Data has been conducted [18].

4. Conclusion

The purpose of this literature review was to provide an overall understanding of how social science literatures and their publication outlets evolve, in relation to AI and Big Data. Three time periods and then four prominent clusters (operation/management, psychology, interdisciplinary social sciences, and geography) have been identified to show both the constant and emerging research fronts and intellectual bases of AI and Big Data related social science work. Although the data source (the WoS database) may be limited in scope, and some of the articles may not be considered as social science work, the findings here nonetheless provide essential groundwork for anyone to begin to explore the application of AI and Big Data in relation to social sciences.

The cross-disciplinary collaboration identified in the article is mostly limited to the level of publication outlets and disciplines. Future research can examine, in greater or and with specific research concerns, how collaboration takes place across disciplines at the level of articles and authors. Notwithstanding its limitations, the article offers a roadmap for researchers, funding bodies, policy-makers and industry professionals to explore and learn from cross-disciplinary collaboration fruitful areas and publication outlets. Also, such a map should help us pay attention to the social, economic and policy needs and challenges facing the adoption of Big Data and AI.

Acknowledgments

The research is funded by the Application-oriented Curriculum Development Projects of "API, Machine Learning and Artificial Intelligence" (NFU 02-40248) and "Big Data" (NFU 02-40249), under the Guangdong Province Department of Education 2018 "Innovation-Strengthening Higher Education Program" Grants, and partly by a project of HCI for Education Development (2018WQNCX283), under the 2018 Major Projects of Guangdong Province Department of Education "Young Innovative Talents" Grants. Han-Teng Liao and Zijia Wang have made equal contributions to the paper, including the data collection and analysis work and have given the approval of the current version for publication.

References

[1] UN News (2019) Artificial intelligence summit focuses on fighting hunger, climate crisis and transition to 'smart sustainable cities.' UN News
[2] PAI (2019) The Partnership on AI. Partnersh. AI PAI
[3] Butler, D. (2017) AI summit aims to help world's poorest. Nat. News 546: 196.
[4] ITU (2019) AI for Good Global Summit, Geneva, Switzerland. AI Good Glob. Summit
[5] Patterson, D. (2016) United Nations CITO: Artificial intelligence will be humanity's final innovation. TechRepublic
[6] UN (2017) Big Data for Sustainable Development. U. N.
[7] GWG (2019) UN Global Working Group (GWG) for Big Data. Big Data UN Glob. Work. Group
[8] Mountrakis, G., Stefanidis, A. (2008) Special issue: Artificial Intelligence in remote sensing - Foreword. Photogramm Eng Remote Sens 74.0: 1199.
[9] Coetzee, S., Rautenbach, V. (2017) A Design Pattern Approach to Cartography with Big Geospatial Data. Cartogr J 54.0: 301.
[10] Akter, S., Wamba, S.F. (2016) Big data analytics in E-commerce: a systematic review and agenda for future research. Electron Mark 26: 173.0.
[11] Guenther, W.A., Mehrizi, M.H.R., Huysman, M., Feldberg, F. (2017) Debating big data: A literature review on realizing value from big data. J Strateg Inf Syst 26: 191.0.
[12] Persson, O. (1994) The intellectual base and research fronts of JASIS 1986–1990. J. Am. Soc. Inf. Sci. 45: 31–8.
[13] boyd, danah, Crawford, K. (2012) Critical Questions for Big Data. Inf. Commun. Soc. 15: 662–79.
[14] Loebbecke, C., Picot, A. (2015) Reflections on societal and business model transformation arising from digitization and big data analytics: A research agenda. J Strateg Inf Syst 24: 149.0.
[15] Mittelstadt, B. (2017) Ethics of the health-related internet of things: a narrative review. Ethics Inf Technol 19: 157.
[16] Cui, J. (2018) Dynamic Migration Algorithm of Marine Big Data in Cloud Computing Environment. J Coast Res 706.

[17] Sung, Y.-T., Chen, J.-L., Cha, J.-H., Tseng, H.-C., Chang, T.-H., Chang, K.-E. (2015) Constructing and validating readability models: the method of integrating multilevel linguistic features with machine learning. Behav Res Methods 47.0: 340.

[18] Cowls, J., Schroeder, R. (2015) Causation, Correlation, and Big Data in Social Science Research. Policy Internet 7: 447–72.

IOP Conf. Series: Materials Science and Engineering **806** (2020) 012020 doi:10.1088/1757-899X/806/1/012020

A New Quick View Retrieval Method for GF-6 Remote Sensing Images

Deng Guang[1,*], Zhang Peng[2], Li Zhiyong[2] and Tian xin[1]

[1]Research Institute of Forest Resource Information Techniques, China Academy of Forestry Sciences, Beijing, 100091, China

[2]Hunan Shenfan Technology Co., Ltd., Changsha, 410000, China

[*]Corresponding author's e-mail: javawsdp@sina.com

Abstract. As of November 3, 2019, all seven satellites for project of the high-resolution earth observation system have been launched. We have ushered in an era of massive remote sensing data. When we do large-scale forest land mapping, such as national and provincial scale, we need to retrieve images that meet the image quality requirements, are as cloudless as possible, cover the research area, and have as few images as possible. The conventional practice is that the application unit searches the web page provided by the remote sensing data provider for the remote sensing data it needs, and orders to obtain the data. It is often time-consuming and laborious to query data that meet the needs. This paper puts forward a new selection process method of GF-6,GF-1 and GF-2 satellite data. This is a step-by-step method. First, the cloud coverage, time and area are selected according to the metadatas provided by the data provider to meet the requirements. Then, according to the image quality and cloud amount set by the program, it is selected again. Finally, in this image set, the final selected image is given according to the requirements of maximum coverage of the research area and minimum overlap of the image itself. The method can greatly save the time for manually selecting images and improve the working efficiency. This paper designs a program, which can be used conveniently by users. The paper introduces the idea of program design in detail.

1. Introduction

China has implemented the project of the high-resolution earth observation system. As of November 3, 2019, all seven satellites in the civil field have been successfully launched. The application department of government has a large amount of remote sensing data that can be used for its own professional application. The challenge of remote sensing application in the era of big data is that the available remote sensing data may no longer be in short supply, but it still takes some time to select the remote sensing images that really meet the needs, and even, as the amount of data increases exponentially, the human workload will double. In the past, some workflows in the era of data scarcity could not adapt to the application of big data remote sensing.

In this paper, a new method of remote sensing data retrieval is proposed, which is oriented to application classification and mapping and implemented step by step and has general significance. This paper takes Gaofen(GF)-6 remote sensing image retrieval as an example to illustrate the workflow. In order to increase usability, our program also supports GF-1 and GF-2 remote sensing images at the same time, which makes joint image data selection possible when applied. This paper first briefly introduces GF-6,GF-1 and GF-2 satellite data, and then gives the algorithm flow. In the implementation part, we described the programming process in detail. Finally, the difference between our program and the existing research methods is given.

2. Data

GF-1 is equipped with two 2-meter spatial resolution panchromatic and 8-meter spatial resolution multispectral cameras and four 16-meter spatial resolution multispectral wide cameras. GF-2 carries two high-resolution 1-meter panchromatic and 4-meter multispectral cameras with sub-meter spatial resolution. GF-6 is an optical satellite similar to the GF-1 satellite, but using a different instrument suit, consisting of a 2/8 m resolution panchromatic/hyperspectral camera with an image swath of greater than 90 km and a 16 m resolution wide angle camera with an 800 km image swath. Both cameras use a three-mirror anastigmat telescope. Both covers visible light to NIR bands (wavelength 450-900 nm). Data of GF-6,GF-1 and GF-2 have great application potential in forest land mapping, forest type classification and other fields.

3. Method

The conventional remote sensing data retrieval method is a quick view based on information such as sensors, imaging time of images, selected areas, cloud cover, etc. Users can select images covering the selected areas through visual interpretation.[1-3]

There are two common problems here. One is that the cloud amount in the image metadata is not accurately judged, and the actual cloud amount of the image exceeds the value in the metadata. Even if such an image is selected, the user will finally discard it. Another problem is that some of the images after the primary election may be repeatedly covered for the selected area, and the user only wants the smallest set of images that can cover the research area as much as possible.

In order to solve the above problems, this paper adds two steps after the conventional data retrieval steps. Conventional data retrieval method is to select data according to sensors, spatial range, time range and cloud amount provided by metadata. After that, the image quality and cloud cover judgment index developed by the authors are used to screen the fast view again. Last, an alternative image is further selected according to the maximum coverage of the image coverage research area and the minimum overlap between the images.

The overall step-by-step algorithm flow is shown in the figure 1. Please refer to 3.2 and 3.3 of [4]

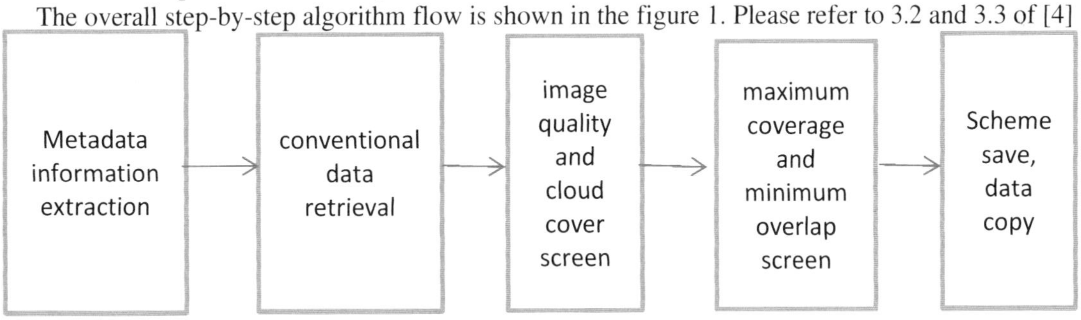

Figure 1. Workflow of GF-6 data retrieval method

for a detailed description of GF-6 fast view image quality and cloud cover judgment and screening method proposed by this algorithm, which will not be described in detail in this paper.

This paper gives the algorithm of the last step of the method. Each quick view can be viewed as a small rectangle. The spatial selection area can be viewed as a large rectangle. The overlap of every two fast views, i.e. two small rectangles, is well calculated. The overlapping degree of multiple fast views, that is, two or more small rectangles, can be calculated in the same way. The calculation method of the overlap between the selected fast view and the spatial selection area is as follows: select the small rectangle and intersect with the large rectangle area first, then divide by the large rectangle area.

How to select the images which the maximum coverage of the research area and the minimum overlap between the images. First, let each small rectangle intersect with the large rectangle area, record the coverage value, and record the overlapping degree of the selected small rectangles. Then let every two small rectangles intersect with the large rectangle area, record the coverage value, and record the overlap degree of each group of selected rectangles. Until finally all the small rectangles and the large rectangle areas are intersected, the coverage value is recorded, and the overlapping degree of the selected small rectangles is recorded at the same time. According to the order of coverage values and overlap degrees, first select the scheme with the largest coverage value, then select the scheme with the smallest overlap degree, and record the corresponding small rectangle numbers.

4. Results and Discussions

The program of the above algorithm is implemented by C# program framework. The program is presented in the form of a combination of a client and a web service, and the calculation is carried out on the client. The overall framework of the program uses GMap.NET. GMap.NET is a powerful, free, cross platform, open source .NET control, which enables use routing, geocoding and maps from Google, Yahoo!, Bing etc in Windows forms and presentation, supports caching and runs on windows mobile. The map service calls Amap by Autonavi Holdings Limited and so on. In the program, we also used OpenCvSharp3-AnyCPU.4.0.0.20181129 for image processing algorithm implementation, and System.Data.SQLite.Core.1.0.109.1 for fast view overlap and coverage calculation and then stored in the database to sort in the database.

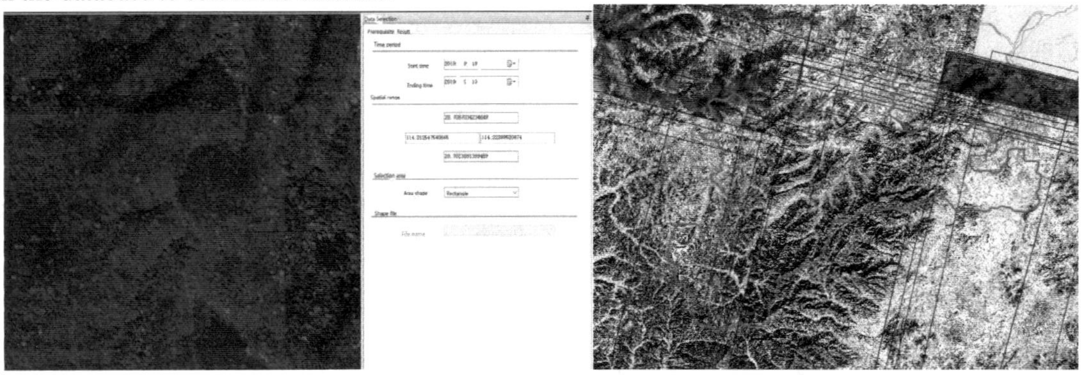

Figure 2. (1) left is interface of GF-6 data retrieval program; (2) right shows the results

Figure 2 (1) shows the interface of software GF-6 remote sensing data query. Some special data selection conditions are set in the parameter setting. Figure 2 (2) shows the results of data selection. The program has a variety of scheme formulation and scheme adjustment functions. It has the capability of cloud judgment, setting up regions and automatic data selection under multi-time period setting. The optimal coverage area has at least repeated coverage adjustment. It has a variety of region visualization methods and has the capability of outputting in various formats such as sensor, quality, batch product number and scene serial number according to scheme.

This paper is the same as and different from the content-based remote sensing image retrieval [5,6] which is currently being studied. The same thing is that they all put forward the goal of further refinement of remote sensing image retrieval. The purpose of this algorithm is to reduce the number of images while covering the research area as much as possible, which is a new problem in the era of remote sensing big data. However, most content-based remote sensing image retrieval methods [6] try to find certain ground objects in remote sensing images, such as buildings, clouds, etc. The algorithm in this paper only selects data at a higher level, but it does not extract objects at a deeper level inside the image.

5. Conclusions

At present, in forest mapping and natural resources survey mapping, satellite data screening mainly relies on manual visual inspection to select fast views, which is time-consuming and laborious. Based on metadatas and fast views of GF-6 data, this paper proposes a fast automatic optimization and screening technology for GF6 satellite data to meet the data quality requirements. The main purpose is to distinguish similar and duplicate data, reduce the workload of manual selection and improve work efficiency. The algorithm also supports the joint data optimization searching of China's high-resolution satellite data on GF-6 ,GF-1 and GF-2.

Acknowledgments
This work was supported by "Research on GF-6 Satellite Optimal Data Selection for Fine Recognition of Forest Types"under Grant No. 21-Y20A06-9001-17/18-1.

References

[1] Li,Y., Wang,F.H., Fan,S.M. (2008) Research of the Browse and Query Technology for Remote Sensing Satellite. Spacecraft Engineering.,17 (1) :100-104.
[2] Li, Y., Wang, F. H., (2009) Query Technology Optimization and Realization of Satellite Data Information . Remote Sensing Information.,24(2) : 64-66.
[3] China Center for Resources Satellite Date and Application. (2019) http://www.cresda.com/CN/
[4] Deng, G., Zhang, P., Li, Zh.Y., Tian, X. (2019) Research on GF-6 Data Selection Technology for Fine Classification of Forest Land. ICGEC 2019.E3S Web of Conferences.
[5] Wei, X., Lv,Y.F. , Cui,Y.Q., Zhang,X.H., Gu,X.Q. (2019) A Discriminative Feature Learning Approach for Remote Sensing Image Retrieval . Remote Sensing., 11(3), 281
[6] Mihai D., Herbert D., Andrea P. (2004) Information mining in remote sensing image archives: System concepts. IEEE Transactions on Geoscience and Remote Sensing., 41(12):2923-2936.

IOP Conf. Series: Materials Science and Engineering **806** (2020) 012021 doi:10.1088/1757-899X/806/1/012021

A Localization Algorithm Based on NLOS Identification with Improved VB-AKF

Hao Chu[1*], Zhibo Fan[1], Fei Wang[1], Peng Ji[1], Chengdong Wu[1]

[1] Faculty of Robot Science and Engineering, Northeastern University, Shenyang, Liaoning, 110000, China.

[*]Corresponding author's e-mail: chuhao_neurobot@163.com

Abstract. As a research hotspot, wireless sensor network (WSN) has a wide range of application prospect. The non-line of sight (NLOS) localization is one of the most basic techniques. In order to mitigate the NLOS influence and increase the localization accuracy, we proposes an improved VB-AKF algorithm in this paper. Firstly, the propagation state of the signal is identified. Then, for the LOS and NLOS mix environments, we reduce the NLOS error by changing the attenuation coefficient. And the maximum likelihood (ML) method is used for localization. Finally, the simulation results show that our method can effectively obtain higher localization accuracy.

1. Introduction

With the rapid development of communication technology, WSN has been widely applied to many fields[1]. Obtaining accurate positional information is the basics to all these applications. In the LOS environment, we can gain the accurate position easily. However, the presence of NLOS propagation and complex environment decrease location accuracy greatly. Therefore, it is of great significance to deliberate the mobile localization methods in the NLOS environment[2].

In recent years, the mobile localization algorithms in NLOS environment can be summarized into two categories. The first class of methods is to depend on all available LOS and NLOS measurement information. In [3], a NLOS mobile location algorithm based on probability data association is proposed. And the algorithm has high location accuracy and good robustness in different environments. In [4], a NLOS location algorithm based on likelihood matrix correction is proposed. The distance is filtered and estimated by KF and HF methods, and the location is determined by the ML method.

The second class of methods includes the NLOS identification and NLOS error mitigation. In [5], a KF algorithm based on variable mean is proposed. The propagation condition is identified by hypothesis test, and KF algorithm with variable mean is used when the measurements are under NLOS environment. Furthermore NLOS error is suppressed by data association method. In [6], a method based on K-means clustering and improved SA algorithm is proposed, which is able to effectively reduce the NLOS error and has low complexity.

Although some progress have made in NLOS localization, but most of them need to foresee the parameters of NLOS error. In practice, the NLOS error model is usually unknown. To solve the problem, we propose an improved VB-AKF algorithm. First, the propagation condition is identified. Then the NLOS error is mitigated by changing the attenuation coefficient of VB-AKF algorithm. Finally, the position of the moving target can be determined by ML method.

2. Construct the measurement model

N beacon nodes and one mobile node are randomly deployed. The coordinates of N beacon nodes are $\theta_1, \theta_2, ..., \theta_N (\theta_i = [x_i, y_i])$. The coordinate of the mobile node at time k is $(x(k), y(k))$. In this paper the TOA based ranging method is selected.

2.1. Range model

At time k, the real distance between the unknown node and the ith beacon node is:

$$d_i(k) = \sqrt{(x(k) - x_i)^2 + (y(k) - y_i)^2} \tag{1}$$

In the LOS environment, the measurement is only affected by the environmental noise, and the measurement distance between the ith beacon node and the mobile node at time k is:

$$\hat{d}_i(k) = d_i(k) + n_i(k) \tag{2}$$

where $n_i(k)$ is the noise with zero mean and variance σ_i^2.

In the NLOS environment, the existence of obstacles will cause NLOS error. Therefore, the measurement distance between the ith beacon node and the mobile node at time k is:

$$\hat{d}_i(k) = d_i(k) + n_i(k) + n_{NLOS} \tag{3}$$

where n_{NLOS} is the NLOS error and the mean value of n_{NLOS} is always positive.

2.2. System model

The state vector of the ith beacon node is defined as:

$$\mathbf{X}_i(k) = [d_i(k), \dot{d}_i(k)]^T \tag{4}$$

where $\dot{d}_i(k)$ is the speed of the mobile node relative to the ith beacon node.

In this paper, it is assumed that the mobile node moves at a constant speed, and the equation of state is:

$$\mathbf{X}_i(k) = \mathbf{F} \cdot \mathbf{X}_i(k-1) + \mathbf{v}_i(k-1) \tag{5}$$

where $\mathbf{F} = \begin{bmatrix} 1 & T_s \\ 0 & 1 \end{bmatrix}$, T_s is the sampling period and $\mathbf{v}_i(k-1)$ is the process noise.

According to Eq.(2) and (3), the measurement equation in LOS and NLOS condition can be shown as:

$$Z_i(k) = \hat{d}_i(k) = d_i(k) + n_i(k) \tag{6}$$

$$Z_i(k) = \hat{d}_i(k) = d_i(k) + n_i(k) + n_{NLOS} \tag{7}$$

3. Proposed method

3.1. Algorithm structure

In VB-AKF algorithm, attenuation coefficient ρ_i represents the correlation degree of the measurement noise variances between two adjacent times. If the propagation condition is the same as before, $\rho = 1$; else $\rho = \rho_m$ ($0 < \rho_m < 0.5$).

The system structure of the proposed algorithm is shown in Figure 1. First, the state and the covariance matrix are calculated using the Kalman prediction, and then the propagation condition is identified. If the propagation condition is the same as the previous time, $\rho = 1$; else $\rho = \rho_m$. And the parameters are updated to estimate the mean and variance of the measurement noise. Finally the ML method is used to determine the positon.

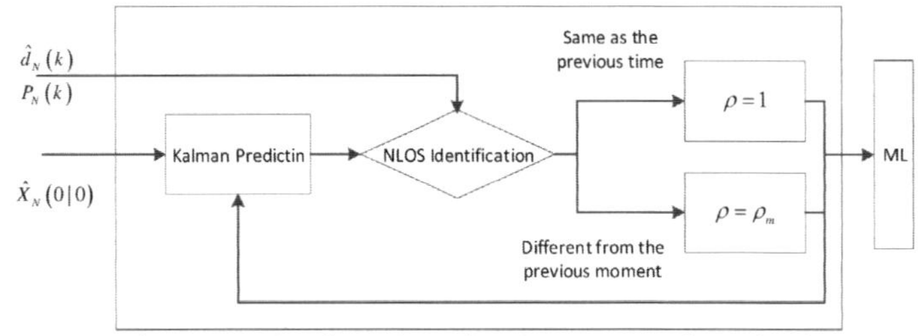

Figure 1. The flowchart for the proposed method

The state vector is initialized as $\hat{\mathbf{X}}_i(0|0)=\left[\hat{d}_i(0),1\right]^T$, and the Kalman prediction equations are:

$$\hat{\mathbf{X}}_i(k|k-1)=\mathbf{F}\cdot\hat{\mathbf{X}}_i(k-1) \tag{8}$$

$$\mathbf{P}_i(k|k-1)=\mathbf{F}\cdot\mathbf{P}_i(k-1)\cdot\mathbf{F}^T+\mathbf{Q}_i(k-1) \tag{9}$$

3.2. Propagation state identification

In this paper, the likelihood ratio test method is used to detect the propagation state. If the propagation state is LOS, the measurement distance of the *i*th beacon node and the probability density function can be expressed as:

$$\begin{cases}\hat{d}_i=d_i+n_i\\ f_{LOS}(\hat{d}_i)=\dfrac{1}{\sqrt{2\pi\sigma_i^2}}\exp(-\dfrac{(\hat{d}_i-d_i)^2}{2\sigma_i^2})\end{cases} \tag{10}$$

If the propagation state of the signal is NLOS, the measurement distance and the probability density function of the *i*th beacon node can be expressed as:

$$\begin{cases}\hat{d}_i=d_i+n_i+n_{NLOS}\\ f_{LOS}(\hat{d}_i)=\dfrac{1}{\sqrt{2\pi(\sigma_i^2+\sigma_{NLOS}^2)}}\exp(-\dfrac{(\hat{d}_i-d_i-\mu_{NLOS})^2}{2(\sigma_i^2+\sigma_{NLOS}^2)})\end{cases} \tag{11}$$

We can establish the generalized likelihood ratio as follow:

$$\lambda(\mathrm{k})=\frac{f_{LOS}(\hat{d}_i|E^i)}{f_{NLOS}(\hat{d}_i|E^i)} \tag{12}$$

If $\lambda(\mathrm{k})>1$, the current propagation state is LOS, else the current propagation state is NLOS.

3.3. Improved VB-AKF algorithm

The improved VB-AKF algorithm can realize continuous estimation of the state and measurement noise. The steps of the improved VB-AKF algorithm are as follows:

(1) Parameter transformation

$$\begin{cases}\alpha_i(k|k-1)=\rho_i\cdot\alpha_i(k-1)\\ \beta_i(k|k-1)=\rho_i\cdot\beta_i(k-1)\end{cases} \tag{13}$$

(2) Parameter update

a. Setting initial value

$$\begin{cases}\hat{X}^{(0)}(k|k-1)=\hat{X}(k|k-1)\\ P^{(0)}(k|k-1)=P(k|k-1)\end{cases} \tag{14}$$

$$\begin{cases}\alpha_i^{(0)}(k|k-1)=1/2+\alpha_i(k|k-1)\\ \beta_i^{(0)}(k|k-1)=\beta_i(k-1)\end{cases} \tag{15}$$

b. Iterative computation

$$\begin{cases} \hat{\mathbf{R}}^{(n)}(k) = diag\left(\beta(k)/\alpha(k)\right) \\ K^{(n+1)}(k\mid k) = \mathbf{P}^{(0)}(k\mid k-1)\mathbf{H}^T(k)\left(\mathbf{H}(K)\cdot\mathbf{P}^{(0)}(k\mid k-1)\cdot\mathbf{H}^T(K)+\hat{\mathbf{R}}^{(n)}(k)\right)^{-1} \\ \hat{\mathbf{X}}^{(n+1)}(k\mid k) = \hat{\mathbf{X}}^{(0)}(k\mid k-1) + K^{(n+1)}(k\mid k)(z(k) - H(k)\hat{\mathbf{X}}^{(0)}(k\mid k-1)) \\ \mathbf{P}^{(n+1)}(k\mid k) = \mathbf{P}^{(0)}(k\mid k) - \mathbf{K}^{n+1}(k\mid k)\cdot\mathbf{H}^T(k)\cdot\mathbf{P}(k\mid k-1) \\ \beta^{(n+1)}(k\mid k,i) = \beta^{(0)}(k\mid k-1) + \frac{1}{2}\left(\mathbf{Z}_i(k) - \mathbf{H}_i(k)\cdot\hat{\mathbf{X}}_i^{(n+1)}(k\mid k)\right)^2 + \frac{1}{2}\left(\mathbf{H}_i(k)\cdot\mathbf{P}(k\mid k)\cdot\mathbf{H}_i^T(k)\right) \end{cases} \tag{16}$$

c. Final value of parameter

$$\begin{cases} \beta(k\mid k-1) = \beta^{(N)}(k\mid k-1) \\ \hat{x}(k\mid k) = \hat{x}^{(N)}(k\mid k) \\ P(k\mid k) = P^{(N)}(k\mid k) \end{cases} \tag{17}$$

The filtering distance of the ith beacon node is:

$$\tilde{d}_i(k) = \mathbf{G}\cdot\hat{\mathbf{X}}_i(k\mid k) \tag{18}$$

where **G**=[1,0], and then the maximum likelihood method is used for location.

3.4. Maximum likelihood localization method

Set the position of the mobile node at time k as $\mathbf{M}(\mathbf{k}) = \left[x^M(k), y^M(k)\right]^T$, and we can establish the following equations:

$$\begin{cases} \left(x_1 - x^M(k)\right)^2 + \left(y_1 - y^M(k)\right)^2 = \left(\tilde{d}_1(k)\right)^2 \\ \vdots \\ \left(x_N - x^M(k)\right)^2 + \left(y_N - y^M(k)\right)^2 = \left(\tilde{d}_N(k)\right)^2 \end{cases} \tag{19}$$

When A and B are as follows, the above equation can be converted to $\mathbf{A}\cdot\hat{\mathbf{M}}(\mathbf{k}) = \mathbf{B}$.

$$\mathbf{A} = 2\begin{bmatrix} (x_1 - x_2) & (y_1 - y_2) \\ \vdots & \vdots \\ (x_1 - x_N) & (y_1 - y_N) \end{bmatrix} \tag{20}$$

$$\mathbf{B} = \begin{bmatrix} \left(\tilde{d}_2(k)\right)^2 - \left(\tilde{d}_1(k)\right)^2 - \left(x_2^2 + y_2^2\right) + \left(x_1^2 + y_1^2\right) \\ \vdots \\ \left(\tilde{d}_N(k)\right)^2 - \left(\tilde{d}_1(k)\right)^2 - \left(x_N^2 + y_N^2\right) + \left(x_1^2 + y_1^2\right) \end{bmatrix} \tag{21}$$

The position of the mobile node can be estimated as:

$$\hat{\mathbf{M}}(k) = \left(\mathbf{A}^T\mathbf{A}\right)^{-1}\mathbf{A}^T\mathbf{B} \tag{22}$$

4. Simulation results

In this section, the effectiveness of the proposed algorithm is verified by simulation experiments. We compare the proposed method with the KF method and the RKF method, and the simulation results are obtained by 500 Monte Carlo experiments.

Figure 2 shows the localization results for the proposed method. It illustrates that the localization errors of KF algorithm and RKF algorithm are large due to the influence of NLOS error, and the proposed algorithm is superior to KF algorithm and RKF algorithm. The average localization errors of KF, RKF and the proposed algorithm are 0.7717m, 0.6325m and 0.3300m.

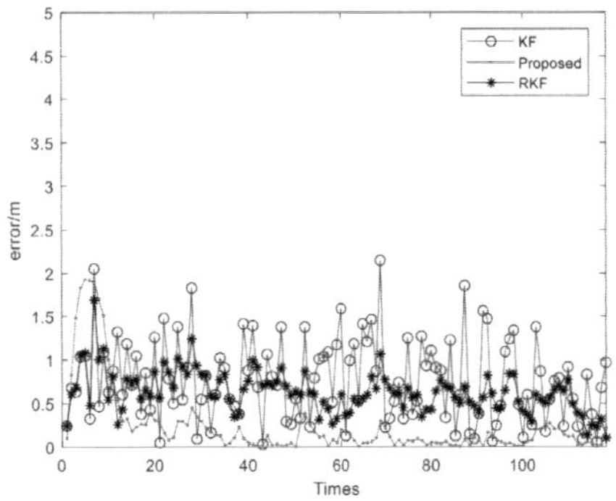

Figure 2. Localization errors at different points

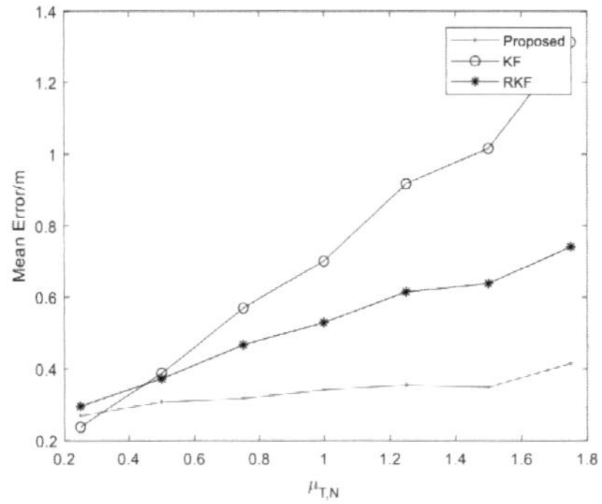

Figure 3. Mean of NLOS versus location error

Figure 3 shows the relationship between the mean value of NLOS error and mean localization error of TOA. It illustrates that the average localization errors of the three algorithms increase with the increase of $\mu_{T,N}$. KF algorithm and RKF algorithm are greatly affected, but the proposed method is influenced little. Therefore, the proposed method has better robustness, and has the highest location accuracy compared with the other two algorithms.

5. Conclusion

In order to reduce the influence of NLOS propagation and improve the localization accuracy, an improved VB-AKF algorithm is proposed in this paper. In complex propagation conditions, the location accuracy can be effectively improved by changing the attenuation coefficient. The simulation results show that the proposed algorithm can effectively reduce the NLOS error, which is better than KF and RKF algorithm.

Acknowledgments

This work was supported in part by the Fundamental Research Fund for the Central Universities of China N172604004, N182613001, N172604003 and N172603001; the National Natural Science Foundation of China under Grant nos. 61901098, 61701101, 61803077, 61603080.

References

[1] Sharma, G., Kumar, A. (2018) Improved DV-Hop localization algorithm using teaching learning based optimization for wireless sensor networks. Telecommunication Systems, 67: 1-16.
[2] Gwo, J. J., Gwo J. H. (2019) Optimal Path Planning for a Biomedical Combined WSN System via RSSI and LQI. Wireless Personal Communications, 108: 957-976.
[3] Yu, X., Hu, N. (2016) A Novel NLOS Mobile Node Localization Method in Wireless Sensor Network. In: International Conference on Communications. Chengdu. pp. 541-549.
[4] Cheng, L., Wu, C. (2013) Indoor Mobile Localization in Wireless Sensor Network under Unknown NLOS Errors. International Journal of Distributed Sensor Networks, 1: 59-64.
[5] Yu, X., Ji, P., Wang, Y. (2017) Mean Shift-Based Mobile Localization Method in Mixed LOS/NLOS Environments for Wireless Sensor Network. Journal of Sensors, 6: 1-8.
[6] Cheng, L., Wu, X. (2017) A non-line of sight localization method based on k-means clustering algorithm. In: 7th IEEE International Conference on Electronics Information and Emergency Communication. pp. 465-468.

IOP Conf. Series: Materials Science and Engineering **806** (2020) 012022 doi:10.1088/1757-899X/806/1/012022

Analysing Gender Bias in IMDB Films Based on Social Networks

Jiayuan Chen[1], Minglu Cui[2] *

[1] Department of English, School of Foreign Languages, Northwest University, Xi'an, Shaanxi, 710000, China

[2]Department of English, School of Foreign Languages, Northwest University, Xi'an, Shaanxi, 710000, China

*Corresponding author's e-mail: mtsuei96@nwu.cn

Abstract. The film industry has a major impact on society. This paper focuses on the gender bias in films. First, the social network analysis (SNA) method is used to construct a movie social network from the online movie dataset IMDb. By analysing these social network data, natural language processing methods are used to analyse movie titles, movie subtitles and casts. Then, this paper constructs a film gender bias classifier. Because the online movie data set IMDb lacks large-scale manual annotation data, and the classifier using traditional deep learning is prone to over-fitting, so this paper proposes a movie gender based on dual learning. The bias feature extractor and classifier enable training of the deep learning model from a small amount of annotated data and classify the bias. Finally, we find that women's roles in the film have improved in all aspects, including the central role of female characters.

1. Introduction

Gender bias is a potential problem in today's society that permeates our culture, learning, workplace, and even our language, often in invisible forms. The first step in changing gender biases is to find out where they exist, and that is where the emerging field of computational linguistics comes in.

The film industry has a significant impact on society, as a microcosm of society, affecting aspects of life such as self-esteem and career choice. It is of great significance to analyse the portrayal of men and women in films and to dig out the gender bias in films.

Most previous studies on gender bias fall into two categories: 1. Provide simple statistics from the data to emphasize the gender gap [1]. 2. Use a small amount of data for data mining [2][3]. Carnes et al. [4] and Terrell et al. [5] analysed the film information retrieval tasks mainly involved, as well as the various previous works involved in this field. Fast et al. [6] focused on gender bias in English fiction, especially in online fiction writing communities. The work involved analysing the behaviour of men and women and describing it in this online novel. In addition, the study showed that men were overrepresented and found that this was traditional in the online fiction data used for analysis, and gender bias was common across all genres. Godden et al. [7] analysed children's books and Millar et al. [8] These studies are useful for revealing trends, but the derivation of these analyses has been done on very small data sets. Kay et al. [9] analysed gender bias in occupations. In the image search results of occupations, compared with male counterparts, women are usually assigned lower level roles.

In the field of gender bias in movies, Lv et al. [10] established a social network of character movies, StoryRoleNet, to analyse gender bias in movies.

This study used machine learning and data mining algorithms to study gender bias in movies. At present, deep learning technology has achieved great success in computer vision, natural language processing and other fields, but it relies on large-scale manual annotation training data, and lacks large-scale manual annotation data set in the field of gender bias in movies. This project creatively used dual learning to mine the gender bias data of movies, and solves the problem of inadequate manual annotation data set in the field of gender bias in movies. The signal used by dual learning exists naturally in artificial intelligence tasks, but it is seldom used by people. We call it the structural duality of artificial intelligence tasks. Structural duality means that the output of one ai task happens to be the input of another, and vice versa. After solving the data set problem, this paper proposes an entity matching method to construct a movie social network from IMDb data of movie titles and credits for gender bias mining. Finally, this paper proposes a novel classifier that can predict whether a film has passed the Bechdel test and evaluate the changes in gender bias in thousands of films over decades.

2. Related Work

2.1. Film Social Network

In recent years, the study of social network has been widely spread. Researchers have found that the analytical techniques of social networks can be used in many areas that do not have explicit data with a network structure. The film industry, for example, is one such area. The researchers used social network analysis to analyze movies, gaining new insights not only about specific movies, but also about the film and television industry. For example, using social networks can empirically analyze the social connections between movie characters. Maintaining the Integrity of the Specifications.

Weng et al. [11] proposed RoleNet, a method for converting movies into social networks. The RoleNet algorithm builds the network by connecting links between characters that appear in the same scene. RoleNet is the basis for scene detection and face recognition based on image processing to discover the appearance of characters. Weng et al. evaluated RoleNet experimentally on 10 films and 3 television shows. RoleNet is used to perform semantic analysis of movies, find communities, detect protagonists, and identify story segments. Park et al. [12] developed character-net, another method for converting movies into networks. Character-net creates a social network for conversations between characters, and USES script caption alignment to extract who is talking to in the scene. Park et al. evaluated their method in 13 films. Character-net is similar to RoleNet, which is used to detect the protagonist and cluster the community.

Tran and Jung [13] developed CoCharNet, which added weights to links in interactive networks, where weights are functions of the number of times two characters occur together. Tran and Jung used CoCharNet to assess the importance of the characters in the movie. They demonstrated that network centrality (e.g., tightness centrality, inter-centrality, and weighting) can be used to categorize secondary and primary characters in films. For example, they used a near central character test with an accuracy of 74.16 percent. Lv et al. [10] proposed an algorithm to improve the accuracy of creating movie social networks. They proposed StoryRoleNet, which combines video and subtitle analysis to build a more accurate movie social network. Subtitles are used to add other links to the video analysis that may be missing. Lv et al are similar to RoleNet and character-net. Use movie social networks to bring communities together and identify key characters.

2.2. Gender Bias Assessment

Many studies have attempted to assess the gender gap between men and women across different fields [14]. Researchers have identified many manifestations of gender bias in our society. Lariviere et al. [15] found that scientific papers in which women held lead author positions were less cited. Wagner et al. [16] observed the unequal coverage of men and women on Wikipedia. The situation for women in the film industry is similar to that in other fields: female characters are underrepresented and poorly portrayed [17]. The Bechdel test, named after Alison Bechdel's 1985 comic Dykes to Watch Out For, aims to address the under characterization of female characters in films and television. Here are the tests:

1. There are at least two female characters in the movie (there are also variations that require female characters with names);

2. There is at least one conversation between two female characters;

3. This conversation can't be about men. Currently, only 57% of films pass this test. Researchers also used the Bechdel test. In recent years, studies have used the test to assess gender bias in movies.

Garcia et al. [16] quantified the Bechdel test and applied it to social media. They added YouTube trailers, movie scripts and Twitter data to produce 704 trailers that could play 493 movies. Garcia et al. [16] used movie data to create a conversation social network. They set up a network of conversations among Twitter users discussing the trailer. The Bechdel test score was calculated using the dialog network. The study found that movie trailers with a male bias were more popular. Similarly, they found that Twitter conversations had a similar bias to movie conversations. Agarwal et al. [18] examined the differences between films that passed and failed the Bechdel test. They created a classifier for the Bechdel test, which was trained in 367 films and evaluated in 90. They found that films that failed the Bechdel test tended to have women play unimportant roles.

2.3. Dual Learning

As a new machine learning paradigm, dual learning was first proposed by He et al. [19] in machine translation tasks. Dual learning solves the problem of insufficient training data in the practical application of machine learning. When we lack sufficient manually labelled large-scale data for effective training, we need to find other signals to drive the training process. The signals that dual learning USES are naturally present in ai tasks, but are rarely used. We call this signal a structural dual property in an artificial intelligence task. Dual learning is proposed based on the observation that many machine learning tasks have dual forms. For example, in machine translation task, English translation into Chinese task and Chinese translation into English task are dual tasks; in image comprehension task, image generation text and text generation image are dual tasks; in speech recognition field, speech recognition and speech synthesis are dual tasks.

This research takes the dual learning in machine translation as an example, as shown in figure 1. In the machine translation task, it is assumed that we only have monolingual data, which are English documents with unmarked information and Chinese documents with unmarked information, as well as two weak initial 6.m and Chinese-English model. The task of dual learning is to use the unmarked monolingual data to continuously learn, improve the translation ability of the two initial models, and finally get a very strong two 6.s model and Chinese-English model. To achieve this, learn to use an English sentence with no information, translate it into Chinese using the original translation model, and then translate it back into English using the original translation model in the opposite direction. By comparing the original English sentence with the translated English sentence, as well as the grammar and morphology of the translated result in the middle, we can get a series of feedback signals to update the original model and make it continuously improve. When we have a large amount of monolingual data, dual learning can continuously improve the performance of the translation model to a high level.

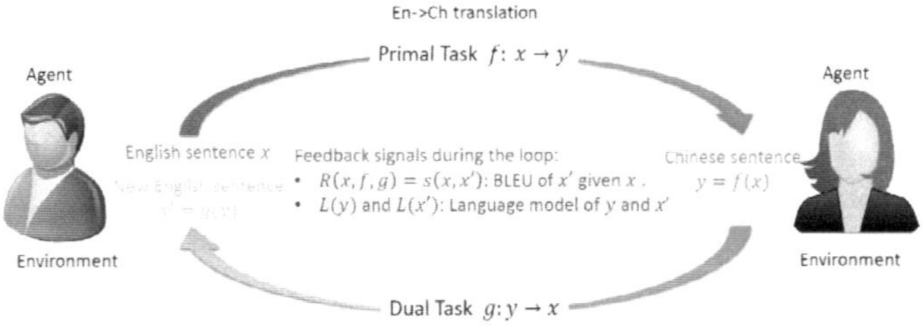

Figure 1. Machine Translation based on Dual Learning

The idea of dual learning is not only limited to artificial intelligence tasks with dual tasks, but also applicable to artificial intelligence models with dual models. Such as automatic encoder (Autoencoder,

AE) against emergent network (Generative Adversarial Networks, GANs). Both the encoder and decoder in AE have structure dual properties, and the generator and discriminator in GANs also have structure dual properties. However, these dual properties of structure are often ignored in current studies. Encoders and decoders in AE are often trained separately in recommendation systems, and their feedback signals are often not Shared. This paper proposes the design of stacked automatic encoder as a dual closed loop, which USES the structure dual property between encoder and decoder to train encoder and decoder simultaneously, and USES feedback signal to update the training process of encoder and decoder, so as to increase its performance of extracting commodity hidden layer features.

3. Gender Bias Approaches Based on Film Social Network

3.1. Film Social Network Construction

The key to constructing the movie social network is to construct an algorithm that can generate the movie social network directly from the input movie. This research used movie names and movie characters' name data to construct movie social network.

First, a movie social network is constructed by entering a movie $M := \langle V, E \rangle$, where V is the set of network vertices and E is the set of edges that connect the vertices. Where each vertex $v \in V$ represents a character in the movie. Each edge $e := (u, v, w) \in E$ represents the interaction between role u and role v and w refers to the times of interactions. As shown in figure 2, a movie social network is constructed using movie subtitles, and figure 2. (a) character names are extracted from movie subtitles using a named entity recognition algorithm. Figure 2. (c) to match character entities to the credits, and figure 2. (b) to link roles and increase edge weights by 1.

The steps to construct a movie social network through the movie credits and movie subtitles are as follows: 1. Use the named entity recognition algorithm to detect and extract individual or organizational entities from the input movie subtitles and store the occurrence time of such entities in the movie. 2. The entities in the movie credits are matched to the list of characters in the credits. It is worth mentioning that it is not possible to map one-to-one between characters in the cast character list and those extracted from the subtitles. In the dark knight, for example, Bruce Wayne is referred to as "Bruce Wayne" three times, Bruce 16 times and Wayne 20 times. 3. In order to solve the problem that the characters in the cast and crew character list and the characters extracted from the subtitle are not mapped one by one. This paper proposes a character entity matching algorithm. As shown in algorithm 1, first, divide all the characters into first and last names and link them to the full names of the actors and characters. Then, link all characters that appear in the role to the subtitle.

Algorithm 1 Character Entity Matching Algorithm

Data: Actor Name, Character Name, Threshold Value;
Results: Role Matching;
1. $Name \leftarrow PersonName.split()$;
2. foreach $N_i \in Names$ **do**
3. if $Role[N_i].lengh = 1$ **then**
4. **return** $Role[N_i]$;
5. end
6. return $MaxWRatio(PersonName, role[N_i], Threshold)$
7. end

In order to solve the matching problem of actors' roles and names, as in algorithm 1. First, divide all the characters into first and last names and link them to the full names of the actors and characters (line 2). Then, if there is only one character with a specific last or first name (a one-to-one match), you need to link all the characters that appear in the subtitle to that character (lines 3-5). Compare strings

IOP Conf. Series: Materials Science and Engineering **806** (2020) 012022 doi:10.1088/1757-899X/806/1/012022

by using WRatio and match part of the name to the full name. With WRatio, the highest matching character with a score above the "threshold" is finally selected (line 6).

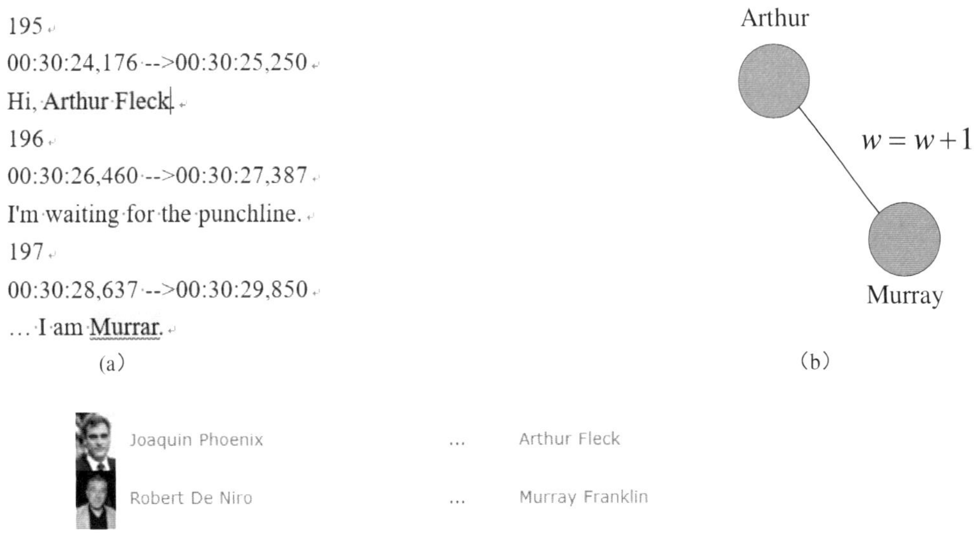

(c)

Figure 2. Movie Social Network Construction Based on Movie Subtitle

3.2. Social Network Features

In order to study gender bias in films, this study calculated five types of features in the film social network: vertex features, network features, film features, gender representation features and actor features. By calculation, we analyse how these characteristics change over time.

In addition, these characteristics are used to construct a machine learning classifier.

First, the definition of these features is given:

Vertex characteristics: the neighbor nodes of a vertex in a movie social network are defined as. The vertex feature is the total weight feature: the total weight of all edges, representing the number of times the character "v" appears in the movie $\text{Total}_w(v) = \sum_{\{(v,u,w)|((v,u,w)\in E\}} w$. Compactness center feature: the reciprocal of the total distance to all nodes in the graph, where, $C_c(v) = \frac{1}{\sum_{v\in V} d(v,u)}$ $d(u,v)$ represents the shortest distance between them.

Network features include network edge number, network vertex number, network node group, and statistical network feature number.

Movie characteristics include: release time, movie IMDb ranking, total movie playing time, movie genre, movie IMDb rating.

Gender represents characteristics: number of genders, number of females in the first 10 roles. Number of actors, number of actresses, gender relations triple.

Actor characteristics include the year the actor was born, the year the actor died, and the age of the actor at the time the film was released.

In order to check the status of gender gap, the gender gap in movies, especially by type, was investigated. This study only analyzed the most popular movies (movies with more than n votes on IMDb). To answer the first research question, is there a genre without gender differences? First, the vertices and character characteristics of all the characters are calculated, and then the data is divided by gender and movie type. Finally, Mann-Whitney U was used to test these characteristics to see if there were statistical differences between different types of male and female characters. To study relationships in movies and answer the second question about what gender relationships reveal, this study calculated all relational triangles in the network and grouped them by the number of women in each triangle. Then, the triangle is subdivided by type and the change of triangle with time.

In order to investigate the role of gender centrality, the third research question of the centrality of female characters, this study calculated the PageRank of nodes in all movie networks, analyzed the number of men and women in the top 10 roles in movies, and studied how this number has changed over the years.

3.3. Gender Bias Feature Extractor

In order to achieve the film from a small amount of film subtitles with the movie ratings gender bias feature extraction. Using dual learning technology, this paper puts forward a kind of automatic encoder movie gender bias feature extraction based on duality, as shown in figure 3.

This paper proposes to use stacked automatic encoders to model movie subtitle data to extract hidden layer feature information from movie subtitle data. Traditional stackable automatic encoders do not consider the structural dual information between encoder and decoder, train encoder and decoder separately, and feedback signals between encoder and decoder are not Shared. This paper presents a new type of automatic encoder structure, dual stack of first design and dual main encoder decoder structure is a kind of dual closed loop.

Then, the main encoder and the dual decoder are trained together, so that the feedback signals of the master encoder and the dual decoder can be Shared, and the performance of the dual stack automatic encoder to extract the hidden layer feature information in the implicit data is improved.

As shown in the main encoder module in figure 3, the binary movie rating $e_i \in \mathbb{R}^n$ is first encoded by the main encoder into the hidden layer representation of the movie rating e_i^r. Where the main encoder code is expressed as:

$$Main\ Encoder: \begin{cases} e_i^1 = a_1(W_1 e_i + b_1) \\ e_i^r = a_2(W_2 e_i^1 + b_2) \end{cases} \tag{1}$$

Dual decoder decoding is expressed as:

$$Dual\ Decoder: \begin{cases} e_i^3 = a_3(W_3 e_i^r + b_3) \\ \hat{u}_i = a_4(W_4 e_i^3 + b_4) \end{cases} \tag{2}$$

Among them, This subscript i of e_i^r represents the specific movie rating information, and the superscript r represents the implicit representation encoded from the movie rating. $W_1 \in \mathbb{R}^{v_1 \times m}$, $W_2 \in \mathbb{R}^{v \times v_1}$, $W_3 \in \mathbb{R}^{v_1 \times v}$, $W_4 \in \mathbb{R}^{m \times v_1}$ represent the weight matrix. m represents the number of comments, v_1 represents the dimension of the first hidden layer, and v represents the dimension of the bottleneck layer.

The model modeling in this paper is based on the data of movie subtitle and movie rating. In order to better model the data information of movie subtitle and movie rating, this paper inserts confidence matrix into the square loss letter, and designs the model loss function of the dual automatic encoder of the gate attention mechanism as follows:

$$L_{GADAE} = \sum_{i=1}^{n}\sum_{j=1}^{m} \| C_{j,i}(D_{j,i} - \hat{D}_{j,i}) \|_2^2 = \| C^T \odot (D^T - \hat{D}^T) \|_F^2 \tag{3}$$

Where $\odot$ represents the element product, $\|\cdot\|_F$ is the F-norm of the matrix, and the confidence matrix is defined as:

$$C_{i,j} = \begin{cases} \rho & \text{IF } D_{i,j} = 1 \\ 1 & \text{ELSE} \end{cases} \tag{4}$$

$C \in \mathbb{R}^{m \times n}$

In this paper, the objective function of the dual automatic encoder for the design of the gate attention mechanism is

$$C_{i,j} = \begin{cases} \rho & \text{IF } D_{i,j} = 1 \\ 1 & \text{ELSE} \end{cases}$$

$$L = L_{GADAE} + \lambda(\| W_* \|_F^2 + \| w_t \|_2^2) \quad (5)$$

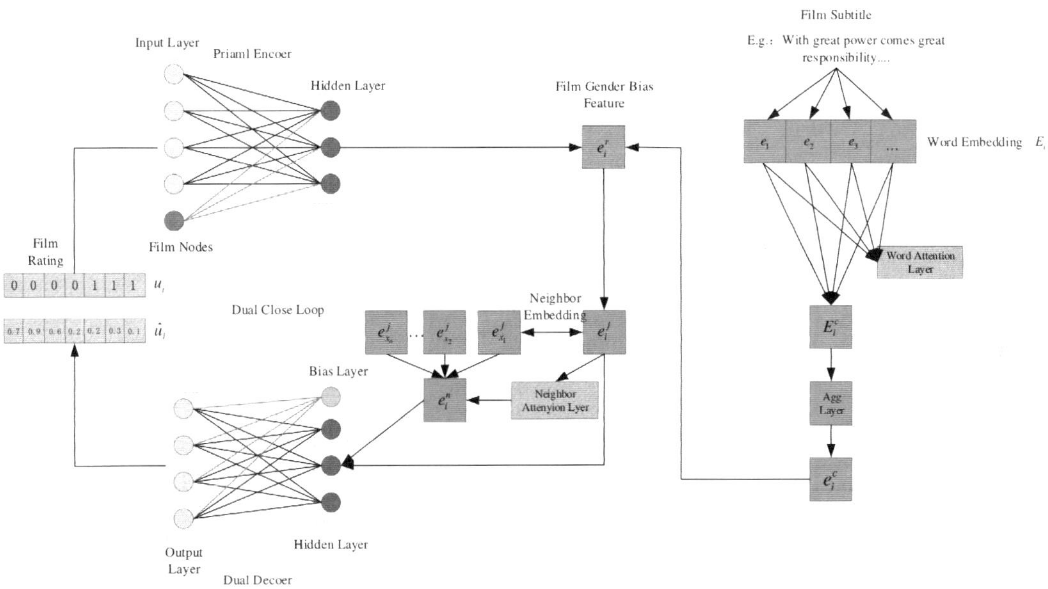

Figure 3. Movie Gender Gap Feature Extraction Based on Dual Autoencoder

Algorithm 2 Dual automatic encoder training process

Input: Movie rating vector u_i, Movie embedding vector e_i^c, Fusion vector e_i^j, Adjacent to the vector e_i^n;

Output: Parameterized θ_E main encoder E,
Parameterized θ_D dual coder D;

Random initialization θ_E and θ_D

repeat

Get a minibatch of m pairs $\{e_i, d_i\}_{j=1}^m$;

Computing the gradient:

$$G_E = \nabla_{\theta_E} (\frac{1}{m}) \sum_{j=1}^m [L_1(f(e_j;\theta_E), y_j]\ ;$$

$$G_D = \nabla_{\theta_D} (\frac{1}{m}) \sum_{j=1}^m [L_2(f(d_j;\theta_D), y_j]\ ;$$

update θ_E and θ_D

$\theta_E \leftarrow Opt_1(\theta_E, G_E), \theta_D \leftarrow Opt_2(\theta_D, G_D)$

until the model of convergence

The optimizer uses the master encoder and the dual decoder in the dual noise reduction automatic encoder adjacent to the attentional mechanism, so that the feedback signals of the master encoder and the dual decoder can be shared between the two models.

3.4. Film Gender Bias Classifier

To explore differences in the representation of men and women in the film industry, the study trained a classifier that used character data (vertex and actor characteristics) and movie data (IMDb characteristics without male and female counts) to predict a character's gender. The basic assumption is that if the representation of men and women is similar, the roles of men and women cannot be distinguished. But once you can predict the gender of a role, you can learn which characteristics are most descriptive.

To establish the data set, this study collected data on the most popular actors from the movies. And define a popular actor as an actor who has participated in at least N films and has more votes than m in IMDb. The idea is to avoid analyzing niche films and unknown actors with secondary roles and no meaningful function. For each actor's movie appearance, the above functionality was extracted in this study (see section 3.4.1) to evaluate the model due to the availability of movie participation data for each actor from multiple movies. This requires splitting all data for the same actor into training sets or validation sets. In order to solve this evaluation problem, this study used the method of setting aside a set of cross-validation. In order to create a true fact for the classifier, the gender of each actor must be determined. For most roles, extract the gender from the IMDb in a similar way. IMDb has actor or actress attributes, which allow us to identify the gender. As mentioned earlier, the IMDb dataset is only partial, so to overcome this problem, we use a dataset that maps the first name to the gender.

4. Experiment Settings

4.1. Data Selection

To evaluate and test our movie social network construction algorithm above on real data, we assembled a large dataset of movie subtitles and movie character lists. In addition, we collected a list of movie characters from the IMDb website and movie subtitles from 15,540 movies. In addition, we used data from 4,658 Bechdel *films.*

To analyze the content of the movie, we extract information from the subtitles. Subtitles are freely available online at many sites. For example, OpenSubtitles alone hosts more than 500,000 English subtitles that are manually created by the community. We collect subtitles using Subliminal 7, a Python library for searching for and downloading subtitles. Subliminal downloads the subtitles from multiple sources and uses internal scoring methods to determine which ones are best for a particular movie. Using Subliminal, we downloaded the subtitles for 15,540 movies.

4.2. Experimental Environment and Settings

The software and hardware environment of the experimental program in this paper is shown in table 1: men.

Table 1. The Experimental Environment

Items	Details
Processor	Intel(R) Core i9-9900k
Graphics Card	Nvidia GeForce GTX Titan X
Development Language	Python 3.6
Development Framework	Pytorch

5. Experimental Results and Analysis

To analyse the gender gap in the film industry, we analysed movie titles that received at least 1,000 votes on IMDb.

This results in a dataset containing 15,540 movies, which is 20 times the largest movie dataset currently available.

First, we predict the gender of the actors from the movie and the vertex features. We evaluated our categorized actors using cross-validation and found that we could predict gender by character significantly better than by chance. The AUC of our classifier is 0.85, and the accuracy and recall rate are 0.76 and 0.79, respectively. Table 2 shows the ranking of the five most important features of the movie gender classifier.

Table 2. The Film Feature Index

Features	Importance	Male Average	Female Average
Actor age	0.20280	44	45
Actor's birth year	0.19913	1950	1960
Film ranking	0.18571	6.43	6.4
Release time	0.03254	1993	1995
Broadcast time	0.02014	109	108

Then, in order to examine the relationships between the characters in the movie social network, we analysed the triangles in the relationship network. We found that most triangles have three people, we found the relationships between the roles, we analysed the role relationship triangles in the network. We find that most triangles have three men, while the triangle with three women is the least common, as shown in table 3, the proportion of women in the film social network triangle in the five film genres.

Table 3. Gender Classifier Top-5 Features

The number of women in the triangle	0	1	2	3
Action	45.85%	41.07%	12.89%	1.45%
Crime	42.56%	40.45%	15.78%	2.10%
Romance	21.29%	43.56%	22.54%	6.04%
War	65.64%	24.67%	25.74%	3.63%
Comedy	56.98%	41.98	20.84%	3.97%

In the five genres, the most common triangle type is 3 men (no women), and in all the others, 2 men and 1 woman. According to the results, romance was the most interactive type among women, while war was the least.

Next, we analysed the gender composition of the top 10 core characters, as shown in figure 4. We found that in most movies, more men than women play male characters. Moreover, as the data show, there are hardly any films without men and women in the top 10. Also, in many movies, the majority of the top 10 most important characters are men.

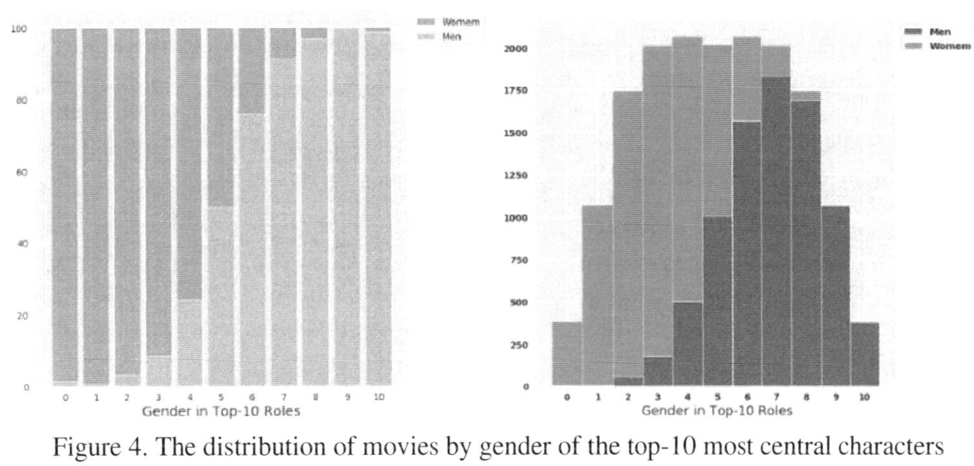

Figure 4. The distribution of movies by gender of the top-10 most central characters

6. Conclusion

In this paper, we proposed a new method of constructing movie social network, an algorithm of constructing movie social network based on character entity recognition, and a feature extraction algorithm of movie social network based on dual automatic encoder. Moreover, we constructed a feature extraction algorithm of movie character gender bias and a feature classifier of movie character gender. Based on the IMDb dataset of online movie data, the experiment results show that gender differences still exist in almost all types of movies. For example, the relationship triangle in the movie is dominated by men. In terms of the top 10 leading movie characters, it's mostly men. However, we also found that the majority of the top 10 most important roles were men. By the ecological analysis of IMDb films, we found that gender bias still exists. Films, as the major social media, female social gender stereotypes are always interpreted, which not only guides the audience, but also negatively affects the expression of female consciousness and the image and interpretation of women in films.

Acknowledgments
This research was part the project "Analysing Gender Bias in IMDb Films based on Dual Learning and Entity Matching", which was supported by the sustentation item of innovation and entrepreneurship training program (S201910697354) for college students in Shaanxi province.

References

[1] Angela Coyle. Behind the Scenes: Women in Television[M]// Women and Work. 1988.
[2] Agarwal A, Zheng J, Kamath S, et al. Key female characters in film have more to talk about besides men: Automating the bechdel test[C]//Proceedings of the 2015 Conference of the North American Chapter of the Association for Computational Linguistics: Human Language Technologies. 2015: 830-840.
[3] Glascock J. Gender roles on prime-time network television: Demographics and behaviors[J]. Journal of Broadcasting & Electronic Media, 2001, 45(4): 656-669.
[4] Carnes M, Devine P G, Manwell L B, et al. Effect of an intervention to break the gender bias habit for faculty at one institution: a cluster randomized, controlled trial[J]. Academic medicine: journal of the Association of American Medical Colleges, 2015, 90(2): 221.
[5] Terrell J, Kofink A, Middleton J, et al. Gender differences and bias in open source: Pull request acceptance of women versus men[J]. PeerJ Computer Science, 2017, 3: e111.
[6] Fast E, Vachovsky T, Bernstein M S. Shirtless and dangerous: Quantifying linguistic signals of gender bias in an online fiction writing community[C]//Tenth International AAAI Conference on Web and Social Media. 2016.
[7] Gooden A M, Gooden M A. Gender representation in notable children's picture books: 1995–1999[J]. Sex roles, 2001, 45(1-2): 89-101.
[8] Millar B. Selective hearing: gender bias in the music preferences of young adults[J]. Psychology of music, 2008, 36(4): 429-445.
[9] Kay M, Matuszek C, Munson S A. Unequal representation and gender stereotypes in image search results for occupations[C]//Proceedings of the 33rd Annual ACM Conference on Human Factors in Computing Systems. ACM, 2015: 3819-3828.
[10] Lv J , Wu B , Zhou L , et al. StoryRoleNet: Social Network Construction of Role Relationship in Video[J]. IEEE Access, 2018:1-1.
[11] Weng C Y, Chu W T, Wu J L. Rolenet: Movie analysis from the perspective of social networks[J]. IEEE Transactions on Multimedia, 2009, 11(2): 256-271.
[12] Park S B, Oh K J, Jo G S. Social network analysis in a movie using character-net[J]. Multimedia Tools and Applications, 2012, 59(2): 601-627.
[13] Tran Q D, Jung J E. CoCharNet: Extracting Social Networks using Character Co-occurrence in Movies[J]. J. UCS, 2015, 21(6): 796-815.
[14] Jia S, Lansdall-Welfare T, Sudhahar S, et al. Women are seen more than heard in online newspapers[J]. PloS one, 2016, 11(2): e0148434.
[15] Larivière V, Ni C, Gingras Y, et al. Bibliometrics: Global gender disparities in science[J]. Nature News, 2013, 504(7479): 211.

[16] Wagner C, Garcia D, Jadidi M, et al. It's a man's Wikipedia? Assessing gender inequality in an online encyclopedia[C]//Ninth international AAAI conference on web and social media. 2015.
[17] Wood J T. Gendered media: The influence of media on views of gender[J]. Gendered lives: Communication, gender, and culture, 1994, 9: 231-244.
[18] Agarwal A, Zheng J, Kamath S, et al. Key female characters in film have more to talk about besides men: Automating the bechdel test[C]//Proceedings of the 2015 Conference of the North American Chapter of the Association for Computational Linguistics: Human Language Technologies. 2015: 830-840.
[19] He D, Xia Y, Qin T, et al. Dual learning for machine translation[C]//Advances in Neural Information Processing Systems. 2016: 820-828.

IOP Conf. Series: Materials Science and Engineering **806** (2020) 012023 doi:10.1088/1757-899X/806/1/012023

Exploring different notions of literacy: a literature review analysis of literacy research related to Artificial Intelligence and Big Data application

Zijia Wang[1*]

[1]New Media Research Center, Sun Yat-sen University Nanfang College, Guangzhou, Guangdong, 510970, China

[*]Corresponding author's e-mail: wang.zijia@outlook.com

Abstract. Artificial Intelligence (AI) and Big Data projects have been rapidly applied by organizations and adopted by younger users, leading to an increasing number of various capacity building applications, as well as significant impacts on the well-being of both designers and users. This research presents a review of literacy-related literature on "Big Data", "Artificial Intelligence", "Machine Learning", based on 73 articles retrieved from the Web of Science Database. The author produced a keyword co-occurrence network map highlighting three main domains: digital literacy, information literacy, and data literacy. This paper concluded that data and information literacy share similar semantic networks whereas digital literacy is related more to pedagogy and innovations, while being the more distant node from both data and information literacy. For stakeholders such as educators, academic researchers, and policymakers, the paper provides reference points for understanding technological literacy and the current relationship among key literacy concepts and domains in relation to AI and Big Data,

1. Introduction

The industrial applications, our economies, and even education are rapidly transformed by rapid technological advancement in big data, AI and machine learning [1]. Such domains of technological development also make great potential to fight hunger, climate crisis, etc. towards a more sustainable future [2], which brings great value to educate younger generations and overcome the digital divide [3]. In the previous literature, Nuortimo (2018) has conducted a review on digital transformation in the Mining industry [4]; MacMillan (2014) has reviewed data management and the sharing of librarians [5]. The existing literature has yet covered the impact of emerging technology trends such as big data, AI or machine learning towards literacy. This paper aims to provide a literature review on the specific topic of literacy and its relations with AI and Big Data applications.

To examine the relationships between the notion of literacy, a key concept for education and educators, and Big Data and AI applications, this paper conducted a literature review of 72 retrieved articles from the Web of Science (WoS) database. The purpose of this article is to explore the potential links and gaps among different domains in literacy and technologies for educators, researchers and policymakers.

2. Data and methods

The paper first describes the query design that defines literacy in relation to big data, AI and machine learning, and then explains the mapping protocols and strategies as follows.

First, based on the entries of "digital literacy" and "literacy" from Oxford Bibliographies [6,7], a series of related terms have been compiled, ranging from data literacy, digital literacy to multiliteracies.

Then from the WoS database, we executed queries below to gather literature data on the various terms of literacy in relation to Big Data and AI on Nov 2019:

- TS = ("Artificial Intelligence" OR "machine learning" OR "Big Data") AND TS = ("data literacy" OR "digital literacy" OR "information literacy" OR "internet literacy" OR "computer literacy" OR "ICT literacy" OR "Information Communication and Technology literacy" OR "Information Communication & Technology literacy" OR "media literacy" OR "new literacy" OR "multiliteracy" OR "multiliteracies")

As a result, 73 articles are retrieved in November 2019. VOSviewer and Python visualization packages was used for mapping.

3. Research mapping results

3.1. Annual trends

Figure 1 shows that the number of publications has been scarcely from 2000 toward 2013. However, the number starts to increase rapidly since 2013 and keeps on growing steadily. Figure 2 shows the annual citation numbers, revealing the heightened interests and perhaps paper quality on the topic.

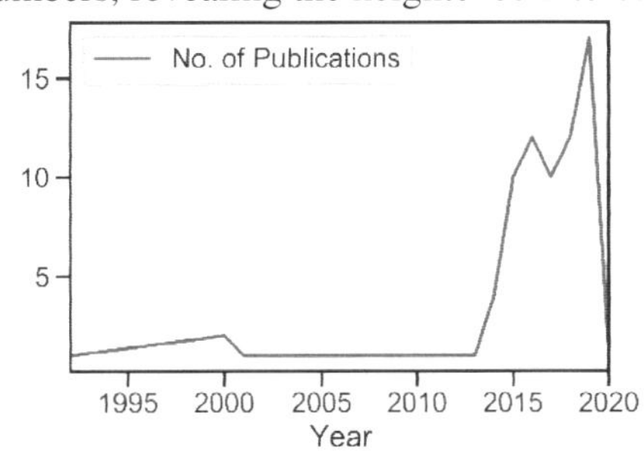

Figure 1. Publications Trend from 2000 to 2019

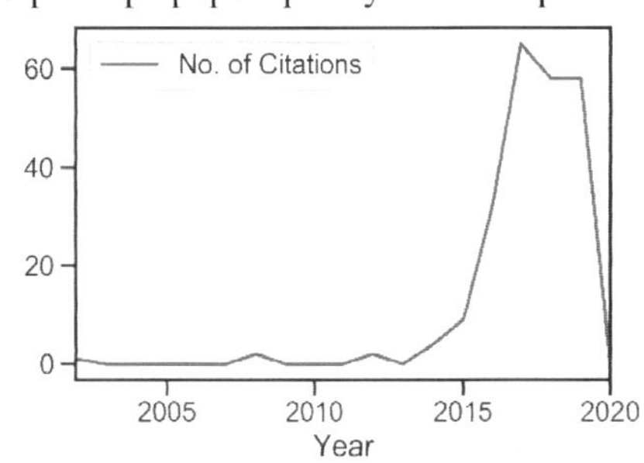

Figure 2. Citations Trend from 2001 to 2019.

3.2. Top WoS categories, research areas and organizations

The top WoS categories are listed in Table 1, with Information Science & Library Science, Education & Educational Research, and Computer Science, Information Systems as the top three. The top three categories make up about one-third of the total publications, indicating clustering in the fields of information, library and education related research. Some other categories include social sciences and interdisciplinary applications.

Table 1. Top WoS categories

WoS categories	No. of Articles	136 in total
Information Science & Library Science	21	15%
Education & Educational Research	15	11%
Computer Science, Information Systems	11	8%
Computer Science, Theory & Methods	7	5%
Social Sciences, Interdisciplinary	5	4%
Computer Science, Artificial Intelligence	5	4%
Computer Science, Interdisciplinary Applications	5	4%
Computer Science, Software Engineering	5	4%
Engineering, Electrical & Electronic	4	3%
Education, Scientific Disciplines	4	3%

The top research areas are shown in Table 2. Of the 13 areas that have more than 2 articles, the top three are the same as WoS categories. Those three same areas make up 52% of total publications, which indicate a more concentrated clustering on computer and information sciences. More diverse research areas have appeared in WoS categories such as business, government & law, public administration, etc.

Table 2. Top research areas

Research areas	No. of Articles	%
Computer Science	23	19%
Information Science & Library Science	21	18%
Education & Educational Research	18	15%
Engineering	9	8%
Business & Economics	5	4%
Social Sciences - Other Topics	5	4%
Telecommunications	3	3%
Government & Law	2	2%
Physical Geography	2	2%
Public Administration	2	2%
Remote Sensing	2	2%
Operations Research & Management Science	2	2%
Communication	2	2%
Others	24	20%

Table 3. Top Countries

Countries	No. of Articles	%
USA	18	23%
China	15	19%
England	10	13%
Germany	5	6%
Canada	4	5%
Indonesia	3	4%
Greece	2	3%
Hungary	2	3%
Netherlands	2	3%
Scotland	2	3%
Spain	2	3%
Others	14	18%

3.3. Top countries
Table 3 lists all countries that have two articles. The top three regions are USA, China and England, which makes up 55% in total. Among them, Indonesia stands out as the main developing country other than China.

3.4. Top publications
Among the top-cited articles, Hargittai argued that the biased sampling frames of social network analysis could lead to methodological challenges of big data research [8]. The other top 10 most cited articles mainly focused on the impact of data/information literacy on various domains such as accounting [9], undergraduate chemistry curriculum [10], nurse shift workers [11], and librarians [12], etc. No clear author clusters were found, implying a very decentralized citation networks without top or star scholars.

3.5. Keyword map and Clusters
This research analysed occurrences and produced a co-occurrence network based on the keyword data, resulting in a concise keyword map due to the relatively small number of total articles from the database. VOSviewer was used to visualize the network, in order to detect the relationships among keywords. Figure 3 shows the overall keyword map based on co-occurrence data. The keywords analysis network is consisted of five clusters: (1) machine learning and digital literacy and, (2) artificial intelligence, data and librarians, (3) big data and data literacy, (4) information literacy and education, and (5) technology. Technology cluster contains only one node at the center of the network. Three types of literacy belongs to different clusters, echoing the complexity of defining terms such as "data literacy"[13]. Among the three types of literacy, digital literacy appears to be peripheral (i.e. further away from the more closely situated data literacy and information literacy). Keywords such as "innovation" and "technology" serve as nodes so that connect digital literacy to the core network.

Figure 3. overall keyword map based on co-occurrence data

As shown in (a), (b), (c) of Figure 4, three technology-related keywords stand out on the map. Among them, "big data" shares the most intensive connections compared with "artificial intelligence" and "machine learning". Three literacy-related keywords stand out as main connections among these technologies, which are "information literacy", "digital literacy" and "data literacy" shown in (e), (f), (g) of Figure 4. Big data shares a connection with all three domains of literacy; machine learning only relates to information and digital literacy, yet AI relates to none. Both AI and machine learning are linked to librarians and pedagogy.

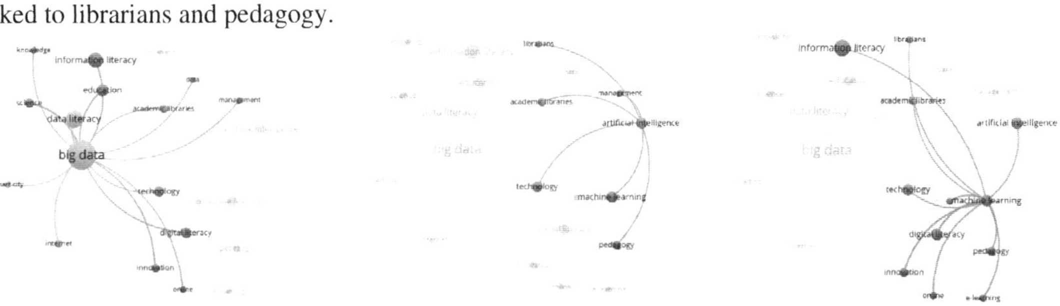

(a) big data as the node | (b) artificial intelligence as the node | (c) machine learning as the node

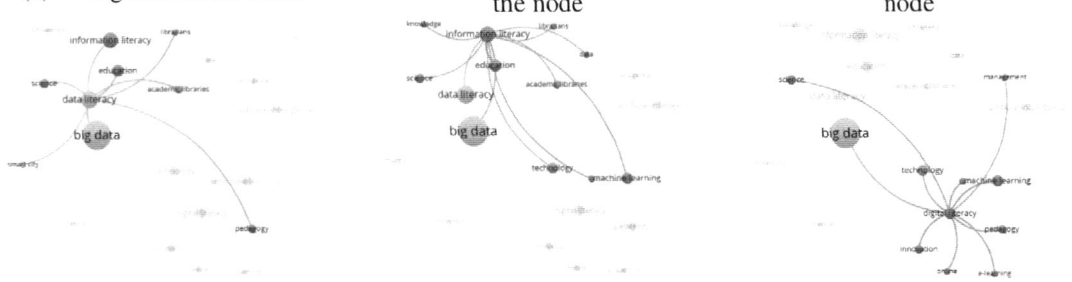

(d) data literacy as the node | (e) information literacy as the node | (f) digital literacy as the node

Figure 4. A keyword map based on co-occurrence data

4. CONCLUSION

The paper intends to contribute to the body of knowledge towards the relationship between domains of literacy and technological keywords. With the aim to understand the links and gaps of literature related to literacy, big data, AI and machine learning, this paper visualizes the keyword analysis networks of key publications from the WoS database. In relation to literacy, big data and machine learning are in more close links compared with AI, implying great potential in the promotion of AI literacy. In addition, data and information literacy share part of networks, indicating some similarity and overlapping use of both concepts. While information literacy relates to knowledge and technology, and data literacy relates to smart city and pedagogy. As a relatively independent cluster from data and information literacy, digital literacy shares more connections with keywords such as management, innovation and e-learning, revealing more separate domains with potential connections with business and online education.

The results of this paper have revealed the gaps and potentials to better understand the dynamics of different literacy with technologies, providing reference points for stakeholders such as educators, academic researchers, and policymakers who are interested in technological literacy and the connections among the key literacy domains in relation to AI and Big Data application.

Acknowledgments

The research is funded by the Application-oriented Curriculum Development Projects of "API, Machine Learning and Artificial Intelligence" (NFU 02-40248) and "Big Data" (NFU 02-40249), under the Guangdong Province Department of Education 2018 "Innovation-Strengthening Higher Education Program" Grants, and partly by a project of HCI for Education Development (2018WQNCX283), under the 2018 Major Projects of Guangdong Province Department of Education "Young Innovative Talents" Grants.

References

[1] UNESCO (2019) Principles for AI: Towards a Humanistic Approach? UNESCO (Paris: UNESCO)
[2] UN News (2019) Artificial intelligence summit focuses on fighting hunger, climate crisis and transition to 'smart sustainable cities.' UN News
[3] Schweitzer, E.J. (2019) Digital divide. SOCIETY
[4] Nuortimo, K. (2018) Measuring public acceptance with opinion mining: The case of the energy industry with long-term coal R&D investment projects. J Intell Stud Bus 8.0: 6.
[5] MacMillan, D. (2014) Data Sharing and Discovery: What Librarians Need to Know. J. Acad. Librariansh. 40: 541–9.
[6] Hicks, T., Baleja, K., Zhang, M. (2019) Digital Literacy. Communication
[7] Rennie, J., Ortlieb, E. (2013) Literacy. Education 9780199756810–0095.
[8] Hargittai, E. (2015) Is Bigger Always Better? Potential Biases of Big Data Derived from Social Network Sites. Ann. Am. Acad. Pol. Soc. Sci. 659: 63–76.
[9] Bhimani, A., Willcocks, L. (2014) Digitisation, 'Big Data' and the transformation of accounting information. Account. Bus. Res. 44: 469–90.
[10] Reisner, B.A., Vaughan, K.T.L., Shorish, Y.L. (2014) Making Data Management Accessible in the Undergraduate Chemistry Curriculum. J. Chem. Educ. 91: 1943–6.
[11] Hobbs, B.B., Farr, L.A. (2004) Assessing Internet Survey Data Collection Methods with Ethnic Nurse Shift Workers. Chronobiol. Int. 21: 1003–13.
[12] Koltay, T. (2016) Data governance, data literacy and the management of data quality. IFLA J. 42: 303–12.
[13] Koltay, T. (2015) Data literacy: in search of a name and identity. J. Doc. 71: 401–15.

IOP Conf. Series: Materials Science and Engineering **806** (2020) 012024 doi:10.1088/1757-899X/806/1/012024

An enhanced artificial bee colony algorithm for numerical function optimization

Yi Yu, Yonggang Wu[*] and Xinglong Liu

School of Hydropower and Information Engineering, Huazhong University of Science and Technology, Wuhan, 430074, China

[*]Corresponding author's e-mail: vehust@126.com

Abstract. In response to the problems of slow convergence, low convergence accuracy and easy to fall in local optimum of the standard artificial bee colony (ABC) algorithm, this paper proposed the following two improvement measures: Firstly, on the basis of using the current optimal solution, the gradient modification strategy was used to reduce the leading role of the current optimal solution, so that the nectar resources could be fully searched to ensure the diversity of the population; Secondly, the crossover and mutation operation in genetic algorithm was used to replace the random search of scout bees, and the prior information was effectively utilized, which not only improved the search efficiency but also ensured the diversity of population to a certain extent. The numerical experiment results show that the proposed algorithm has good convergence accuracy and robustness, and also presents an excellent performance in multi-dimensional optimization.

1. Introduction

Artificial bee colony (ABC) algorithm is a global and random optimization algorithm based on bees foraging, which is proposed by Karaboga in 2005[1]. The algorithm has a few parameters, simple operation and strong robustness. Since the algorithm has been put forward, it has attracted great attention from many scholars around the world and they have conducted in-depth research on the theoretical research and application of algorithms. So far, the algorithm has been successfully applied to artificial neural network training, combinatorial optimization, power system optimization, data mining and image recognition and so on [2-6].

As a novel swarm intelligence algorithm, ABC has obviously been widely recognized and affirmed by domestic and foreign academic circles for its excellent performance for solving high-dimensional problems. However, as other optimization algorithms, traditional ABC algorithm also has the disadvantages of premature convergence and slow convergence and tending to fall into the local optimal solution [7-9]. In response to these, researchers have put forward a large number of effective methods for achieving a better balance of the ABC algorithm's exploration and exploitation [10-18]. GABC algorithm [19] is an example of many successful improved ABC algorithms, which is mainly inspired by the current optimal solution guidance mechanism in particle swarm optimization (PSO) algorithm. It introduces the current optimal solution in the local search formula, which obviously improves the global search ability of the algorithm. The convergence of this algorithm has also been proved by literature [19]. However, research [20] points out that the improved measure may reduce the algorithm's local optimization ability to some extent. This is due to it will quickly guide the employees to the vicinity of the current optimal solution, while other nectar sources of mining value have not been fully searched. In order to optimize the performance of the algorithm further, this paper improves it from two aspects. On one hand, the gradient information of the global optimal solution is introduced into the search formula for properly weakening the global optimal solution's guiding effect, which results in a fully search in the neighbourhood under a lower convergence speed. On the other hand, inspired by the genetic algorithm, the mechanism of crossover and mutation is introduced in the scouters' global search, in which crossover can make full use of prior information and mutation can ensure the diversity of individuals, which make the random searchers obtain extra nectar information thus improve their search efficiency. Prior to this, some mixed algorithms about ABC and GA have been proposed in references [20], [21] and [22]. In these references, their genetic operators are all implemented to manipulate all the populations in each generation so as to prefer superior species to the next generation. These may lead to unsatisfactory optimization results due to incomplete local search and premature optimization. This paper tried to use genetic operator to generate new nectar resource in the case of that any nectar resource has not been updated exceeding the limit, which can effectively avoid the premature convergence, and make full use of the prior information and ensure the diversity of the population. These measures can better achieve the balance between the exploration and exploitation ability of the ABC algorithm in theory and a series of classical test functions are used to attest the improved algorithm's rationality and superiority in this paper at last.

2. Overview of artificial bee colony algorithm

The standard ABC algorithm which simulates behaviors of gathering nectar by bees, searches out the nectar resource which contains the maximum nectar through sharing information among the different divisions of bees. The bees are divided into three categories based on division of labor: employees, onlookers and scouters. The employees share the information of nectar resource to onlookers by their "8" type of dance in dance-floor, onlookers wait in hive and search around the self-considered ideal nectar resource since employees sharing the information and scouters randomly search for a new valuable nectar resource near the hive. The process of ABC algorithm for solving problem actually is the process of searching in potential solution space. The position of each nectar resource represents a feasible solution to the problem and the amount of nectar indicates the corresponding solution's fitness. In each generation, the number of employees is equal to the number of nectar resources, and there is a one-to-one relationship between the employees and nectar resources.

In order to ensure the diversity, employees are required to carry on a local search for better nectar resources around the corresponding resources in each generation based on the following formula,

$$\overline{x_{ij}} = x_{ij} + \text{rand}(-1,1)\cdot\left(x_{ij} - x_{kj}\right) \tag{1}$$

Where $\overline{x_{ij}}$ is the value of generated nectar resource in jth dimension, x_{ij} is the value of ith nectar resource in jth dimension, x_{kj} is the value of kth nectar resource in jth dimension, in which k is a random number that is less than population quantity and not equal to i. Comparing the generated nectar resource with the original one, the one with higher fitness is retained. The fitness is calculated as follow.

$$Fit_i = \begin{cases} \dfrac{1}{1+f_i} & ,if\ f_i > 0 \\ 1+\left|f_i\right| & ,if\ f_i \le 0 \end{cases} \tag{2}$$

According to the information of nectar resources transmitted by employees, each onlooker will choose a nectar resource based on roulette strategy. The formula of possibility of being selected is as follow.

$$p_i = \frac{Fit_i}{\sum_{j=1}^{n} Fit_j} \tag{3}$$

Where Fit_i is the fitness of ith nectar resource and n is the number of nectar resources. The onlookers search for new nectar resources according to the Eq. (1) after selecting nectar resources by roulette strategy. Meanwhile, employees update the nectar resources by fitness on basis of the greedy strategy. If any nectar resource has not been updated within a given *limit* of generation, the corresponding employee gives up the nectar resource, and searches for a new one according to the following formula,

$$x_{id} = x_{id}^{min} + \text{rand}(0,1)\cdot\left(x_{id}^{max} - x_{id}^{min}\right) \quad (4)$$

Where x_{id} is the value of *i*th nectar resource in *d*th dimension. x_{id}^{min} and x_{id}^{max} are the lower and upper bounds of *i*th nectar resource in *d*th dimension, respectively.

3. Enhanced artificial bee colony algorithm

The exploitation and exploration are two important performances of swarm intelligent algorithms [23]. The former is reflected in algorithms' local search capability and the latter is reflected in the global search capability. The ABC algorithm's local search is performed by employees and onlookers and global search is mainly reflected in the search process of scouters. In order to achieve a better balance of exploitation and exploration, the standard ABC algorithm is improved from the below two aspects.

3.1. Global optimal solution guidance and gradient adjustment strategies

Local search is based on Eq. (1) in standard artificial bee colony algorithm. It will be found by analyzing the Eq.(1) that the standard ABC algorithm's local search is to select one dimension from one nectar resource as its local optimization variable, take the selected dimension of this nectar resource as the center and regard projection distance between the nectar resource and another one in this dimension as the search scope. In Eq. (1), the coefficient is a totally random number in [-1,1], $x_{k,j}$ is a random individual in the population and the possibility of selecting a good solution is nearly equal to that of selecting a bad solution. Therefore, the new candidate solution is not promising to be a solution better than the previous one and it also confirms the statement that the artificial bee colony algorithm is tired of local search. To solve this problem, the literature [19] presents GABC to introduce the global optima into the search formula for improving the exploitation which refers to particle swarm optimization and the specific is shown in Eq. (5).

$$\overline{x_{id}} = x_{id} + \varphi_{id}\cdot\left(x_{id} - x_{kd}\right) + \beta\cdot\left(x_{d}^{Global} - x_{id}\right) \quad (5)$$

Where x_{d}^{Global} is the d-dimension value of the current optimal solution, β is a random number which subjects to a uniformly distributed in the range [0, C], and the study found that the performance of the algorithm is the best when C is equal to 1.5. However, the research [20] has shown that the GABC algorithm inherits the feature of fast convergence speed of PSO, but leads to incomplete local search of other valuable individuals because of the effects of quick guide to the local optima. Aimed at the problem, this paper proposes a gradient adjustment strategy which reduces the speed of the bees falling into the local optimum through the reverse effect of the current optima's gradient, which can make full search of the other individuals and ensure a good species diversity.

It is assumed that the N-ary function has first-order continuous partial derivative in the space region G, where $g(x_1, x_2, \cdots x_N) \in G$, the gradient of point *g* can be expressed as follow.

$$grad^{g} = \frac{\partial F}{\partial x_1}\overline{x_1} + \frac{\partial F}{\partial x_2}\overline{x_2} + \cdots + \frac{\partial F}{\partial x_N}\overline{x_N} \quad (6)$$

Similarly, if the current spatial solution of the optimal solution is marked as $x^{global}=(x_1,x_2,\ldots,x_D)$, the gradient of the current optimal solution is expressed below $grad^{Global}$.

$$grad^{Global} = \left(\frac{\partial f}{\partial x_1}, \frac{\partial f}{\partial x_2}, \cdots \frac{\partial f}{\partial x_D}\right) \quad (7)$$

$$\frac{\partial f}{\partial x_i} = \lim_{\Delta x \to 0} \frac{f(\cdots x_{i-1}, x_i + \Delta x, x_{i+1} \cdots) - f(x_1, x_2, \cdots x_D)}{\Delta x}, i = 1,2\cdots D \quad (8)$$

After introducing the current optimal solution and gradient adjustment strategy into the local search of the ABC algorithm, the local search of the algorithm is transformed from formula (1) to formula (9).

$$\overline{x_{id}} = x_{id} + \varphi_{id} \cdot (x_{id} - x_{kd}) + \beta \cdot (x_d^{Global} - x_{id}) + (-1)^k \cdot \gamma(t) \cdot \left| x_j^{Global} - x_{id} \right| \cdot \left| grad_{e,d}^{Global} \right| \tag{9}$$

Where k is the direction control parameter, if $x_{i,d} > x_d^{Global}$, k is even; otherwise, k is odd. $\gamma(t)$ is the adaptive parameter, which changes with the number of iterations; $\left| x_j^{Global} - x_{id} \right|$ is the control item of step size; $\left| grad_{e,d}^{Global} \right|$ is the scalar of gradient unit vector in d dimension, and it is the control parameter of step size change. The gradient unit vector is expressed as follow.

$$grad_e^{Global} = \left(\frac{\partial f}{\partial x_1}, \frac{\partial f}{\partial x_2}, \cdots \frac{\partial f}{\partial x_D}\right) / \sqrt{\left(\frac{\partial f}{\partial x_1}\right)^2 + \left(\frac{\partial f}{\partial x_2}\right)^2 + \cdots \left(\frac{\partial f}{\partial x_D}\right)^2} \tag{10}$$

3.2. Crossover and mutation strategy

In ABC algorithm, the nectar source can be considered to be exhausted if the number of nectar source that has not been updated reaches the limit. At this time, the employee corresponding to the nectar source will change its role as a scout bee, abandon the mined nectar source, and then randomly search in the search space to generate a new one. The mechanism makes the algorithm possible jump out of the trap when it falls into the local optimal state, which is also a highlight in the design of ABC algorithm. In the standard ABC algorithm, global random search is conducted according to formula (4), but it is undoubtedly inefficient to search for nectar sources in the complex solution space, especially in the later stage of optimization.

As is known to all, the GA has the characteristics of self-organization, self-adaptation and self-learning [24], and it realizes the continuous evolution of population through selection, crossover and mutation. Selection is to reserve some preferred individuals directly to the next generation, or to provide an operational basis for crossover operators and mutation operators, so as to protect the good genes. Crossover generates two new individuals through the recombination of the selected parent generation, which is used for the crossover combination to generate new individuals to search the solution space effectively and reduce the probability of destroying the effective mode. Mutation randomly selects an individual from the population to exchange its gene position to increase the diversity of the population. It has good global search ability, and can quickly search out the approximate optimal solution in the solution space [24-25].

Considering the inefficiency of scout's random search in the standard ABC algorithm and the good global search capability of GA, the idea of crossover and mutation is introduced, instead of the original random searching. On one hand, crossover can make full use of the prior information of the population and improve the search efficiency of scout bee. On the other hand, mutation retains the original character of random search which can still guarantee the diversity of population with a certain probability. So the formula is changed from equation (4) to equation (11).

$$x_{id} = \begin{cases} \lambda \cdot x_{kd} + (1-\lambda) \cdot x_{ld} & if\ 0 \le p \le p_c \\ L_d + \text{rand}(0,1) \cdot (U_d - L_d) & if\ p_c < p \le 1 \end{cases} \tag{11}$$

In the formula, λ is a uniformly distributed random number within the range of [0, 1], k and l are two nectar sources randomly selected by roulette based on sorting probability; p_c is the probability of cross, p_m is the probability of mutation, p_m=1- p_c.

4. Enhanced artificial bee colony algorithm flow

1) Total number of bees is initialized N, in which one half are employees and the others are onlookers. The number of nectar resources is initialized $N/2$. Maximum residence times is initialized *limit*. Iteration is marked iter=0, Maximum iteration is initialized *Max_Cycle*.

2) The bees randomly select N nectar resources in solution space. Calculate fitness values with Eq.(3) and record the initial sign the res_times(1)=0.

3) Each employer randomly search around its located nectar resource according to the Eq.(9). Calculate each fitness value, if it is better than the original source, update the position of the employed bees and res_times(i)=0. Otherwise, res_times(i+1)= res_times(i)+1.

4) Calculate selection probability according to the Eq.(2), each onlooker selects a nectar reource to follow according to the Eq.(3), update the nectar resource and res_times(i).

5) Determine whether res_times(i) is greater than *limit*. If it is, continues. Otherwise, switch to 7.

6) The ith employee gives up its nectar resource and searches for a new one according to Eq.(11).

7) Record the current optimal solution and iterations *iter*=*iter*+1.Determine whether iterations is greater than *Max_Cycle*. If it is, termination. Otherwise, skip to 4.

5. Experiments

5.1. Benchmark functions

In order to test the performance of EABC algorithm on numerical function optimization, a series of benchmark functions shown in table 1 are used here.

Table 1. Benchmark functions

Functions	Search space	The minimum solution
$f_1(x)=\frac{1}{4000}\sum_{i=1}^{D}x_i^2-\prod_{i=1}^{D}\cos(\frac{x_i}{\sqrt{i}})+1$	$[-600,600]^D$	$f_1(\vec{0})=0$
$f_2(x)=\sum_{i=1}^{D}[x_i^2-10\cos(2\pi x_i)+10]$	$[-5.12,5.12]^D$	$f_2(\vec{0})=0$
$f_3(x)=\sum_{i=1}^{D-1}[100(x_{i+1}-x_i^2)^2+(x_i-1)^2]$	$[-50,50]^D$	$f_3(\vec{0})=0$
$f_4(x)=-20\exp(-0.2\sqrt{\frac{1}{D}\sum_{i=1}^{D}x_i^2})-\exp(\frac{1}{D}\sum_{i=1}^{D}\cos(2\pi x_i))+20+e$	$[-32,32]^D$	$f_4(\vec{0})=0$
$f_5=D\times 418.9829+\sum_{i=1}^{D}-x_i\sin\left(\sqrt{\lvert x_i\rvert}\right)$	$[-500,500]^D$	$f_5(\overrightarrow{420.9867})=0$

5.2. Parameter setting

The experiment in this paper includes two parts: one is to evaluate the improvement effect of EABC algorithm by comparing with the results of ABC and GABC algorithm in the reference [19]; the other one is to evaluate the optimization performance of EABC algorithm by comparing with the results of the representative optimization algorithms, such as GA, PSO and PS-EA in the reference [26]. In order to ensure the objectivity, the parameters in this paper are set as consistent with the relevant literatures.The population size is 80 and the maximum generations is 5000 in the first part, the population size is 125 in the other one. Each experiment was repeated 30 times independently.

5.3. Experimental results

The comparison with the results in the reference [19] is shown in table 2. As can be seen that EABC is obviously superior to the traditional ABC algorithm in the optimization results of different dimensions of each test function, which proved that the optimization performance of EABC has been greatly improved compared with the traditional ABC algorithm. Compared with GABC test results, EABC is also significantly better than other algorithms in terms of mean and standard deviation, except for that in the optimization of 30-dimensional test Griewank function. Therefore, compared with ABC and GABC, the optimization performance of EABC algorithm has been greatly improved, indicating the effectiveness of the improvement measures in this paper. For the sake of intuition, figures 1 and 2 respectively show the convergence of the optimal solutions of ABC, GABC and EABC algorithm in

30-dim and 60-dim Griewank, Rastrigin, Ackley and Schwefel functions. It is easy to find that the optimization ability of the EABC algorithm is obviously better than that of ABC and GABC.

Table 2. Comparison with ABC and GABC algorithms

Function	Alg.	ABC		GABC		EABC	
	Dim	Mean	SD	Mean	SD	Mean	SD
f_1	30	1.27306E-15	1.46400E-15	2.96059E-17	4.99300E-17	1.66531E-16	1.56354E-16
	60	2.51040E-13	7.51400E-13	7.54952E-16	4.12700E-16	3.70000E-16	4.12432E-16
f_2	30	1.34529E-13	7.96600E-14	1.32635E-14	2.44500E-14	3.78953E-15	1.41791E-14
	60	2.06479E-08	1.12100E-07	3.52429E-13	1.24300E-13	1.44004E-13	1.63259E-13
f_3	2	9.93136E-03	8.14300E-03	1.68497E-04	1.45400E-04	3.66933E-08	1.97600E-07
	3	6.44947E-02	4.85200E-02	2.65914E-03	2.22000E-03	1.09957E-04	3.66540E-04
f_4	30	4.69550E-14	5.95400E-15	3.21521E-14	3.25200E-15	1.45785E-14	2.35015E-15
	60	1.66089E-13	2.21700E-14	1.00009E-13	6.08900E-15	9.97120E-14	1.81829E-14
f_5	30	-	-	-	-	3.81830E-04	1.62630E-19
	60	-	-	-	-	7.65998E-04	9.93060E-06

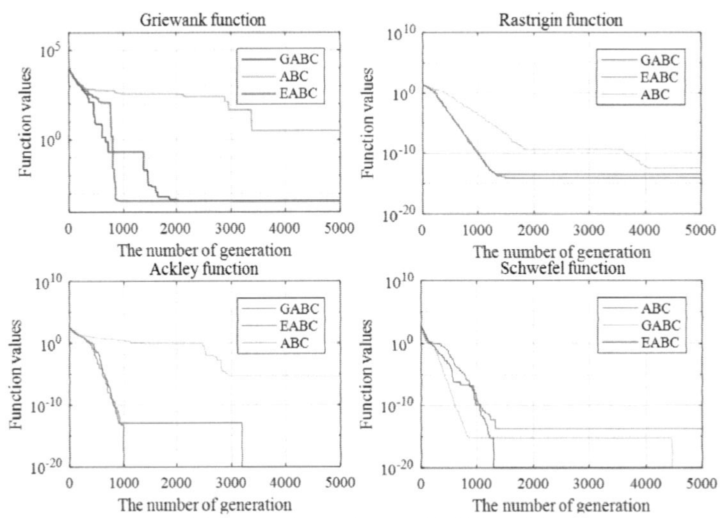

Figure 1. Functions with Dim=30

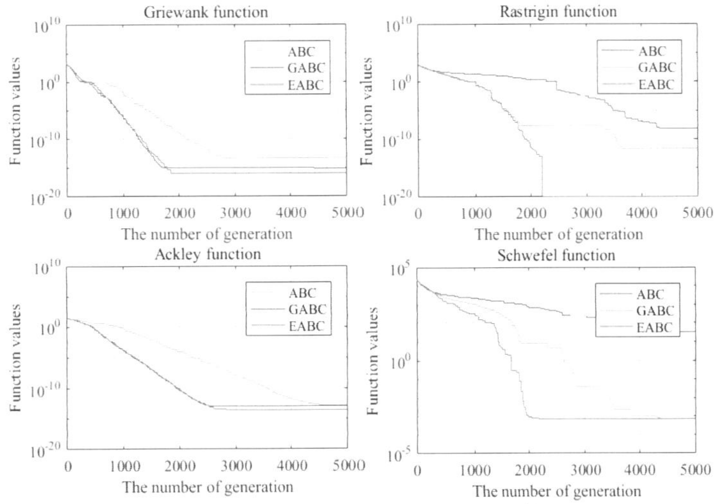

Figure 2. Functions with Dim=60

In order to further verify the superiority of the EABC algorithm, the algorithm is compared with representative algorithms such as GA, PSO and PS-EA, and the results are shown in table 2. Experimental data of GA, PSO and PS-EA algorithms in table 3 are extracted from reference [26]. As can be seen from table 3, compared with other algorithms, EABC algorithm has high convergence accuracy and strong robustness in different dimensions in each test functions, and the results of EABC algorithm are obviously superior to other algorithms, reflecting the superiority of EABC algorithm.

Table 3. Comparison with GA, PSO and PS-EA algorithms

Function	Alg. / Dim	GA Mean	GA SD	PSO Mean	PSO SD	PS-EA Mean	PS-EA SD	EABC Mean	EABC SD
	10	5.02280E-02	2.95230E-02	7.93930E-02	3.34510E-02	2.22366E-01	7.81000E-02	1.36607E-03	2.47313E-03
f_1	20	1.01390E+00	2.69660E-02	3.05650E-02	2.54190E-02	5.90360E-01	2.03000E-01	9.45744E-04	2.92831E-03
	30	1.23420E+00	1.10450E-01	1.11510E-02	1.42090E-02	8.21100E-01	1.39400E-01	3.93297E-04	1.78842E-03
	10	1.39280E+00	7.63190E-01	2.65590E+00	1.38960E+00	4.34040E-01	2.55100E-01	0	0
f_2	20	6.03090E+00	1.45370E+00	1.20590E+01	3.32160E+00	1.81350E+00	2.55100E-01	2.55795E-14	8.56904E-14
	30	1.04388E+01	2.63860E+00	3.24760E+01	6.95210E+00	3.05270E+00	9.98500E-01	1.02489E-02	1.85949E-02
	10	4.63184E+01	3.38217E+01	4.37130E+00	2.38110E+00	2.53030E+01	2.97964E+01	2.43003E+01	2.60310E+00
f_3	20	1.03930E+02	2.95050E+01	7.73820E+01	9.49010E+01	7.24520E+01	2.73441E+01	9.79914E+00	2.88693E+00
	30	1.66283E+02	5.95102E+01	4.02540E+02	6.33650E+02	9.84070E+01	3.55791E+01	2.43003E+01	2.60310E+00
	10	5.92670E-01	2.24820E-01	9.84990E-13	9.62020E-13	1.92090E-01	1.95100E-01	5.39032E-15	6.70333E-15
f_4	20	9.24130E-01	2.25990E-01	1.17780E-06	1.58420E-06	3.23210E-01	9.73530E-02	5.79992E-12	1.63625E-11
	30	1.09890E+00	2.49560E-01	1.49170E-06	1.86120E-06	3.77100E-01	9.87620E-02	2.43905E-04	2.34852E-04
	10	1.95190E+00	1.30440E+00	1.61870E+02	1.44160E+02	3.20370E-01	1.61850E+00	1.27280E-04	0.00000E+00
f_5	20	7.28500E+00	2.99710E+00	5.43070E+02	3.60220E+02	1.49840E+00	8.46120E-01	2.54726E-04	7.05002E-07
	30	1.35346E+01	4.95340E+00	9.90770E+02	5.81140E+02	3.27200E+00	1.61850E+00	3.21173E-03	1.48244E-02

The values obtained by GA, PSO, PS-EA and EABC after 500, 750 and 1,000 cycles for dimensions 10, 20 and 30

6. Conclusion

An enhanced artificial bee colony algorithm is proposed in this paper, which is optimized by using the gradient correction strategy and the crossover and mutation of genetic algorithm. The experimental results show that the improved algorithm has high convergence accuracy, strong robustness, and remarkable effect. And compared with the current advanced algorithms, the EABC algorithm has some superiority, and has certain advantages for the high dimensional optimization problems.

Acknowledgment
This work is partially supported by the National Natural Science Foundation of China (No.51379081) and Project of Science and Technology of SGCC (DKJS-13-00220).

References

[1] Karaboga D. (2005) An idea based on honey bee swarm for numerical optimization. Erciyes University Technical Report.
[2] Lučić P, Teodorović D. (2003) Transportation Modeling:An Artificial Life Approach. In:IEEE International Conference with Artificial Intelligence.
[3] Teodorović D. (2009) Bee colony optimization. Innovations in Swarm Intelligence.
[4] H D, S S, S Y. (2005)Cooperative Bees Swarm for Solving the Maximum Weighted Satisfiability Problem[Z]. Vilanova i la Geltrú.
[5] Wedde H F, Farooq M, Yue Z. (2004) BeeHive: An Efficient Fault-Tolerant Routing Algorithm Inspired by Honey Bee Behavior. In:4th International Workshop Ant Colony Optimization and Swarm Intelligence.
[6] Shaik S, Xu F Y, Lai C S, Sinha N, Lai L L. (2012) Short-term hydro -thermal scheduling with Artificial Bee Colony. In:2012 International Conference on Machine Learning and Cybernetics (ICMLC) IEEE.
[7] Yi Y, Yonggang W, Binqi H, Xinglong L, Quan Z. (2018) An enhanced artificial bee colony algorithm (EABC) for solving dispatching of hydro-thermal system (DHTS) problem. Plos One, 13: e189282.
[8] Karaboga D, Akay B. (2009) A comparative study of Artificial Bee Colony algorithm. Applied Mathematics & Computation, 214: 108-132.
[9] Yuan X, Ji B, Chen Z, Chen Z. (2014) A novel approach for economic dispatch of hydrothermal system via gravitational search algorithm. Applied Mathematics & Computation, 247: 535-546.
[10] Cui L, Li G, Zhu Z, Lin Q, Wen Z, Lu N, Wong K, Chen J. (2017) A novel artificial bee colony algorithm with an adaptive population size for numerical function optimization. Information Sciences: S1915193703.
[11] Akbari R, Mohammadi A, Ziarati K. (2010) A novel bee swarm optimization algorithm for numerical function optimization. Communications in Nonlinear Science & Numerical Simulation, 15: 3142-3155.
[12] Bouaziz A, Draa A, Chikhi S. (2013) A Quantum-inspired Artificial Bee Colony algorithm for numerical optimisation. In:2013 11th International Symposium on Programming and Systems.
[13] Peng G, Cheng W, Jian L. (2011) Global artificial bee colony search algorithm for numerical function optimization. In:2011 Seventh International Conference on Natural Computation IEEE.
[14] Bi X J, Wang Y J. (2011) Niche artificial bee colony algorithm for multi-peak function optimization. Systems Engineering and Electronics, 33: 2564-2568.
[15] Gao W, Liu S, Fei J, Zhang J. (2011) Hybrid artificial bee colony algorithm. Systems Engineering & Electronics, 33: 1167-1170.
[16] Bao L, Zeng J. (2010) Self-adapting search space chaos-artificial bee colony algorithm. Application Research, 27: 1330-1334.
[17] Luo J, Li Y. (2010) Artificial bee colony algorithm with chaotic-search strategy. Control & Decision, 25: 1913-1916.
[18] Bi X, Wang Y. (2011) An improved artificial bee colony algorithm. In:International Conference

on Computer Research & Development.
[19] Zhu G, Kwong S. (2010) Gbest-guided artificial bee colony algorithm for numerical function optimization. Applied Mathematics & Computation, 217: 3166-3173.
[20] M J, D Y. (2014) Artificial bee colony algorithm and its application. Science Press,Beijing.
[21] Yanhua Y. Genetic bee colony algorithm and its application [D]. Nanjing: Nanjing University of Science and Technology, 2012.
[22] Gunasekaran L, Subramaniam S. (2016) FAGA: hybridization of fractional order ABC and GA for optimization. Int Arab J Inf Techn, 13: 1045-1053.
[23] Shi Y. (2014) Diversity and Mechanisms in Swarm Intelligence. International Journal of Swarm Intelligence Research, 5: 1-12.
[24] Richardson J T. (1989) Some guidelines for genetic algorithms with penalty functions. In:Proceedings of the third international conference on Genetic algorithms.
[25] Orero S O, Irving M R. (1998) A genetic algorithm modelling framework and solution technique for short term optimal hydrothermal scheduling. IEEE Transactions on Power Systems, 13: 501-518.
[26] Karaboga D, Basturk B. (2007) A powerful and efficient algorithm for numerical function optimization: artificial bee colony (ABC) algorithm. Global Optimization, 39: 459-471.

Review on Application of Binocular Vision Technology in Field Obstacle Detection

Qiao Wang[1], Zhijun Meng[1*] and Hui Liu[2]

[1] National Engineering Research Center of Intelligent Equipment for Agriculture, Beijing Academy of Agricultural and Forestry Sciences, Beijing, China

[2] Information Engineering College, Capital Normal University, Beijing, China

[*]Corresponding author's e-mail: mengzj@nercita.org.cn

Abstract: Obstacles within the field are the main external source of safety hazard in field operation of agricultural machineries. Autonomous obstacle avoidance is a critical technology to be solved for the robotization of field operation by agricultural machineries. For this technology, a critical premise is to timely and accurately perceive static and dynamic information of field obstacles. Firstly, this paper describes main detection parameters of field obstacles. Whereafter, by compared with other detection technologies, applicability of binocular vision-based detection technology in field obstacle detection is analyzed. Afterwards, this paper summarizes the whole process of binocular vision-based obstacle detection into three steps. For step one, main detecting methods for locating obstacles in images, main principles of these methods and their applicability in field operation scene are discussed. For step two, main existing methods for detecting spatial position parameters of obstacles based on detection results of step one and their application status in field operation scene are summarized. For step three, typical target tracking methods for obtaining the motion state of obstacles are compared and summarized, and their reference value for future research is pointed out. At last, the main challenges of detecting obstacles in field scene based on binocular vision are summarized.

1. Introduction

The robotization of field operation is an indispensable part of the robotization of agricultural production, which is of great significance for solving the shortage of agricultural labor force, promoting targeted and intellectualized agricultural machineries and improving efficiency and economic benefits of agricultural production. From the current view, the whole-process autonomous navigation walking technology of agricultural machinery in the non-structured field operation environment is one of the key technical bottlenecks for achieving the transition from automation to robotization of field operation.

For the whole-process autonomous navigation walking of agricultural machinery (mainly referring to the large-scale conventional agricultural machinery in this paper) during field operation, it mainly includes the following three parts of walking process: autonomous navigation walking during normal operation in the field, autonomous turning at the end of field, and autonomous obstacle avoidance when encountering obstacles in field. Among them, the walking process of autonomous obstacle avoidance, which is usually manifested as conditional stress behavior requiring real-time response and has random characteristics, is an important feedback mechanism responding to external safety hazards for agricultural machineries during field operation. Obviously, in order to realize this part of autonomous walking process, the primary task is to timely sense obstacles in the field environment, especially obstacles within the coverage area of operation path in front of the current agricultural machinery, so as to obtain their overall dimensions, spatial position relative to the current agricultural machinery and their state of motion.

Besides, binocular stereo vision, which is similar to human binocular view and conforms to human habit of perception [1], can provide depth distance or three-dimensional information of obstacles in scenes based on the principle of distance measurement using binocular disparity by stereo matching, and its application in obstacle detection draws increasing attention of scholars worldwide. In view of this, this paper mainly focuses on an overview of the application status and development tendency of binocular stereo vision technology in field obstacle detection, so as to provide a systematic and comprehensive technical reference and information support for further development and improvement of related algorithms.

2. Main detection parameters of field obstacles

For a single target obstacle, whether static or dynamic obstacle, main detection parameters include two types of static parameters - its spatial position relative to the machinery described by two parameters of depth distance and azimuth angle, and its dimensions of two-dimensional projected contour described by two dimensions of height and width/length – and dynamic parameters - magnitude and direction of its movement, of which, for static obstacles, the magnitude of their movement is regarded as 0. For a single target obstacle group merged from multiple target obstacles, main detection parameters are similar. As shown in Figure 1(a), when the machinery operates normally and arrives at the point O_p along the crop rows, the distance O_{cE} shown in the figure is the depth distance of the current obstacle, the angle θ is the azimuth angle of the current obstacle, and the height and width of circumscribed rectangle as shown in Figure 1(b) are the dimensions of two-dimensional projected contour.

3. Advantages and disadvantages of binocular vision technology in detecting field obstacles

Current sensing technologies for detecting obstacles mainly include ultrasonic sensing [2] [3], laser radar sensing [4] [5], millimeter-wave radar sensing [6] [7], and binocular vision sensing [8] [9] and multiple-sensor fusion sensing[10][11]. Most of the current obstacle detection technologies focus on indoor environment or simple outdoor environment with known structure (e.g. structured road/highway or urban streets environment with favorable conditions), while the study on field environment is less performed [12]. Applicability in using a single type of sensing technology described above to detect obstacles in outdoor scenes [12] [13] [14] is compared, as shown in Table 1.

Relative to laser radar, millimeter-wave radar and ultrasonic sensor, binocular vision features light load and low cost, with passive measurement method and without emitting signals interfering each other. Moreover, it can provide abundant and dense 3D information (e.g., dimensions, colors, texture, depth distance, spatial position, etc.), and can detect obstacles in wider view, obstacles lower than the ground surface (that is, negative obstacles such as ditches, pits and pools) and obstacles lower than top of the crops but not completely covered. However, the current study on binocular vision-based obstacle detection mainly focuses on indoor environment or simple outdoor environment. In the actual unstructured and complicated field environment, affected by weather (e.g., dust, rain, snow or fog), natural light change and blowing dust during field operation, as well as random appearance of obstacles with various types, using binocular vision to rapidly identify obstacles in the field has been a research hotpot for scholars worldwide.

In accordance with Table 1, using a single type of sensor to sense obstacles in outdoor environment shows imperfections and can barely satisfy requirements of real-time capability and reliability. Meanwhile, multi-sensor fusion system usually taken vision sensor as one of main sensors for detecting obstacles in outdoor scenes draws increasing concern. However, although multi-sensor fusion technology may provide comprehensive and precise information, it brings comparatively complex structure, requires high costs and can’t ensure real-time capability.

4. Application of binocular vision technology in field obstacle detection

Different from obstacle recognition in whether structured or unstructured road scene, recognition of obstacles in field operation scene usually doesn't require the recognition of the road area before detecting obstacles. In general, the field area in front of agricultural machinery during its operation is considered the area of interest with obstacles to be recognized by default, and obstacle detection may be directly performed in this area. In view of this, road recognition is not taken into consideration in this paper when discussing binocular vision-based obstacle detection methods in various scenes. Binocular vision-based obstacle detection mainly consists of three detection steps: step one – to locate target obstacles in images; step two – to locate target obstacles in three-dimensional space; step three – to obtain the motion state of target obstacles.

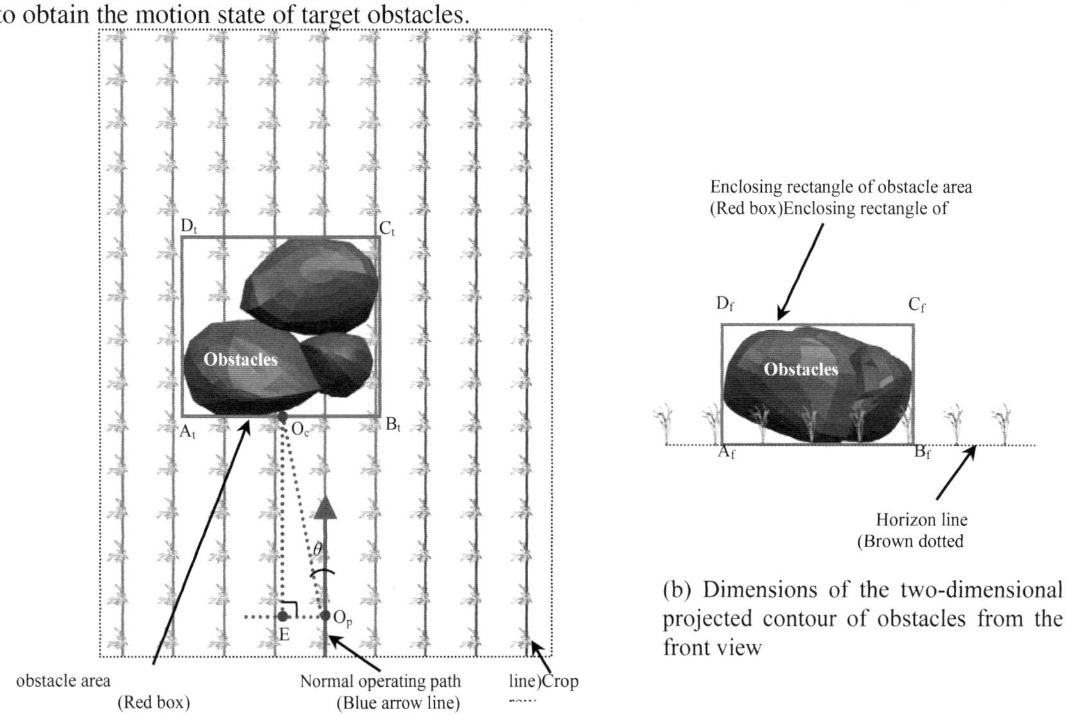

(b) Dimensions of the two-dimensional projected contour of obstacles from the front view

(a) Azimuth of obstacles relative to the machinery from the top view

Figure 1 Schematic view of static detection parameters of target obstacles in the field

Notes:

A_t, B_t, C_t, D_t: four apexes of circumscribed rectangle of obstacle area in the top view. Where, the left and right sides of the rectangle are parallel to crop rows.

O_c: midpoint of the side A_tB_t.

O_p: a normal operating position point of the machinery before encountering obstacles, at which the operation direction is forward along the crow rows.

θ: azimuth angle of the current obstacle, which is the angle of O_c deviated from normal operation direction when the machinery is at the position point O_p during normal operation.

O_{cE}: distance between O_c and the operation point O_p along normal operation direction, i.e. along the crop rows. When the agricultural machinery operates normally and arrives at the position point O_p, the distance O_cE is the depth distance of the current obstacle relative to the machinery.

A_f, B_f, C_f, D_f: four apexes of circumscribed rectangle of obstacle area in the front view. Where, the top and bottom sides of the rectangle are parallel to the horizon line. The actual dimensions of this circumscribed rectangle are the dimensions of the two-dimensional projected contour of the obstacles projected toward the machinery.

Table 1 Comparison of applicability of obstacle detection technologies in outdoor scenes, especially in unstructured and complicated field environment

Technology	Weight, cost and detection method	Weather (dust/rain/snow/ fog) and light (sunny/cloudy, day/ night) influence	Detectable obstacles types, detectable parameters and detection distance	Timeliness and accuracy	Reference
Ultrasonic	•Light weight •Low cost •Active sensor, with interference among multiple sensors, not well-suited for collaborative operation of multiple machineries	•Limited performance measurement. It may be influenced by air temperature and humidity •Operation in any light	•Affected by the types of obstacle surface, it can't detect obstacles mixed with crops, especially those lower than top of the crops. It can hardly identify obstacles lower than the ground surface. •It can't detect information such as dimensions of two-dimensional projected contour and accurate azimuth angle. •Limited, generally within 10m	•Speed of data acquisition and processing is fast enough for field scene. •Poor accuracy due to poor directionality, frequent misreadings and specular reflections	[15][16]
Millimeter-wave radar	•Light weight •High cost, •Active sensor, with interference among multiple sensors, not well-suited for collaborative operation of multiple machineries	•Operation in any weather •Operation in any light	•Unable to detect obstacles mixed with crops, especially those lower than top of the crops. It may miss obstacles with small volume. •For negative obstacles lower than the ground surface, it can't ensure accuracy in distance detection. It can't detect information such as dimensions of two-dimensional projected contour and accurate azimuth angle •Enough	•Speed of data acquisition and processing is fast enough for field scene. •Accurate enough	[13][14]
Laser radar	•Heavy weight •Highest cost, especially three-dimensional laser radar •Active sensor, with interference among multiple sensors, not well-suited for collaborative operation of multiple machineries	•Decreased performance measurement •Operation in any light	•Unable to detect obstacles mixed with crops, especially those lower than top of the crops. Significantly influenced by specular reflection of smooth surface of objects. It may miss obstacles with small volume. •For negative obstacles lower than the surface, it can't ensure accuracy in distance detection. It can't detect information such as dimensions of two-dimensional projected contour and accurate azimuth angle by two-dimensional laser radar, while the above static and dynamic parameters can all be detected by three-dimensional laser radar. •Enough	•For field scene, speed of data acquisition and processing is fast enough for two-dimensional laser radar, but not well-suited for three-dimensional laser radar with low imaging speed. •Accurate enough for three-dimensional laser radar	[17] [12]

Binocular vision	•Light weight •Low cost •Mainly passive sensors, with no interference among multiple passive sensors, provide abundant and dense 3D information, applicable to collaborative operation of multiple machineries	•Limited performance measurement in dust or fog environment, while less affected by rain or snow •Affected by light change	•It is difficult to detect obstacles similar to the environment. It can detect those lower than the ground surface or top of the crops but not completely covered •The above static and dynamic parameters can all be detected. •Limited, with satisfied detection accuracy within 10m, generally not exceeding 30m	•Short data acquisition cycle. Efficiency of data processing depends on the capabilities of image processing algorithm. •Depend on image processing algorithm and decrease with the increase of distance.	[18][19]

4.1. Step one: to locate target obstacles in images

Binocular vision-based obstacle detection usually requires to locate target obstacles in the images first, that is, to extract target obstacle regions or their edge contour. Methods for such extraction work mainly include the following three types:

Type 1: locating obstacles in images based on monocular vision. To be specific, there are mainly three approaches. The principles of these approaches and their limitations in field operation scene are shown in Table 2.

Type 2: locating obstacles in images based on binocular disparity. To be specific, there are currently three approaches. The principles of these approaches and their applicability in field operation scene are shown in Table 2.

Type 3: locating obstacles in images by combining monocular and binocular vision. The principle and main advantages of this method in field operation scene are shown in Table 2.

Table 2 Methods used for step one and their applicability in field operation scene

Main methods	Main approaches and principles	Applicability in field operation scene	Reference
Monocular vision-based method	**Approach 1:** •Based on traditional characteristic analysis of colours, texture, morphology and edges •Extract obstacle area from the image background	•Applicable to detect specific types of obstacles with specific characteristics under a specific field scene, season, weather or light conditions.	[20][21]
	Approach 2: •Based on motion characteristics, such as methods of inter-frame difference, optical flow analysis, and motion compensation etc. •Obtain the area position of obstacles	•Affected by changeful posture and motion state of obstacles, light change, complicated and changeful field operation scene, etc.	[22][23]
	Approach 3: •Based on learning characteristics of neural network-based or support vector machine-based machine learning or deep learning •Locate obstacles in the testing image	•In order to be applicable in complicated and changing field operation environment with diverse obstacle characteristics, it is necessary to collect a large number of training sample sets of obstacles with different characteristics under different field scenes, seasons, weather and light conditions.	[24][25]
Binocular disparity-based method	**Approach 1:** •Based on inverse perspective transformation •Obtain non-zero disparity regions (i.e. obstacle area)	•Applicable to detect obstacles on flat surface rather than in field operation scene.	[26][27]
	Approach 2: •Based on plane-line projection characteristics of UV-disparity •Obtain height of vertical line segments in V-disparity maps (i.e. height of obstacle area) •Obtain length of horizontal line segments in U-disparity maps (i.e. width of obstacle area)	•Applicable to detect obstacles with plane characteristics in normal structured road scene. •Not well suitable in unstructured field operation scene without noticeable and fixed road characteristics but with complicated types of obstacles, especially obstacles lower than top of the crops and negative obstacles lower than the ground surface.	[28][9]
	Approach 3: •Based on binocular stereo matching •Combined with edge and texture feature of disparity or depth map, or point cloud density feature of 3D point cloud map •Combined with limiting conditions such as distance range (e.g., taking nearby obstacles as target obstacles) and size dimensions determined by practical application •Combined with methods such as region clustering, etc. •Obtain the area position or edge contour of the obstacles	•Precise stereo matching of image pairs in complicated and unstructured field scene is the most complicated and time-consuming part. •Sensitive to regions with ambiguous or repeated texture which commonly appear in field scene, or occlusion by crops or other obstacles in the field •Difficult to detect obstacles with their characteristics similar to field background •Affected by geometric distortion, imaging distortion, view difference and natural light change, etc.	[29][30]
Method of combining monocular and binocular vision	•Based on the combination of the above two methods •Preliminarily and respectively locate suspected obstacles in image pairs with the monocular vision-based method •Finally locate target obstacles of interest after removing false obstacles or currently subordinate obstacles with binocular disparity-based method	•Influence factors restricted its applicability in field scene cover the above related content. •Determining candidate target zones in early stage can narrow down searching range during subsequent stereo matching, reduce complexity of matching algorithm, decrease computation load and improve real-time capability.	[31][32]

4.2. Step two: to locate target obstacles in three-dimensional space

Obviously, it is far from being enough by just obtaining the position information of obstacles in images, and their location information relative to the subject of obstacle avoidance in three-dimensional space is actually the critical information truly needed in practical obstacle avoidance. On the basis of the principle of disparity in binocular stereo imaging system, especially extensively used parallel binocular stereo imaging system, depth distance of the characteristic points on the target relative to the camera can be easily obtained or their three-dimensional coordinate information can even be regained by combining triangulation method after obtaining internal and external parameters of the camera by binocular camera calibration and obtaining disparity information of matched pairs of the characteristic points by stereo matching. Therefore, on this basis, according to disparity information obtained by stereo matching of the overall target obstacle region [33] or its edge contour [34] or its characteristic points [35]), three-dimensional reconstruction for the target obstacles may be performed to regain their spatial location and three-dimensional geometrical information.

Although three-dimensional point cloud data of the overall obstacle region or its edge contour can be obtained via three-dimensional reconstruction to provide more abundant and complete spatial azimuth information and three-dimensional contour information of obstacles for planning the obstacle avoidance path. Nonetheless, for obstacle avoidance during field operation of agricultural machineries, considering the regular field operation route of the machineries along the rows and the practical requirement for preventing damage to crops in their growing period, no matter for static or dynamic obstacles, only the following three spatial position parameters need to be detected as described in Section 2.2 in order to locate obstacles in three-dimensional space: depth distance, azimuth angle, dimensions of two-dimensional projected contour.

4.2.1. Detection of depth distance. In the above three parameters, detection of depth distance is fundamental and critical. Currently, detection of depth distance of obstacles has been extensively discussed in many papers.

If in step one, the target obstacle region or the edge contour of obstacles is located in respective left and right image pair with the monocular vision-based method, stereo matching may be performed for characteristic points (e.g. randomly selected points, angular points, centroid or center of mass) on the target region pairs [36] [37] or characteristic points (e.g. randomly selected points, discrete sampling points, vertexes or angular points) on the edge contour pairs [34] in step two to obtain respective disparities, and then the depth distance of the obstacles can be detected based on triangulation method.

If local position matching point pairs of target obstacles or area position or edge contour of target obstacles in the disparity map is obtained with the binocular disparity-based method or with the method of combing monocular and binocular vision in step one, the corresponding disparity information may be directly used to measure the depth distance of the target obstacles by triangulation method in step two [38]. If the area position or edge contour of obstacles in the depth map or three-dimensional map of point cloud are obtained in step one, the depth distance of the obstacles may be directly obtained in step two based on corresponding depth data or three-dimensional data [39].

4.2.2. Determination of azimuth angle. For determination of azimuth angle, generally, once three-dimensional coordinate information at the center of the obstacle region or center of its edge contour or center of the characteristic points on it is regained based on the principle of distance measurement by binocular disparity, information about the azimuth angle as well as the depth distance may be simply estimated [40].

4.2.3. Detection of dimensions of two-dimensional projected contour. For detection of dimensions of two-dimensional projected contour, if three-dimensional point cloud data of the overall obstacle region or its edge contour is regained based on the principle of distance measurement by binocular disparity, its three-dimensional contour dimensions, including dimensions of the two-dimensional projected contour and thickness information in the depth direction, can be obtained based on the difference between the maximum and the minimum values of point cloud data along each coordinate axis.

For the regular-shaped obstacle, only three-dimensional coordinate information of vertexes or angular points on obstacle region or its edge contour need to be regained in order to obtain three-dimensional contour dimensions of the obstacle [33][36].

In addition, if the circumscribed rectangle of obstacle region in original image is known, actual dimensions of the circumscribed rectangle may be calculated by combining depth distance with the triangulation method, or by regaining the three-dimensional coordinate information of 2 diagonal vertexes of the circumscribed rectangle or 4 points in obstacle region closest to four sides of the circumscribed rectangle based on the principle of distance measurement by binocular disparity [40]. Of which, if the vertexes are not within the obstacle region, disparity value within obstacle region will be regarded as disparity value of the vertexes. The dimensions of the circumscribed rectangle are approximate to the required dimensions of the two-dimensional projected contour.

Thus it can be seen that calibration of camera parameters, accuracy of locating target characteristics (e.g. point, line or region) and precision of obtaining disparity by stereo matching are main factors influencing the detection accuracy of the above parameters, of which, the precision of stereo matching is particularly critical. About locating obstacles in three-dimensional space based on binocular vision, study of its application in indoor scenes or outdoor structured road scenes is rather extensive, while its application in complicated and unstructured field scene is still at the exploration stage [41][22][31][32]. Meanwhile, the related research work is mainly about the detection of depth distance and contour dimensions, and the detection of azimuth angle is usually omitted. In addition, relevant study results indicate that the absolute value of detection errors tends to increase with the increase of detection distance [41] [42]. In the wide field environment, the effective detection distance is relatively limited. To be specific, detection distance with satisfactory detection accuracy is within 10m and the maximum distance usually does not exceed 30m.

4.3. Step three: to obtain motion state of target obstacles

For dynamic obstacles, as described in Section 2.2, in addition to detection of the above static parameters, dynamic parameters such as the magnitude and direction of motion speed are also required to be obtained in real time. The above mentioned obstacle detection methods, no matter based on monocular or binocular vision, and no matter used for locating moving obstacles in images or in three-dimensional space, can only realize the location of all target obstacles at a specific moment but lack correlation of position changes of the same target obstacle at different moments. From the point view of target detection and tracking, all these methods are within the study scope of dynamic obstacles detection. In order to detect dynamic parameters, the same target obstacle should be focused on to determine the change rule of its position in three-dimensional space based on consecutively captured images. Meanwhile, the difference between target measurement value and actual value should be minimized, and the motion state of the target at the next moment should be predicted according to the estimation of its current motion state. On this basis, from the point view of target detection and tracking, detection in this step is within the study scope of dynamic obstacle tracking.

Overall, the current algorithm for target tracking consists of three types. The typical methods of each type and the advantages and disadvantages of these three types are shown in Table 3.

Unfortunately, current study on application of the above tracking algorithms in tracking dynamic obstacles in unstructured field scene is insufficient. In some researches [18], to detect the motion state of target obstacles in field operation scene was mainly based on three-dimensional position coordinates of obstacles at different moments and the corresponding time interval. Specifically, the average motion speed in this period is calculated and directly regarded as the motion speed of obstacles in this period, with the motion direction pointed from the position at previous moment to the position at the current moment. Obviously, when measuring and calculating the speed, such researches may evade the problem of matching target obstacles in different image frames by considering that there is only one target obstacle within the field of view and it is the same target obstacle by default. If multiple obstacles appear simultaneously within the field of view, matching target obstacles in different image frames will be required before measurement and calculation. In addition, although this measurement and calculation method is simple, it doesn't take into consideration of measurement errors of the three-dimensional position coordinates caused by interference of noise as well as the final deviation in speed calculation resulting from the measurement errors. In order to eliminate the deviation between the measured and calculated value and the actual value, data processing such as filtering or denoising is required in subsequent work according to change rules of distribution of measured and calculated speed in a continuous time slice.

Table 3 Typical methods used for target tracking and comparison between them

Main types	Principles	Typical methods	Comparison
Generative method	•Establish a model of target appearance in the current frame based on generative models •Search candidate areas with the maximum similarity to the current model in the next frame by using certain search strategy	•Mean shift [43] •Continuously adaptive mean shift [44] •Particle filter [45] •Kalman filter [46][8] •Subspace-based representation [47] •Sparse representation [48]	•Generative models belong to matching models, with heavy computation load and poor real-time capability when obtaining similarity between candidate areas and target template. •Only target itself is considered when forming the model, while background information is ignored. •When appearance of target varies drastically, target is occluded or target information is similar to interference information, processing results may not be that ideal.
Discriminative method	•Regard target tracking as a dichotomy problem based on discriminative models •Obtain position information of target objects by using online learning or offline trained classifier to distinguish target objects from background	•SVM-based discriminative method [49] •Boosting-based discriminative method [50] •Method based on online learning detection, e.g. TLD for tracking a single target in a long term [51] •Deep learning-based method [52] •Correlation filtering-based method [53]	•Compared with generative methods, its classifier utilizes background information during training so as to clearly distinguish target objects from background, which brings higher robustness. •Become a mainstream method for target tracking.
Hybrid method	•Combine different generative methods [54] •Combine different discriminative methods [55] [9] •Combine the above two types of methods [56]		•Helpful to increase accuracy, real-time capability or robustness of tracking algorithm by incorporating advantages of different algorithms.

5. Conclusions and future perspectives

Obstacles within the field are the main external source of safety hazard in field operation of agricultural machineries. In order to realize the robotization of field operation by agricultural machineries and completely replace manual work in field operation, autonomous obstacle avoidance becomes one of the critical technologies to be solved. And for autonomous obstacle avoidance by agricultural machineries during field operation, it is a critical premise to timely and accurately perceive static and dynamic information of field obstacles.

Overall, no matter for locating obstacles in images or in three-dimensional space, detection by stereo matching using binocular disparity shows significant advantages. It does not need priori knowledge of obstacles, and it is not restricted by the motion of obstacles. In addition, based on binocular parameter calibration and the principle of binocular distance measurement, information of the actual position of obstacles can be simply obtained. However, due to complexity of practical application scenes, especially field operation scene of agricultural machineries, complexity and variance of a number of factors such as types of field operation environment, types, positions and postures of obstacles in the field, light and weather conditions, challenges still exist when using this method to appropriately detect and locate obstacles in images with complicated field background, of which, precise stereo matching of image pairs is the most complicated and time-consuming part.

In addition, limited by complexity of field scene and variance of dynamic obstacles in the field, current tracking algorithms can hardly satisfy the practical application requirements simultaneously in accuracy, real-time capability and robustness required by obstacle avoidance during field operation of agricultural machineries. Although the practical operation scene within the field is relatively simple, to track obstacles in the field with strong robustness and good real-time performance still faces the following challenges: change of forms of target obstacles (e.g. change of postures of pedestrians in the field such as sitting, lying, standing and bending, as well as adjustment of azimuth and attitude of agricultural implements), change of dimensions varying with the change of depth distance, indistinct motion caused by quick movement of target obstacles or by camera trembling with vibration of the camera carrier during operation in the field, target obstacles partially or completely occluded by crops or other static or dynamic obstacles, interference caused by similar background regions (e.g. when tracking targets based on the characteristic of colour, background regions similar in colour to the targets may cause tracking shift), mutual interference between similar target obstacles, change of natural light and real-time capability in practical application. For study in the future, the above object tracking algorithms need to be proactively and boldly used and integrated to discover methods applicable to detecting dynamic target obstacles in field operation scene of agricultural machineries.

Acknowledgements
This research was based upon the project supported by Beijing Postdoctoral Research Foundation and the project funded by China Postdoctoral Science Foundation.

References
[1] Hu, J., Gao, L., Bai, X., Li, T., Liu, X., 2015. Review of research on automatic guidance of agricultural vehicles. Transaction of the Chinese Society of Agricultural Engineering 31(10), 1-10.
[2] Zhang, S., 2013. Local Obstacle Avoidance Algorithm and Application of Mobile Robot Based on Ultrasonic. Master Thesis, Harbin Institute of Technology, Harbin, China.
[3] Zhang, M., 2016. Research and Implementation of Obstacle Pattern Recognition for Mobile Robot. Master Thesis, Jinan University, Jinan, China.
[4] Shang, E., An, X., Wu, T., Hu, T., Yuan, Q., He, H., 2016. LiDAR Based Negative Obstacle Detection for Field Autonomous Land Vehicles. Journal of Field Robotics 33(5): 591-617.
[5] Asvadi, A., Premebida, C., Peixoto, P., Nunes, U., 2016. 3D Lidar-based Static and Moving Obstacle Detection in Driving Environments: An Approach Based on voxels and Multi-region Ground Planes. Robotics and Autonomous Systems 83(C): 299-311.
[6] Pang, C., 2015. Detection System of Ahead Vehicle Based on Data Fusion of Ranging Radar and Vision. Master Thesis, Southeast University, Nanjing, China.
[7] Na, T., 2016. Detection of Ahead Vehicle Obstacle Based on Radar and Computer Vision. Master Thesis, Hefei University of Technology, Hefei, China.
[8] Huh, K., Park, J., Hwang, J., Hong, D., 2008. A Stereo Vision-based Obstacle Detection System in Vehicles. Optics and Lasers in Engineering 46(2): 168-178.
[9] Dairi, A., Harrou, F., Senouci, M., Sun, Y., 2018. Unsupervised Obstacle Detection in Driving Environments Using Deep-learning-based Stereovision. Robotics and Autonomous Systems 100: 287-301.

[10] Budzan, S., Kasprzyk, J., 2016. Fusion of 3D Laser Scanner and Depth Images for Obstacle Recognition in Mobile Applications. Optics and Lasers in Engineering 77(7): 230-240.

[11] Mocanu, B., Tapu, R., Zaharia, T., 2016. When Ultrasonic Sensors and Computer Vision Join Forces for Efficient Obstacle Detection and Recognition. Sensors 16(11): 1807.

[12] He, Y., Jiang, H., Fang, H., Wang, Y., Liu, Y., 2018. Research Progress of intelligent obstacle detection methods of vehicles and their application on agriculture. Transactions of the Chinese Society of Agricultural Engineering 34(9): 21-32.

[13] Gray, K.W., 2000. Obstacle Detection and Avoidance for an Autonomous Farm Tractor. Master Thesis, Utah State University, Logan, American.

[14] Jiang, C., 2016. Unmanned Aerial Vehicle (UAV) Obstacle Detection Based on Radar and Vision Sensor. Shenyang Ligong University, Master Dissertation.

[15] Li, Z., Lv, J., Hong, T., Song, S., Huang, J., Chen, H., 2011. Obstacle Avoiding System of Mountain Orchard Steel Wire Rope Traction Cargo Vehicle Based on Ultra-sonic Sensing. Transactions of the Chinese Society for Agricultural Machinery 42(10): 116-120.

[16] Zhang, T., Li, H., Chen, D., Huang, P., Zhuang, X., 2015. Agricultural Vehicle Path Tracking Navigation System Based on Information Fusion of Multi-source Sensor. Transactions of the Chinese Society for Agricultural Machinery 46(3): 37-42.

[17] Kragh, M., Jørgensen, R.N., Pedersen, H., 2015. Object Detection and Terrain Classification in Agricultural Fields Using 3D Lidar Data. 10th International Conference on Computer Vision Systems, 9163: 188-197.

[18] Wei, J., Rovira-Mas, F., Reid, J.F., Han, S., 2005. Obstacle Detection Using Stereo Vision to Enhance Safety of Autonomous Machines. Transactions of the American Society of Agricultural Engineers 48(6): 2389-2397.

[19] Ball, D., Upcroft, B., Wyeth, G., Corke, P., Engligh, A., Ross, P., Pattrn, T., Fitch, R., Sukkarieh, S., Bate, A., 2016. Vision-based Obstacle Detection and Navigation for an Agriculture Robot. Journal of Field Robotics 33(8): 1107-1130.

[20] Han, Y., Wang, Y., Kang, F., Zhao, Y., 2013. Detection of Obstacles in Farmland Based on Wavelet Multi-resolution Transform. Transactions of the Chinese Society of Agricultural Machinery 44(6): 215-221.

[21] Campos, Y., Sossa, H., Pajares, G., 2016. Spatio-temporal Analysis for Obstacle Detection in Agricultural Videos. Applied Soft Computing 45(C): 86-97.

[22] Yang, L., Noguchi, N., 2012. Human Detection for a Robot Tractor Using Omni-directional Stereo Vision. Computers and Electronics in Agriculture 89(3): 116-125.

[23] Zhou, J., Cheng, J., 2011. Moving Obstacle Detection Based on Machine Vision for Agriculture Mobile Robot. Transactions of the Chinese Society for Agricultural Machinery 48(2): 154-158

[24] Stentz, A., Dima, C., Wellington, C., Herman, H., Stager D., 2002. A System for Semi-Autonomous Tractor Operations. Autonomous Robots 13(1): 87-104.

[25] Christiansen, P., Nielsen, L.N., Steen, K.A., Jørgensen, R.N., Karstoft, H., 2016. DeepAnomaly: Combining Background Subtraction and Deep Learning for Detecting Obstacles and Anomalies in an Agricultural Field. Sensors 16 (11): 1904.

[26] Chen, W., 2009. Research of Road Identification and Obstacle Detection for Intelligent Vehicle Based on Binocular Vision. Master Thesis, Xi'an University of Technology, Xi'an, China.

[27] Qu, S., Meng, C., 2014. Statistical Classification Based Fast Drivable Region Detection for indoor Mobile Robot. Internal Journal of Humanoid Robotics 11(01): 14500010.

[28] Sun, Y., Zhang, L., Leng, J., Luo, T., Wu, Y., 2018. An Obstacle Detection Method Based on Binocular Stereovision. Advances in Multimedia Information Processing, PCM 2017: 571-580.

[29] Gou, Q., 2013. Research on Obstacle Detection in Unknown Farmland Based on Binocular Vision. Northwest A & F University, Master Dissertation.

[30] Ji, C., Shen, Z., Gu, B., Tian, G., Zhang, J., 2015. Obstacle Detection Based on Point Clouds in

Application of Agricultural Navigation. Transactions of the Chinese Society of Agricultural Engineering 31(7): 173-179.

[31] Ross, P., English, A., Ball, D., Upcroft, B., Wyeth, G., Corke, P., 2014. Novelty-based Visual Obstacle Detection in Agriculture. Proceedings of the 2014 IEEE International Conference on Robotics and Automation: 1699-1705.

[32] Ball, D., Upcroft, B., Wyeth, G., Corke, P., English, A., Ross, P., Patten, T., Fitch, R., Sukkarieh, S., Bate, A., 2016. Vision-based Obstacle Detection and Navigation for an Agricultural Robot. Journal of Field Robotics 33(8): 1107-1130.

[33] Chen, Z., Luo, X., Dai, B., 2017. Design of Obstacle Avoidance System for Micro-UAV Based on Binocular Vision. 2017 International Conference on Industrial Informatics – Computing Technology, Intelligent Technology, Industrial Information Integration: 67-70.

[34] Shen, Z., 2014. Research on Variable Angle of Robot Obstacle Detection Methods Based on Contour Recognition. Master Thesis, Zhejiang University of Technology, Hangzhou, China.

[35] Fu, H., 2018. Research on 3D Reconstruction of Binocular Vision System Based on Assembly Robot. Master Thesis Xi'an University of Technology, Xi'an, China.

[36] Zhao, L., 2014. Study on the Vision-based Detection and Planning Techniques for Hexapod Walking Robot under Environment with Obstacles. Doctoral Thesis, Zhejiang University, Hangzhou, China.

[37] Yang, C., Wu, H., Li, Z., He, W., Wang, N., Su, C., 2017. Mind Control of a Robotic Arm with Visual Fusion Technology. IEEE Transactions on Industrial Informatics 14(9): 3822-3830.

[38] Li, G., Li, G., Chen, P., 2016. The Method of Detecting Nearest Distance Between Obstacles and Vehicle Tail Based on Binocular Vision System. 2016 IEEE Vehicle Power and Propulsion Conference: 1-5.

[39] Ge, X., 2018. Design and Implementation of Aircraft Routing System Based on Heterogeneous Multicore Embedded Platform. Master Thesis, University of Electronic Science and Technology of China, Chengdu, China.

[40] She, H., 2016. The AGV Guidance Control and Obstacle Detection Based on Vision. Master Dissertation, Zhejiang University of Technology, Hangzhou, China.

[41] Zhang, L, 2007. Study of the Detection Method for Farm Area Boundary and Obstacle Based on Machine Vision. Doctoral Thesis, China Agricultural University, Beijing, China.

[42] Ji, C., Shen, Z., Gu, B., 2015. Obstacle Detection Based on Point Clouds in Application of Agricultural Navigation. Transactions of the Chinese Society of Agricultural Engineering 31(7): 173-179.

[43] Hu, J.S., Juan, C.W., Wang, J.J., 2008. A Spatial-color Mean-shift Object Tracking Algorithm with Scale and Orientation Estimation. Proceedings of the IEEE International Conference on Robotics and Automation 29(16): 2165-2173.

[44] Li, Y., Shen, X., Bei, S., 2011. Real-time Tracking Method for Moving Target Based on an Improved Camshift Algorithm. Proceedings of the 2011 International Conference on Mechatronic Science, Electric Engineering and Computer: 978-981.

[45] Sun, H., Zou, H., Zhou, S., Wang, C., EI-Sheimy., N., 2013. Surrounding Moving Obstacle Detection for Autonomous Driving Using Stereo Vision. International Journal of Advanced Robotic Systems 10(36): 526-531.

[46] Reid, D., 2013. An Algorithm for Tracking Multiple Targets. IEEE Transactions on Automatic Control 24(6): 843-854.

[47] Ross, D.A., Lim, J., Lin, R.S., Yang, M.S., 2008. Incremental Learning for Robust Visual Tracking. International Journal of Computer Vision 77(1-3): 125-141.

[48] Huang, H., Bi, D., Hou, Z., Hu, C., Gao, S., Zha, Y., Ku, T., 2018. Research of Sparse Representation-based Visual Object Tracking: A Survey. Acta Automatic Sinica 44(10): 1747-1763.

[49] Hare, S., Saffari, A., Torr, P.H.S., 2012. Struck: Structured Output Tracking with Kernels. IEEE International Conference on Computer Vision: 263-270.

[50] Grabner, H., Leistner, C., Bischof, H., 2008. Semi-supervised On-line Boosting for Robust Tracking. In: Proc of European Conference on Computer Vision: 234-247.

[51] Kalal, Z., Mikolajczyk, K., Matas, J., 2012. Tracking-Learning-Detection. IEEE Transactions on Pattern Analysis & Machine Intelligence 34(7): 1409-1422.

[52] Zhang, K., Liu, Q., Wu, Y., 2016. Robust Visual Tracking via Convolutional networks Without Training. IEEE Transactions on Image Processing 25(4): 1779-1792.

[53] Henriques, J.F., Caseiro, R., Martins, P., Batista, J., 2015. High-speed Tracking with Kernelized Correlation filters. IEEE Transactions on Pattern Analysis and Machine Intelligence 37(3): 583-596.

[54] Zhang, Z., Xu, H., Chao, Z., Li, X., Wang, C., 2015. A Novel Vehicle Reversing Speed Control Based on Obstacle Detection and Sparse Representation. IEEE Transactions on Intelligent Transportation Systems 16(3): 1321-1334.

[55] Valmadre, J., Bertinetto, L., Henriques, J.F., Vedaldi, A., Torr, P.H.S., 2017. End-to-end Representation Learning for Correlation Filter Based Tracking. Proceedings of the IEEE Conference on Computer Vision and Pattern Recognition: 2805-2813.

[56] Zhang, T., Xu, C., Yang, M., 2017. Multi-task Correlation Particle Filter for Robust Visual Tracking. Proceedings of the IEEE Conference on Computer Vision and Pattern Recognition: 4335-4343.

IOP Conf. Series: Materials Science and Engineering **806** (2020) 012026 doi:10.1088/1757-899X/806/1/012026

A Scientometric Analysis of Artificial Intelligence and Big data for well-being and human potential

Zonghe Zhang[1], Han-Teng Liao[1*], Xue Wu[1], Zhichao Xu[1]

[1] New Media Research Center, Sun Yat-sen University Nan-fang College, Guangzhou, Guangdong, 510970, China

[*]Corresponding author's e-mail: h.liao@oxon.org

Abstract. Technology has a significant impact on our way of life. Research on artificial intelligence and big data and the cross-field of human well-being plays an important role in guiding the application of science and technology to the people-oriented and good application direction. The reviews in this field focused on the achievements of artificial intelligence and big data in a specific field， such as subjective well-being/biobanks/Translational medicine (TM)/connected health (CH) and medical Internet of Things(mIoT). There is no bibliometric study on the use of artificial intelligence and big data to promote human happiness and health. To provide a systematic review of the literature on Positive Artificial Intelligence and Big data, this paper presents the main institutions, authors, funding agencies and disciplines, based on a scientometric analysis of 243 articles collected from Web of Science.

1. Introduction

Technology has a significant impact on our way of life. It is of great significance to study artificial intelligence and big data for well-being and human potential because it can point out the direction of the application of science and technology. Current technology practitioners are increasingly concerned about how to improve well-being and human potential through Artificial Intelligence and Big data. For example, people are using the Internet of things [1,2], social networking data [3], and artificial intelligence [4], to improve our Healthcare capabilities, and use Internet of things to build smart cities and Healthy Cities [5].

The reviews on this topic focused on the achievements of artificial intelligence and big data in a specific field， such as subjective well-being [6], biobanks [7], Translational medicine (TM) [1], connected health (CH) and medical Internet of Things(mIoT)[8]. There is no bibliometric study on the use of artificial intelligence and big data to promote human happiness and health. To provide a systematic review of the literature on Positive Artificial Intelligence and Big data, this paper presents the main institutions, authors, funding agencies and disciplines, based on a scientometric analysis of 243 articles collected from Web of Science.

2. Data and methods

Following the conventions of scientometric analysis, the paper first describes the query design and then explains the mapping protocols and strategies as follows.

2.1. Data query design

Gathering literatures from the platform of Web of Science, we deployed the queries as below:

IOP Conf. Series: Materials Science and Engineering **806** (2020) 012026 doi:10.1088/1757-899X/806/1/012026

- TS = ("AI" OR "Artificial Intelligence" OR "machine learning" OR "Big Data") AND TS=("well-being" OR "Positive Technology" OR "Positive Psychology" or "Positive Computing")

About 283 articles were retrieved in November 2019. After deleting the irrelevant articles, there are 243 left. To help us with exploratory analysis, we used VOSviewer to draw the resulting graph.

3. RESEARCH MAPPING RESULTS

This paper aims to provide a literature review on the specific topic of well-being and human potential.

3.1. Annual trends

As is shown in the figure1, the publications on the topic began to appear in 1999 and increased rapidly in 2016, and the growth trend has been consistent.

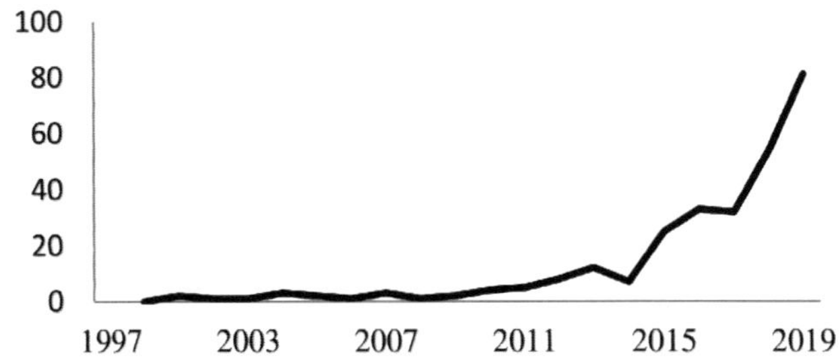

Figure 1. Annual Publication Trend from 1997 to 2019.

3.2. Top disciplines

The top four disciplines are Computer Science, Psychology, Engineering, and Health Care Sciences & Services. We can see that the main research direction of this topic is to use big data and artificial intelligence as tools to study human happiness, Ecosystem services and Health care.

Table 1. Top disciplines

Disciplines	%	Main topics
Computer Science	11%	Data mining, machine learning, artificial intelligence
Psychology	8%	well-being, happiness, mental health
Engineering	7%	Ecosystem services
Health Care Sciences & Services	5%	Health care, mHealth,

3.3. Top institutions

The University of California System published the most literature on this topic. Apart from the Consiglio Nazionale delle Ricerche, a national research institution in Italy, the University of California system, the University of Pennsylvania and the Massachusetts Institute of Technology are all universities and are all based in us.

Table 2. Top institutions

Institutions	No. of Articles
University of California System	20
University of Pennsylvania	15
Harvard University	11
Consiglio Nazionale delle Ricerche	7
Massachusetts Institute of Technology	7

3.4. Top funding agencies

The National Institutes of Health and the United States Department of Health Human Services funded the most published papers on the topic, followed by the National Natural Science Foundation of China.

Table 3. Top funding agencies

Funding Agencies	No. of Articles
National Institutes of Health	19
United States Department of Health Human Services	19
National Natural Science Foundation of China	10
National Science Foundation	7
National Institute of Mental Health	7

3.5. Top published countries
Research on this topic is dominated by countries with established technology, the main literature is from North America and Europe. The United States has the most research on this topic, followed by the UK.

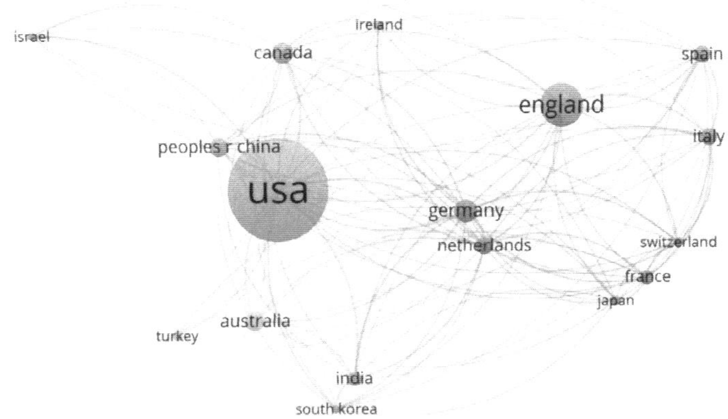

Figure 2. Top published countries.

3.6. Top cited authors
Breiman and Diener are the most cited expert on this topic. As can be seen from figure 3, cooperation between researchers on this topic is not high.

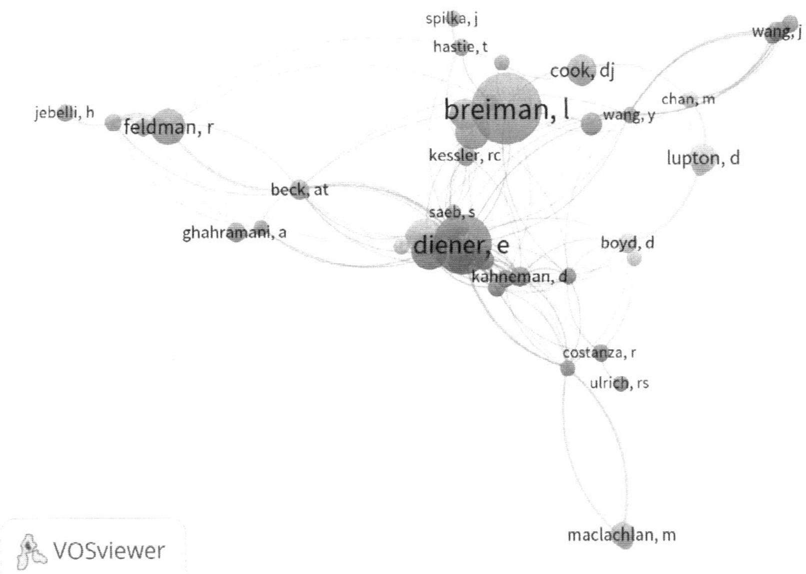

Figure 3. Top cited authors.

4. CONCLUSION

The purpose of this paper is to understand the research situation of Artificial Intelligence and Big data for well-being and human potential, and the results show that the heat of this research field has gradually increased in recent years, and North America and Europe are the main contribution areas. More and more researchers use big data and artificial intelligence as tools to study human well-being, ecosystem services and Health care. This article provides a reference point for researchers, funding bodies, policy-makers, and industry professionals to study the application of artificial intelligence and big data for well-being and human potential.

With the increase in the convenience of data acquisition such as wearable devices, the Internet of things, music data and network usage data, there are more and more things we can do to promote human physical and mental health for well-being and human potential. Not just using smart wear to get physical health data to understand our health [1-4], using music data to understand and promote human mental health [9], future research can combine a variety of these data to research in the field of psychological problem identification and intervention.

References

[1] Dimitrov Dimiter V. (2016) Medical Internet of Things and Big Data in Healthcare. Healthcare Informatics Research 22: 156-163.

[2] Monekosso, D.N., Remagnino, P. (2010) Behavior Analysis for Assisted Living. IEEE Transactions on Automation Science & Engineering 7: 879–886.

[3] Anon (2015) Psychological Language on Twitter Predicts County-Level Heart Disease Mortality. Psychological Science 26:159-169.

[4] Nguyen, T., Khosravi, A., Creighton, D., Nahavandi, S. (2015) Classification of healthcare data using genetic fuzzy logic system and wavelets. Expert Systems with Applications 42: 2184–2197.

[5] Kamel Boulos, M.N., Al-Shorbaji, N.M. (2014) On the Internet of Things, smart cities and the WHO Healthy Cities. International Journal of Health Geographics 13: 10.

[6] Luhmann, Maike (2017) Using Big Data to study subjective well-being. Current Opinion in Behavioral Sciences 18: 28–33.

[7] Kinkorová, J., Topolčan, O. (2018) Biobanks in Horizon 2020: sustainability and attractive perspectives. The EPMA Journal 9: 345–353.

[8] Jagadeeswari, V., Subramaniyaswamy, V., Logesh, R., Vijayakumar, V. (2018) A study on medical Internet of Things and Big Data in personalized healthcare system. Health Information Science and Systems 6:14

[9] Greenberg, D.M., Rentfrow, P.J. (2017) Music and big data: a new frontier. Current Opinion in Behavioral Sciences 18: 50–56.

IOP Conf. Series: Materials Science and Engineering **806** (2020) 012027 doi:10.1088/1757-899X/806/1/012027

Research on Virtualization Security in Cloud Computing

Lei Chen[1*] , Ming Xian, Jian Liu and Huimei Wang

[1]National University of Defense Technology College of Electronic Science and Technology Changsha, China

*corresponding author's email: 18299927375@163.com

Abstract. Cloud computing services, with its advantages of flexible configuration, efficient utilization of resources and easy management, have achieved rapid development and wide application, in recent years. However, the security issues in cloud computing are also becoming increasingly prominent, at same time, cloud computing security has become an important factor hindering the development of cloud computing. The virtualization technology is one of the core technologies of cloud computing, meanwhile, virtualization security is the core issue of cloud computing security as well. This article introduces the mainstream virtualization technology, reviews the security threats of virtualization in cloud computing, and puts forward the corresponding countermeasures.

1. Introduction

With the rise of cloud computing[1], its technology and concept are becoming mature and popular. In addition, more and more service providers are realizing the great advantages of cloud computing in information processing and use. As a new business service model, it has been recognized and developed rapidly. Virtualization technology is the key technology of cloud computing, but with the wide application of virtualization technology, security issues are also gradually emerging. Its harm degree is far higher than the traditional stand-alone environment, seriously restricted the development of cloud computing industry[2]. The security protection of virtualization in cloud computing is an important measure to promote the safe and orderly development of cloud computing.

2. Virtualization technology

Virtualization[3] is the process by which a computer is virtualized into multiple logical computers, and each logical computer can run a different operating system. What's more, the applications can run in separate Spaces without affecting each other, thus significantly improve the work efficiency of the computer. VMM is the core software for virtualization and a kind of middle layer software that runs between the physical server and the operating system. It has access to all physical devices on the server, including CPU, memory, disk, and network card. The VMM not only coordinates access to these hardware resources, but also provides protection between the various virtual machines. When the server starts up and executes the VMM, it loads all the virtual machine client operating systems and allocates an appropriate amount of memory, CPU, network, and disk to each virtual machine. The virtualization model is shown in Figure 1.

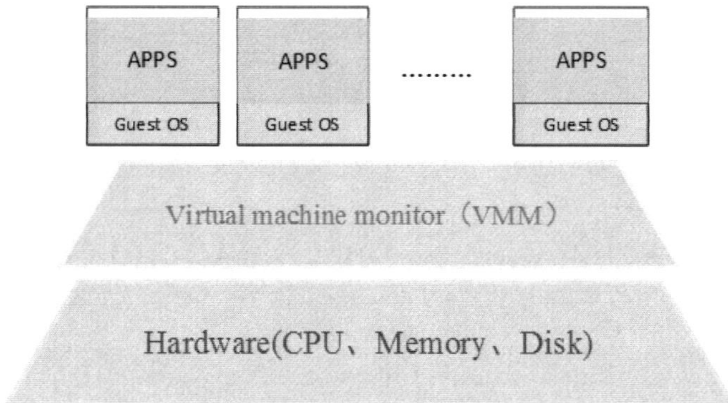

Figure 1. Virtualization technology

Depending on the degree and level of virtualization, there are software virtualization and hardware virtualization, full virtualization and paravirtualization. Virtualization technology refers to the technology of realizing virtualization at the software level. On the whole, it can be divided into two camps: open source virtualization and commercial virtualization. Typical examples are Xen, KVM, VMware, Hyper-V and Docker container. Their respective characteristics are shown in Table 1.

Table 1. comparison of mainstream virtualization technology

Types	Characteristics explanation
Xen	XEN is one of the fastest growing, most stable, and least resource-intensive open source virtualization technologies based on X86 architecture
KVM	KVM is open source full virtualization that USES QEMU to provide device virtualization, which belongs to the Hypervisor model, supports a variety of hardware platforms.
VMware	VMWare is a global leader in virtualization and cloud computing infrastructure, providing customer-proven solutions that improve IT efficiency by reducing complexity and delivering services more flexibly and nimbly.
Hyper-v	Hyper-v is a hypervisor virtualization technology proposed by Microsoft,can achieve desktop virtualization
Docker container	Docker containers are more portable than traditional virtual machines. Each container has its own file system, processes between containers do not affect each other, and it can distinguish computing word resources

3. Security threats to virtualization in cloud computing

3.1. Virtual machine migration

Virtual machine hot migration is done without shutting down the virtual machine, migrate the virtual machine to another physical machine. There are many reasons for performing virtual machine migration, such as maintaining fault tolerance or load balancing. During the migration, the contents of the virtual machine are exposed to the network, which may result in the loss of data privacy and integrity.

3.2. virtual machine escape

With virtual machines, users can share the resources of the host and achieve mutual isolation.Ideally, a program running in a virtual machine should not affect other virtual machines. However, due to technical limitations and some loopholes in virtualization software, in some cases, programs running in a virtual machine can bypass isolation restrictions and run directly on the host machine, which is called virtual machine escape. When the virtual machine escape attack is successful, it poses a great threat to both the Hypervisor and the host.

3.3. Rookit attack

Rootkit is a special type of malware that hides itself and specified file processes and network links on the installation target, what is more common is that rootkits are commonly used in combination with other malicious programs such as Trojan and backdoors. Rootkit hides information by loading special drivers and modifying the system kernel.

3.4. Denial of service attack

Virtual machines on the same physical machine share resources. If an attacker USES one virtual machine to obtain all the resources of the host machine, the resources of other virtual machines will also be disturbed, or even crash because there are no resources available, which will result in a denial of service attack in the virtual environment.

3.5. Virtual machine monitor problem

Virtual machine monitor is the key module of virtualization, virtual machine monitors are responsible for both managing and isolating virtual machines and generating or managing virtual resources. If the virtual machine monitor is compromised, all the virtual machines it manages will be controlled by the attacker, and the virtual machine metadata stored by the virtual machine monitor will be exposed to the attacker.

3.6 Decoupling attacks on virtualization platforms

Under the traditional working environment, a single vulnerability can only attack the physical machine with system vulnerability, and the attack scope is very narrow and limited. Multiple rent users based on cloud computing virtualization platform are able to share information of the entire system, and then attack the entire virtual vulnerability platform through a single virtual machine.

4. Countermeasures for virtualization security problems in cloud computing

4.1. Security regime of Hypervisor

At present, the mainstream virtualization software[4], such as Xen, VMware and KVM, all have security holes, It can destroy the Hypervisor's integrity and defenses, the security of the Hypervisor is critical to the security of the virtualization platform.Currently, there are three main aspects of Hypervisor security hardening, the first is to build a lightweight Hypervisor, the second is to protect the integrity of the Hypervisor based on trusted computing technology, the third is to improve the defense capability of Hypervisor by configuring virtual firewall and reasonably allocating host resources.

4.2. Virtual machine security isolation

Through the virtual machine security isolation mechanism, the virtual machines of the users with interests run independently and do not interfere with each other. It is mainly realized from two aspects: first, SMM security memory management model is applied to encrypt the memory through SMM controller, second, SIOM security I/O management model is used to weaken Dom 0 for Xen.

4.3. Virtual machine access control

Security models such as sHype, Chinese wall and BLP are used to reduce the risk of hidden information flow in the virtual machine system, and to control the use of resources and event behavior inside the virtual machine to improve the security of the system.

4.4. Virtual machine security monitoring

To ensure the healthy operation of each virtual machine, virtual machine security monitoring is indispensable[5]. There are two mainstream virtual machine security monitoring architectures. One is the monitoring architecture based on virtual machine introspection technology, which puts the monitoring module in the Hypervisor and detects other virtual machines through virtual machine introspection technology, The other is the security active monitoring architecture based on virtualization, which intercepts the change of system state by inserting some hook functions into the monitored virtual machine and jumps to a separate secure virtual machine for monitoring and management.

4.5. Virtual trusted computing technology

Through the trusted measurement trusted storage and trusted report mechanism provided by trusted computing, the computing environment of terminals is purified, the trusted connection between terminals is built, and the virtual space of honest mutual trust is built. The dynamic integrity of the virtual machine is guaranteed by the trusted measurement mechanism, the trust communication between different virtual environments is realized by the trusted report mechanism, and the data migration storage and access control is guaranteed by the trusted storage mechanism.

Strengthen the virtualization security protection in cloud computing from the above five aspects can effectively promote the virtualization security of cloud computing, so as to improve the security and reliability of cloud computing services. The overall virtualization security protection model is shown in Figure 2.

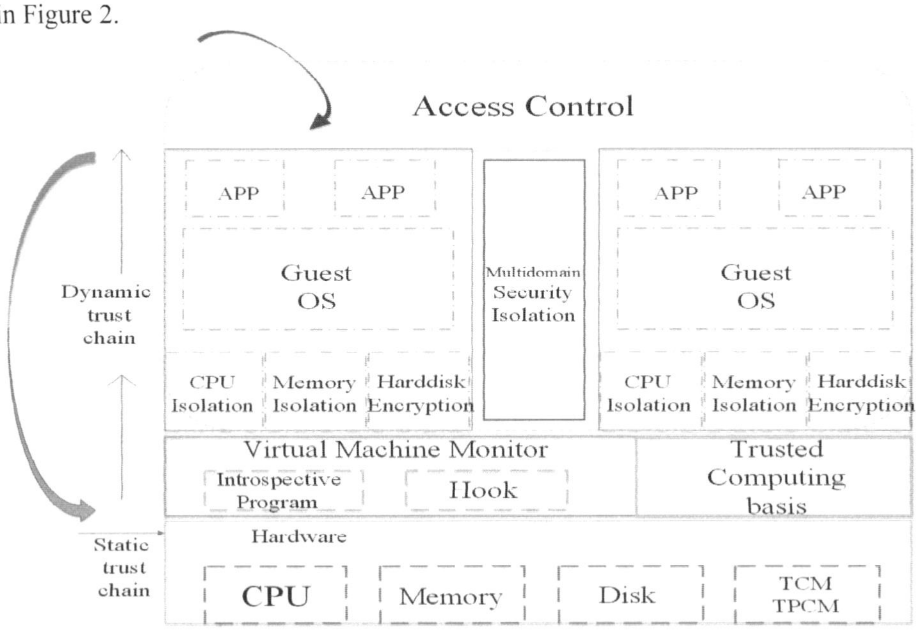

Figure 2. Virtualization security protection model

5. Conclusion

Virtualization plays a major role in reducing administrative costs and increasing the availability and flexibility of hardware resources. As the core of the cloud computing technology, virtualization is facing security threats hindered the rapid development of cloud computing and widely used. It is of great significance to analyze the main security threats faced by virtualization technology and strengthen the security protection of virtualization.

References:

[1] Sharma, S., Soni, S., Sengar, S., (2012) Security in cloud computing National Conf. on Security Issues in Network Technologies, 1-6.

[2] Sen, J., (2013) Security and privacy issues in cloud computing Retrieved from arxiv.org/pdf/1303.4814.

[3] Khurana, S., Verma, A.G., (2013) Comparisons of cloud computing service model: SaaS, PaaS, IaaS International Journal of Electronics & Communication Technology (IJECT), 4 3 29-32.
[4] Ashraf, I., (2014) An overview of service model of cloud computing Int. J. of Multidisciplinary and Current Research 2, 779-783.
[5] Zhang, T., Lee, R.B., (2016) Monitoring and Attestation of Virtual Machine Security Health in Cloud Computing [J]. IEEE Micro, 36(5): 28-37.

IOP Conf. Series: Materials Science and Engineering **806** (2020) 012028 doi:10.1088/1757-899X/806/1/012028

The Science and Technology Innovation Report on the Ocean Big Data Application along the ice silk road

Bo Lin [1*]

[1] The National Marine Environmental Forecasting Center (NMEFC) of Ministry of Natural Resources of the People's Republic of China, Beijing 100081

[*] Corresponding author's e-mail: linbo@nmefc.cn

Abstract: Over the past twenty years, the National Marine Environmental Forecasting Center (NMEFC) has carried out safety and security tasks for a total of 9 batches of Arctic science and technology navigations and 22 voyages of merchant ships along the North East Arctic waterway. The paper summarizes their successful experience and technology innovation achievements of applying on big data, focusing on the infrastructure innovation construction of the ocean forecast big data: creating an autonomous Arctic observation system, exploring the Arctic international cooperation model, participating in the global regional observation plan; scientific and technological business innovation: research method innovation, security technology innovation, super-computing innovation; as well as science innovation achievement: science research achievements, application technology achievements, computer technology achievements. Finally, a brief prospect for the future of the Arctic is given from the perspective of prediction.

1. Introduction

The Silk Road on Ice (hereinafter referred to as the Ice Silk Road) extended the traditional Maritime Silk Road of the eastbound ocean lane [1], and it starts from Vladivostok, Russian, goes across the Bering Strait northward into the Arctic Ocean, and flows westward through the North-East Arctic channel [2] and the Barents sea to the coastal ports of northwestward Europe . The accelerated melting of Arctic ice and snow caused by global warming, the modernization of marine ships brought about by the rapid development of science and technology, the rich resources to be developed under the Arctic ice sheet, the international obligations of energy conservation and emission reduction undertaken by our country consciously, the scientific cognition and the responsibility of developing the Arctic people jointly bear, etc., all foreshadow that the ice road will become a sea highway. The National Marine Environment Forecasting Center (hereinafter referred to as the Forecast Center) has realized the innovation of big data infrastructure and innovation of scientific and technological methods along the ice silk road in the long-term practice of business scientific research and foreign cooperation by applying big data and other technologies.

2. Ice Silk Road Ocean Forecast Big Data Infrastructure Innovation

2.1. *Creating a fully autonomous Arctic Ocean observation system*

Before 2012, satellite remote sensing monitoring data and Arctic science test data were the main sources of data for Arctic Ocean forecasting at that time. Since 2012, the forecasting center, along with China's fifth and sixth Arctic science and technology team undertake safety and security tasks of science and technology, use the ocean hydrologic-sea ice-meteorological-atmospheric boundary layer stereoscopic observation model consisting of air-sighting and sounding instrument, ocean buoys and submarine, as well as automatic weather stations to collect and accumulate the ocean observation data, and make application demonstration products of marine forecast on site through participation in navigation observation, ice station operations, aerial exploration and other work processes [3].The two business innovations of on-site observation and forecasting of Arctic science examination are realized. During the seventh and eighth Arctic science examinations in 2016-2017, the forecasting center team with supporting service personnel crossed the Central Arctic Passage for the first time (Figure 1.) and obtained a large amount of valuable key information on ocean hydrology and space atmosphere in the main functional area of the Arctic marine environment on the basis of completing the Arctic Science and Technology Survey on Ice Survey.

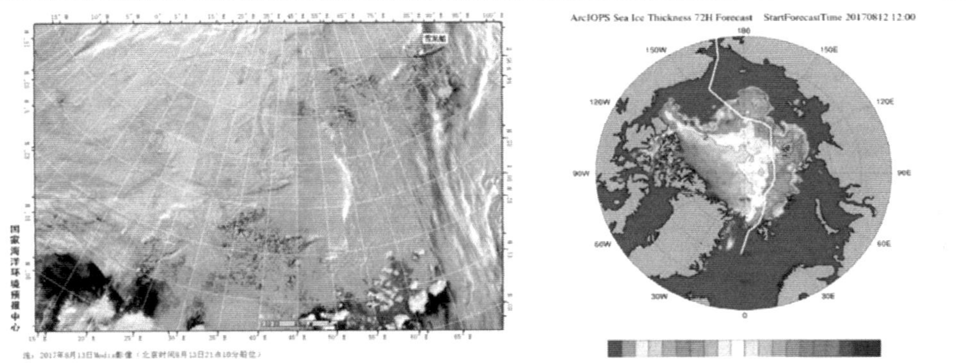

Figure 1. Ice-based drift weather station of the prediction centre for Arctic scientific research.

2.2. *Exploring the inter-State cooperation mode in Arctic Ocean observations*

Through cooperation with the Finnish National Meteorological Service on science research in polar sea ice and meteorological field observation, data analysis and data assimilation, numerical simulation, the "China/Finland Polar Atmosphere-Snow Sustainable Joint Monitoring and Numerical Simulation Research" was approved for implementation by the National Oceanic Administration in 2013 [4], and a series of important results were achieved in the cooperation. In December 2014, a memorandum of understanding on polar forecasting cooperation was signed with the Alfred Wegner Polar Research Institute of Germany, agreeing to strengthen cooperation in the field of marine and polar research, to jointly optimize the use of polar observation networks, develop data assimilation techniques and improve polar forecasting models [5]. In October 2018, the national key R & D plan "The intergovernmental key project of international innovation cooperation-Research on multi-factor space observation and information service of Arctic environment" [6] participated by the prediction center and cooperated with the EU 2020 plan "integrated Arctic observation system (INTAROS) " entered the substantive research stage.

2.3. *Participation in the global ocean observation regional plan*

As a member of the Scientific Steering Group of the Polar Prediction Program (PPP) organized by the World Meteorological Organization (WMO), the Centre collaborates with other member countries in polar forecasting studies based on observational data (Figure 2.) and numerical models [7]. As a member of Northeast Asia Regional Ocean Observing System (NEAR-GOOS) of the Global Ocean Observing System (GOOS), we will work with Russia, Japan, South Korea and other member countries to promote the free sharing of ocean observation data and other international cooperation in the field of ocean [8]. We have participated in the International Arctic Rafting Ice Station Program, relying on the German "Polar Star" icebreaker, with sea ice, ocean and atmosphere within 50 kilometers of the icebreaker as the main observation object, and rafted along the direction from the East Siberian Sea to the North Pole under the action of "crossing the pole flow". Through a year of artificial observation, the understanding of the central Arctic ocean atmosphere-sea ice-ocean-ecosystem coupling process is deepen during the collection of a large amount of ocean data at the same time, to investigate the physical characteristics of arctic sea fog and the mechanism of extinction, thus exploring the Arctic sea fog numerical simulation [9].

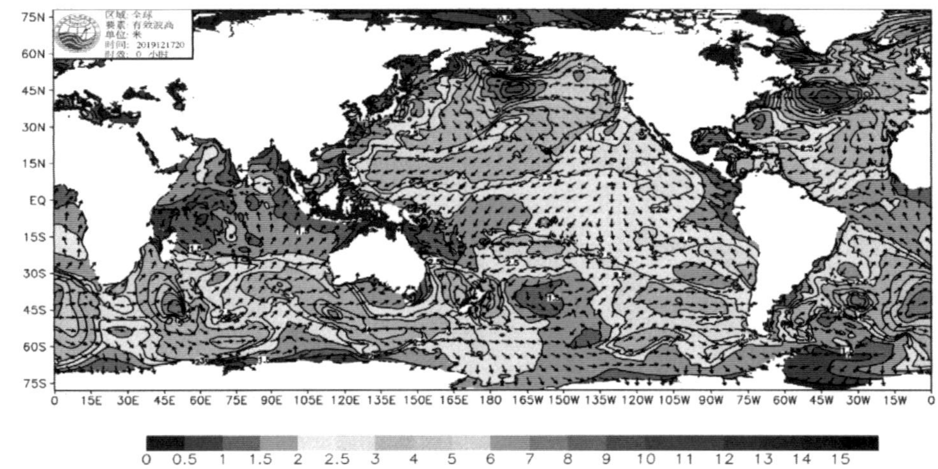

Figure 2. Real-time observation data of the north-polar waves of the global ocean.

3. Ocean Forecast Business Innovation on Big Data Technology along the Ice Silk Road

3.1. Innovation in the research method of ocean forecast along the Ice Silk Road

In 2013, the Center was successfully elected as a member of the Arctic University Organization (UArctic) [10], which opens the path of Arctic Ocean forecasting innovation in the combination of "production, learning and research". The UArctic includes eight Arctic countries (Canada, Denmark, Finland, Iceland, Norway, Sweden, Russia, and the United States) and China, the United Kingdom, Belgium, France and Japan, the strong teaching and research team of which promotes the simultaneous development of research and education in Arctic Ocean prediction. In 2015, senior experts from the research and development of polar sea ice numerical simulation and forecasting systems of the National Environmental Prediction Center of the United States were invited to lecture in Beijing to offer a series of lectures on polar climate models, sea ice models, data assimilation, collection forecasting, predictive testing, and to cooperate in the use of the Climate Prediction System (CFSv2) to carry out the mid- and long-term numerical prediction tests of Arctic sea ice, and the Arctic scientific research and merchant ship's first voyage was successfully applied [11]. Six Arctic prediction seminars were jointly organized with foreign scientific research institutions to discuss and develop the Arctic sea ice ocean integrated prediction system and the current-atmosphere-wave-sea ice coupling system.

3.2. Technology Innovation of Safety and Security Services along the Icy Silk Road

In December 2018, the China Ocean Forecast net developed by the Forecasting Center will be operational [12]. This is China's only national marine forecasting professional website, marking China's future-oriented marine forecasting services enters the "Internet +" and "mobile Internet +" era, this website is China's professional information platform of marine forecasting and early warning, which is responsible for issuing storm surge, waves, sea ice (Figure 3.), tsunami, red tide, green tide and sea temperature, sea currents and other marine disaster warning (forecasting) information to the external in authority, and gathers information, public science, professional marine forecasting and protection services. Experts' replies and question-and-answer interaction are achieved through mobile applications and WAP services, and the service surface seamlessly covers the global oceans and poles. Immediately thereafter, the Marine Environment Prediction and Guarantee System of the Maritime Silk Road was put into operation and the Marine Environment Forecast products along the Maritime Silk Road were released in Chinese and English (more languages will be added in the future) through the China Marine Forecasting Network's "Marine Silk Road Special Channel" [13].

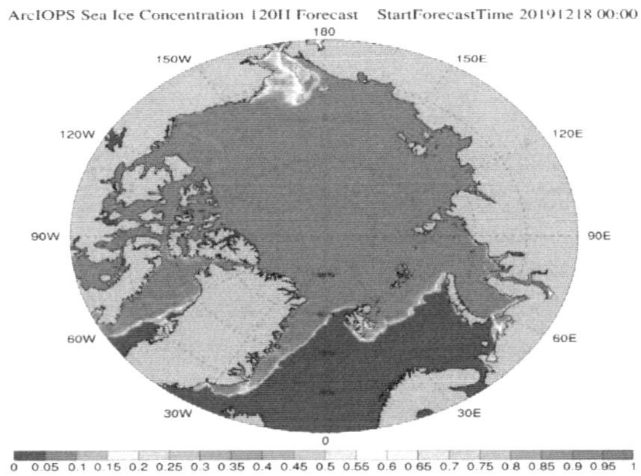

Figure 3. Big data for 120-hour numerical forecast of North-polar sea ice.

3.3. High-performance Computing Clusters and Supercomputing Innovations

Ocean observation data is traditional data; ocean forecasting products are big data. Global ocean forecasting, especially Arctic ocean forecasting, is depending more and more on high-performance computing, it can be said that without high-performance computing there is no modern ocean forecast, high-performance computing technology and research has changed from the past ocean forecasting auxiliary role to the same role as the ocean forecast science and technology. Driven by innovation, the forecasting center has formed a computing resource mine consisting of Sunway 3000 + high-performance computing system, IBM blade cluster, ocean real-time data center, ocean supercomputing center, ocean forecast numerical model software set, global ocean monitoring and forecasting data lake, global marine environment forecasting business network, ocean cloud computing and big data visualization intelligent platform, etc., which is sufficient to ensure the sustainable development of ocean forecasting technology innovation along the Ice Silk Road. In August 2017, the Forecasting Center joined the National Supercomputing Innovation Alliance and served as a member of the Technology and Standards Group of the Alliance, since then it has joined the ranks of China's top-level supercomputing innovation teams.

4. Innovation results on ocean forecast big data application along the Ice Silk Road

4.1. It has won the first prize of China Marine Science and Technology Award in the Global Business Oceanographic Forecasting System

The global operational oceanographic forecasting system, which is developed mainly by the forecasting center, is the first independent marine environment numerical forecasting system nested by the global-ocean and polar-offshore (Figure 4.). It achieves the full coverage of the integrated ocean forecasting business of space resolution from 100 km to kilometer, and the advent and application of the forecasting system has completely reversed the long-term use of foreign ocean forecasting products in China, thus becoming one of the few countries after the United States and Europe to carry out the global business-oriented ocean forecast. In December 2017, the project passed application evaluation, and it's believed that the scientific and technological achievements have reached the international advanced level as a whole, and is in the international leading position in the physical parameterization scheme, the assimilation of autonomous power satellite data, and the application of environmental forecasting technology [14]. It is evaluated as one of China's top ten marine science and technology progress in 2013, the first prize of the 2015 China Marine Science and Technology Award, and in June 2016, it was selected to join the exhibition of the national "12th Five-Year Plan" scientific and technological innovation achievement exhibition with the theme of "Innovation-driven development and science and technology leading the future".

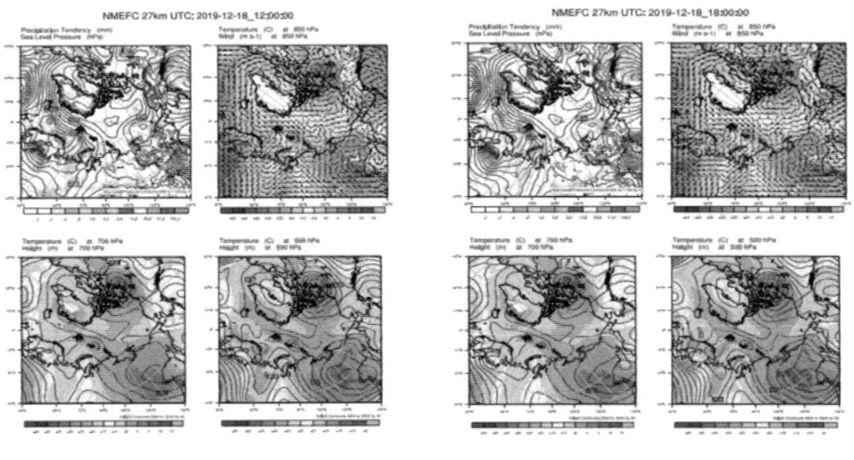

Figure 4. Big data for near-sea 500hPa forecast of the North Pole.

4.2. The key technologies of China's merchant ship polar routes won the China Marine Science and Technology Award

After entering the 21st century, the Forecasting Center took the initiative to face climate change, intervened in the Arctic seaway trade navigation marine forecast ingress and guarantee service research in advance, successfully completed 9 batches of China's Arctic science and technology ocean forecast safeguard mission, and has provided forecast guarantee for China Ocean Shipping Group and other 15 merchant ships and a total of 22 voyages of the Arctic Northeast Seaway ever and again (Figure 5.). In 2016 alone, the "Yongsheng Wheel +" project of six Arctic voyage ships saved a total of 32,137 sea miles, 108 days of sailing time, 4077 tons of power fuel, which greatly reduced the operating costs of enterprises, carbon emissions and environmental pollution [15]. The normalization of commercial shipping in the Northeast Arctic Seaway has made the traditional "Sea Silk Road" route extend naturally northward and directly to Europe via the Arctic Ocean. The "Research and Application of Polar Marine Meteorological Support Technology" project the Forecasting Center participated in, won the first prize of the 2008 China Marine Science and Technology Award and the project "Research and Application of Key Technologies of The Polar Route of Chinese Merchant Ships", as well as the Special Prize of the 2018 China Maritime Science and Technology Award.

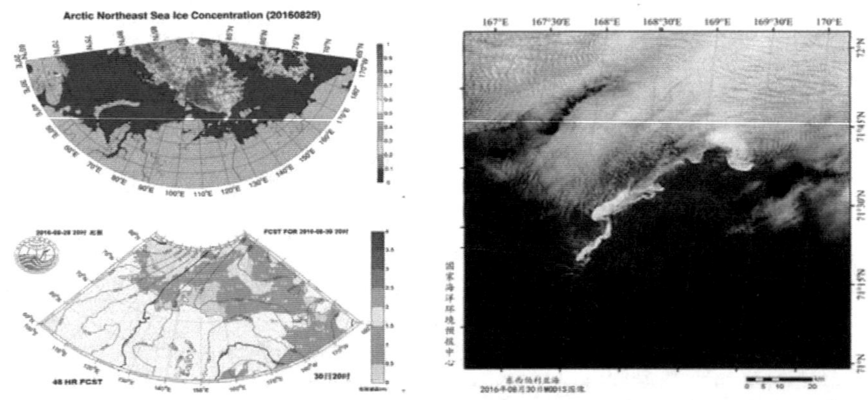

Figure 5. Big data for sea wave and sea ice forecast in the eastern section along the Northeast Passage of the North Pole.

4.3. Numerical forecast over-calculation results has won the National PAC FinalBronze Award for four consecutive years

The Forecasting Center is a member of the International Operational Ocean Forecasting Organization (GOV) in the field of numerical models, and has achieved new progress and achievements in data assimilation, business process optimization and other fields of global operational ocean forecasting system (CGOFS), especially in the field of polar science and ice silk road commercial shipping, such as the research and application of "sea current-atmosphere-wave-sea ice coupling forecasting system", marine monitoring forecast cloud, big data application and global ocean forecast visualization human-machine dialogue. The research results of “Research on the application of NEMO in the global marine environment forecasting model”, “the optimization of the computational performance of the NEMO-based global marine environment forecast ingress in the supercomputing cluster”, “the coupling line of the global atmospheric-wave numerical calculation model” (Figure 6.), “the parallel optimization of the typhoon-coupled numerical calculation model” and 4 related theme papers have participated in the 2016, 2017, 2018 and 2019 National Parallel Application Challenge (PAC) respectively and won the Bronze Award of the Challenge Final for four consecutive years, and all the theme papers were awarded the outstanding thesis award in the competition at the same time.

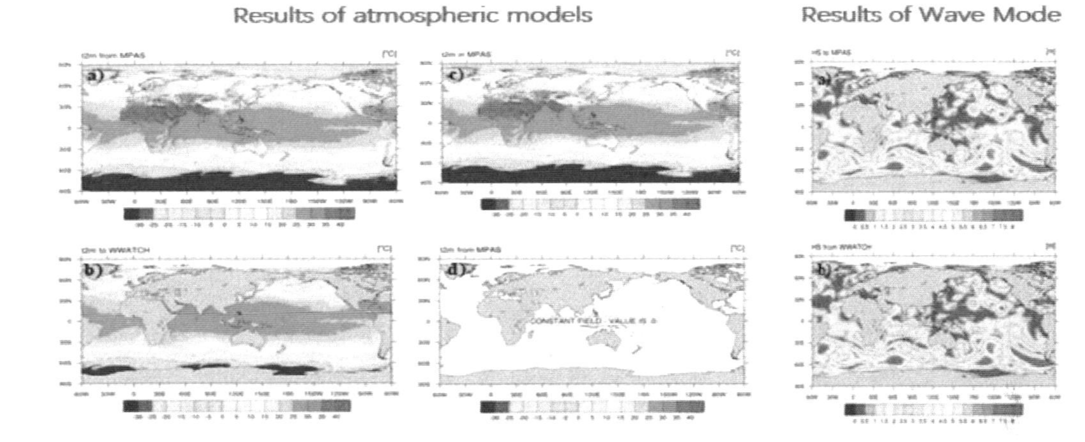

Correctness test of 2m temperature variable in atmospheric mode,Where a) and b) are the results of the atmospheric mode output variables and the wave mode received variables in the restart file written by the coupler, respectively; c) the output variable results for the atmospheric mode in the coupled mode; d) is the difference between a) and c).

Correctness test of the effective wave height (HS) variable in the wave mode,a) and b) are the results of receiving variables in the atmospheric mode and the results of output in the wave mode in the restart file written by the coupler.

Figure 6. Effect diagram of Global Atmosphere-Sea Wave optimization coupling model.

5. Conclusions

According to the latest estimates by scientists, the global warming rate of the upper ocean within 2000 meters is 0.36-0.39wm-2 from 1971 to 2010, which accelerated significantly in the 1990s and increased to 0.55-0.68 Wm-2 after 1991, and the phenomenon continues [16]. At the same time, studies have found that between 1901 and 2012, the rate of warming in the Arctic was twice the global average, and it is noted that Arctic warming occurred mainly in the winter of the 1990s [17]. As with the growth of the Ice Silk Road (North-East Arctic Channel), the Northwest and Central Passage of the Arctic are also ushering in the dawn of commercial shipping. We will strive to "improve the level and capacity of scientific research in the Arctic, deepen scientific cognition and understanding of the Arctic, explore objective laws of arctic change and development, and create favorable conditions for enhancing human capacity to protect, utilize and govern the Arctic [18]."

Acknowledgments

Thanks for the following funded projects: the national key research and development plan- national key project of high-efficiency computing “Development of high-efficiency numerical simulation software for the marine environment [2016YFB0201100]; the national key plan of research and development "key project of innovation and cooperation for the intergovernmental international science" [2017YFE0111700].Thanks for the strong supports from departments such as the polar environment research and forecast office of the national marine environment forecasting centre as well as numerous experts on the marine forecast data of the North Pole.

References

[1] Yao Q.H. Hu X.P. (2018) "The Silk Road at the 21st Century" and the New Mechanism for Regional Cooperation. Shanghai Academy of Social Sciences Press, Shanghai.
[2] Li Z.F. (2016) A Strategic Study of the Arctic Route of the Silk Road, Dalian Maritime University Press, Dalian.
[3] National Marine Environmental Forecasting Center, 2014. News_2014.07.15. http://www.nmefc.cn/
[4] National Marine Environmental Forecasting Center, 2014. News_2014.10.23. http://www.nmefc.cn/
[5] National Marine Environmental Forecasting Center, 2014. News_2014.12.08. http://www.nmefc.cn/
[6] National Marine Environmental Forecasting Center, 2018. News_2018.10.26. http://www.nmefc.cn/
[7] National Marine Environmental Forecasting Center, 2016. News_2016.05.30. http://www.nmefc.cn/
[8] National Marine Environmental Forecasting Center, 2017. News_2017.12.05. http://www.nmefc.cn/
[9] National Marine Environmental Forecasting Center, 2018. News_2018.08.15. http://www.nmefc.cn/
[10] National Marine Environmental Forecasting Center, 2013. News_2013.06.14. http://www.nmefc.cn/
[11] National Marine Environmental Forecasting Center, 2015. News_2015.07.10. http://www.nmefc.cn/
[12] National Marine Environmental Forecasting Center, 2018. News_2018.12.14. http://www.nmefc.cn/
[13] National Marine Environmental Forecasting Center, 2018. News_2018.12.26. http://www.nmefc.cn/
[14] National Marine Environmental Forecasting Center, 2017. News_2017.12.21. http://www.nmefc.cn/
[15] National Marine Environmental Forecasting Center, 2016. News_2016.10.17. http://www.nmefc.cn/
[16] Cheng Lijing(the official website of the Institute of Atmospheric Physics of the Chinese Academy of Sciences), 2019. What do you know about global ocean warming? http://www.iap.cas.cn/xwzx/tpxw/201901/t20190111_5227062.html
[17] Zhang Jingjing(the official website of The Chinese Science Daily), 2018. Is the polar bear extinct? http://news.sciencenet.cn/htmlnews/2018/8/416667.shtm
[18] Information Office of the State Council(Official website of the Central People's Government of the People's Republic of China), 2018. China's Arctic Policy (January 2018) White Paper. http://www.gov.cn/xinwen/2018-01/26/content_5260891.htm

IOP Conf. Series: Materials Science and Engineering **806** (2020) 012029 doi:10.1088/1757-899X/806/1/012029

Comparison of Machine Learning Models for Sentimental Analysis of Hotel Reviews

Xiaoyu Li [1,a], **Cai Liu** [2,b]

[1] Department of Automotive Engineering Tongji University Shanghai, 200092, China

[2] Lvmama Co., Ltd Shanghai, 200033, China

[a] 1352116@tongji.edu.cn

[b] liucai@lvmama.com

Abstract — It is widely acknowledged that the enormous boost of Internet technology has greatly changed tourism industry, especially for hotel booking. A plenty of research have addressed that most tourists do decision-making based on online hotel reviews. With the goal of predicting satisfaction of hotel customers precisely and efficiently, in this study ten thousand hotel reviews across US with one to five stars are collected from TripAdvisor, and the performance of a variety of machine learning models, namely Logistic Regression, Naive Bayesian, Decision Tree, Random Forest, Support Vector Machine (SVM) and Neural Network, are compared through two 10-folder-cross-validation experiments. The result indicates that models directly trained on review titles which have only 25.7 byte lengths on average could achieve promising classification accuracies ranging from 84% to 87%, and with review contents and irrelevant word filtering, the SVM model is able to reach maximal predictive accuracy nearly 92%. Furthermore, this paper demonstrates that classical machine learning models, such as SVM and Naive Bayesian, are computationally efficient and perform well in terms of accuracy, recall and precision, while neural network requires careful structure design as well as parameter tuning.

1. INTRODUCTION

With rapid development of Internet technology, tourism industry, specifically hotel booking, has been transformed greatly [1]. According to statistics, approximately 150 million tourist bookings have been made online at year 2016 [2], and this tendency has grown more and more heated so far. Moreover, researchers have found out that review rating from social media can explain hotel performance metric better than traditional customer satisfaction evaluations [3]. Furthermore, through survey method, a study [4] has revealed that 70 percent of potential customers worldwide trust hotel reviews from website for decision-making, especially for the cases where tourists have never been to the travelling spots [5]. Therefore, extracting textual information and analysing sentiments from hotel reviews is beneficial for the whole tourism industry chain.

There have been excellent research works on textual analytics of hotel guest experience according to online reviews [6][7]. However, the existing tourist experience evaluation system have the following problems: (1) customer reviews can be informal and fuzzy, and traditional factor analysis and linear models are not able to capture the majority of high-dimensional data patterns. (2) since deep neural network and the domain of Natural Language Processing gain unprecedented popularity during this decade, neural network models with millions, even billions of parameters are able to achieve brilliant performance [8]. However, the training process could take up to several months, and the models fail to be guaranteed that they do not overfitting.

In this study, ten thousand hotel reviews were collected from TripAdvisor, and 1-3 star feedbacks were labelled as 'negative' while 4-5 star reviews were regarded as 'positive'. To investigate the performance of classical and up-to-date machine learning models, a variety of machine learning models, namely Logistic Regression, Naive Bayesian, Decision Tree, Random Forest, Support Vector Machine (SVM) and Neural Network, are compared through two 10-folder-cross-validation experiments which can reduce the risk of overfitting. This study aims to compare the predictive power and training efficiency of widely-used machine learning models, and discuss how text processing impact the model performances.

2. Data Preparation

2.1. Data Collection

The data was collected from the database: data.world [9] and this database acquired as well as formatted 10,000 hotel comments from TripAdvisor into csv files. As **Fig.1** illustrates, approximately 7800 reviews are 4 to 5 stars and were labelled as 'positive (1)' while around 2200 reviews are 1-3 stars and were treated as 'negative (0)'.

2.2. Experimental Preparation

To overcome the problem of overfitting, the data is randomly shuffled and divided into ten trunks. Each time 9 chunks (9,000 data) was utilized for training and the labels of the rest chunk (1000 data) was applied for testing. This design

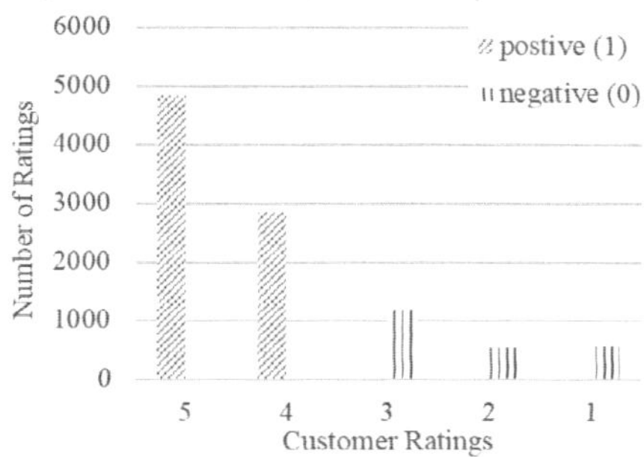

Figure 1. Customer rating data

is called 10-folder-cross-validation and it helps to reduce the risk of excessively fitting and local optimal solution.

As Table 1 shows, because in the first experiment only titles were used, each single data vector merely consists of a few words and the size of bytes (1 English letter or numerical digit requires 1 byte for storage) has mean 25.7 and standard deviation 14.6. For the second experiment, the contents of review were firstly filtered by punctuations (comma, period, semi-colon, colon, hyphen, etc) and neutral emotional words with high frequency ('hotel','inn','motel', etc). After adding filtered review contents for training, the average bytes of each sample increase from 25.7 to 442.4 and the sample size varies more intensively with the standard deviation being 394.0.

TABLE I. EXPERIMENTS PREPARATION

Designs	Details		
	Training/Testing dataset size	***Mean of bytes***	***Standard deviation of bytes***
Experiment 1	9000/1000	25.7	14.6
Experiment 2	9000/1000	442.4	394.0

3. EXPERIMENTS

3.1. Analysis on Experiment 1

As shown in **Table 2**, the six machine learning models are imported from Python module 'sklearn' [10] and the parameters for models are listed in detail for readers' experimental reproduction. For instance, for Neural Network a 3-layer structure with 5,4,2 neurons on each layer was applied, relu function was chosen as activation function and the learning rate was 10^{-5}.

The predictive performance of six models, namely Decision Tree, Naive Bayesian, Logistic Regression, SVM, Neural Network and Random Forest, are depicted in **Fig.2** as boxplot format which contains five horizontal lines to indicate minimum, first quartile, median, third quartile, and maximum of data. First of all, neural network performs worst among all five models, and this phenomena attributes to small training data which is just 25.7 bytes on average. Furthermore, due to the fact that Random Forest consists of average results of 100 decision trees given specific randomness, Random Forest outperforms Decision Tree in terms of accuracy. Thirdly, as mentioned before, the training data is tiny and monotonous, linear model such as Logistic Regression performs above non-linear models such as neural network. Apart from that, SVM has the advantage of dealing with unstructured data and scales relatively well to high dimensional data, therefore for textual analysis SVM has the best predictive performance. Last but as significant, each review titles merely consists of a few words and feeding them into Machine Learning models only takes time in the magnitude of seconds, but five of six models are able to achieve prediction accuracy ranging from 84% to 87%. This result indicates that without dataset in the scale of gigabytes or deep neural network trained for months with billions of parameters, classical models with relatively tiny structures have the power to achieve satisfying predictive performance.

TABLE II. 'SKLEARN' MODULE PARAMETERS

Models	Parameters
Decision Tree	criterion='gini', min_samples_leaf=1, min_samples_split=2
Naive Bayesian	alpha=1.0, fit_prior=True
Logistic Regression	max_iter=100, multi_class='warn', penalty='l2', solver='lbfgs'
SVM	degree=3, gamma='scale', kernel='rbf'
Neural Network	activation='relu', alpha=10-5, hidden_layer_sizes=(5, 4, 2)
Random Forest	criterion='gini', n_estimators=100

IOP Conf. Series: Materials Science and Engineering **806** (2020) 012029 doi:10.1088/1757-899X/806/1/012029

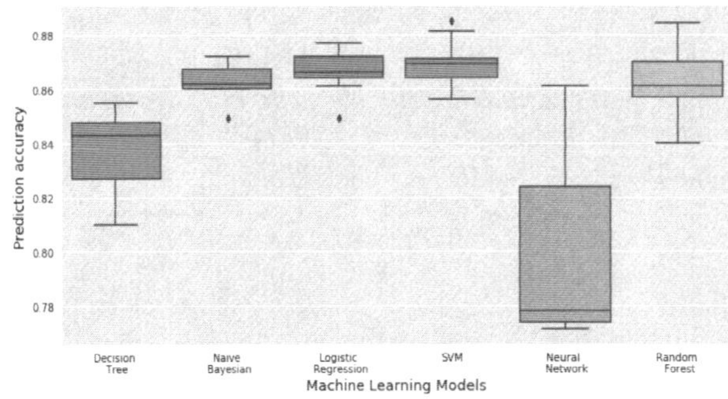

Figure 2. Accuracy performance of ML models on title texts

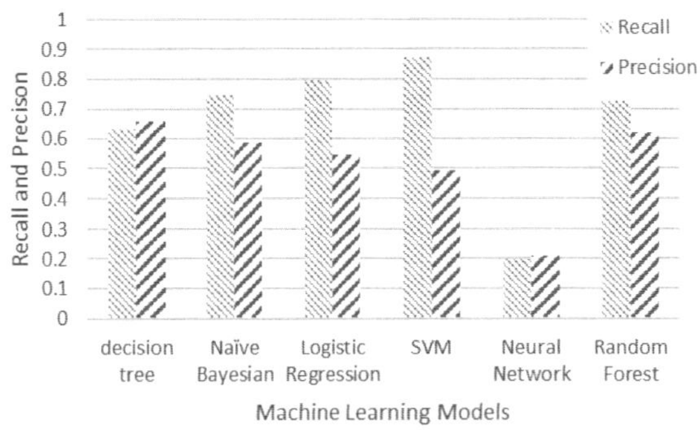

Figure 3. Recall and Precision of ML models on title texts

When it comes to **Fig.3**, even though SVM achieves optimal predictive accuracy, this methodology behaves apparent imbalance between precision and recall. Also, in terms of Logistic Regression which reaches secondary accuracy performance, the approach still has non-negligible distinction between precision and recall. Therefore, it can be concluded that although SVM as well as Logistic Regression models obtain satisfying accuracy based on small training sample bytes, there is a disturbing tendency for these two models to predict positive customer reviews as negative. In real practices, the two models would result in unwilling bias.

3.2. Analysis on Experiment 2

As described in Section 2.2, a lot more textural information was supplemented for training and disruptive bytes such as punctuations and neutrally common words have been filtered before the experiment.

Judging from the discrepancies between **Fig.4** and **Fig.5**, first and foremost, SVM and Naive Bayesian achieves nearly 92% maximal prediction accuracy and around 89% median predictive accuracy. Hence, by feeding more relevant data into machine learning models there will be distinguishable improvement on performance. Secondly, neural network reached 86% accuracy which raised by 8% from 78% during which merely titles were utilized for training. This finding corresponds to the property of Neural Network that with larger related datasets, the approach will behave steadily better predictive power.

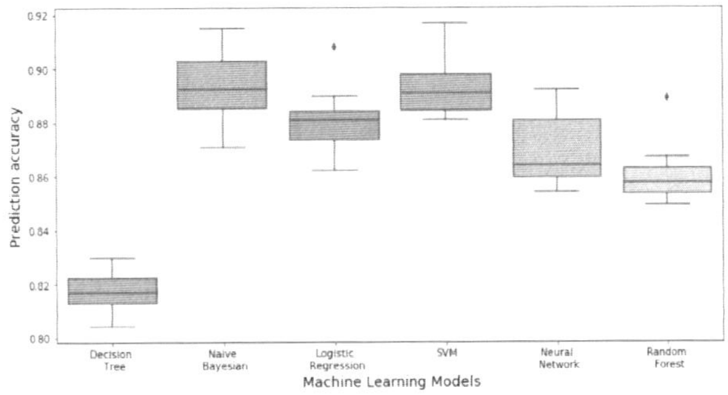

Figure 4. Accuracy performance of ML models on filtered review contents

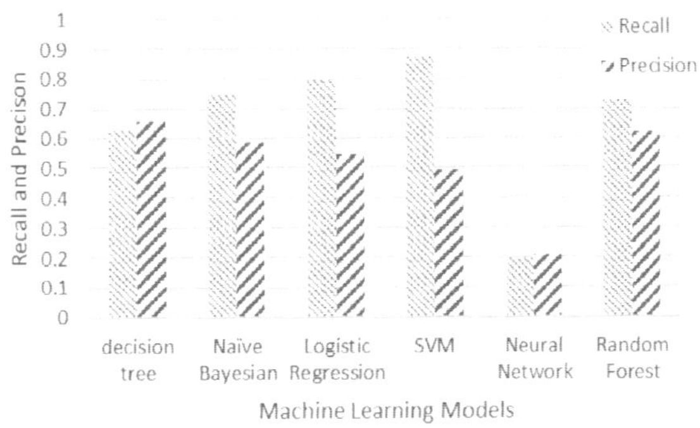

Figure 5. Recall and Precision of ML models on filtered review contents

In contrast to **Fig.3**, according to **Fig.5**, both Logistic Regression and SVM display harmonious differences between recall and precision. Hence, the two models will behave less bias towards True positive and have few errors to mistakenly treat actual positive reviews as negative.

4. CONCLUSION

In this paper, ten thousand hotel reviews in English were collected and labelled as negative or positive, and then review titles as well as filtered review contents were trained by Logistic Regression, Naive Bayesian, Decision Tree, Random Forest, SVM and Neural Network models through 10-folder-cross-validation scheme respectively. The conclusions are summarized below:

- When training samples have small byte sizes, the 3-layer Neural Network model distinctly performs worse than other five classical models. With more textual information, predictive accuracy of Neural Network model improves massively. However, Neural Network still cannot surpass traditional models such as SVM and Logistic Regression in the two experiments, which, on the contrary, indicates that neutral network requires careful structure design as well as parameter tuning for excellent performance.
- Among all six models, SVM outperforms others in terms of accuracy, precision and recall, due to its model strengths in dealing with unstructured data and high-dimensional features.
- When models were directly trained on review titles which have only 25.7 byte lengths on average, five out of six models are able to achieve predictive accuracy ranging from 84% to 87%. Considering the fact that training merely takes a few seconds, this finding casts light on business application of classical machine learning models for individual hotels to identify customer satisfaction.
- With more textual data and irrelevant punctuations as well as neutral words filtered, SVM model could reach maximal predictive accuracy nearly 92%.

IOP Conf. Series: Materials Science and Engineering **806** (2020) 012029 doi:10.1088/1757-899X/806/1/012029

REFERENCES

[1] A. Emir, H. Halim, A. Hedre, D. Abdullah, A. Azmi, and S. Kamal. Factors influencing online hotel booking intention: A conceptual framework from stimulus-organism-response perspective. International Academic Research Journal of Business and Technology, 2016, 2(2), 129-134.

[2] Statistic Brain, 2016. Internet Travel & Hotel Booking Statistics. https://www.statisticbrain.com/ internet-travel-hotel-booking-statistics/

[3] W. G. Kim, and S. A. Park. Social media review rating versus traditional customer satisfaction: which one has more incremental predictive power in explaining hotel performance?. International Journal of Contemporary Hospitality Management, 2017, 29(2), 784-802.

[4] A. J. Flanagin, and M. J. Metzger. Trusting expert-versus user-generated ratings online: The role of information volume, valence, and consumer characteristics. Computers in Human Behavior, 2013, 29(4), 1626-1634.

[5] R. K. Nielsen, and K. C. Schrøder. The relative importance of social media for accessing, finding, and engaging with news: An eight-country cross-media comparison. Digital journalism, 2014, 2(4), 472-489.

[6] X. Tian, W. He, R. Tao, and V. Akula. Mining online hotel reviews: a case study from hotels in China, 2016.

[7] Z. Xiang, Z. Schwartz, Jr, J. H. Gerdes, and M. Uysal. What can big data and text analytics tell us about hotel guest experience and satisfaction?. International Journal of Hospitality Management, 2015, 44, 120-130.

[8] Z. Jianqiang, G. Xiaolin, and Z. Xuejun. Deep convolution neural networks for twitter sentiment analysis. IEEE Access, 2018, 6, 23253-23260.

[9] data.world, 2019. Hotel Reviews – dataset in DATAFINITY. https://data.world/datafiniti/hotel-reviews.

[10] J. Hao, and T. K. Ho. Machine Learning Made Easy: A Review of Scikit-learn Package in Python Programming Language. Journal of Educational and Behavioral Statistics, 2019, 44(3), 348-361.

Robust Multi-View Feature Selection Method

Yufang Liu[1], Sibin Xu[2], He Huang[1], Fei Zhao[2*], Linyu Wang[3], Xin Liang[2], Ruixin Tang[2]

[1]State Grid Jiangsu Electric Power Company Limited

[2]Guangdong University of Technology

[3]State Grid (Suzhou) City & Energy Research Institute Co., LTD

[*]Corresponding author's e-mail: 503188839@qq.com

Abstract. Recently, Multi-view learning has attracted widely attention. Inspired by the idea of RFS [6], we propose to utilize the Multi-View features to develop a Multi-View feature selection method. In our work, the weight of each view can be adaptively adjusted. An efficient algorithm is designed to solve the proposed method. Numerical study confirms the effectiveness of our proposed method.

1. Introduction

Feature dimension explosion is one of the critical issues in efficiency improvement of machine learning model. Too many features significantly decrease the accuracy, training speed and nearly all other efficiencies in model adjustment [1][2]. Thus feature selection becomes a necessary steps before model learning with large feature dimension [3][4].

The feature selection plays a key role in machine learning, effective selection is the basis of accuracy of prediction. Feature Selection refers to an art of extracting a subset of features from an original data space. Feature selection process will try to find the smallest input set for a predictor that produces the best results concerning accuracy and model size [5]. The most recent approaches have been reported in [6]-[8]. L1 regularization can produce sparse weight matrix used for feature selection, in [6], the author proposed a new robust feature selection method with emphasizing joint L_{2-1}-norm minimization on both loss function and regularization, which makes the feature selection process more efficient. PCA(Principle Component Analysis) is also a powerful tool for analysis data, in [7], the article shows that features selection can be done effectively using combination of thresholding-based ROI template and PCA (Principle Component Analysis) methods. Pearson product-moment correlation coefficient can easure the correlation between two variables, in [8], the author proposed sentiment analysis on movie review using ensemble features and Bag of Words and selection Features Pearson's Correlation to reduce the dimension of the feature and get the optimal feature combinations.

The development of data collection technology allows people to describe things through more complex and diverse means, which also leads to more data features. Among the features have different attributes, so it is not suitable to use the same model for learning, Multi-View method can solve this problem effectively. The most recent approaches have been reported in [9]-[11]. In [9], the author proposed a Multi-View semi-supervised sparse feature selection method based on graph Laplacian, the results show that the proposed Multi-View method outperforms the single-view methods. In [10], the paper present a new semi-supervised sparse feature selection framework based on Multi-View Hessian regularization to obtain better performance, experimental results show that the proposed method can realize feature selection well. In [11], the author applies robust Multi-View K-means to obtain the robust and high quality pseudo labels for sparse feature selection in an efficient way.

This paper is an attempt on combination between Multi-View structure and model in [6]. The new model selects the $L_{2,1}$-norm from [6] across all samples with joint sparsity for feature significance presentation. Different from model in [6], the proposed model is established on Multi-View structure with new parameters. With this structure, feature selection could be not only unitedly processed but also evaluated under different views.

Content of this paper is introduced below. Section 2 introduced the detail of the proposed model. Section 3 introduces the training process of the proposed model. Section 4 selects a numerical study to verify performance of the proposed model.

2. Feature Selection Modelling

This paper selects Least Square Regression as based model. Equation (1) introduces the objective function in [6] and equation (2) introduces the objective function of proposed model.

$$\min_{W}: Obj_{NIPS10} = \frac{1}{\gamma} \| X^T W - Y^T \|_{2,1} + \| W \|_{2,1} \tag{1}$$

$$\begin{cases} \min\limits_{\alpha, W}: Obj_{NEW} = \frac{1}{\gamma} \sum\limits_{n=1}^{N} \{ \alpha_n^r \| X_n^T W_n - Y^T \|_{2,1} \} + \| W \|_{2,1} \\ s,t: \sum\limits_{n=1}^{N} \alpha_n = 1 \end{cases} \tag{2}$$

$$\| W \|_{2,1} = \left(\sum_{j=1}^{J} (w_{tj})^2 \right)^{\frac{1}{2}} \tag{3}$$

Difference between equation (1) and (2) is the term representing accuracy insurance. In equation (1), the accuracy insuring term is a typical least square regression. In equation (2), it becomes a least square regression under Multi-View data. In equation (2), feature space is separated into N views. Each view is marked as X_n. The corresponding weights matrix to each X_n is W_n. Since different views capture the different levels of information about the object, we introduce an independent parameter α_n^r to reflect the weight of each view. Moreover, the sparsity constraint is added in equation (2) to select the reduced features of each view. Equation (3) is the expression of $L_{2,1}$-norm.

From equation (3), $L_{2,1}$-norm is the summary of L_2-norm of each row in W. When separates W under Multi-View data, W can be directly separated into several W_n by telling apart rows only. So equation (2) can be rewritten as equation (4).

$$\begin{cases} \min\limits_{\alpha, W}: Obj_{NEW} = \frac{1}{\gamma} \sum\limits_{n=1}^{N} \{ \alpha_n^r \| X_n^T W_n - Y^T \|_{2,1} + \gamma \| W_n \|_{2,1} \} \\ s,t: \sum\limits_{n=1}^{N} \alpha_n = 1 \end{cases} \tag{4}$$

3. Model Learning Algorithm

Model learning is the optimization process on equation (4). α and W are two sets of parameters requiring optimization. This section chooses to optimize them sequentially. First to update α and then W.

3.1. Optimization Step for α_n Only

In the step for α_n, W is recognized as constant value. The Lagrangian function of equation (4) can be rewritten into equation (5).

$$\begin{cases} L = A + B \\ A = \dfrac{1}{\gamma}\sum_{n=1}^{N}\{\alpha_n^r \|X_n^T W_n - Y^T\|_{2,1} + \gamma \|W_n\|_{2,1}\} \\ B = \lambda\left(1 - \sum_{n=1}^{N}\alpha_n\right) \end{cases} \tag{5}$$

Consider the KKT condition, equation (6) can be obtained.

$$\begin{cases} \dfrac{\partial L}{\partial \alpha_n} = \dfrac{r\alpha_n^{r-1}}{\gamma}\|X_n^T W_n - Y^T\|_{2,1} - \lambda = 0 \\ \sum_{n=1}^{N}\alpha_n = 1 \end{cases} \tag{6}$$

Solution of the function in equation (6) is shown in equation (7).

$$\alpha_n = \frac{\left(\dfrac{1}{\|X_n^T W_n - Y^T\|_{2,1}}\right)^{\frac{1}{r-1}}}{\sum_{n=1}^{N}\left(\dfrac{1}{\|X_n^T W_n - Y^T\|_{2,1}}\right)^{\frac{1}{r-1}}} \tag{7}$$

With equation (7), the update of α_n can be obtained in equation (8). In equation (8), $W_{n,(i)}$ means W_n in the ith step. $\alpha_{n,(i+1)}$ means α_n in the i+1 step.

$$\alpha_{n,(i+1)} = \frac{\left(\dfrac{1}{\|X_n^T W_{n,(i)} - Y^T\|_{2,1}}\right)^{\frac{1}{r-1}}}{\sum_{n=1}^{N}\left(\dfrac{1}{\|X_n^T W_{n,(i)} - Y^T\|_{2,1}}\right)^{\frac{1}{r-1}}} \tag{8}$$

3.2. Optimization Step for W Only

In the step for W, α_n is recognized as constant. As the constraint in equation (4) is independent to W, this step can ignore the constraint in equation (4). Following similar spirit in [6], objective function can be rewritten into equation (9).

$$\begin{cases} Obj_{NEW} = \sum_{n=1}^{N}\{\|E_n\|_{2,1} + \gamma\|W_n\|_{2,1}\} \\ Y^T = \alpha_n^r X_n^T W_n + \gamma E_n \end{cases} \tag{9}$$

To further simplification, equation (9) can be rewritten into equation (10).

$$U_n = \begin{bmatrix} W_n \\ E_n \end{bmatrix} \Rightarrow \begin{cases} Obj_{NEW} = \sum_{n=1}^{N}(\|U_n\|_{2,1}) \\ s.t.\ A_n U_n = Y^T, A_n = [\alpha_n^r X_n^T, \gamma I_n] \end{cases} \tag{10}$$

Based on equation (10), the Lagrangian function can be obtain in equation (11).

$$\begin{cases} L_n = \|U_n\|_{2,1} - Tr(\Lambda_n^T(A_n U_n - Y^T)) \\ \dfrac{\partial L_n}{\partial U_n} = 2D_n Un - A_n^T \Lambda_n \\ D_n\ is\ diagonal\ matrix, d_{ii} = \dfrac{1}{2\|u_n^i\|_2} \end{cases} \tag{11}$$

Consider that the gradient is equal to 0, equation (12) can be obtained.

$$U_n = D_n^{-1} A_n^T (A_n D_n^{-1} A_n^T)^{-1} Y^T \tag{12}$$

Due to matrix D_n is dependent to U_n, the optimization can firstly updates matrix D_n and then updates the matrix U_n. Equation (13) finally reveals the updates of W_n.

$$\left\{\begin{array}{c} U_{n,(i)}=\begin{bmatrix} W_{n,(i)} \\ E_{n,(i)} \end{bmatrix} \\ \Rightarrow D_{n,(i+1)}\ is\ \text{diag}onal\ matrix, d_{jj}=\dfrac{1}{2\|u^{j}_{n,(i)}\|_2} \\ \Rightarrow U_{n,(i+1)}=D^{-1}_{n,(i+1)}A^T_n(A_nD^{-1}_{n,(i+1)}A^T_n)^{-1}Y^T \\ \Rightarrow \begin{bmatrix} W_{n,(i+1)} \\ E_{n,(i+1)} \end{bmatrix}=U_{n,(i+1)} \end{array}\right. \tag{13}$$

3.3. Learning Algorithm

The sequential learning algorithm is constructed by expression in equation (8) and equation (13). Table I reveals the optimization process of model learning.

Table 1. The Entire Optimization Process

Data Import: X_n, Y
Result: W_n, α_n
Set $t = 0$. **Initialize** W_n, α_n, D_n
for i←0 to k
  calculate α_n with equation (8)
  calculate U_n with equation (13)
  calculate the diagonal matrix D_{t+1}, **where the** i-th **diagonal element is** $\dfrac{1}{2\|u^{i}_{t+1}\|_2}$
  Obtain W_n from U_n
end for

In Table 1, the necessary data X_n and Y. W_n, α_n, D_n are initialized. A limited maximum step is set to end the algorithm. With in each iteration, the algorithm firstly fix W_n and D_n, but update α_n with equation (8). Secondly, the algorithm fix α_n and update U_n, D_n with equation (13). W_n can be obtained from the upper half of U_n.

4. Numerical Study

4.1. Background

A numerical study is implemented to verify the capability of the proposed method. MSRC-v1 and Caltech 101-7 are the two sets of data selected for model training. In MSRC-v1, there are 6 views and the total feature dimension is 2428 [12]. In Caltech 101-7, there are 6 views and the total feature dimension is 3766 [13]. The proposed method will be compared with the model in [RFS]. The selected features from both models will be sent to a typical KNN model to examine the effect of feature selection. Matlab 2016 is used as the platform for model learning.

4.2. Experimental Result of MSRC-v1

Table II reveals the result on MSRC-v1. It compares the results from RFS and the proposed model under two dimensions, which includes different value of r and different quantity of selected features.

IOP Conf. Series: Materials Science and Engineering **806** (2020) 012030 doi:10.1088/1757-899X/806/1/012030

Table 2. Experimental Result of RFS and The Proposed Model on MSRC-v1

	F5		F10	
	New Model	RFS	New Model	RFS
r = 0.1	0.9	0.890476	0.933333	0.933333
r = 0.2	0.92381	0.890476	0.961905	0.933333
r = 0.4	0.919048	0.890476	0.971429	0.933333
r = 0.8	0.928571	0.890476	0.942857	0.933333
r = 1.2	0.857143	0.890476	0.947619	0.933333
r = 1.4	0.885714	0.890476	0.928571	0.933333
r = 1.6	0.866667	0.890476	0.933333	0.933333
r = 1.8	0.857143	0.890476	0.92381	0.933333
r = 2	0.809524	0.890476	0.895238	0.933333
r = 4	0.733333	0.890476	0.871429	0.933333
r = 8	0.733333	0.890476	0.871429	0.933333
r = 10	0.733333	0.890476	0.871429	0.933333
	F15		F20	
	New Model	RFS	New Model	RFS
r = 0.1	0.947619	0.952381	0.985714	0.961905
r = 0.2	0.966667	0.952381	0.97619	0.961905
r = 0.4	0.966667	0.952381	0.971429	0.961905
r = 0.8	0.961905	0.952381	0.961905	0.961905
r = 1.2	0.933333	0.952381	0.952381	0.961905
r = 1.4	0.92381	0.952381	0.942857	0.961905
r = 1.6	0.928571	0.952381	0.947619	0.961905
r = 1.8	0.914286	0.952381	0.928571	0.961905
r = 2	0.904762	0.952381	0.92381	0.961905
r = 4	0.880952	0.952381	0.880952	0.961905
r = 8	0.880952	0.952381	0.880952	0.961905
r = 10	0.880952	0.952381	0.880952	0.961905
	F25		F30	
	New Model	RFS	New Model	RFS
r = 0.1	0.97619	0.957143	0.957143	0.947619
r = 0.2	0.961905	0.957143	0.961905	0.947619
r = 0.4	0.966667	0.957143	0.966667	0.947619
r = 0.8	0.97619	0.957143	0.985714	0.947619
r = 1.2	0.957143	0.957143	0.957143	0.947619
r = 1.4	0.966667	0.957143	0.947619	0.947619
r = 1.6	0.952381	0.957143	0.957143	0.947619
r = 1.8	0.947619	0.957143	0.933333	0.947619
r = 2	0.933333	0.957143	0.942857	0.947619
r = 4	0.904762	0.957143	0.933333	0.947619
r = 8	0.904762	0.957143	0.933333	0.947619
r = 10	0.904762	0.957143	0.933333	0.947619

Table 3. Experimental Result of RFS and The Proposed Model on Caltech 101-7

	F5		F10	
	New Model	RFS	New Model	RFS
r = 0.1	0.901934	0.892964	0.916786	0.918726
r = 0.2	0.905925	0.892964	0.924266	0.918726
r = 0.4	0.908665	0.892964	0.930354	0.918726
r = 0.8	0.926089	0.892964	0.932893	0.918726
r = 1.2	0.913909	0.892964	0.922088	0.918726
r = 1.4	0.919942	0.892964	0.936325	0.918726
r = 1.6	0.915964	0.892964	0.927509	0.918726
r = 1.8	0.909853	0.892964	0.92829	0.918726
r = 2	0.903839	0.892964	0.920105	0.918726
r = 4	0.879577	0.892964	0.895976	0.918726
r = 8	0.879577	0.892964	0.895976	0.918726
r = 10	0.879577	0.892964	0.895976	0.918726
	F15		F20	
	New Model	RFS	New Model	RFS
r = 0.1	0.924869	0.922097	0.923443	0.924814
r = 0.2	0.93103	0.922097	0.932304	0.924814
r = 0.4	0.935757	0.922097	0.936364	0.924814
r = 0.8	0.94174	0.922097	0.945873	0.924814
r = 1.2	0.937062	0.922097	0.943748	0.924814
r = 1.4	0.937767	0.922097	0.943064	0.924814
r = 1.6	0.934314	0.922097	0.943782	0.924814
r = 1.8	0.932365	0.922097	0.945124	0.924814
r = 2	0.937707	0.922097	0.940475	0.924814
r = 4	0.921407	0.922097	0.921385	0.924814
r = 8	0.921407	0.922097	0.921385	0.924814
r = 10	0.921407	0.922097	0.921385	0.924814
	F25		F30	
	New Model	RFS	New Model	RFS
r = 0.1	0.930346	0.928874	0.931004	0.933642
r = 0.2	0.935049	0.928874	0.937072	0.933642
r = 0.4	0.93772	0.928874	0.937729	0.933642
r = 0.8	0.951965	0.928874	0.954587	0.933642
r = 1.2	0.950536	0.928874	0.953861	0.933642
r = 1.4	0.948481	0.928874	0.949911	0.933642
r = 1.6	0.949161	0.928874	0.953879	0.933642
r = 1.8	0.94781	0.928874	0.952513	0.933642
r = 2	0.948513	0.928874	0.955239	0.933642
r = 4	0.93566	0.928874	0.942411	0.933642
r = 8	0.932231	0.928874	0.942411	0.933642
r = 10	0.932231	0.928874	0.942411	0.933642

From angle of dimension r, it is obvious that new model performs better than RFS when r is smaller than 1. Results from Table 2 shows that the new model obtains 1% accuracy improvement than RFS averagely. Fig.1 introduces the average improvements at different optimal feature quantity with $r < 1$. As the following, we explain why setting $r < 1$ can obtain the better performance. In Eq.(8), we can find that the weight α_n is proportional to the regression residual of each view when $r < 1$. The large (small) residual regression can lead to a large (small) weight. In the next iteration, this large (small) weight can enhance (reduce) the residual regression. Hence, this updating rule can help to fully utilize the complementary information of Multi-View data.

4.3. Experimental Result of Caltech 101-7

Table III reveals the result on Caltech 101-7. It compares the results from RFS and the proposed model under two dimensions, which includes different value of r and different quantity of selected features.

From angle of dimension r, new model performs better than RFS in most of the cases, especially for case with r = 0.4. Results from Table 3 show that the new model obtains about 1% accuracy improvement than RFS averagely. Fig.2 introduces the average improvements at different optimal feature quantity with r = 0.4.

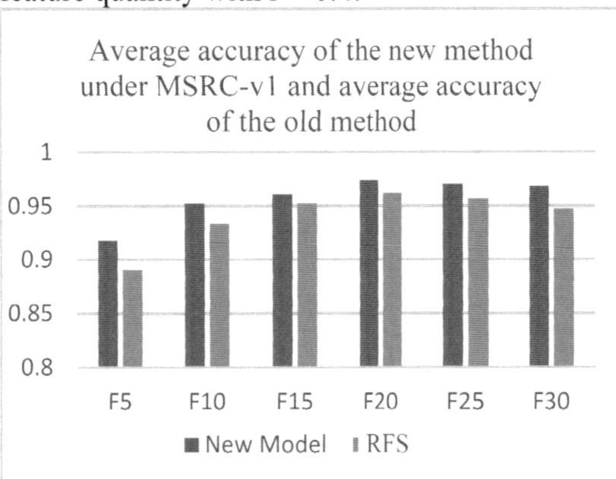

Fig.1 Average Accuracy (r = 0.8) of New Model and RFS Under MSRC-v1

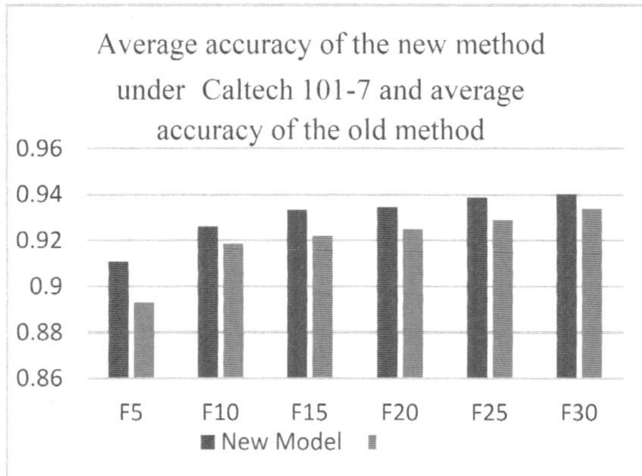

Fig.2 Average Accuracy (r = 0.4) of New Model and RFS Under Caltech 101-7

5. Conclusions

This paper proposed a feature selection method for classification under Multi-View data. The idea is derived from RFS and is combined with Multi-View data. It evaluates the effect of all features together under several views with α_n. A two stages Lagrangian optimization is proposed for this model. In each iteration, one stage is to optimize W and fix α_n and the other is to optimize α_n and fix W. Simulation result shows that the proposed model performs better when r is less than 1.

Acknowledgment

This research work was supported in part by the National Science Foundation of China: A Time-Based-Demand-Response Program of Compensated Multiple-Shape Pricing Scheme, Grant No. 51707041; in part by the State Grid Energy Research Institute Project (Penetration Capability Research of Distribution Grid and The algorithm of Relevant Analyse Optimization Model); in part by the State Grid (Suzhou) City & Energy Research Institute project (Critical Data Acquisition and Obtaining Method research on City Energy Research); in part by the science and technology project of State Grid Corporation of China (The key technologies and applications of State Grid's social image improvement based on social media big data, Grant no. 1400-201957293A-0-0-00); in part by the young talents project of State Grid Energy Research Institute Co. Ltd. (Research on recognition and processing technology of news hotspots in energy industry based on the natural language processing, Grant no.526700180001).

References

[1] J. Shawe-Taylor, N. Cristianini. (2004) Kernel Methods for Pattern Analysis. Cambridge University Press, Cambridge.
[2] Y. Sun, D. Wu. (2008) A relief based feature extraction algorithm. Proc. SIAM Int. Conf. Data

Mining, pp. 188-195.
[3] R. Kohavi, G. H. John. (1997) Wrappers for feature subset selection. Artif. Intell., vol. 97, no. 1/2, pp. 273-324.
[4] P. A. Mundra, J. C. Rajapakse. (2010) SVM-RFE with MRMR filter for gene selection. IEEE Trans. NanoBiosci., vol. 9, no. 1, pp. 31-37.
[5] S. Visalakshi, V. Radha. (2014) A literature review of feature selection techniques and applications: Review of feature selection in data mining. IEEE International Conference on Computational Intelligence and Computing Research (ICCIC), pp. 1-6.
[6] Nie, Feiping, Huang, Heng, Cai, Xiao , Ding, Chris. (2010) Efficient and robust feature selection via joint ℓ2;1-norms minimization. 24th Annual Conference on Neural Information Processing Systems, pp. 1-6.
[7] N.R. Ratnasari, Adhi Susanto, Indah Soesanti, Maesadji. (2013) Thoracic X-ray features extraction using thresholding-based ROI template and PCA-based features selection for lung TB classification purposes. 2013 3rd International Conference on Instrumentation, Communications, Information Technology and Biomedical Engineering (ICICI-BME), pp. 65-69.
[8] Fachrul Rozy Saputra Rangkuti, M. Ali Fauzi, Yuita Arum Sari, Eka Dewi Lukmana Sari. (2018) Sentiment Analysis on Movie Reviews Using Ensemble Features and Pearson Correlation Based Feature Selection. 2018 International Conference on Sustainable Information Engineering and Technology (SIET), pp. 88-91.
[9] Caijuan Shi, Qiuqi Ruan, Gaoyun An, (2014) Multi-view Laplacian sparse feature selection for web image annotation. 2014 12th International Conference on Signal Processing (ICSP), pp. 1026-1029.
[10] Caijuan Shi, Jian Liu, Liping Liu, Xiaodong Yan, (2015) Semi-supervised sparse feature selection based on multi-view Hessian regularization. 6th International Conference on Wireless, Mobile and Multi-Media (ICWMMN 2015), pp. 158-161.
[11] Hongfu Liu, Haiyi Mao, Yun Fu, (2016) Robust Multi-View Feature Selection. 2016 IEEE 16th International Conference on Data Mining (ICDM), pp. 281-290.
[12] Winn J M, Jojic N. (2005) LOCUS: Learning Object Classes with Unsupervised Segmentation[C]. 10th IEEE International Conference on Computer Vision (ICCV 2005), Beijing, China.
[13] Fei-Fei L , Fergus R , Perona P. (2007) Learning generative visual models from few training examples: An incremental Bayesian approach tested on 101 object categories[J]. Computer Vision and Image Understanding, 106(1):59-70.

Research on Development Strategy of News App under the Background of Artificial Intelligence

Wei Guo[1], **Bo Zhang**[2,*]

[1]School of Literature and Law, Wuhan Donghu University, Wuhan, Hubei, 430212, China

[2]School of Computer Science, Wuhan Donghu University, Wuhan, Hubei, 430212, China

*Corresponding author's e-mail: bob.cheung@ovspark.com

ABSTRACT: With the rapid development of the mobile Internet, a large number of Internet users have access to current news through the mobile phone. The market performance of the media Applictipon(referred to as App) is strong, and it has become increasingly prominent in the public opinion. This article analyzes the present situation of the development of media APP by consulting literature, market research, comparative research and other methods, aiming at exploring and studying the future development trend of media APP and its development countermeasures.

1. Introduction

With the continuous development of China 's mobile Internet, the level of people 's consumption has increased dramatically, and China 's mobile phone user base has continued to expand. In the 44th "Statistical Report on the Development of the Internet in China" released in August 2019, as of June 2019, the number of mobile Internet users in China reached 847 million, an increase of 29.84 million from the end of 2018. The proportion of Internet users using mobile phones to access the Internet reached 99.1%, an increase of 0.5 percentage points from the end of 2018[1]. Internet penetration rate exceeds 60%, and mobile Internet use continues to deepen.

The above data fully proves that with the advent of the smartphone era and the continuous improvement of communication technology, the mobile Internet era is the general trend. In the media field, because a large number of netizens have switched from PC to mobile clients, news media have aimed at the mobile network market and started to develop news Applictipon (referred to as App).

1.1. News media App industry scale

"2019Q1 China Mobile News App Market Monitoring Report" authoritative release by iiMedia Research in May 2019. With the increasing penetration of Internet, mobile news App has gradually become the main source of information for Chinese. According to iiMedia Research, the scale of mobile news App users in China is expected to reach 693 million in 2019, which indicates huge market potential.

In terms of news App, in 2019Q1, the pattern of China's mobile phone news App market remained stable. The monthly active users of Tencent news and Toutiao news both exceeded 200 million, which is at the leading position of the industry. The specific data is shown in 'figure 1'.

IOP Conf. Series: Materials Science and Engineering **806** (2020) 012031 doi:10.1088/1757-899X/806/1/012031

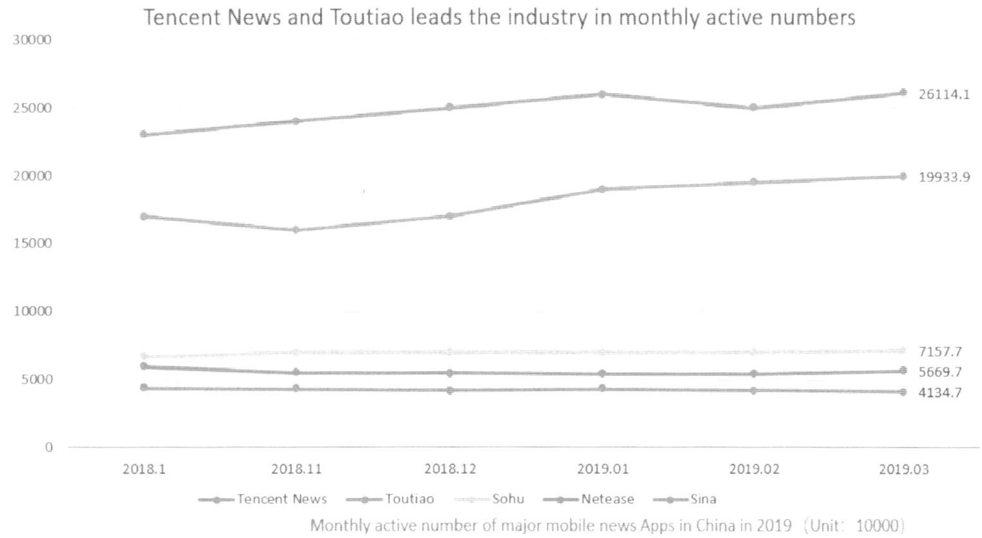

Figure 1. Monthly active number of major mobile news Apps in China in 2019（Unit：10000）（Data from iiMedia Research）

In terms of users, male accounts for 60% of the news App users, whereas high-income groups prefer traditional media news App. Under the background of strict network supervision, news App will stick to truth and start quality improvement. At the same time, the sinking market began to wake up, news App will present news that is more close to the public life to obtain users in third and fourth tier cities, leading to the intensification of industry competition[2].

1.2. Comprehensive market performance of various news media Apps

According to incomplete statistics, there are more than 1,300 news Apps at the national level. The media App industry is fiercely competitive, and the problems that accompany it are gradually increasing. From the perspective of market share, the three major portals of NetEase, Tencent and Sohu, as well as Toutiao and Surging News occupy most of the market share. News Apps created by traditional portals have an advantage in market traffic, and have been put into practice in cutting-edge areas such as the use of big data, intelligent push, and news customization.

1.3. Tencent News and Toutiao App occupy most of the market share

From the perspective of the industry market, according to statistics from Aurora Big Data in May 2019, in terms of monthly active users, Tencent News and Toutiao performed strongly as the first legion, far exceeding other rivals Sohu, Netease and Sina[3].

In the market competition, traditional media Apps mainly meet different user needs with their rich news resources and high-quality original content. According to data analysis, among information App users, high-income groups prefer traditional media Apps, and high-income groups (average monthly income of more than 10,000 yuan) have a high proportion. The specific data is shown in 'figure 2'.

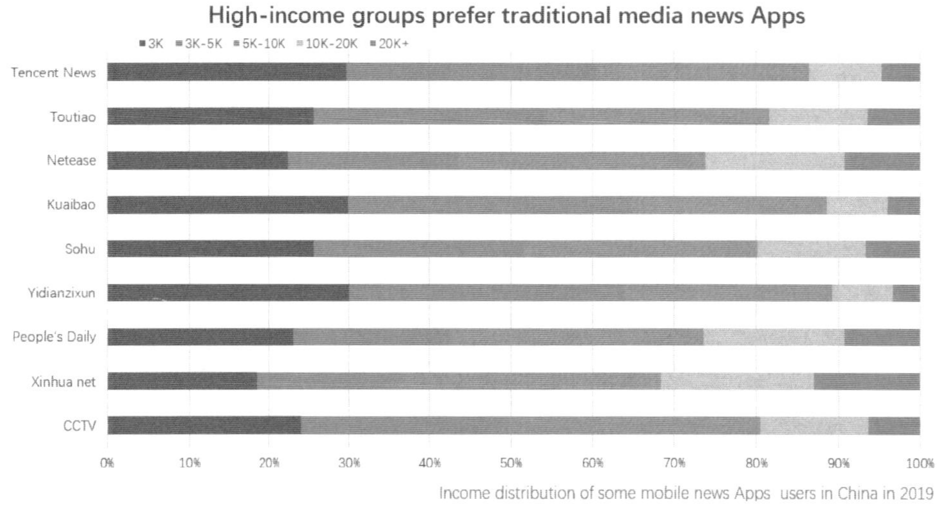

Figure 2. Income distribution of some mobile news Apps users in China in 2019（Data from iiMedia Research）

2. Development Dilemma of News App under the Background of Artificial Intelligence

Although the emerging online media App has technological and market competitive advantages, there are still many problems that need to be solved urgently: the "title party" phenomenon, the tendency of information content to be vulgar, content disputes, and algorithm recommendations to "filter bubbles" gap and so on.

2.1. Poor journalism

As the current market online media App mainly releases information from the media, in order to ensure that the media can upload information in a timely manner, the audits are mostly machine audits, which are mainly based on sensitive word filtering and content rechecking, and there are fewer manual audits. Without the "gatekeeper" of the manual review type, it is difficult to ensure the professionalism of the news.

In response to the widespread problems of poor professional content of news content and the spread of "yellow news" in information media apps, relevant departments are increasing the supervision and review of information and other applications.

2.2. Algorithm recommendation disadvantages

Algorithmic technology determines what content is recommended to users. One measure is clicks, comments, and retweets.

Toutiao algorithm recommendation mode changes the way information is obtained, allowing information to actively "find" users, greatly improving people's efficiency in obtaining news information, saving time and costs, and helping Toutiao to quickly rank among the top news client industries[4]. However, while the algorithm is prevailing, problems such as headline party, fake news, and information cocoon room exposed by the algorithm are becoming increasingly serious, which not only causes aesthetic fatigue of users, but also damages the image of the platform and hinders the development of the platform. At the same time, the effectiveness of advertisements recommended by algorithms has been repeatedly questioned.

2.3. Content copyright issues

Compared with traditional media Apps, the algorithm-recommended Apps are more biased toward quality and quantity.

Take Toutiao as an example, Toutiao refused to pay reprints to traditional media, writers, independent authors, photographers, etc., which led to a series of conflicts between traditional media and today's headline infringement of rights. Copyright disputes are inevitable.

3. Breakthrough and Innovation of News App

In the future, as users' requirements for information content continue to increase and algorithm technology continues to upgrade, media apps will experience the resurgence of traditional mainstream media and the in-depth cooperation between artificial intelligence and content creators.

3.1. News content expertise will continue to increase

After several years, various industries have gradually completed the digital transformation and entered the era of "big content". The contradiction between the wireless expansion of content and the limited attention of users has become increasingly prominent.

Taking CCTV News App as an example, the product mainly has the following outstanding features to solve the problem of some users' poor professionalism of today's news content and a single way of presenting information by mainstream media.

3.1.1. Deep fusion production

CCTV News App leads other platforms in terms of timeliness and originality of news content. By improving the layout of all media, CCTV news App traffic resources are abundant.

3.1.2. Focus on exclusive authoritative content

CCTV News App focuses on combining powerful resource advantages. CCTV has absolute authority in news content and value, and has a first-rate interview resource in the central media background.[5] Through innovative content creation, the form of news presentation is novel.

3.1.3. News content returns to true

When the market enters the stage of stock competition, meeting the fundamental needs of users to obtain information quickly and efficiently becomes the key to success. This means that news content will return to the true nature, and the news client platform will start the upgrade path from "quantity" to "quality".

3.2. Artificial intelligence assists media industry

The impact of artificial intelligence on the traditional media industry is still relatively small, but its impact may not be limited to the field of content distribution, but fundamentally affects everything from content production to consumption forms.

3.2.1. Content production moves towards human-machine collaboration

In terms of media content production, the future will be the road of human-machine collaboration. The media App industry that pays attention to timeliness, journalists are freed from simple and repetitive event reports, so as to carry out in-depth reports, topic design and other content.

3.2.2. Content distribution moves towards single-user communication

In addition to news professionalism, inaccurate customized content has also become a major problem for users using the App. With the help of artificial intelligence, machine recommendation can extract, sort and transfer large amounts of information to users based on their behaviors, characteristics, and demands, and overcome this big pain point through single-user communication.

4. Conclusion

In the context of the rapid development of the mobile Internet, traditional media in the process of transforming new media urgently need to abandon the original paper media creation methodologies, fully integrate the contemporary new media context, and develop a unique mainstream media transformation road.

In the future, algorithm recommendation will become the mainstream of the media app industry. From news creation, review, to distribution, people will work alongside artificial intelligence to participate in it. As a result, algorithm-based recommendation of this cutting-edge technology will face numerous tests from journalists, social ethics, laws and regulations. Related technical specifications, regulations and provisions should be gradually improved.

Acknowledgement

This article is part of the research results of the scientific research project "Research on the Legal Reg ulation of Freedom of Speech from the Media" of the Scientific Research Project of Hubei Province E ducation Department (No : B2017302)

References

[1] The 44th Statistical Report on Internet Development in China, CNNIC, http://www.cnnic.cn/hlwfzyj/hlwxzbg/hlwtjbg/201908/t20190830_70800.htm

[2] 2019Q1 China Mobile News APP Market Monitoring Report, iiMedia Research, https://www.iimedia.cn/c400/64308.html

[3] Aurora Big Data. http://www.shujuju.cn/data/detail/36

[4] Peng Guibing. (2017) A Discourse Study on Copyright Controversy in the "Toutiao" Event—— Concurrently Discussing the Path of the Reform of News Copyright Protection[J], Modern communication, 55-60.

[5] Jiang Lan, Zhang Yan, (2018) Research on the Development of New Media App for TV News in the Context of Converged Media——Taking CCTV News App as an Example, Communication power research, 99

IOP Conf. Series: Materials Science and Engineering **806** (2020) 012032 doi:10.1088/1757-899X/806/1/012032

Research on Legal Protection System of Personal Information in Big Data Era

Xianghua Deng[1] **and Bo Zhang**[2,*]

[1] Department of Law, Wuhan Donghu University, Wuhan, Hubei, 430212, China

[2] Department of Computer Science, Wuhan Donghu University, Wuhan, Hubei, 430212, China

[*]Corresponding author's e-mail:bob.cheung@ovspark.com

Abstract. In the era of big data, the problem of leakage of personal information is becoming more and more serious. With the citizens' increasing legal awareness, people pay more and more attention to the protection of personal information. In the context of big data, this article is based on the impact of big data on the protection of personal information. By studying the current status of the domestic and international legal protection of personal information, this paper puts forward some proposals for the protection of information, combined with the current status and existing problems of the legal protection of personal information,

1. Introduction

In the era of big data, the development and utilization of data has brought huge challenges to citizens' personal information. There are great security risks in the collection and use of personal information, and the information subject is in a weaker position than the information manager in terms of litigation ability. In daily life, when personal information is unreasonably commercialized, it has brought great trouble to our lives. Some personal information has even been used for illegal and criminal activities, which is extremely harmful to society and affects social stability. However, China does not currently have specific legal provisions to deal with this dilemma, so the issue of how to effectively protect personal information in the era of big data needs to be resolved.

2. Overview of the legal protection of personal information in the era of big data

2.1. Features of the Big Data Era

Big data is a collection of information. The collection and management of big data are not achieved through ordinary software, but through a new management model. Only in this way can the processing capacity of information resources be improved. In the past, traditional data processing methods have failed in the storage and processing of big data, because of its large volume, rapid growth, and many types. They exist in the Internet in various forms. The era of big data is a modern era of informationization that relies on big data and integrates data collection, processing, and decision-making. In the process of using big data, it is possible to collect information at a higher speed, process the information to make scientific and reasonable decisions, and apply it to production and life, bringing significant economic and scientific value to society.

2.2. Impact of the era of big data on personal information protection

The Internet has brought convenience to our lives and made it easier for us to collect information, but it is also often used by criminals, bringing huge losses to the lives and work of citizens. As shown in the Figure 1 of the leakage of personal information, websites, including e-commerce platforms, search engines, portals, etc., is considered to be the most likely to leak personal information, which accounts for 14%; followed by Apps on personal information terminals such as mobile phones, PAD, smart watches, and sports bracelets, which accounts for 13%; the third is the automotive industry, such as 4S shops and telecommunications service providers, both reaching 12%; the fourth is communication software such as e-mail, WeChat, QQ, etc. And real estate industry, including real estate industry and communication software, reaching 11%; the fifth is express companies, accounting for 9%; the sixth is the banking and insurance industry, with 8%; the seventh is medical, education, electricity, water, gas, etc. The public service industry takes up about 5%; the other industries are considered to be 3%, and the least is airlines and administrative agencies including air ticket agents and train ticket agents, each about 1%.

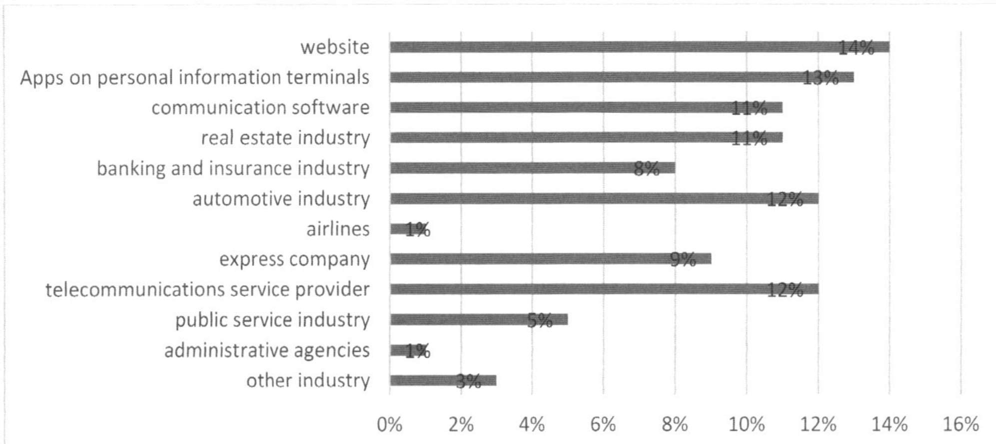

Figure 1. Personal information leakage channels

In the era of big data, citizens' information can be stolen through technologies such as big data screening and mining. As a result, the legal protection of personal information in the context of big data has become extremely difficult.

2.3. The importance of legal protection of personal information in the era of big data

Big data brings a lot of convenience to people's production and life, and also brings people's personal information into the public eye. The content of this information is very extensive and is closely related to the identity, property and security of citizens. If this information is leaked, it will cause significant damage to citizens' personal rights and property rights. In the era of big data, the Internet can be said to run through all aspects of our lives, work and study. Any one of us can do nothing without the Internet. While the Internet brings convenience to our lives, it also puts us on the brink of danger. When people browse the website or shop online, the website will remember our relevant information. Even if you do not log in to the website, you can also remember the user's preferences. If this information cannot be kept secret or is used by criminals, it would cause us losses.

In the context of economic globalization, there are more and more opportunities for exchanges between countries. The exchanges between various organizations and enterprises have also exhibited diversified characteristics. In the cooperation, it is inevitable that with the dissemination and exchange of information, the cross-border transmission of information brings new challenges to the legal protection of personal information. The cross-border transmission of information makes the diffusion of information larger. The people who contact the information may be multiple uncertain people, which makes information protection more difficult[1].

At present, China does not have a unified personal information protection law, and personal information protection work has passed specific laws. Although China has a "Network Information Security Law," it is mainly aimed at network security considerations, and the protection of personal information is not specific enough. It cannot protect personal information in all aspects. The Internet has the characteristics of convenience and speed. Traditionally, relying on social norms such as morals to protect the personal information of citizens is far from enough. In this way, it is even more urgent to protect the personal information of citizens.

3. Status and Defects of Legal Protection of Personal Information in China in the Big Data Era

3.1. Lack of targeted legislation

Although China's "Civil Law", "Criminal Law", "Constitution" and other laws have stipulated the content of personal information protection, but these laws and regulations are lack of pertinence, and many laws stipulate the content of personal information is roughly equivalent, the content Complete, and the protection is not comprehensive enough. In civil law, the right to personal information is considered as a separate civil right. The protection of personal information is not clear and specific. The protection of privacy seems to be too narrow and not comprehensive enough. In criminal law, There are specific provisions on the crime situation, but the provisions on the grasp of the crime are too vague and there are great problems in their application; China's Constitution stipulates the personal information of citizens as a right of personality, and personal information is not only personal information, but also an interest related to the person, but this regulation is still relatively broad, lacking specificity and guidance.

3.2. Imperfect judicial remedy system

When our personal information rights are violated, we may first think of infringement damages. The issue of compensation is divided into material compensation and spiritual compensation. China's laws provide specific provisions for material compensation, but the provisions on spiritual compensation are too vague, and the conditions that constitute spiritual compensation are very high. This is very unfavourable to the protection of personal information. If you encounter a crime as stipulated in the provisions of the criminal law, you need to bear criminal responsibility. Article 253 of the Criminal Law makes specific provisions on this, but it is still difficult to apply. Formatting author affiliations

Please ensure that affiliations are as full and complete as possible and include the country.The addresses of the authors' affiliations follow the list of authors and should also be indented 25 mm to match the abstract. If the authors are at different addresses, numbered superscripts should be used after each surname to reference an author to his/her address. The numbered superscripts should not be inserted using Word's footnote command because this will place the reference in the wrong place—at the bottom of the page (or end of the document) rather than next to the address. Ensure that any numbered superscripts used to link author names and addresses start at 1 and continue on to the number of affiliations. Do not add any footnotes until all the author names are linked to the addresses. For example, to format[2].

3.3. Difficulties in the protection of a single domestic law

In the era of big data, the speed of dissemination of personal information is unprecedented, and personal information can easily be transmitted abroad in a very short period of time. However, the legal regulations on the protection of personal information at home and abroad are different, which has led to the difficulty of transnationalization of personal information Protection.

3.4. Citizens' legal awareness is weak

In recent years, China attaches great importance to the construction of the rule of law, and many units have carried out law popularization activities. These activities cover a wide range of areas. However, due to the large population of our country, the education level of different citizens varies greatly, and the law popularization work is not completely covered. Therefore, some citizens do not have a strong legal awareness and will not take precautions when they ask for personal information. This poses a major threat to the security of their personal information.

4. Legal protection of foreign personal information in the era of big data

4.1. Legal analysis of EU personal information protection

The EU has earlier protection of personal information, forming a unique EU model. With regard to the protection of personal information, the European Union protects all aspects of the collection, processing and use of personal information through laws. The law applies uniformly to the public and non-public sectors, protects personal information through various links, and sets up a comprehensive Supervision by the regulatory authorities. The EU's Personal Data Protection Directive has been recognized by many countries, and its unique legislative model and advanced legislative technology have made it widely applicable in Europe. The EU 's personal information protection model has an early origin, strong applicability, and clear enforcement mechanisms. This law raised the EU's protection of personal information to a new level, and this law entered foreign companies, played an important role and was widely disseminated, making Europe a model for global personal information protection.

4.2. Legal Analysis of Personal Information Protection in the United States

The protection of personal information in the United States is mainly based on the right to privacy. Not only decentralized legislation, but also corresponding industry self-regulation. In the public domain, the United States takes the right to privacy as the basis of the Constitution and administrative law, and adopts a decentralized legislative model to legislate one by one. In the private sector, the United States relies on self-regulatory mechanisms to protect personal information, and according to the specific content of personal information, it is supervised by the corresponding regulatory authority. Although the United States has a unique protection model, the protection based on the right to privacy is not comprehensive, and it is also a question for contemporary personal information protection[3].

4.3. Legal Analysis of Personal Information Protection in Japan

Japan's legal protection of personal information started late and was made by drawing on European and American legislation. Japan attaches great importance to the legal protection of personal information. Japan passed the Personal Information Protection Law in 2005. After the implementation of this law, personal information will be fully protected. For the protection of personal information in the era of big data, Japan also pays attention to industry self-regulation and community participation. A multi-pronged approach has formed a unique personal information protection model in Japan.

4.4. The experience of foreign personal information legal protection in the era of big data and its inspiration to China

With the development of China's Internet technology, modern information technology in the era of big data has brought challenges to the protection of personal information in China. In the public sector, many industries are collecting personal information, but related departments have not fulfilled their obligations of confidentiality, resulting in the leakage of personal information of many citizens; in the non-public sector, personal information leakage is more common, and many industries are on the Internet Platforms conduct transactions, and these platforms do not protect customer information, and even some platforms collect customer information through this channel, and use it for advertising or sales. Although the relevant laws of China have also provided relevant provisions for the protection of personal information, there is no special law on personal information protection, nor has a complete legal system been formed, and the protection of personal information is not comprehensive. Therefore, on the basis of existing laws, China should fully draw on foreign experience to form a personal information legal protection system that is in line with China's national conditions, and implement targeted and comprehensive legislation on the protection of personal information. Personal information is most effectively protected.

5. Suggestions on China's Personal Information Legal Protection System in the Big Data Era

5.1. .Accelerate the legislation of personal information protection

China needs to fully absorb and draw on the legislation of the European Union, the United States, and Japan. Model, formulate the Personal Information Protection Law and improve the relevant legal system. In the context of the era of big data, it is necessary to adopt a more flexible jurisprudence system based on the civil trust law, in order to complete the construction of the order for the use of personal information during continuous adjustment and adaptation, and to give full play to the last of the criminal law under the principle of moderation. Protective effects. Based on the existing laws, it should be based on the top-level design. The protection of personal information must be made specifically for the big data environment. The problem of lagging legislation should be solved, and a systematic and targeted law should protect personal information. Play a greater role.

Under the big data environment, data collection, processing and publication are more convenient, which is both an opportunity and a challenge for the protection of personal information. In the context of the era of big data, the law should also make corresponding provisions on the methods and methods of collecting data information. First, the collection of personal information must be lawful and proper. It must comply with both laws and policies, respect social morality, observe business ethics, and be honest and trustworthy. Illegal collection, use, processing, and transmission of personal information of others are all inconsistent with the law. The purchase, sell, or disclose of personal information must be within the permits of the law. Secondly, when collecting and using personal data of citizens, the data information collector needs to indicate to the collected person the purpose of collecting personal information. It is best to leave a written certificate. This can properly restrict the scope of use of the collector. On the one hand, it also leaves a basis for the later rights protection, in case the information is used by others. It is convenient for citizens to safeguard their rights. Especially for the purpose of collecting data information for the purpose of commercial transactions, the user's personal data cannot be used without the consent of the user. Employers, state agencies, etc. should clearly understand the purpose of collecting personal information. If the collected person does not know the purpose of the information, and may adversely affect the collected person, or the collector has no right to collect, it is a manifestation of violation of personal information. When collecting data and information, it is necessary to tell the citizens who have collected the information whether the information is required by state agencies, is used for commercial activities, or is used in teaching and research. In particular, those engaged in Internet business must fulfill their obligation to keep customers confidential. The collected customer information cannot be used for purposes other than service. Institutions such as schools and hospitals cannot use the collected personal information for profit-making purposes. Otherwise, it will not only infringe, but also affect the credibility of these units and cause significant adverse effects. Finally, the collectors and users of information should keep the information secure, and they should protect the security of the information. They must not leak the information to others or cause the information to be lost due to their own reasons. The integrity of collected personal information should be protected, and cannot be added, deleted, or modified at will.

5.2. Improve judicial protection procedures and judicial remedies

Although China has relevant legal protections for personal information, there are still many problems in its implementation. This requires further improvement of judicial protection procedures and judicial relief measures. The government should strengthen the timeliness of supervision and law enforcement, and establish an authoritative and efficient law enforcement agency. So far, China does not have a specialized data protection agency. When people's personal information is violated, it is relatively difficult to remedy it. The state shall set up a special personal information protection agency to manage personal information and issue strict policies to maintain information security. In China's future personal information protection law, personal information should be standardized and embodied as a separate right.

If the legislation is simple and the implementation is not emphasized, then the legal provisions will be a dead letter. On the basis of legislation, we must strictly enforce the law and strengthen supervision over all aspects of law enforcement. If personal information is infringed, legal remedies can be taken to resolve the problem through litigation procedures. Therefore, judicial institutions and social organizations are required to monitor the market environment and strive to reduce the leakage of personal information at the source and in all aspects of information collection and processing, rather than seeking solutions and approaches after the leakage[4].

5.3. Deepening International Personal Information Security Cooperation

The security of personal information is closely related to the life of each citizen, and also related to the security of the country and the stability of society. In the era of big data, the collection and dissemination of information have become more convenient. Information is not only disseminated within a country, but also easily spread abroad. Because the social nature and laws and regulations of different countries are different, it is difficult to protect personal information across countries. It is no longer possible to meet the needs of society simply relying on domestic power to protect personal information security. Under such circumstances, it is necessary to deepen state cooperation, strengthen communication between countries, establish institutional mechanisms to protect information security, jointly protect information security, and maintain social harmony and stability.

5.4. Raising awareness of citizens' legal protection of personal information

As an ordinary citizen, we should also have the awareness of protecting personal information security. Pay attention to the use of social tools, avoid the annotation of personal information as much as possible, and block the key information when drying photos on QQ space, friends circle, microblog and other platforms. Don't use or use irregular websites or apps. Don't fill in the questionnaire or vote at will online or offline. Be careful to click the link of free gift collection or free fortune telling. When installing mobile software, choose "allow" carefully. Try not to set the mobile phone software as automatic login, and set the password every time you log in; the password should be changed regularly, with high security factor, and do not use a simple number group.

6. Conclusion

Compared with traditional personal information protection, it is more difficult to determine the subject of tort in the era of big data, for the acts of tort are more diverse, and personal information is more likely to be illegally obtained and sold. However, China's personal information protection is obviously insufficient. There are problems of scattered laws, poor operability, vague scope of personal information, imperfect informed consent rules, immature rights relief mechanism, insufficient administrative protection, and insufficient criminal law protection. In view of this, it is necessary to improve the legal protection of personal information in the era of big data by accelerating the legislative work on personal information protection, improving judicial protection procedures and judicial remedies, and deepening international personal information security cooperation, raising the awareness of citizens' legal protection of personal information, etc.

Acknowledgement

This article is part of the research results of the scientific research project "Golden Courses" of the Research Project of Wuhan Dong hu University (No : 201906)

References

[1] Van der Geer, J., Hanraads, J.A.J., Lupton, R.A. (2010) The art of writing a scientific article. J. Sci. Commun., 163: 51–59.

[2] Landau S.（2015）Control use of data to protect privacy. Science. 347: 504-506.

[3] Kuner C, Cate F H, Millard C, et al. (2015) Risk management in data protection .International Data Privacy Law.28:91-92.
[4] Tene O, Polonetsky J. (2013) Big Data for All: Privacy and User Control in the Age of Analytics.Social Science Electronic Publishing.239:36.

The Visualization Analysis of Artificial Intelligence Research in Recent 10 Years in China

Wei Zhang[1], Liling Li[2], Jialiang Liu[1*]

[1]Office of Educational Administration, Foshan Polytechnic, Foshan, Guangdong, 528137 China

[2]College of Business Administration, Foshan Polytechnic, Foshan, Guangdong, 528137 China

[*]Corresponding author's e-mail: 124521287@qq.com

Abstract. Based on the research literature of artificial intelligence in 2010-2019 in CNKI database, this paper uses the econometric analysis method and CiteSpace software to draw the knowledge map and chart of literature distribution, authors, institutions, discipline distribution, keywords and other dimensions. The paper analyzes the current situation and process of artificial intelligence research in China in the past decade from the perspective of visualization, and abstracts six research topics of artificial intelligence.

1. Introduction

According to CNKI, the first artificial intelligence research document was published in the journal philosophy cluster translation in 1978. It has been 40 years since then. A number of researchers have summarized the research status of artificial intelligence, such as Wei Lv. With the rapid rise of artificial intelligence in recent three years, the research topic of artificial intelligence has changed. From the perspective of data visualization, this paper analyzes the current research situation in the field of artificial intelligence in China by means of econometric analysis and big data analysis technology.

2. Data sources

In CNKI database, this paper searches artificial intelligence research literature from 2010 to 2019 with Su = "artificial intelligence" as the search condition, and obtains 13265 literatures in total. The distribution of literature in each year is shown in Figure 1.

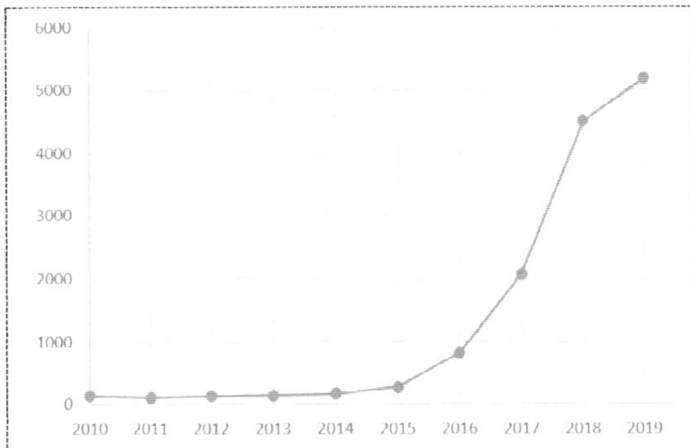

Figure 1. Distribution of artificial intelligence research literature in recent ten years.

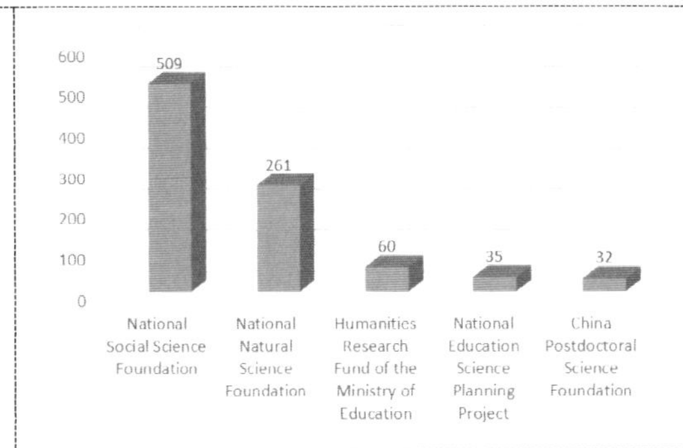

Figure 2. Statistics of artificial intelligence research funding.

3. Research ideas and methods

3.1. Research ideas

This paper analyzes the research literature of artificial intelligence from 2010 to now by using the measurement method. The research dimensions include research team, research institution, research level, and research discipline. The paper uses CiteSpace software to draw keyword knowledge map based on all the research literature, and forms the research cluster of artificial intelligence. Finally, the paper summarizes the research topics and development trends in the field of artificial intelligence in China.

3.2. Research method

3.2.1. Knowledge map analysis. Mapping knowledge domain is a series of different graphs that show the relationship between the development process and structure of knowledge. It uses visualization technology to describe knowledge resources and their carriers, excavate, analyze, construct, draw, and display knowledge and their interrelations[1]. Knowledge map can reveal the general development of a field and discipline from macro, meso and micro levels. It enables people to comprehensively examine the structure and research hotspot of a discipline from various perspectives[2].

3.2.2. Co - word Analysis. Co - word analysis is a method of content analysis. It mainly analyzes the phenomenon that professional terms appear in a document together. These professional terms can express the research topic or research direction of a subject field. Then it can judge the relationship between subjects in the discipline field and show the research structure of the discipline[3]. The more frequently words appear in the same document, the closer the relationship between the two themes is[4].

4. Analysis on the research stage of artificial intelligence

The number of literature can reflect the heat and breadth of research topics. As can be seen from Figure 1, the development of artificial intelligence research in China is slow before 2014, with less than 200 annual research literatures. Since 2015, artificial intelligence has developed rapidly. In 2018, 4504 papers have been published, 34 times higher than that in 2010. It can be seen that artificial intelligence has been widely concerned. Therefore, the research of artificial intelligence in China has gone through two stages. The first stage is the exploration and development period before 2014. During this period, the development of artificial intelligence has experienced many twists and turns. 9 June 2014, General secretary Jinping Xi put forward that "Robot Revolution" is expected to become a breakthrough point and an important growth point of the third industrial revolution, which will affect the global manufacturing industry, and China will become the world's largest robot market[5]. Subsequently, the state issued "*China made 2025*", "*Robot industry development plan (2016 to 2020)*" and "*'Internet +'*

artificial intelligence three year action plan". The government creates a good environment for the development of artificial intelligence in China. From 2015, China's artificial intelligence research has entered a rapid development stage.

5. An analysis of artificial intelligence research funding

Among the documents searched, some of them are supported by national and provincial funds. As shown in Figure 2. The top three funds for supporting literature are National Social Science Foundation (509 articles), National Natural Science Foundation (261 articles) and Ministry of Education Humanities and Social Science Research Fund (60 articles). There are 15 project funds supporting more than 10 documents. By comparing the total number of funded documents with the total number of retrieved documents, it is found that the number of funded documents is not enough.

6. Research level and discipline distribution of artificial intelligence

The research level of artificial intelligence literature is relatively concentrated. There are 3058 (23%) literatures belong to engineering technology level, 2152 (16%) literatures belong to basic research level, and 2004 (15%) literatures belong to industry guidance level, as shown in Figure 3.

By analyzing the subjects of artificial intelligence literature, it is found that control engineering (7015 papers), communication economy (1437 papers) and Computer Science (1280 papers) are the subjects with more literature distribution, and the number of literature in the three subjects accounts for 73% of the total literature. It can be seen that the research subjects of artificial intelligence are relatively concentrated, as shown in Figure 4.

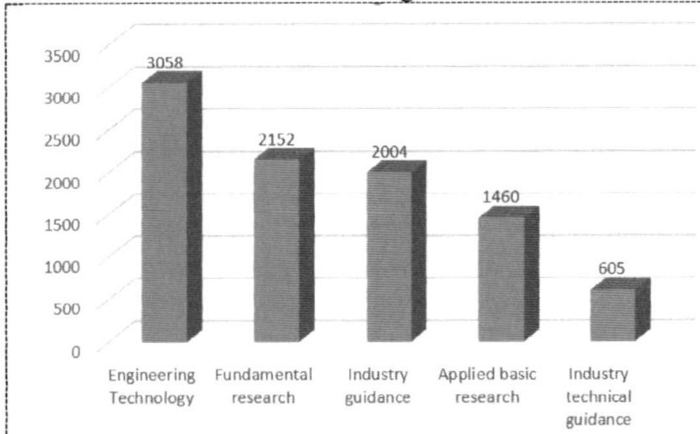

Figure 3. Research level distribution of artificial intelligence.

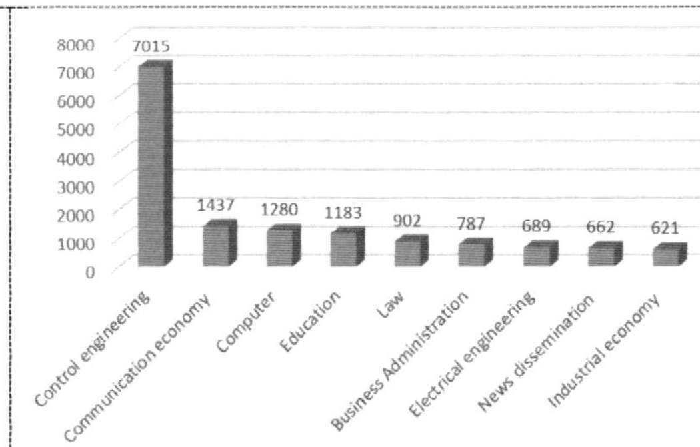

Figure 4. Discipline Distribution of the top ten documents.

7. Distribution of artificial intelligence research institutions

This paper uses CiteSpace to analyze the data of institution. In the CiteSpace software, we set the time from 2010 to 2019 in the "Time Slicing" function area, set the "Year Per Slice" to 1, select the "Institution" function in the "Node Types" function area, select the "Top 50 levels" in the "Selection Criteria" policy area, and other parameters default. Then form the research organization map as shown in Figure 5. It can be seen that the three research institutions with more publications are East China University of political science and Law (108 papers), Tsinghua University (101 papers) and Beijing Normal University (87 papers). Of all the research institutions, there are 37 institutions have published more than 30 articles, all of which are universities and research institutions.

IOP Conf. Series: Materials Science and Engineering **806** (2020) 012033 doi:10.1088/1757-899X/806/1/012033

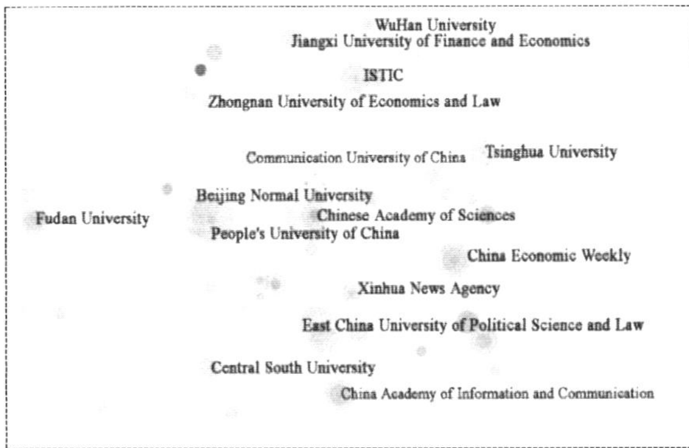

Figure 5. Artificial intelligence research organization map.

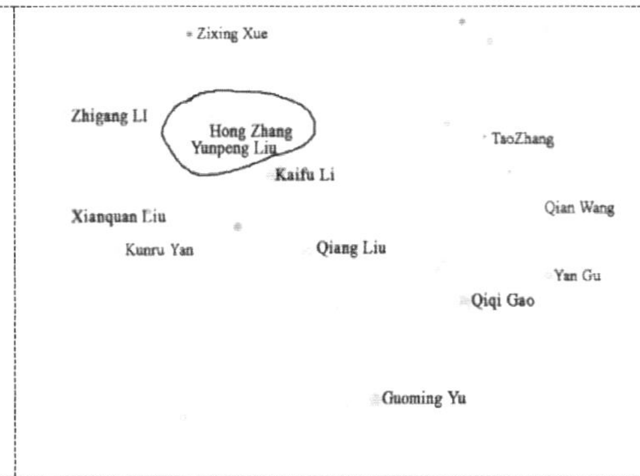

Figure 6. Collaboration map of artificial intelligence research authors.

8. Analysis of artificial intelligence research authors

We use CiteSpace to analyze the author data. In the CiteSpace software, we set the time from 2010 to 2019 in the "Time Slicing" function area, set the "Year Per Slice" to 1, select the " Author " function in the "Node Types" function area, select the "Top 50 levels" in the "Selection Criteria" policy area, and other parameters default. Then form the research organization map as shown in Figure 5. It can be seen that there are 32 authors who have published more than 5 articles. Qiqi Gao (19 papers) and Xianquan Liu (19 papers) are the authors who have published a lot of literature. We find several research teams, such as Hong Zhang and Yunpeng Liu.

9. Analysis on the research topic of artificial intelligence

9.1. Keywords analysis

High frequency keywords can reflect the research topic of a field. In this paper, we use bicomb2 software to obtain 13512 keywords. he standard of high frequency words selection is that the cumulative frequency of high frequency words is about 40% of the total frequency[6]. Therefore, the article selects 52 high-frequency keywords with the word frequency threshold of 51, and the cumulative percentage of word frequency accounts for 40.0%, as shown in Table 1. It can be seen that the research on "artificial intelligence" is relatively extensive, with word frequency reaching 9562, accounting for 21.23% of the cumulative percentage of keyword word frequency. The second key word is "application", which shows that artificial intelligence pays more attention to application research. The related application fields include electrical automation, AI, education, financial accounting, mechanical and electronic engineering, etc. In addition, other high-frequency keywords include "artificial intelligence technology", "intelligent robot", "big data", "AI technology", "network technology" and "machine learning". These high-frequency keywords also represent the main research topics in the field of artificial intelligence.

Table 1. Artificial intelligence high frequency keywords with threshold of 51(partial display).

keywords	threshold	Percentage	Cumulative percentage
Artificial intelligence	9562	21.23	21.23
Application	814	1.81	23.04
Artificial intelligence technology	708	1.57	24.61
Intelligent robot	688	1.53	26.14
PLC	485	1.08	27.22
Big data	440	0.98	28.20
AI	413	0.92	29.11
Network technique	365	0.81	29.92
Machine learning	336	0.75	30.67
Field of artificial intelligence	326	0.72	31.39
Computer	304	0.68	32.07
Deep learning	240	0.53	32.60
Personnel training	208	0.46	33.06
Financial accounting	172	0.38	33.45
Neural network	161	0.36	33.80
Human	148	0.33	34.13
IOT	114	0.25	34.39
Mechanical and Electronic Engineering	111	0.25	34.63
Development	109	0.24	34.87
Algorithm	107	0.24	35.11
Computer network	104	0.23	35.34
Influence	103	0.23	35.57
Library	102	0.23	35.80
The State Council	99	0.22	36.02
Human intelligence	99	0.22	36.24

9.2. High frequency keyword social network analysis

After extracting high-frequency keywords, BICOMB software is used to generate high-frequency keyword co-word matrix, as shown in Table 2. It can be seen that the co occurrence frequency of artificial intelligence, intelligent robot, big data and other key words is high. In order to further analyze the logical relationship between the data, UCINET is used to generate a visualization map of social network relationship for high-frequency keywords, as shown in Figure 7, and then the following conclusions are drawn.

Table 2. Artificial intelligence high frequency keyword co word matrix (partial display).

	Artificial intelligence	Application	Artificial intelligence technology	Intelligent robo	PLC	Big data
Artificial intelligence	9562	577	18	651	314	389
Application	577	814	200	12	192	14
Artificial intelligence technology	18	200	708	22	156	23
Intelligent robot	651	12	22	688	0	17
PLC	314	192	156	0	485	1
Big data	389	14	23	17	1	440

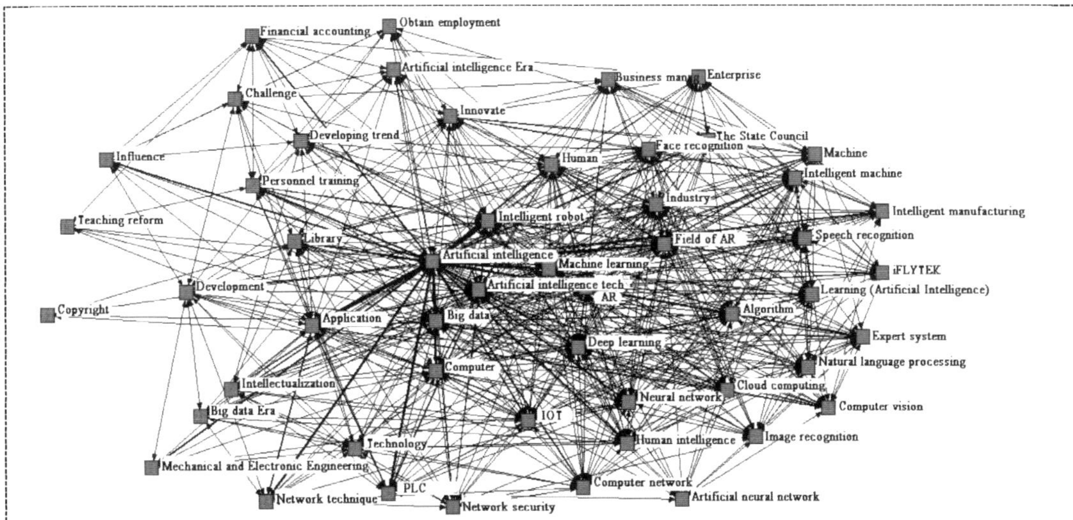

Figure 7. High frequency keyword network relationship map.

9.2.1. The key word "artificial intelligence" is at the center of the research map, which is the core of the research category. Other key words have direct or indirect connection with "artificial intelligence".

9.2.2. This map forms a core area, with the key words of artificial intelligence, intelligent robot, machine learning, AI, artificial intelligence technology, deep learning, big data and computer in the core area. These keywords are basically consistent with the high-frequency keywords mentioned above, and also basically represent the main research topics of artificial intelligence research in China.

9.2.3. The key words "image recognition", "iFLYTEK", "expert system", "natural language processing", "speech recognition", "intelligent machine", "learning (Artificial Intelligence)" are at the edge of the relationship map. These key words have not been paid enough attention, but there is a lot of research space.

9.2.4. The connection and distance between keywords in the social network relationship map represent the closeness between keywords. It can be seen from Figure 7 that "artificial intelligence" is the core keyword as well as the search term. It is close to the three key words of "artificial intelligence technology", "big data" and "intelligent robot". It can be considered that the four key words are closely related.

9.3. Analysis and conclusion of high frequency keyword map

We use CiteSpace software to draw keywords knowledge map. In the CiteSpace software, we set the time from 2010 to 2019 in the "Time Slicing" function area, set the "Year Per Slice" to 1, select the " keyword " function in the "Node Types" function area, select the "Top 50 levels" in the "Selection Criteria" policy area, and other parameters default. CiteSpace uses TF * IDF weighting algorithm to automatically cluster and visualize, and obtains the keyword co-occurrence map as shown in Figure 8. The modularity value of the map (Q value for short) is 0.5665 > 0.3 (Q value > 0.3 means that the structure of the map is significant), and the mean contour value of the map (S value for short) is 0.5585 > 0.5 (S value > 0.5 means that the clustering is reasonable), so the map is effective. It can be seen that the research of artificial intelligence in China mainly forms 8 clustering maps. After the merging of similar clusters, this paper proposes 6 main research topics of artificial intelligence research in China.

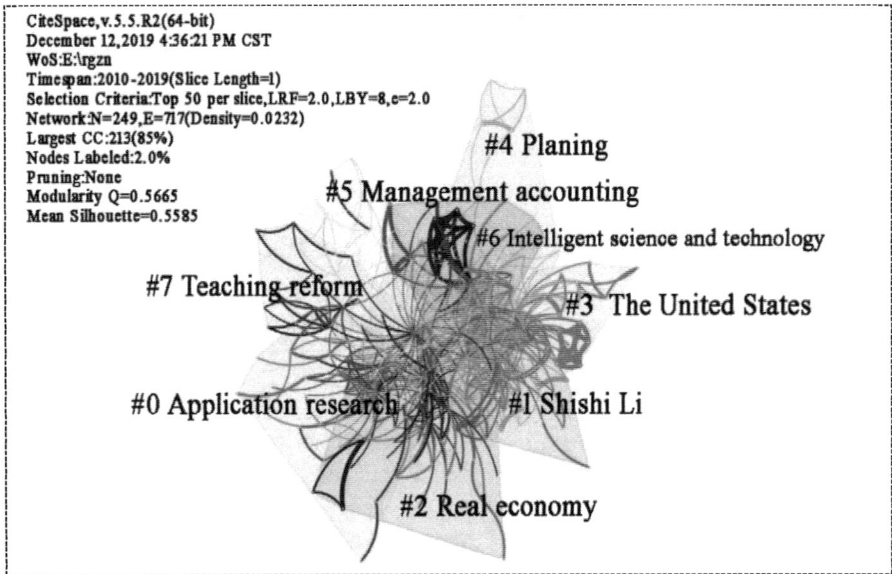

Figure 8. Artificial intelligence keyword knowledge map.

9.3.1. The application of artificial intelligence in various fields. This topic corresponds to two clusters of "Application Research" and "management accounting", represented by high-frequency keywords such as "application", "computer", "electrical automation", "mechanical and electronic engineering", "financial accounting" and "law". This topic analyzes the role of artificial intelligence in the transformation of various fields and promotes the transformation of various fields, especially in accounting, machinery, electrical and other intelligent manufacturing fields. At present, the mature artificial intelligence industry includes pattern recognition (gait recognition, identity recognition, etc.), speech recognition (smart home, smart car, etc.), expert system (intelligent robot, intelligent driving, etc.), human-computer game (online game, etc.) .There are artificial intelligence applications in management accounting, legal consulting, modern medicine, e-commerce shopping guide and other fields.

*9.3.2. The study of alphago and Shishi Li.*This cluster mainly studies the results of alphago and Shishi Li competition，The research contents include alphago algorithm, machine learning technology, deep reinforcement learning and the Enlightenment of alphago to the development of artificial intelligence in various fields。

9.3.3. Research on the challenge and opportunity of artificial intelligence to real economy. This topic is proposed for clustering "#2 real economy", In the short term, artificial intelligence will first bring impact and challenges to the real economy, but in the long term, AI will promote economic growth and social welfare, resulting in intelligent economy, digital economy, etc. Qingqing Cheng put forward the way to integrate artificial intelligence and real economy, that is, to widely apply artificial intelligence equipment to all industries of real economy, widely apply artificial intelligence technology to all links of real economy, so as to realize the intelligence of real economy[7]. The research purpose of this clustering is to promote the transformation of the real economy with the aid of artificial intelligence technology, promote the linkage development of big data, artificial intelligence and the real economy, and realize the "unmanned operation" mode.

9.3.4. The study on the reference of artificial intelligence research in the United States and other countries. This topic corresponds to the cluster "#3 the United States ". This clustering mainly refers to the experience of artificial intelligence research in the United States and other countries, and analyzes the current situation of artificial intelligence research in the world. Zhiming Yan analyzed the reports of "preparing for the future of artificial intelligence" and "national strategic plan for artificial intelligence research and development". On this basis, he expounded the connotation, key technology and application trend of educational artificial intelligence (EAI)[8]. Based on the analysis of artificial intelligence reports such as "preparing for the future of artificial intelligence" in the United States, Wei Hu proposed that the direction of education reform is to cultivate the labor force needed in the era of artificial intelligence[9]. In addition, Shuwang Zhao. Tianchi Lu, etc. elaborated the research of artificial intelligence in American publishing, journalism, intelligence, criminal law, national defense.

9.3.5. Research on artificial intelligence in planning of various fields. This topic is derived from the clustering "# 4 planning"， The research contents mainly include the application of artificial intelligence in urban planning, power grid planning, publishing planning, aerospace planning, and the development planning of artificial intelligence itself. Managers use big data analysis technology of artificial intelligence to find business rules by analyzing current data, which is the basis of planning and decision-making.

9.3.6. Research on Artificial Intelligence Education. This topic corresponds to two clusters of "#6 intelligent science and technology" and "#7 teaching reform". The core keywords include "deep learning", "machine learning", "expert system", "AR technology", "VR technology", "personnel training", "artificial intelligence education", "teaching reform", "intelligent science and technology", etc. The research content includes two aspects: first, take artificial intelligence as a means to carry out the research of artificial intelligence and learning; second, take artificial intelligence as a content to carry out the research of artificial intelligence education. China's high-end IT enterprises have set up research institutions centered on artificial intelligence technology, such as Huawei Noah Ark laboratory, Baidu deep learning research, etc.

10. Analysis of burst term in artificial intelligence research

Burst terms are one or more high-frequency keywords, which represent the new research direction in this field and may be a new potential research cluster. Therefore, extracting emergent words can predict the research direction of artificial intelligence. Based on the key words knowledge map in Figure 3, this paper extracts the burst terms by comparing the centrality (the position of key words in the key words map network) and burst value, as shown in Table 4. It can be predicted that the research on the relationship between artificial intelligence and human beings, the application of artificial intelligence in various fields and the research on artificial intelligence expert system will become new research direction.

Table 3. Keywords with the strongest citation bursts（partial display）.

Keywords	Year	Strength	Begin	End	2010 - 2019
bilingual education	2010	3.295	2010	2012	
Power System	2010	5.876	2010	2015	
data mining	2010	9.4062	2010	2015	
genetic algorithm	2010	16.4771	2010	2015	
expert system	2010	22.0242	2010	2015	
curriculum construction	2010	3.9543	2010	2012	
knowledge base	2010	2.9368	2010	2015	
fault diagnosis	2010	9.9949	2010	2015	
teaching reform	2010	9.4009	2010	2014	
neural network	2010	14.8334	2010	2015	
postgraduate teaching	2010	2.6818	2011	2012	
application field	2010	2.6818	2011	2012	
CAI	2010	2.6818	2011	2012	
practice teaching	2010	2.6818	2011	2012	
PLC	2010	13.8611	2011	2015	
artificial neural network	2010	8.3905	2011	2015	

Table 4. Ranking of burst value of high frequency keywords.

Keywords	Burst	Frequency	centrality	Year
human	52.99	141	0.08	2015
field	32.76	105	0.06	2015
expert system	22.02	68	0.09	2010
industry	21.39	61	0.04	2015

11. Analysis of the research process of artificial intelligence

The Timeline view of CiteSpace software knowledge map can show the development process of keyword clustering, as shown in Figure 9. Clustering "#0application research", "#2 real economy" and "#5 management accounting" run through the starting and ending time of literature retrieval (from 2010 to 2019). These three research clusters have strong development momentum, and may continue to develop, and become the main research topics of artificial intelligence in the future. Clustering "#1 Shishi Li" started in 2010, but ended in 2016, indicating that the research on "Shi shi Li and alphago" gradually weakened after 2016.

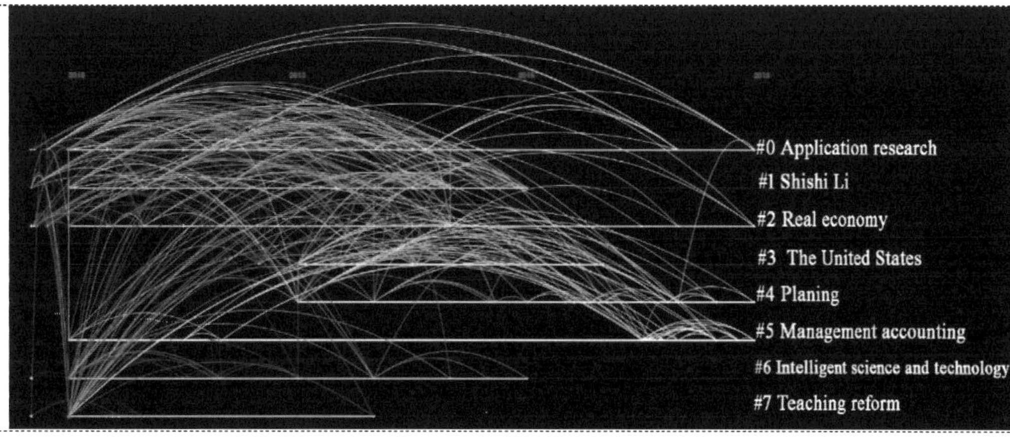

Figure 9. Timeline view of artificial intelligence keyword map.

The connection between clusters reflects the close relationship between clusters. From figure 8, it can be seen that there are many connections between cluster "#0 application research" and other clusters, indicating that cluster "#0 application research" is the main research cluster of artificial intelligence research system, and its research process is synchronous and staggered with other clusters or themes.

As shown in Table 3, in the research process of artificial intelligence, new keywords with high salience value are constantly emerging, such as "human", "industry" and other keywords in 2015, with high burst value, which will also be recorded in the research process of artificial intelligence.

12. summary

This paper analyzes the development trace of artificial intelligence in China in the past ten years through the research of data visualization from many angles, such as the author and cooperative organization of artificial intelligence research. Since 2015, artificial intelligence has developed rapidly and has become the productivity of various industries, playing an important role in financial accounting, real economy and new industries. Based on the analysis of keyword knowledge map, this paper summarizes six research topics of artificial intelligence. Based on the above research conclusions and related research results, this paper puts forward two thoughts.

12.1. We should strengthen the integration of artificial intelligence and education.

As shown in Figure 9, the time line of clustering "Chen 7 teaching reform" stopped in 2014, which is not conducive to the cultivation of artificial intelligence talents. The lack of artificial intelligence talents will inevitably lead to slow development of other research clusters. Therefore, the cluster "Chen 7 teaching reform" still needs to continue to develop. We should vigorously develop artificial intelligence education, develop artificial intelligence teaching courses, and serve the " artificial intelligence +" strategy of various industries. Artificial intelligence can promote the transformation of education, but its strength is not enough. We should introduce the elements of artificial intelligence into the teaching concept, teaching content, teaching method, teaching mode and teaching assessment, so as to train the excellent talents in line with the future industry.

12.2. Develop artificial intelligence standards.

Hawking said: the all-round development of artificial intelligence may lead to the extinction of human beings[10]. The state highly supports the development of artificial intelligence, but also restricts its products. Without standards, it will inevitably lead to product flooding and affect human health and survival. Benign artificial intelligence products contribute to the formation of artificial intelligence culture.

References

[1] Chen,Y..Liu.Z.R.(2005) The rise of mapping knowledge domain. scientific research,23:149-154.
[2] Ren.H.J.,Zhang.Z.Q.(2009) Research on the development of scientific knowledge map based on Bibliometrics. Journal of Information,12: 86-90.
[3] Zhang.Q.,Ma.F.C.(2007) Research paradigm of knowledge management in foreign countries- Using coinage analysis. Journal of Management Sciences in China,6:65-66.
[4] Zhong,W.J.,Li,J.,Yang,X.J.(2008) Research on the method of CO word analysis (3) - the principle and characteristics of the method of CO word cluster analysis. Journal of Information,7:118-120.
[5] CNTV. (2014) Speech by Xi Jinping at the 17th academician conference of the Chinese Academy of Sciences and the 12th academician conference of the Chinese Academy of Engineering. http://news.cntv.cn/2014/06/09/ARTI1402316419354858.shtml.
[6] Chen,Y.L.(2012) The historical evolution of the main research fields of educational technology in China-clustering analysis of key words and subject words based on CNKI. E-education Research,8:36-42.
[7] Cheng,Q.Q.(2018) Research on the integration path of artificial intelligence and real economy. Journal of Beijing Institute of Economics and Management,4:3-9.
[8] Yan,Z.M.,Tang， X.X.,Qin,X.,Zhang,F.,Duan,Y.M.(2017) The connotation, key technology and application trend of EAI- analysis of the report "*Prepare for the future of AI*" and "*National AI research and development strategic plan*" in the United States. Journal of Distance Education,1:26-35.
[9] Hu.W.(2019) Education reform in the age of artificial intelligence: background, direction and path.-analysis based on American AI report. Modern Educational Technology,7:12-17.
[10] CYOL.(2014) Artificial intelligence may cause human extinction. http://zqb.cyol.com/html/2014-12/04/nw.D110000zgqnb_20141204_2-04.htm.

IOP Conf. Series: Materials Science and Engineering **806** (2020) 012034 doi:10.1088/1757-899X/806/1/012034

Modelling and Simulation of Facility Planning Problem Based on Improved SLP Method

Bo Zhang[1]*, Xianlin Wang[2]

[1] National Research Base of Intelligent Manufacturing Service, Chongqing Technology and Business University, Nan'an District. Chongqing, China

[2] Department of Literature and Journalism, Sichuan University. Sichuan, China

*Corresponding author's e-mail: arrow_zb@outlook.com

Abstract. Facility planning plays a critical role in various fields. An excellent facility layout can minimize the losses caused by the connection between facility units in production operations. In this paper, we improved the traditional Systematic Layout Planning (SLP) method, combine it with the multi-objective decision optimization model. The Improved SLP (ISLP) method is described in detail in this paper, including its principles, process of building model, and calculation methods. To make the ISLP more suitable for research and practical problems, we established a mathematical model for it. This optimization model was also applied to the existing facility layout problem to realistically validate the effectiveness of the ISLP and the model. Results on the established model were found first, then the Flexsim simulation software was used to simulate the layout of the facility before and after the improvement. Finally, the results of the experiment were analysed and compared.

1. Introduction

Facility planning plays an increasingly significant role in all walks of life applications, for instance, from assembly line processing factory, logistics sorting centre, experimental base, testing centre, etc., to kitchen layout even. The research content of facility planning mainly focuses on the layout planning of production or laboratory equipment, so as to reduce the loss or waste caused by the interrelationships among each facilities unit such as handling and transmission. A manufacturing company or factory with a well-designed facility layout can at least reduce operating costs by more than 50% [1]. Therefore, more and more attention has been paid to the study of facility planning.

However, it is not easy to design an excellent facility layout in practical applications, either it is difficult to make sure whether the designed new layout is good enough, or it can't be easily moved the existing design layout to validate new design. Therefore, establishing an optimized model for the facilities to be planned through a systematic approach, finding the optimization result with the related method, and verifying the optimization results with the help of the simulation software are of remarkable research value.

The loop layout is one well known general types of design used in production systems [2], the only downside is that it does not take into account the human factors in the facility planning [3]. Ganesharajah et al. [4] reported and discussed an approach called Automated Guided Vehicle (AGV) system that provides greater flexibility, better space utilization, and lower operating costs, but ignores the connections between non-functional areas. The SLP method was proposed by Muther in 1961 [5], which has pushed the study of facility layout issues from the qualitative research stage to the quantitative research stage. The chart analysis and the graphical model were used as main technical means, at the same time, a new concept of Mathematically Quantified Relational Levels(MQRL) [5] was defined to describe the logistics relationship and the non-logistics relationship between operations units, where the former is the direct relationship between facility units, such as the flow relationship between the upper and lower links on the assembly line, the latter is the indirect relationship between facility units, such as the influence of units, the connection of functional areas, which achieved a quantitative analysis of the entire process of facility layout and established a quantitative evaluation model of facility layout design.

However, there is a clear defect in the SLP method. It is relatively easy to be affected by the subjective factors of the layout planner when solving the problem of facility layout, correspondingly, the layout plan for the design varies from person to person and varies. And a host of tedious manual adjustments are needed to meet the requirements of actual conditions in practical applications, the workload of manual adjustment is extremely large especially when there are many partitions.

To overcome the foregoing disadvantages, in this paper, we have improved the traditional SLP method, combine it with the Multi-Objective Optimization Model (MOOM) [6][7]. MOOM can connect the logistics relationship and the non-logistics relationship mentioned above, making the algorithm and ISLP simpler without losing any information. We introduced in detail the ISLP and established a mathematical model for it. Moreover, the existing facilities planning was relocated through the ISLP method, the layout of the facilities before and after the improvement were simulated based on the Flexsim simulation software, and the experiment records are compared and analysed to verify the effectiveness of the algorithm and model.

2. Methodology

In the vast majority of facility planning problem, the reason why the logistics relationship and the non-logistics relationship between units and units generally are considered as the main influence and study factors is that the latter already contains all the factors that cannot be directly quantified except for the former that can be quantified [8][9]. This paper also focuses on these two factors as the optimization goal. Based on the combination of the traditional SLP method and the multi-objective decision optimization model, the static layout problem model of the multi-objective optimization system is established, which makes us propose a new method to improve the static facility planning scheme, so as to minimize the defects caused by the traditional SLP method. Only a small amount of unquantifiable optimization goals is directly scored by experts as non-logistic factors, and the scoring results are quantified in conjunction with multi-objective decision algorithms. It not only simplifies the complexity of the model operation, but also ensures the diversity of the static layout of the facility system for the optimization objectives. The following is a step-by-step explanation of the method and its principles.

1) Firstly, the traditional SLP method is still needed to describe and get the non-logistics relationship among the facility units. At the same time, quantify the logistics relationship among each unit to obtain the flow rate table, as shown in the part 1 of figure 1. In this part, input data is critical, including product details, quantity, route planning, service area, time description, which is the basis of everything and requires an amount of argumentation and design to ensure that an excellent design layout can be designed.

2) Secondly, after obtaining the flow rate table and the non-logistics relationship, the mathematical model method completely replaces the manual layout adjustment to quantify non-quantifiable non-logistic factors. And then set up the objective functions of non-logistics relations and the logistics relations respectively, as shown in the part 2 of figure 1.

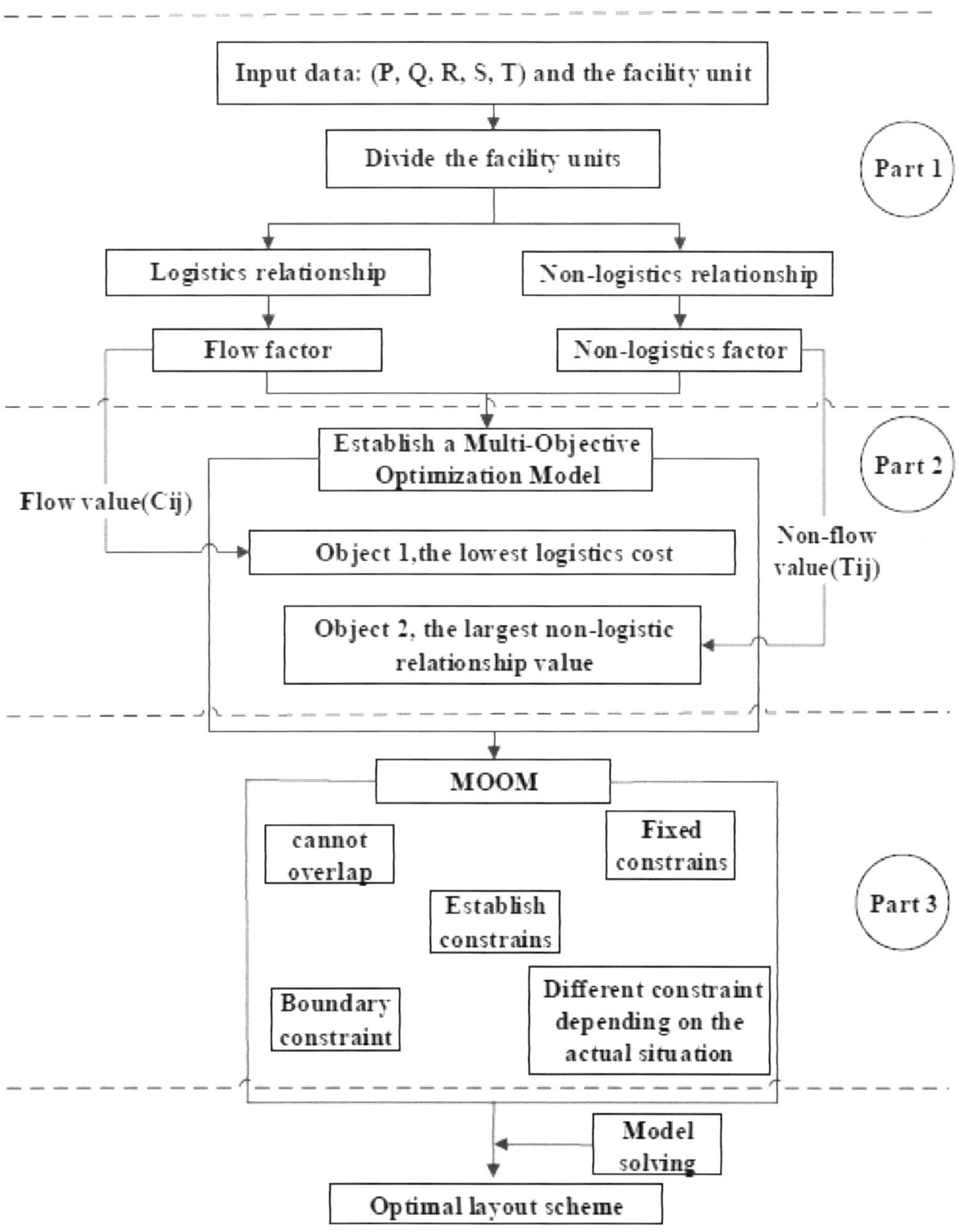

Figure 1. The ISLP structure diagram.

In figure 1. **P**, **Q**, **R**, **S**, **T** in the part 1 represent the five basic elements of SLP method, product, quantity, route, service, time. The data of the part 1 can be adjusted according to actual demand. The objective function of the part 2 mainly considers the logistics factor and the non-logistics factor in this paper. If the realistic situation is more complicated, the objective function can be increased accordingly, which also applies to the part 3. This paper and diagram only considers the general situation.

3) Finally, some other requirements that need to be met in the facility planning layout are considered to be the constraint and then appear in the model in this way, as shown in the part 3.

The entire mathematical model contains two optimization objectives. The MOOD can be used to solve the optimal layout and find the results. The specific improved SLP method structure diagram is shown in figure 1.

3. Mathematical Model

3.1. Establish the objective function

In this model, the goal to be achieved is divided into two parts. The first part is the minimization of quantifiable logistics costs, including planning the shortest logistics path and the functional area with large material flow as close as possible. The second part is the maximization of non-quantitative non-logistics parts, maximizing the convenience of personnel contact between different facilities, and achieving the greatest correlation between office area and production operation. We established the following two objective functions.

1) The objective function of the lowest cost of logistics between facility units can be described by the (1).

$$\text{Min } Z_1 = \sum_{i=1}^{N-1} \sum_{j=i+1}^{N} C_{ij} * Q_{ij} * D_{ij} \tag{1}$$

2) The largest non-logistic relationship value can be described by the (2).

$$\text{Max } Z_2 = \sum_{i=1}^{N-1} \sum_{j=i+1}^{N} T_{ij} * D_{ij} \tag{2}$$

Where **i** and **j** are the facility unit numbers, **N** is the total number of facility units, $\mathbf{C}_{ij}$ is the unit logistics cost per unit distance of facility units **i** to **j**, $\mathbf{Q}_{ij}$ is the average traffic between facility units **i** and **j**, $\mathbf{T}_{ij}$ is the non-logistic correlation value between facility unit **i** and facility unit **j**, which is determined after comprehensively considering the closeness of the non-logistic relationship between facility unit **i** and facility unit **j**, $\mathbf{D}_{ij}$ is the distance between facility units **i** and **j** in the facility layout, showing in (3).

$$D_{ij} = |X_i - X_j| + |Y_i - Y_j| \tag{3}$$

Where $\mathbf{X}_i$, $\mathbf{Y}_i$ is the coordinate values of facility unit **i** in the layout plan of the facility.

3.2. Convert multi-objective optimization decisions to single

The above multi-objectives can be converted into single objective decisions, which is for simplify the target model. Considering the different dimensions of (1) and (2), the unitary factors $\boldsymbol{\mu}_1$and $\boldsymbol{\mu}_2$ are added to unify their dimensions. Meanwhile, the proportion of the two objective functions in the actual application will also be different. Therefore, the weights $\boldsymbol{\theta}_1$ and $\boldsymbol{\theta}_2$ are respectively assigned to the objective function, thereby obtaining the single objective function (4).

$$\text{Min Z} = \mu_1\theta_1 \sum_{i=1}^{N-1} \sum_{j=i+1}^{N} C_{ij} * Q_{ij} * D_{ij} - \mu_2\theta_2 \sum_{i=1}^{N-1} \sum_{j=i+1}^{N} T_{ij} * D_{ij} \tag{4}$$

$$\mu_1 = 1/\sum_{i=1}^{N-1} \sum_{j=i+1}^{N} C_{ij} * Q_{ij} * D \tag{5}$$

$$\mu_2 = 1/\sum_{i=1}^{N-1} \sum_{j=i+1}^{N} T_{ij} \tag{6}$$

$$\theta_1 + \theta_2 = 1 \tag{7}$$

Where **D** is the sum of the length and width of the general planning area. The weights $\boldsymbol{\theta}_1$ and $\boldsymbol{\theta}_2$ depend on the actual application.

3.3. Establish constraints

Whether the SLP method before or after the improvement, it is to better solve practical problems. Therefore, it is critical and necessary to establish the complete constraints according to the requirements of the facility layout.

1) All facility units cannot overlap and there should have a minimum spacing between each two facility units. Get constraint expressions (8), (9).

$$|X_i - X_j| - (L_i + L_j)/2 \geq D_{X_{ij}} \tag{8}$$

$$|Y_i - Y_j| - (W_i + W_j)/2 \geq D_{Y_{ij}} \tag{9}$$

Where $\mathbf{L_i}$ is the length of facility unit **i**, $\mathbf{W_i}$ is the width of the facility unit **i**, $\mathbf{D}_{X_{ij}}$ is the minimum spacing of facility **i** and **j** in the **X**-axis direction and $\mathbf{D}_{Y_{ij}}$ is the minimum spacing of facility **i** and **j** in the **Y**-axis direction.

2) There shall be a certain distance between the facility unit and the perimeter of the layout range. Get constraint expressions (10), (11).

$$L_i/2 + H_i - X_i \leq 0 \tag{10}$$

$$W_i/2 + K_i - Y_i \leq 0 \tag{11}$$

where $\mathbf{H_i}$ is the minimum distance of the facility **i** from the boundary in the **X**-axis direction and $\mathbf{K_i}$ is the minimum distance of the facility **i** from the boundary in the **X**-axis direction.

3) All facility units must be within the layout, which is the size of the facility does not exceed the total length and total width. Get constraint expressions (12), (13).

$$X_i - (H - L_i/2 - H_i) \leq 0 \tag{12}$$

$$Y_i - (K - W_i/2 - K_i) \leq 0 \tag{13}$$

Where **H** is the total length of the planning area in the **X** axis direction and **K** represents the total length of the **Y**-axis direction.

4) Fixed constraints. Some facility units must be fixed in a certain location. Get constraint expressions (14), (15).

$$X_{i*} = X_s \tag{14}$$

$$Y_{i*} = Y_s \tag{15}$$

Where $\boldsymbol{X_{i*}}, \boldsymbol{Y_{i*}}$ is a specific device unit and $\boldsymbol{X_s}, \boldsymbol{Y_s}$ is a fixed value.

5) Non-negative constraints. All factors in the facility planning problem are positive.

6) There are other constraints that exist in the actual problem category. This completely depends on the actual problem and the designer.

4. Model Simulation and Evaluation

According to the general model established in part three, it is applied to the planning of a city's logistics sorting centre facilities. First establish a goal decision model. Then, calculate the optimization result in MATLAB, simulate the result within Flexsim software. Finally, compared the experimental results between before and after optimization to validate the model.

4.1. Describe the experimental background

The experimental object selected in this paper is a sorting centre of a large logistics company. The main task is to classify the goods from various places. These goods are sorted and distributed to all corners of the city. Then, from the moment the goods arrive, the shorter the time the goods stay in the sorting centre, the higher the customer satisfaction.

From the time of the goods arrive to the end of sorting, the goods pass through multiple functional areas. Due to the initial shortage of facilities planning and design, the goods have a residence time in each functional area, and there is also a phenomenon of accumulation of goods, and even time-consuming and labour intensive phenomena, such as manual handling.

Therefore, this paper conducts this experiment by collecting field operation data of this sorting centre in the field, including cargo volume, cargo route, function area function and existing facility layout.

4.2. Build a model based on actual problems.

$$\mathbf{Min}\ Z_1 = \sum_{i=1}^{N-1}\sum_{j=i+1}^{N} C_{ij} * Q_{ij} * D_{ij} + \sum_{i=1}^{N-1}\sum_{j=i+1}^{N} T_{ij} * D_{ij} \tag{16}$$

$$\begin{cases} |X_i - X_j| - (L_i + L_j)/2 \geq D_{X_{ij}} \\ |Y_i - Y_j| - (W_i + W_j)/2 \geq D_{Y_{ij}} \\ L_i/2 + H_i - X_i \leq 0 \\ W_i/2 + K_i - Y_i \leq 0 \\ X_i - (H - L_i/2 - H_i) \leq 0 \\ Y_i - (K - W_i/2 - K_i) \leq 0 \\ X_{i*} = X_s \\ Y_{i*} = Y_s \\ \mu_1 = 1/\sum_{i=1}^{N-1}\sum_{j=i+1}^{N} C_{ij} * Q_{ij} * D \\ \mu_2 = 1/\sum_{i=1}^{N-1}\sum_{j=i+1}^{N} T_{ij} \\ \theta_1 + \theta_2 = 1 \\ D_{ij} = |X_i - X_j| + |Y_i - Y_j| \\ X_{ij}, Y_{ij} \geq 0,\ i,j = 1,\dots,N \end{cases} \tag{17}$$

Among them **H, K,** C_{ij}, **D,** $D_{X_{ij}}$, $D_{Y_{ij}}$, H_i, K_i, θ_1 **and** θ_2 are all known. All data for this experiment was taken from actual sampling.

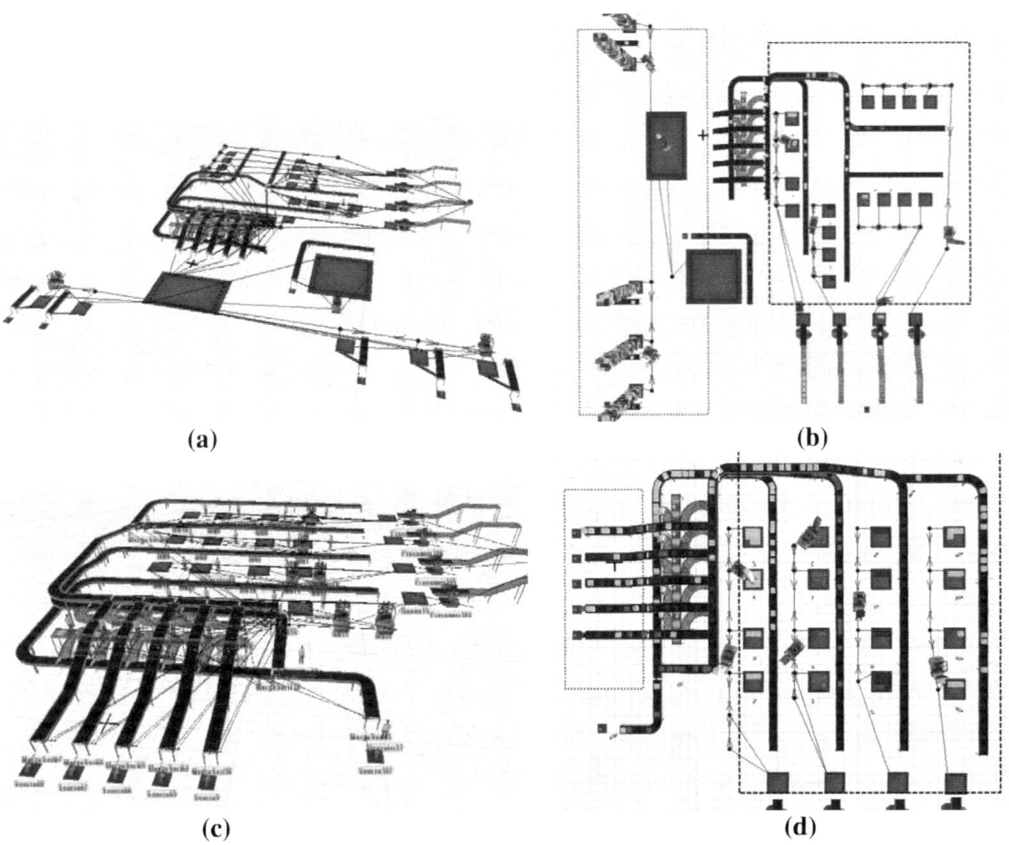

Figure 2. The result of software simulation using Flexsim

In figure 2. The small blue area is the starting position of the goods, the large black area is the cargo temporary storage area, the black long strip is the conveyor belt, and the white, yellow, blue and red small blocks represent goods in different regions. figure 2(a) is the layout of the 3D facility before the improvement, which is the record when the running time is 30 seconds, and the figure 2(b) is the 2D layout before the improvement, which is the record when the running time is 60 seconds. figure 2(c) is the improved 3D facility layout, which is also the record at 30 seconds of runtime. figure 2(d) is the improved 2D layout, which is also the record at 60 seconds.

Calculate the final result through MATLAB and use AutoCAD to draw a plan topology diagram of the facility layout before and after improvement, which is to prepare for the simulation experiment.

4.3. Simulation

Flexsim is a software for logistics system simulation. It can help the users establish and plan the simulation model of process design. The facility planning scheme obtained using the SLP method cannot be implemented on site. Only simulate the actual situation with Flexsim software.

All the data are from the actual site records, and the record time is 30 days; the distribution of sorting and receipt obedience is the simulation of the actual situation. And under the same conditions and at the same time, the quantity of the goods sorted before and after the improvement was compared, and the purpose of verifying the improvement plan and model was finally achieved.

Table 1. Experiments data of simulation

Subject	Time(3600s)		Time(36000s)	
	Before	After	Before	After
Total Sorting	2323	10582	20646	109985
After/Before	4.55		5.33	

[a] Where 'Before' is before the improvement, 'After' is after the improvement.

Under the condition that other parameters are the same, the only change is the layout of facility units. Run simulation models before and after improvement, 3600s and 36000s respectively. Record the experiments outcome and compare it, as is shown in table 1.

From table 1. We can know that when the simulation time is 3600s, the total amount of sorting after improvement is 4.55 times before improvement and it was improved to 5.33 when the simulation time is 36000s.

5. Conclusion

This paper focuses on the modelling and simulation based on the improved SLP facility planning method. By combining the multi-objective decision model with the traditional SLP method, the defects of the traditional SLP method are largely compensated. Based on the existing facility layout, the layout was re-arranged using the improved optimization model, and the facility layout before and after the improvement was simulated using Flexsim software. Experimental results show that the improved facility layout has an overwhelming advantage over the previous facility layout, which also verifies the effectiveness and availability of the algorithm and model.

However, there are still some deficiencies in this paper. For instance, the improved SLP method still has a small number of artificial scoring links. Although it is small, it will still affect the final layout plan. This unquantifiable problem will cause deviations in the layout of the facilities. Although this deviation will be small.

Subsequent research will continue to focus on how best to quantify the relationship between facilities, while focusing more on the establishment of the model.

Acknowledgment

The author sincerely thanks professor Fangyan Yang for her suggestions and comments on the paper. Thanks to professor Zhaochu Yang for his great support. In particular, the first author would like to thank Ms. Wang for her encouragement and support, as well as Mr. Wang Peng and professor Jiafu Su for their help.

References

[1] J. Tompkins, and J. White. "Facilities planning," 1996
[2] R. E. Schofield, "Plant Layout and Materials Handling." Production Engineer 55.4(2009):208.
[3] A. Asef-Vaziri, G. Laporte, "Loop based facility planning and material handling," European Journal of Operational Research, 2005 vol: 164 (1) pp: 1-11
[4] T. Ganesharajah, N. G. Hall, and C. Sriskandarajah. "Design and operational issues in AGV-served manufacturing systems." Annals of Operations Research 76.1(1998):109-154.
[5] G. Cachon, M. Lariviere. "Supply Chain Coordination with Revenue- Sharing Contracts: Strengths and Limitations." Management Science, 2005 vol: 51(1) pp:30-44
[6] Hu, M, and H. Cho. "A probability constrained multi-objective optimization model for CCHP system operation decision support." Applied Energy 116.116(2014):230-242.
[7] Su J.F, Wei M, Liu A.J. A robust predictive-reactive allocating approach, considering random design change in complex product design process. International Journal of Computational Intelligence Systems. 2018, 11(1): 1210-1228.
[8] Z. X. Huang, and H. Luo. "Study on Planning and Design of Production Facility for Bearing Manufacturing Industry Based on SLP." Advanced Materials Research 542-543(2012):428-432.
[9] Su J.F, Li C, Zeng Q, et al. A green closed-loop supply chain coordination mechanism based on third-party recycling[J]. Sustainability, 2019, 11(19): 5335.

IOP Conf. Series: Materials Science and Engineering **806** (2020) 012035 doi:10.1088/1757-899X/806/1/012035

Human Posture Recognition Method Based On Skeleton Vector With Depth Sensor

Jian Wang[1], Xue Hua liu[2 *]

[1]Network Technology Department, South China Institute of Software Engineering, Guangzhou, GuangDong, 510900, China

[2]Computer Department Department, South China Institute of Software Engineering, Guangzhou, GuangDong, 510900, China

[*]Corresponding author's e-mail: wjn@sise.com.cn

Abstract. This paper proposes a human body gesture recognition method. The method collects bone information of the human body through the Kinect depth sensor, and then uses the direction cosine method for feature extraction.Finally, the feature vector is sent to the BP neural network for training and recognition. In the case of less sample training, the system can still accurately identify the five postures of standing, sitting, leaning forward, leaning backward and underarm. The input is composed of low-dimensional bone information, so the speed and recognition response speed during network training is better than the conventional RGB image recognition method. Experiments show that the recognition rate and real-time performance of the system can meet the needs of practical applications.

1. Introduction

Human posture recognition has broad application prospects in many fields such as human-computer interaction, medical health, augmented reality, sports analysis and robot control[1].The identification of human action began in the 1990s[2], the related research and papers increased year by year. There are generally two methods: one is to collect human body posture data through a wearable sensor, and then use a machine learning method such as a support vector machine to classify[3][4]. The advantages are high accuracy and high real-time performance. The disadvantage is that the user must wear the sensor at a specified position on the body, which will reduce the user's comfort, so the attention is not high and the development is slow[5]; the other is based on A non-invasive detection method for computer vision. This method processes and analyzes raw RGB image data collected by a color camera, and uses a deep learning method to recognize a person's posture[6]. The feature is that the recognition rate is high, but the network level is deeper and deeper[7]. The real-time performance in recognition is not good.

2. The realization of ideas

We solve this problem by reducing the dimension of the input features. The depth image sensor can provide the third-dimensional depth data[8][9][10], as shown in Figure 1. First, the image segmentation technique is used to peel off the human body in the image from the background environment, and then according to The "skeletal tracking" human joint points generate a skeleton system, obtain the feature information needed for machine learning, and finally use the machine learning method to construct the classifier, and optimize the classifier to improve the accuracy. The kinect depth sensor used in this study has provided three kinds of data stream output of color image, depth image data and human bone data.

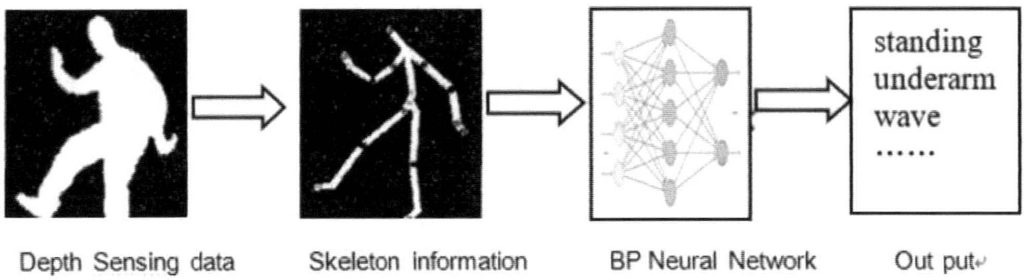

Figure 1 Body motion recognition based on depth sensor

3. Bone vector feature extraction

In order to reduce the feature data dimension and real-time response rate, compared with the commonly used convolution feature extraction, this paper adopts a feature extraction method based on the skeletal vector direction cosine. The specific ideas are as follows: stratify the 20 joint points of the human body, such as Figure 2 shows: first layer: body torso joint points. This layer mainly includes 8 joint points (number 1) such as head, left and right shoulders and spine. The second layer: joint points of the limbs. Includes 8 joint points for the left and right elbows, wrists, left and right knees, and ankle (number 2). The third layer: hands and feet. The remaining 4 joint points of the left and right hands and feet are classified into the third layer (number 3). The role of the palms and feet in the human posture studied in this paper is very small, so it can be ignored.

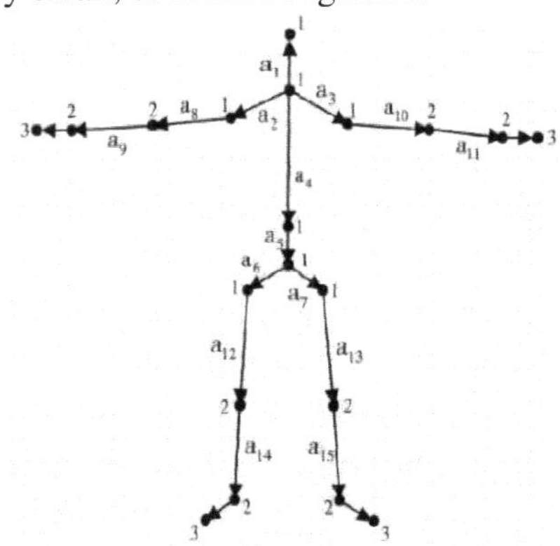

Figure 2 bone information

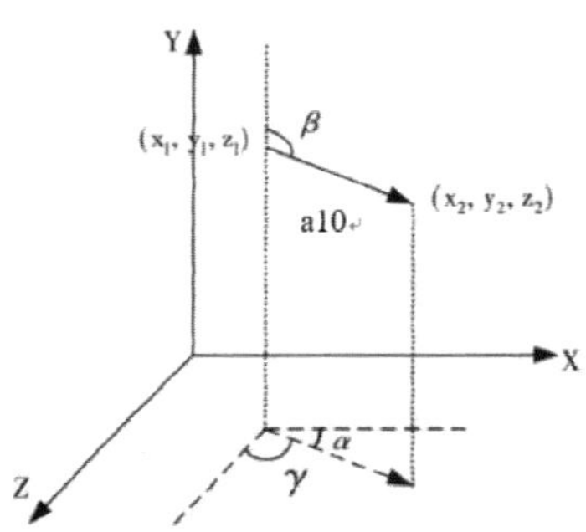

Figure 3 Direction cosine calculation

For Connect each of the 16 joint points to form a skeleton, define it as a vector, mark the 7 vectors of the first layer as { $\mathbf{a_1, a_2, ..., a_8}$}; the second layer is marked as { $\mathbf{a_9, a_9, ..., a_{16}}$}. When people make different postures, they have different position and angle information for each segment of the human body. Define the direction of the cosine of the 16 skeletal vectors to represent a certain type of posture. Now take the shoulder joint of the right hand to the elbow joint (the vector labeled a_8) for detailed algorithm analysis: Set Kinect to obtain the three-dimensional coordinates of the shoulder joint. For ($\mathbf{x_1, y_1, z_1}$), the three-dimensional coordinates of the elbow joint are ($\mathbf{x_2, y_2, z_2}$), as shown in Figure 3. Thus, the skeletal vector can be expressed as (here, rotated image representation, so $\mathbf{a_{10}}$, not $\mathbf{a_8}$): $\mathbf{a_{10}=\{x_2-x_1, y_2-y_1, z_2-z_1\}}$, set vector a_8 and Kinect coordinate system. The three directions are: α, β, γ. then the feature vector of the bone can be defined as $\cos\alpha$, $\cos\beta$, $\cos\gamma$. The formula is as in equation (1) ,(2), (3).

$$\cos\alpha = \frac{x_1-x_2}{\sqrt{(x_2-x_1)^2+(y_2-y_1)^2+(z_2-z_1)^2}} \quad (1)$$

$$\cos\beta = \frac{y_2-y_1}{\sqrt{(x_2-x_1)^2+(y_2-y_1)^2+(z_2-z_1)^2}} \quad (2)$$

$$\cos\gamma = \frac{z_1-z_2}{\sqrt{(x_2-x_1)^2+(y_2-y_1)^2+(z_2-z_1)^2}} \quad (3)$$

Through the above algorithm, the direction cosine of 16 bone vectors can be obtained, which can be sent as a feature value to the neural network for training and recognition.

4. BP neural network design

The structure of the BP neural network is shown in Figure 4. It consists of a 16-node input layer, two 1024-nodes hidden layers, and 5-nodes output layer, where {$\mathbf{x_1, x_2, x_3...x_{16}}$} is the skeletal feature vector input,{$\mathbf{o_1,o_2,o_3, o_4,o_5}$}is the output.

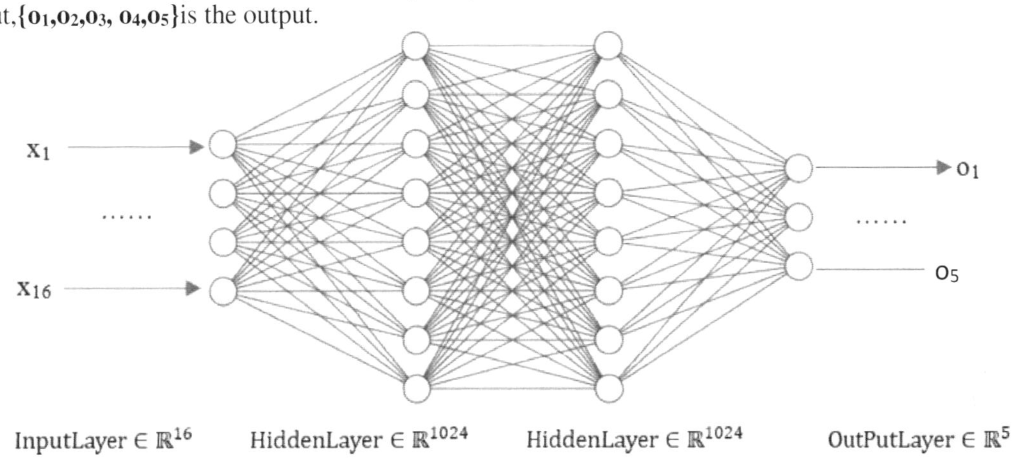

Figure 4. Figure with short caption (caption centred).

The output of the hidden layer is as in equation (4).

$$b_j = f\left(\sum_{i=1}^{N} \omega_{ij} x_i - \theta_j\right) \quad (4)$$

The output of the output layer neuron is is as in equation (5).

$$c_t = f\left(\sum_{j=1}^{p} v_{ij} b_i - r_t\right) \quad (5)$$

Where $\boldsymbol{\omega_{ij}}$ represents the connection weight between the i-th neuron of the input layer and the j-th neuron of the hidden layer;vjt represents the connection right between the j-th neuron of the hidden layer and the t-th neuron of the output layer; $\boldsymbol{\theta_j}$, $\boldsymbol{r_t}$ represents the activation threshold of the corresponding neuron; **f(x)** represents the activation function of the neuron. The network uses the softmax activation function, which maps the output of multiple neurons to the (0,1) interval and can be interpreted as a probability to make multiple classifications. The loss function is as in equation (6).

$$\varepsilon = \frac{1}{K} \sum_{k=1}^{K} \sum_{t=1}^{T} (y_t - c_t)^2 \quad (6)$$

$\boldsymbol{y_t}$ represents the expected output value of the t-th neural network; $\boldsymbol{c_t}$ represents the actual output value of the t-th neural network; when the total number of training samples is **K** the global output error of the network is calculated by the minimum variance. The weight w_{ij} adjustment here uses the gradient descent algorithm. The gradient value d of the connection v between the hidden layer and the output layer neurons is as in equation (7).

$$d_t^k = \left(y_t^k - c_t\right) \cdot (1 - c_t), (t = 1,2, \ldots., q) \quad (7)$$

Use $\boldsymbol{d_t^k}$, $\boldsymbol{b_j}$, $\boldsymbol{v_{jt}}$, and $\boldsymbol{r_t}$ to calculate the new connection weight and threshold between the next hidden layer and the output layer is as in equation (8). (9).

$$v_{jt}(N) = v_{jt}(N-1) + a \cdot d_t^k \cdot b_j \quad (8)$$

$$r_t(N) = r_t(N-1) + a \cdot d_t^k \quad (9)$$

The correction amount e of the connection right between the input layer and the hidden layer as in equation (10).

$$e_j^k = \sum_{t=1}^{q} (d_t v_{jt}) \cdot b_j\left(1 - b_j\right), (j = 1,2, \ldots., p) \quad (10)$$

Calculate the new threshold and connection weight between the next input layer and the middle layer with $\boldsymbol{e_j^k}$, $\boldsymbol{a_i^k}$, $\boldsymbol{\omega_{ij}}$, and $\boldsymbol{\theta_j}$ as in equation (11). (12).

$$\theta_j(N) = \beta \cdot e_j^k + \theta_j(N-1) \quad (11)$$

IOP Conf. Series: Materials Science and Engineering **806** (2020) 012035 doi:10.1088/1757-899X/806/1/012035

$$\omega_{ij}(N) = \beta \cdot e_j^k \cdot a_i^k + \omega_{ij}(N-1), (i = 1,2, \ldots, n) \quad (12)$$

After each sample is learned, adjust the corresponding connection weight until the K samples are all finished learning, then determine whether the global output error function reaches the set convergence limit value until the error function reaches the limit value, and the network training ends; otherwise, if the error is still greater than the set value when the maximum number of learning is reached, the training is also over, then the network training fails.

5. Experiment and result

We experiment our way in Microsoft's kinect second-generation sensor with Microsoft's kinect second-generation sensor, the camera resolution is color 1080p, and the depth camera is 512*424. The software running environment is windows10, VisualStudio2016, python 2.7,Kinect SDK 1.5, and the development language adopts C#.Net. to analyze the results in more detail, we show the confusion matrix (Tab. 1)and rate of activity recognition (Tab. 2) .the results show that the method achieves the expected goal.

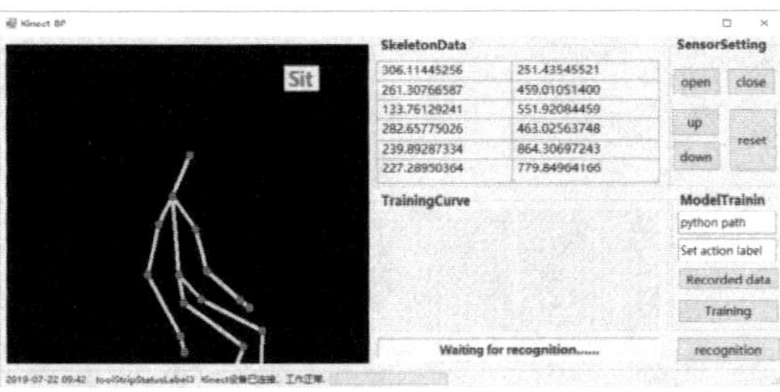

Figure 5 Experimental demonstration system

Table 1. Confusion Matrix for result of experiment.

Activity	accuracy rates
standing	91.57%
sitting	90.30%
leaning forward	91.68%
leaning down	92.71%
squat	96.03%

Table 2. Recognition rate for result of experiment.

		Predict Class				
		Standing	sitting	leaning forward	leaning down	squat
ActualClass	standing	91	0	4	5	0
	sitting	1	90	0	7	2
	leaning forward	6	0	91	1	2
	leaning down	7	0	1	92	1
	squat	0	3	0	1	96

6. Conclusion

In this paper, we propose a human posture recognition method based on bone vector.Compared with existing studies, it has the advantages of high recognition rate and simple network structure.To evaluate the method, we design and implement a set of experiments.the results show that the recognition rate of various actions can reach more than 90%,which is effective in various application scenarios. We intend to expand our work by refining the data set and improving recognition performance.

Acknowledgments

Research and Education Teaching Research Project of South China Institute of Software Engineering in 2018（KY201806），School-level quality engineering project of South China Institute of Software Engineering in 2018（JXTD201804）, 2019 Research Team Project (HR201901) of South China Institute of Software Engineering in 2019

References

[1] Wang, J. , Liu, Z. , Wu, Y. , & Yuan, J. . (2013). Learning actionlet ensemble for 3d human action recognition. IEEE Transactions on Software Engineering, 36(5), 914-927.

[2] Poppe, R. . (2010). A survey on vision-based human action recognition. Image and Vision Computing, 28(6), 976-990.

[3] Ali, S. , & Shah, M. . (2010). Human action recognition in videos using kinematic features and multiple instance learning. IEEE Transactions on Software Engineering, 32(2), 288-303.

[4] Chaudhry, R. , Ofli, F. , Kurillo, G. , Bajcsy, R. , & Vidal, R. . (2013). Bio-inspired dynamic 3d discriminative skeletal features for human action recognition.

[5] Liu, M. , Liu, H. , & Chen, C. . (2017). Enhanced skeleton visualization for view invariant human action recognition. Pattern Recognition, 68, 346-362.

[6] Nasiri, J. A. , Charkari, N. M. , & Mozafari, K. . (2014). Energy-based model of least squares twin support vector machines for human action recognition. Signal Processing, 104(6), 248-257.

[7] Chen, C. , Jafari, R. , & Kehtarnavaz, N. . (2016). Fusion of depth, skeleton, and inertial data for human action recognition. 41st IEEE International Conference on Acoustics, Speech and Signal Processing (ICASSP 2016). IEEE.

[8] Dan, Z. , Xingwen, C. , Shuying, Z. , Jiwei, L. I. , Yu, B. , & Center, I. E. , et al. (2014). Posture recognition method based on kinect predefined bone. Journal of Computer Applications.

[9] Li, G. S. , & Luo, J. H. . (2014). Somatosensory application study based on kinect skeletal tracking. Applied Mechanics and Materials, 635-637, 901-905.

[10] Abiddin, W. Z. W. Z. , Jailani, R. , Omar, A. R. , & Yassin, I. M. . (2016). Development of MATLAB Kinect Skeletal Tracking System (MKSTS) for gait analysis. 2016 IEEE Symposium on Computer Applications & Industrial Electronics (ISCAIE). IEEE.

IOP Conf. Series: Materials Science and Engineering **806** (2020) 012036 doi:10.1088/1757-899X/806/1/012036

Exploring research fronts and topics of Big Data and Artificial Intelligence application for cultural heritage and museum research

Man Zhao[1], Xue Wu[1], Han-Teng Liao[1*], and Yu Liu[1]

[1] New Media Research Center, Sun Yat-sen University Nanfang College, Guangzhou, Guangdong, 510970, China

[*]Corresponding author's e-mail: h.liao@oxon.org

Abstract. Modern technologies have supported the preservation, curation and reuse of human cultural heritage assets and resources, the latest development in Artificial Intelligence (AI) and Big Data technologies are expected to have great potentials in the datafication, digitization, and reuse of such resources. With the aim to reveal the research fronts and topics of such potential development, the paper conducted a bibliometric analysis based on 206 related articles from the Web of Science (WoS) database. The paper presents the findings on publication trends, main publication outlets and disciplines, main keyword clusters and research gaps. The paper also finds that the current research scope of AI and Big Data in cultural heritage and museum applications is still in the early stage of storage, calculation and presentation of Big Data. The paper help researchers, funding agencies and practitioners to understand the current status of AI and Big Data applications in cultural heritage and museum.

1. Introduction

Modern technologies have supported the preservation, curation and reuse of human cultural heritage assets and resources. For instance, the use of Information Communication Technologies (ICT) has been incorporated into educational activities of museums, and recommendations have been made to creatively apply ICTs to engage museum visitors with digital educational resources, use various educational ICTs[1].Recent research has also observed that the ICT methods and tools have helped museums to transform their framework of cultural communication, therefore important part of museum strategies to promote arts and cultural heritage activities[2].The global dialogues also explore the ways in which ICT and education innovation can contribute to the sustainable development goal for education (SDG4), as confirmed in the 2017 Qingdao Statement [3,4], promoting the use of ICT to achieve education targets in sustainable development goals.

Artificial Intelligence (AI) and Big Data, as part of the latest development in the ICTs, are expected to have great potentials in the datafication, digitization, and reuse of such educational and cultural resources [5,6], including interconnecting the knowledge structures of digital heritage with the social web using Big Data[7]. Using AI and Big Data may enhance museum collections management, visitor engagement, etc., thereby curatorial decisions (which collections to display, how to display them, and where to place them) can be made based on evidence[8]. Specific literature reviews have been conducted for cultural heritage and museum (e.g. recommender systems [9]), but no systematic review has been conducted to assess the current status and research frontiers of AI and Big Data in cultural heritage and museum applications.

2. Data and methods

To fill the gap, the paper conducted a bibliometric analysis based on articles collected from the Web of Science (WoS) database. The paper used the advanced query of the WoS as below:

TS = (AI OR Artificial Intelligence OR machine learning OR Big Data) AND TS = (museum OR heritage)

In November 2019, the query above gave us 206 articles. To overcome the analytical challenges of different terms in author keywords, we also built a thesaurus that treat terms such as "cultural heritage (CH)", "cultural-heritage", etc. as the same concept of "cultural heritage". Then the research conducted various bibliometric analysis using VOSviewer.

3. Research findings

Annual trends Annual trends of publications and citations reveal that the related research has experienced growth, with less than 8 articles per year before 2015 and about 34 articles on average per year after 2015.

3.1. Top publication and bibliographic coupling network and clusters

Since both co-citation and bibliographic coupling networks can reveal research fronts and intellectual bases [10], we visualized such networks at the level of sources (journals and conference proceedings). Figure 1 shows the relationship and distance among different sources, and their respective disciplines. The top journal, *Remote Sensing*, mostly related to cultural heritage via the disciplines of physical geography, geology, etc., has several research articles applying machine learning and feature extraction to understand cultural heritage. The second journal, *ACM Journal on Computing and Cultural Heritage*, largely an interdisciplinary publication outlet among humanities and computer science, has articles on the use of technologies for presenting and curating cultural heritage. There are two journals at the third ranking. While *In Situ-Revue De Patrimoines* is also interdisciplinary, *Journal of Cultural Heritage* is largely multi-disciplinary, mostly archeology and art, also chemistry geology, material science and spectroscopy. Categorized as "green & sustainable science technology", *Sustainability* and *Journal of Cultural Heritage Management and Sustainable Development*, has articles utilizing Big Data for better environment understanding.

Revealing another aspect of citations (i.e. intellectual bases), several clusters have identified as the important cited outlets (see Figure 2): tourism studies (e.g. *Tourism Management*, in blue), computer science (e.g. *Lecture Notes in Computer Science*, in green), remote sensing (in yellow), and other science journals (e.g. *Nature and PLOS One*, in red).

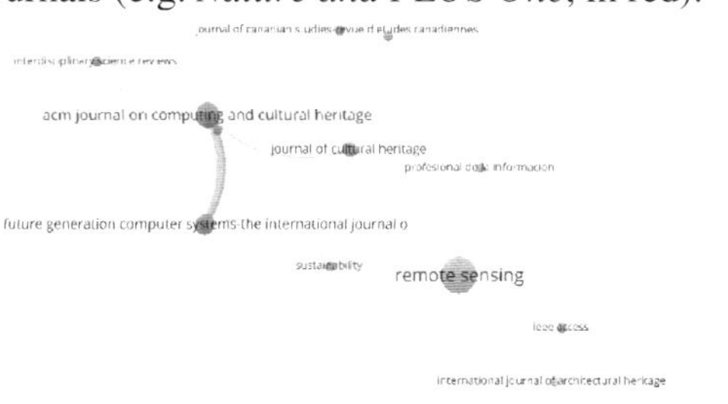

Figure 1. Bibliographic coupling network

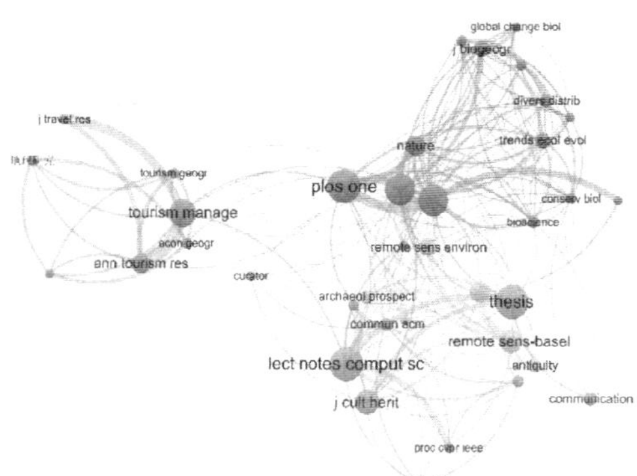

Figure 2. Co-citation network

These above findings reveal the research fronts and intellectual bases of Big Data and AI in the fields of heritage and museum research: while they are interdisciplinary, they nonetheless are related to geography, geology, archeology, environment, sustainability, etc.

3.2. Keyword map and Clusters

Keyword data is visualize based on the interconnection of the keywords resulting in a keyword co-occurrence network. Figure 3(a) show 4 clusters: (a) of: Big Data and cultural heritage (in blue color), AI (in red color), museum (in yellow color) and machine learning (in green color).

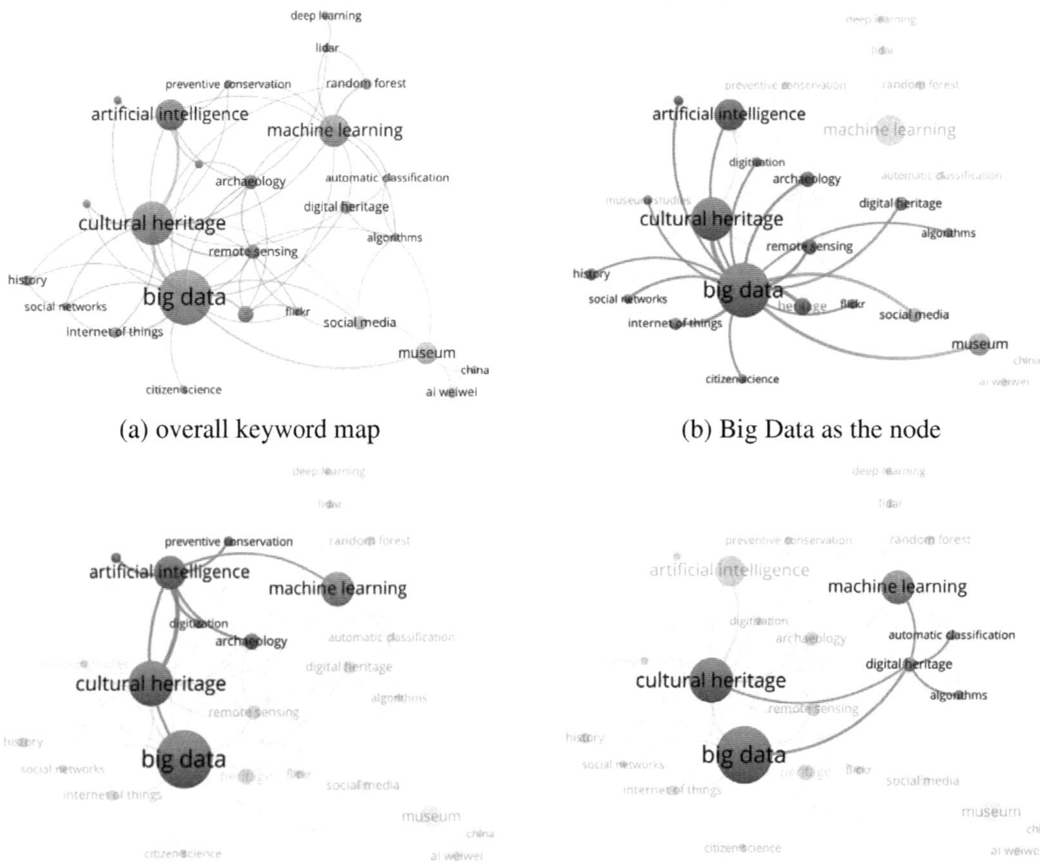

(a) overall keyword map (b) Big Data as the node

(c) AI as the node (d) digital heritage as the node

Figure 3. A keyword map based on co-occurrence data

As shown in Figure 3(b), Big Data is more prominent than AI. Well-connected to AI, cultural heritage and museums, Big Data is at the core in the discussions. As shown in Figure 3(c) and 3(d), although connected with Big Data, cultural heritage and machine learning, AI applications appear to be at the peripheral, suggesting more research needed in filling such gaps.

Indeed, more research appears to focus on the storage, calculation, and reproduction of data to enhance digital experiences and digital services (e.g. [8], [9],[11]) than the application of AI.

3.3. Importance of Internet platforms

One emerging research area we found relates to the potentials of AI and Big Data services provided by Internet platform companies. Empirical research includes the remote sensing Big Data using the Google Earth Engine Cloud Platform [12][13], the Google Arts & Culture projects, the Shrewsbury Museum & Art Gallery project based on Microsoft Azure Search Cognitive Search.

4. Conclusion

The purpose of this literature review is to provide a comprehensive, albeit preliminary, understanding of the current status of AI and Big Data applications in cultural heritage and museum applications.

Although the articles visualized and discussed here are limited to the 206 articles we have analyzed, they nonetheless reveal specific trends, main publication outlets and disciplines, main keyword clusters and research gaps. Future research must address such gaps.

The paper help researchers, funding agencies and practitioners to understand the current status of AI and Big Data applications in cultural heritage and museum. Given the scale and importance of the Internet platforms, future research on this topic can explore how AI and Big Data enhance the digital management, digital services and digitization of cultural heritage and museum through open collaboration and AI and Big Data services on the Internet platforms.

Acknowledgments

The research is funded by a project of HCI for Education Development (2018WQNCX283), under the 2018 Major Projects of Guangdong Province Department of Education "Young Innovative Talents" Grants, and partly by the Application-oriented Curriculum Development Project of "Information Visualization Design" (NFU 02-40250), under the Guangdong Province Department of Education 2018.

Man Zhao, Xue Wu, and Han-Teng Liao made equal contributions to the paper, including the data collection, analysis, and interpretation. Man Zhao, and Han-Teng Liao have been involved in drafting and revising the manuscript and have given the approval of the current version for publication.

References

[1] UNESCO IITE (2011) ICTs in Museum Education. (UNESCO Institute for Information Technologies in Education (IITE))
[2] Sylaiou, S., Papaioannou, G. (2019) ICT in the Promotion of Arts and Cultural Heritage Education in Museums. Strategic Innovative Marketing and Tourism Springer Proceedings in Business and Economics ed A Kavoura, E Kefallonitis and A Giovanis (Cham: Springer International Publishing) pp 363–70
[3] UNESCO IITE (2018) Global Dialogue on ICT and Education Innovation – Towards Sustainable Development Goal for Education (SDG 4). Ministerial Forum Proceedings (Moscow, Russian: UNESCO Institute for Information Technologies in Education (IITE))
[4] UNESCO (2017) Leveraging information and communication technology to achieve Education 2030: report of the UNESCO 2017 International Forum on ICT and Education 2030. (Qingdao, China: UNESCO)
[5] UNESCO (2019) Artificial intelligence for Sustainable Development: synthesis report, Mobile Learning Week 2019. (Paris, France: UNESCO)
[6] UNESCO (2019) Artificial intelligence in education, compendium of promising initiatives: Mobile Learning Week 2019. (Paris, France: UNESCO)
[7] Bonacchi, C., Krzyzanska, M. (2019) Digital heritage research re-theorised: ontologies and epistemologies in a world of big data. Int. J. Herit. Stud. 25: 1235–47.
[8] Tavčar, A., Csaba, A., Butila, E.V. (2016) Recommender System for Virtual Assistant Supported Museum Tours. Informatica 40:
[9] Pavlidis, G. (2019) Recommender systems, cultural heritage applications, and the way forward. J. Cult. Herit. 35: 183–96.
[10] Persson, O. (1994) The intellectual base and research fronts of JASIS 1986–1990. J. Am. Soc. Inf. Sci. 45: 31–8.
[11] Bogdanovych, A., Rodríguez-Aguilar, J.A., Simoff, S.J., Cohen, A.S. (2010) Authentic interactive reenactment of cultural heritage with 3D virtual worlds and artificial intelligence. Appl. Artif. Intell. 24: 617–47.
[12] Agapiou, A. (2017) Remote sensing heritage in a petabyte-scale: satellite data and heritage Earth Engine© applications. Int. J. Digit. Earth 10: 85–102.
[13] Luo, L., Wang, X., Guo, Q., Lasaponara, R., Shi, P., Bachagha, N., li, li, Yao, Y., Masini, N., Chen, F., Ji, W., Cao, H., Li, C., Hu, N. (2018) Google Earth as a Powerful Tool for Archaeological and Cultural Heritage Applications: A Review. Remote Sens. 10: 1558

A Novel Fuzzy Evaluation Method on Teaching Quality of College English based on Preference Selection Index

Guojiaozi Liu[1,*], Qing Cai[2] and Haiping Ren[2]

[1]School of Foreign Languages, East China Jiaotong University, Nanchang, Jiangxi, 330013, China

[2]Teaching Department of Basic Subjects, Jiangxi University of Science and Technology, Nanchang, Jiangxi, 330013, China

[*]Corresponding author's e-mail: 3103729351@qq.com

Abstract. The evaluation of teaching quality is an important part of the quality management of teaching in colleges and universities. Many evaluation systems exist shortcomings, which are not suitable for the evaluation of English class. Because English subject has its own characteristics, in which the evaluation subject taken on students is more suitable than on teachers. This paper will propose a college English evaluation index system on the basis of students as the evaluation subject. Since the evaluation indicators are mainly qualitative indicators, which makes the evaluation value fuzzy, this paper proposes a new evaluation method of college English classroom teaching evaluation based on linguistic terms, triangular fuzzy numbers and preference selection index method. Finally, an algorithm case is analyzed to illustrate the effectiveness and feasibility of the proposed method.

1. Introduction

The evaluation of teaching quality is a value judgment made by using the theory and technology of education evaluation on whether the teaching process and its results meet certain quality requirements [1]. Many scholars have done a lot of in-depth and perfect research on theory, evaluation content, evaluation subject, evaluation content, evaluation method and even influencing factors [2]. With the deepening of College English teaching reform, the evaluation of College English teachers can not only be limited to the traditional classroom teaching. Second, the connotation of some evaluation indicators is beyond the students' understanding ability. At present, the main body of University rating is students. The rating table is also designed for students. The evaluation index should be within the scope of students' understanding ability. Looking at the existing English teaching evaluation system, some indicators are obviously beyond the scope of students' understanding [3]. For example, in the index of "the content is rich, which can reflect the new achievements and new trends of the subject", obviously, students don't understand the new achievements and new trends, so they can't make an objective and fair judgment. Reference [4] is aware of these problems and try to use fuzzy comprehensive evaluation method to combine qualitative evaluation and quantitative evaluation, which provides a new perspective for the study of College English teaching quality evaluation.

However, most of these studies focus on other subjects, some of which only make a fuzzy evaluation of oral English, grammar, or multimedia. This paper attempts to establish a practical teaching quality evaluation system based on students. First, the evaluation values of indicators are firstly transformed into triangular fuzzy numbers (TFNs), and then we put forward a new evaluation method to evaluate English teaching quality based on Preference Selection Index method, first proposed by Maniya and Bhatt in [5], which is a well-known multi-attribute decision making method.

2. Evaluation Model of College English Teaching

In the great wave of college English teaching reform, extra-curricular counselling has gradually become the new normal of college English teaching. It is not only a supplement to classroom teaching but also a supplement[6]. As a first-level indicator, it highlights its function and function in College English teaching. The specific evaluation indicators are described in Table 1.

Table 1. Evaluation indicators of teaching quality evaluation

Indicators	Context of each indicator
Teaching attitude (o_1)	Enthusiastic and willing to communicate with students; attends and leaves classes on time and makes effective use of class time.
Teaching content (o_2)	Appropriate integration of English cultural background knowledge; appropriate guidance of learning strategies and methods; content enrichment, highlighting key points and difficulties, etc.
Teaching methods (o_3)	Proficient in explanation and in depth; proficient in using multimedia and other teaching equipment; classroom teaching forms are various; inspires students' thinking and pays attention to the development of students' critical thinking; good interaction and active classroom atmosphere; takes into account the differences in students' English level.
Teaching language (o_4)	English pronunciation standard; uses a lot of English and Chinese properly; fluent, clear and simple.
Extracurricular tutoring (o_5)	Homework is properly arranged, which can consolidate the effect of classroom learning; after class (online) tutoring is detailed and timely; actively guide and participate in all extracurricular English activities.
Teaching effect(o_6)	Gained a lot in classroom learning and improved its English performance; built up the confidence to learn English well; improved its oral English.

The language evaluation set can be used to evaluate the quality of English teaching. Now, the set of evaluation indicators are divided into seven grades and the relationship of linguistic terms with TFNs is[4]: When Linguistic terms set is { Very Poor (VP), Poor (P), Medium Poor (MP), Fair (F), Medium Good (MG), Good (G), Very Good (VG)}, the corresponding TFNs set is {(0,0,1), (0,1,3), (1,3,5), (3,5,7), (5,7,9), (5,7,9), (7,9,10), (9,10,10)}. The definition of TFN can be found in [7], and the operational law of two any TFNs $A=(a_1,a_2,a_3)$ and $B=(b_1,b_2,b_3)$ is:

$$A+B=(a_1+b_1,a_2+b_2,a_3+b_3) \text{ and } \quad kA=(ka_1,ka_2,ka_3), k\in R.$$

Considering the problem of classroom teaching evaluation of English Majors in Colleges and universities, a set of teachers $X=\{x_1,x_2,\cdots,x_m\}$ is to be evaluated, and the set of evaluation indicators is $O=\{o_1,o_2,\cdots,o_n\}$ .The importance of the jth indicators is w_j .A set of evaluation experts is $D=\{D_1,D_2,\cdots,D_s\}$. The evaluation value is a linguistic term s_{ij}^k , provided by D_k , is the rating of x_i on o_j, so the evaluation decision matrix is obtained with the form $\tilde{S}^k=(s_{ij}^k)_{m\times n}$,, which requires the evaluation of teachers' teaching quality according to $\tilde{S}^k, k=1,2,...,s$. Now, we give the evaluation model, and the calculation steps are given as follows:

Step1. Collect $\tilde{D}^k=(\tilde{a}_{ij}^k)_{m\times n}$ into a crisp decision making matrix $R=(\tilde{r}_{ij})_{m\times n}=((r_{ij}^l,r_{ij}^m,r_{ij}^u))_{m\times n}$ as in [8], where $\tilde{x}_{ij}=(\tilde{a}_{ij}^1+\tilde{a}_{ij}^2+\cdots+\tilde{a}_{ij}^s)/s$.

Step2. Transform $\tilde{R}=(\tilde{r}_{ij})_{m\times n}$ into decision matrix $X=(x_{ij})_{m\times n}$, where $x_{ij}=(r_{ij}^l+r_{ij}^m+r_{ij}^u)/3$.

Step 3. Determine weights of the index using entropy weight method as in [9].

Step 4. Compute the preference variation value (PV_j) and overall preference value (Ψ_j) as:

$PV_j = \sum_{i=1}^{m}(w_j r_{ij} - w_j \bar{r}_j)^2$, $\Psi_j = \Phi_j / \sum_{j=1}^{n}\Phi_j$, where $\bar{r}_j = \sum_{i=1}^{m} r_{ij} / m$ and $\Phi_j = 1 - PV_j$

Step 5. Calculate the preference selection index (I_i): $I_i = \sum_{j=1}^{n} r_{ij} \times \Psi_j$ and Give the evaluation result according to the rule: The larger the value of I_i with the better teaching quality of the teacher x_i .

3. Numerical case study

A foreign language college plans to evaluate the teaching quality of three English teachers x_1, x_2, x_3 in the college. The evaluation indicators are as in the second section. The college employs 10 peer experts to form an expert group. The expert group evaluates the four teachers, and the evaluation values are shown in Table 2.

Table 2. Language evaluation value provided by experts

Indicators	Teachers	Linguistic terms and corresponding expert numbers				
		P	F	G	VG	E
	x_1	0	1	3	4	2
o_1	x_2	0	2	2	4	2
	x_3	1	3	3	2	1
	x_1	0	1	4	2	3
o_2	x_2	0	2	4	1	3
	x_3	0	2	2	5	1
	x_1	0	2	2	2	4
o_3	x_2	1	1	4	2	2
	x_3	0	1	3	3	3
	x_1	0	3	3	3	1
o_4	x_2	0	1	3	4	2
	x_3	0	2	5	2	1
	x_1	0	1	2	3	4
	x_2	1	2	4	2	1
o_5	x_3	0	1	2	3	4
	x_1	0	3	4	2	1
o_6	x_2	0	1	2	5	2
	x_3	0	2	4	3	1

The following uses the methods proposed in this paper to evaluate the classroom teaching quality of these four teachers. The specific steps are as follows:

Step 1. Gather expert evaluation information into a TFN decision matrix $\tilde{A} = (\tilde{a}_{ij})_{3\times 6}$, where

$$\tilde{A} = \begin{pmatrix} (42.5,67.5,87.5) & (42.5,67.5,85) & (45.00,70.00,85.00) & (30,55,77.5) & (50,75,90) & (27.5,52.5,75) \\ (40,65,85) & (37.5,62.5,80) & (35,57.5,77.5) & (42.5,67.5,87.5) & (27.5,50,72.5) & (45,70,90) \\ (25,47.5,70) & (37.5,62.5,85) & (45,70,87.5) & (30,55,77.5) & (50,75,90) & (32.5,57.5,80) \end{pmatrix}$$

Step 2. The attribute weights obtained by entropy weight method are:

$w_1 = 0.2136$, $w_2 = 0.0135$, $w_3 = 0.1233$, $w_4 = 0.1146$, $w_5 = 0.3369$, $w_6 = 0.1982$.

Step 3. Calculate the preference variation value (PV_j), deviation (Φ_j), and overall preference variation value (Ψ_j) , and illustrate them in Table 3.

Table 3. Values of preference variation, deviation and overall preference variation

	o_1	o_2	o_3	o_4	o_5	o_6
PV_j	18.0440	0.0073	3.9010	2.9322	81.3428	15.8633
Φ_j	-17.0440	0.9927	-2.9010	-1.9322	-80.3428	-14.8633
Ψ_j	0.1468	-0.0086	0.0250	0.0166	0.6921	0.1280

Step 4. The preference selection indexes are obtained as follows:

$$I_1 = 20.3864, I_2 = 15.6713, I_3 = 19.9414$$

Then the results of English teaching quality evaluation of these four teachers are $x_1 > x_3 > x_2$.

4. Conclusions

In view of the problems in classroom teaching quality evaluation of College English teachers, this paper puts forward an evaluation index system of English teaching quality. This index system takes the students as the main body of evaluation, and its content is comprehensive. The language is concise and easy to understand, which can better monitor and guide college English teaching. On this basis, this paper proposes a triangular fuzzy number based assessment method of English classroom teaching quality. Compared with the conventional weighted average method, the fuzzy evaluation method proposed in this paper effectively reduces the interference brought by the fuzziness of indicators and makes the evaluation results have higher reliability and validity.

Acknowledgments

The author thanks for the support of the Science and Technology Research Project of Jiangxi Educational Committee (No. GJJ170496) and School Level Teaching Reform of JUST (No. XJG-2018-35).

References

[1] Warman, S.M. (2015) Challenges and issues in the evaluation of teaching quality: How does it affect teachers' professional practice? a UK perspective. Journal of Veterinary Medical Education, 42(3):1-7.

[2] Cai, Q. (2019) Evaluation on teaching quality of college English based on triangular fuzzy numbers. Journal of Guizhou Education University, 35(1): 67-71.

[3] Tu, L.H. (2017) Teaching quality evaluation of college English classroom based on interval-valued fuzzy sets. Journal of Zhejiang Ocean University (Humane Science), 34(5):83-88.

[4] Gu, S.L., Xiao, T.T.,Wang, M.M. (2015) A Study on the College Teaching Quality Assessmen. Mathematics in Practice and Theory, 45(18):287-292.

[5] Maniya, K., Bhatt, M.G. (2010) A selection of material using a novel type decision-making method: Preference selection index method. Materials and Design, 31: 1785-1789.

[6] Zhao, X.R., Zhou, Y.Q., Qu, L.Y., et al. (2018) College physics MOOC teaching quality evaluation based on fuzzy synthesis estimate method. Physics and Engineering, 28: 190-195.

[7] Xu, Z.S. (2002) A method for priorities of triangular fuzzy number complementary judgement matrices. Fuzzy Systems and Mathematics, 16(1): 47-50.

[8] Chen, C.T. (2000) Extensions of the TOPSIS for group decision-making under fuzzy environment, Fuzzy Sets & Systems, 114(1):1-9.

IOP Conf. Series: Materials Science and Engineering **806** (2020) 012038 doi:10.1088/1757-899X/806/1/012038

Deep Metric Learning for text data based on Triplet Network

Xin Deng [1,a], Wei Wu [2,b], Fei Wang [3,c]

[1] College of Computer Science and Software Engineering, Shenzhen University Shenzhen, 518060 P.R. China

[2] Software Development Department, Yunce Company Shenzhen, 518060 P.R. China

[3] College of Computer Science and Software Engineering, Shenzhen University Shenzhen, 518060 P.R. China

[a] 1062388964@qq.com

[b] wuwei0955@163.com

[c] wangfei2017@email.szu.edu.cn

Abstract—Text data has not only the characteristics of high dimension and serialization, but also most of the data are not labeled. This makes it difficult for the traditional method to achieve good results only relying on the training class mark to solve the problem of text classification. Later, some people put forward the method of metric learning to solve the problem of text classification. Its principle is to learn the measurement distance function for a specific task according to different tasks, that is to say, samples are classified by "distance measurement". The triplet network used in this paper is an algorithm based on "distance measurement". By optimizing the original triplet network, we can learn a non-linear measurement space with stronger data representation ability, so as to achieve better results than the traditional metric learning. Finally, we analyze the performance and efficiency of the method on IMDB movie review data set, and prove the superiority of the method for text classification.

1. INTRODUCTION

Triplet network is a method of deep metric learning, which is mainly used in image field at first. The principle is to learn an embedding function, which maps samples from feature space to measure space[1]. Then through the training parameters, the distance between the same kind of samples is close, and the distance between different kinds of samples is far away, so as to complete the classification[2]. Inspired by this method, this paper improves triplet network according to the characteristics of high dimension and serialization of text data. The model structure,loss function,selection of triplet and distance function which are more suitable for text classification scene are used. Finally, the experimental analysis of the method is carried out on IMDB movie review data set, and the experimental results are better than the traditional metric learning method.

2. RELATED WORK

Metric learning is an algorithm based on distance to realize classification[3]. As early as 2006, some people proposed the research method of text classification based on metric learning. The paper metric learning for text documents by Lebanon et al. has good reference significance and research value[4].

Later, as deep learning becomes more and more popular, there have been many researches on the classification of composition texts by deep metric learning. For example, The paper "Survey on distance metric learning and dimensionality reduction in data mining"[5] and "Low rank similarity metric learning in high dimension" published around 2015[6]. These researches make the deep metric learning method more and more popular.

Triplet network is a popular deep metric learning algorithm in recent years, which has been widely concerned by researchers all over the world. As early as 2014, Elad Hoffer first proposed the concept of triple network depth measurement learning in the paper "Deep metric learning using triplet network", defined the network structure of triple network and proved the advantages of triplet network method compared with traditional methods[7].

Later, during 2015-2017, Florian Schroff and Jiawei Liu et al. applied the triplet network method to face recognition and computer vision image classification[8][9]. They put forward some varieties of triple network and completed the end-to-end training combined with convolutional neural network. Finally, the accuracy of face recognition is higher than the traditional method, so triplet network is mostly used to study image problems.

Compared with triplet network in the field of image research, there are relatively few achievements in the field of text. In recent years, only Stefan Thaler have proved the feasibility of triplet network for text classification research[10]. Based on the previous research, this paper explores the general idea, network structure and application value of triplet network to solve the problem of text classification.

3. Triplet Network

Here we present our all experimental detail, including the model structure. We want to show that our optimized triplet network model has stronger representation ability and is more suitable for text type data.

3.1. Data source

Our dataset is IMDB movie review dataset. IMDB data set is a well-known data set, which is mostly used in the field of text classification.The dataset is a binary text dataset containing 50000 movie reviews from the Internet. The data set is divided into 25000 reviews for training and 25000 reviews for testing. Both the training set and the test set contain 50% positive and 50% negative reviews.

3.2. Data Pre-processing

In order to make the data adapt to the model better, we need to preprocess the data. First of all, we should remove the stop words that contain mood particles, conjunctions and special characters, etc. Next, we need to use word2vector to process sentences into the following formats:

$$\mathbf{Sent_1=\{W_{11},W_{12},...,W_{1n}\}}$$

$$\mathbf{Sent_2=\{W_{21},W_{22},...,W_{2n}\}}$$

...

$$\mathbf{Sent_{n-1}=\{W_{n1-1},W_{n2-1},...,W_{nn-1}\}}$$

$$\mathbf{Sent_n=\{W_{n1},W_{n2},...,W_{nn}\}}$$

Where $sent_n$ stands for sentences, W_{nn} stands for words. And when the vocabulary is built, every word has a unique id, so the sentences can be transferred to vectors.We define word's id as k_i in the upper form.

$$\mathbf{Sent_1=\{k_{11},k_{12},...,k_{1n}\}}$$

$$\mathbf{Sent_2=\{k_{21},k_{22},...,k_{2n}\}}$$

...

$$\mathbf{Sent_{n-1}=\{k_{n1-1},k_{n2-1},...,k_{nn-1}\}}$$

$Sent_n=\{k_{n1},k_{n2},...,k_{nn}\}$

In addition, it should be noted to prevent the length inconsistency of the sequence,we also need to truncate the sentence to a uniform length.

Finally, we need to divide the training data into 80% training set and 20% verification set. We also need to scramble the data and repeat the experiment many times, which is also to reduce the influence of data sampling imbalance on the experimental results.

3.3. Model

As we know, the model is the core of an algorithm and plays an important role. Our model is an improved triplet network. The structure of our triplet network model includes the following parts: the triplet selection, loss function, measure space and final classification model.The model structure is shown in Figure 1 below:

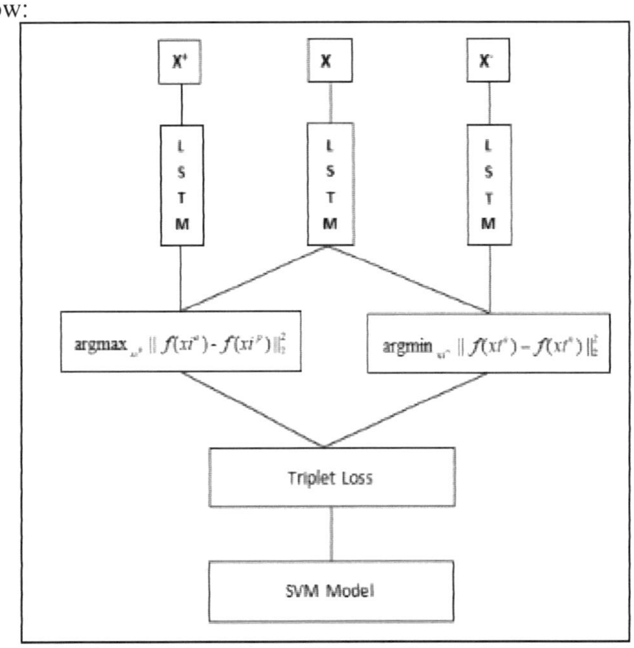

Figure 1. The model structure diagram of our triplet network

In the above picture,<X^+, X, X^-> represents the triplet sequence of the input.Then, we define an LSTM network structure and output an 8-dimensional vector space. Then in the metric space, we use the Mahalanobis distance to calculate the distance before different sample points, so that the semantically similar sequences are close to each other and the dissimilar sequences are far away from each other. Then we define the triple loss function and use Adam as the optimizer to minimize the loss function. At this time, the samples of different categories can be separated. Finally, we use SVM model to complete the final classification.

Compared with the traditional triplet network, in order to better adapt to the text type data, our model uses the LSTM model instead of the traditional CNN model, uses the Mahalanobis distance instead of the traditional European distance and finally uses SVM as the classifier to complete the text classification.In addition, we can try some unsupervised methods to optimize the selection of triplet. Some outliers are excluded by outlier detection methods such as Isolated Forest and OneClassSvm, and the rest normal samples are used as anchor points. This method can effectively avoid the impact of outlier samples on the results, so as to achieve better classification results.

3.4. Triplet selection

Before training the model, we need to choose different triplet pairs (A, P, N).Where A represents the anchor point, P refers to the same sample as class A, and N refers to a different sample from class A.

Generally, we have two methods to traverse triples. One method is called hard_triplet_loss, which means that after randomly selecting an anchor point, the nearest heterogeneous sample and the farthest similar sample are selected to form a triplet. Another one is called all_triplet_loss, which means that after randomly selecting an anchor point, it traverses all samples of the same type and different types of anchor point to form n triples.

Obviously, the two ways have their own characteristics. Hard_triplet_loss can form the optimal triplet, but it has higher time and space complexity. Although the time complexity of all_triple_loss is relatively low, but because there are many easy triplets that are not the best choice, the average loss is very small. Therefore, we should choose to use the triple loss according to the actual problems.

3.5. Triplet Loss

The calculation method of triple loss is different from that of traditional loss. It is not to adjust the loss by predicting the difference between the real class and the real class, but to minimize the loss function by adjusting the distance between the anchor point and the positive and negative samples.The calculation method is shown in Formula 1.

$$L = \max(d(a,p) - d(a,n) + m\ argin, 0) \tag{1}$$

Where a is the anchor point, p is the same sample, n is the different sample, d is the distance between sample points, margin is the manually set distance.

From the above formula, we find that the traditional triplet loss only considers the relative distance between positive and negative sample pairs, but not the absolute distance between positive sample pairs. Therefore, we propose an improved triple loss and add a driving term d (a, p). In this way, not only the distance between positive and negative samples in the feature space is far enough, but also the distance between positive samples is close enough. Experiments proves that our improved loss method has better results than the traditional triplet loss method.The specific formula is as follows:

$$L = \max(d(a,p) + d(a,p) - d(a,n) + m\ argin, 0) \tag{2}$$

3.6. Distance metric

The traditional triplet network uses the Euclidean distance to measure the similarity between two unknown samples, but the biggest problem of the Euclidean distance is to treat the differences between different attributes of the samples equally, which can only be applied to the case where the measurement standards of each component are unified, and the text does not fit the characteristics of the text data.

In this paper, we use a more suitable Mahalanobis distance. Mahalanobis distance represents the covariance distance of data. Compared with Euclidean distance, it takes into account the relationship between various characteristics of samples and is not affected by dimensions, so it can more truly reflect the distance between different samples. The specific calculation method is shown in the formula:

$$D_{\varkappa}(x) = \sqrt{(x-\mu)^{T} \sum^{-1} (x-\mu)} \tag{3}$$

3.7. Algorithm pseudo code

The following algorithm pseudo code shows the process of our triplet model training.

Algorithm1 Triplet network training

1: Load data to X,Y
2: Update X,Y by word2vector
3: Split TrainX,TestX from X

4: Split TrainY,TestY from Y
5: Build LSTMNetwork
6: Update Y,Y' by input TrainX,TrainY into LSTM Network
7: Generate A,P,N by all_triplet_loss or hard_triplet_loss
8: Update AP by d(A-P)
9: Update AN by d(A-N)
10: Update L by max[AP-AN+margin, 0]
11: Train LSTMNetwork by L
12: Test LSTMNetwork use TestX, TestY

4. EXPERIMENT

4.1. Data distribution map

Figures2 and 3 below reflect the distribution of test data before and after using the triplet network model. Because of the large sample size of IMDB full data set, we randomly selected 1000 samples to replace the full data set.

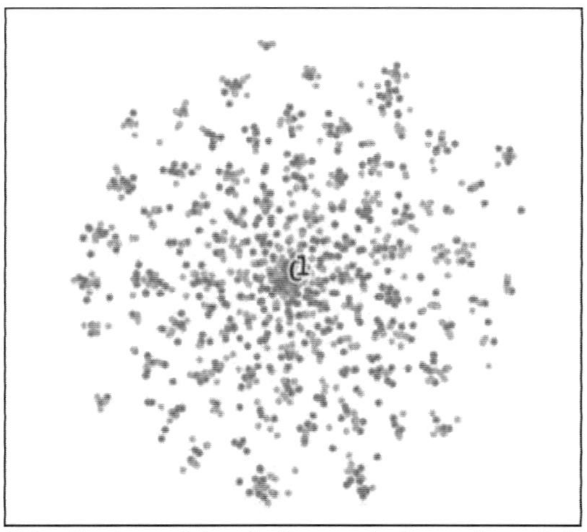

Figure 2. The Distribution of data before and after using triple network

Figure 3. The Distribution of data before and after using triple network

From the figure above, we can see that after the training of triple network, the data samples of different categories can be more distributed to achieve the effect of classification.

4.2. Confusion matrix

Figure 3 and 4 below shows the results of confusion matrix and normalized confusion matrix after using our improved triplet network model for 25000 training datasets. From the figure, We can see that triplet network can distinguish two types of text clearly.

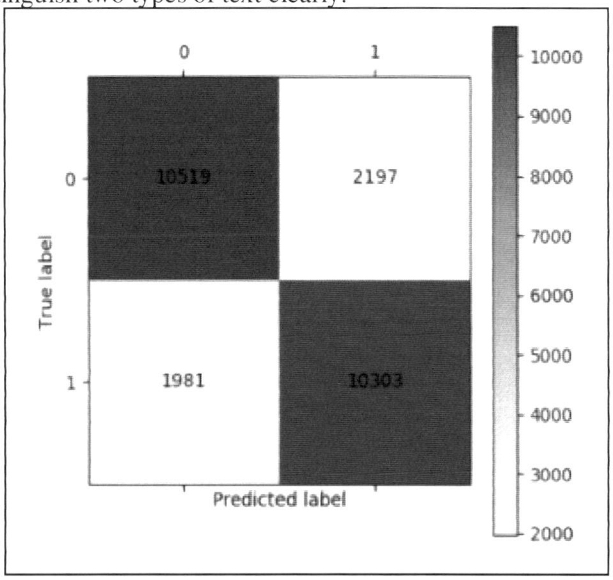

Figure 4. The confusion matrix of experimental results

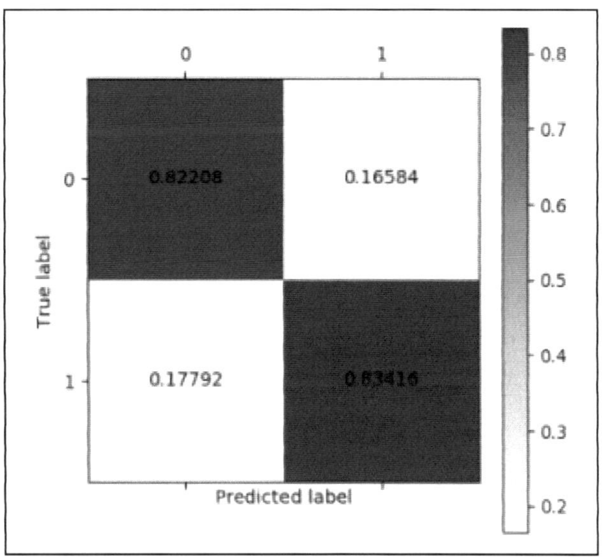

Figure 5. The normalized confusion matrix of experimental results

4.3. Comparative experiment results

Table 1 shows the experimental results of triplet network compared with other traditional metric learning algorithms and deep metric learning algorithms in IMDB dataset. From the results, it can be seen that triplet network is superior to the traditional metric learning algorithms in terms of the performance of several key indicators, such as accuracy, precision, recall and F1_score.

In the experiment in the table below, we choose two traditional metric learning methods and two deep metric learning methods to compare with our triplet network. From the experimental results, we can see that,our improved triple network algorithm has obvious advantages in dealing with text classification.

TABLE I. COMPARISON OF TRIPLE NETWORK AND OTHER MEASUREMENT LEARNING METHODS

Evaluation	**Metric Learning Method**				
	LMNN	**ITML**	**DDML**	**Siamese**	**Triplet**
Accuracy	0.538	0.582	0.698	0.783	0.833
Precision	0.549	0.580	0.704	0.792	0.839
Recall	0.582	0.578	0.712	0.785	0.832
F1_score	0.596	0.604	0.705	0.776	0.824

In addition, in the experiment, there are some key variables that affect the experimental results. such as: the number of anchor points selected, the number of layers of neural network, the setting of positive and negative sample margin, etc.We can optimize these indicators for better training results.

5. CONCLUSIONS

This paper explores a new method of text classification using deep metric learning, and proves that deep metric learning method based on triplet network can also achieve good results in text classification. Based on the traditional triplet network, this paper optimizes and innovates the triplet selection, triplet loss function, triplet network structure and distance measurement function, which makes the classification effect of triplet network better than that of traditional metric learning method in text field. We provide important academic value and application value for triplet network research in the field of text in the future.

In the future, There is still a large optimization space for triplet network applied in text field. we can optimize the triplet anchor selection and the distance function to achieve better experimental results. In addition, the research methods of this paper can also be extended to other fields.

REFERENCES

[1] Aggarwal C C, Zhai C X. A survey of text classification algorithms[M] Mining text data. Springer, Boston, MA, 2012: 163-222.

[2] Kulis B. Metric Learning: A Survey[J]. Foundations & Trends® in Machine Learning, 2013, 5(4):287-364.

[3] Nguyen B, Morell C, De Baets B. Large-scale distance metric learning for k-nearest neighbors regression[J]. Neurocomputing, 2016, 214: 805-814.

[4] Lebanon G. Metric learning for text documents[J]. IEEE Transactions on Pattern Analysis and Machine Intelligence, 2006, 28(4): 497-508.

[5] Weinberger K Q, Saul L K. Distance metric learning for large margin nearest neighbor classification[J]. Journal of Machine Learning Research, 2009, 10(Feb): 207-244.

[6] Davis J V, Dhillon I S. Structured metric learning for high dimensional problems[C] Proceedings of the 14th ACM SIGKDD international conference on Knowledge discovery and data mining. ACM, 2008: 195-203.

[7] Hoffer E, Ailon N. Deep metric learning using triplet network[C] International Workshop on Similarity-Based Pattern Recognition. Springer, Cham, 2015: 84-92.

[8] Schroff F, Kalenichenko D, Philbin J. Facenet: A unified embedding for face recognition and clustering[C] Proceedings of the IEEE conference on computer vision and pattern recognition. 2015: 815-823.

[9] Liu J, Zha Z J, Tian Q I, et al. Multi-scale triplet cnn for person re-identification[C] Proceedings of the 24th ACM international conference on Multimedia. ACM, 2016: 192-196.

[10] Thaler S, Menkovski V, Petkovic M. Deep metric learning for sequential data using approximate information[C] International Conference on Machine Learning and Data Mining in Pattern Recognition. Springer, Cham, 2018: 269-282.

Towards the eco-design of Artificial Intelligence and Big Data applications: a bibliometric analysis of related research

Zijia Wang[1] **and Han-Teng Liao**[1*]

[1]New Media Research Center, Sun Yat-sen University Nanfang College, Guangzhou, Guangdong, 510970, China

[*]Corresponding author's e-mail: h.liao@oxon.org

Abstract. Given the impacts of information and communications technologies on the planet and people, researchers from all disciplines need to pay attention to the environmental footprints and sustainable opportunities associated with the Artificial Intelligence (AI) and Big Data applications. In order to provide a preliminary discussion on the eco-design of AI and Big Data applications, the paper aims to map out the sources and topics to explore (1) the intersections of eco-design with AI and Big Data, and (2) the sustainable and eco-friendly design. Based on a bibliometric analysis of 236 articles gathered from the Web of Science (WoS) database, the paper has identified main research topics, publication outlets, research fronts, and intellectual bases and explored their inter-relationship and grouping. The leading application disciplines include construction & building technology, automation & control system, computer & information science, and business & economics. Multi-disciplinary and inter-disciplinary journals such as *Sustainability*, *Journal of Cleaner Production*, and *Sustainable Cities and Society* have been proven to be more receptive to such eco-design research outputs. As the notion of eco-design of AI and Big Data applications is expected to be pertinent and relevant for designing greener strategies, products and services for green digital transformation, we argue that more concerted efforts must be made to advance both the theoretical and empirical research on the nascent topic among researchers, funding bodies, policy-makers and industry professionals.

1. Introduction

The environment impact of Information and Communication Technologies (ICT), including its use and assets, has been an important agenda for research activities such as the eco-design of ICT, computing within limits, tech for good, etc. Indeed as often associated with the notion that 80% of all product-related impact is defined during the design phase of a product[1,2], eco-design of digital services[3] and green ICT[4] has become important areas for policy-making and innovation, as an important piece for sustainable growth[5]. For instance, the European Framework Initiative for Energy & Environmental Efficiency in the ICT Sector, or ictfootprint.eu, has gathered 4,000 community members, 21 ICT sustainable suppliers, etc., promoting low-carbon ICT standards and communities.

As ICT technologies have advanced the socio-economic development of human societies, we expect that the latest iterations of Artificial Intelligence (AI) and Big Data technologies and applications will have or already have had impacts on human environmental footprints, be them positive or negative. Indeed, as major institutions such as Facebook, Google, Microsoft, UNICEF, etc., founded the Partnership on Artificial Intelligence to Benefit People and Society in 2016 [6,7], researchers and policy makers must examine both negative and positive impacts of AI and Big Data applications on sustainable development. We need to "transform ourselves as innovators", as stated by the UN Chief Information

Technology Officer [8]. We need to leverage Big Data for sustainable development[9]. Thus, any Big Data and AI application may need to pay attention to environmental and sustainable challenges.

Some efforts have been made to review the design of Big Data or AI applications for sustainable development, including smart sustainable cities [10], green building [11,12], sustainable industry[13,14], etc. Yet none of them has provided a comprehensive enough overview of the state-of-art research. In order to provide one such overall roadmap for exploring the related research development on the eco- and sustainable designs of Big Data and AI, the paper aims to map out how the sources and topics are involved in such efforts to explore the intersections of AI and Big Data on one hand, and the sustainable and eco-friendly designs on the other.

2. Data and methods

To capture relevant the literature on the topic of eco-design of Big Data and AI applications, this bibliometric analysis starts with the Web of Science (WoS) "Advanced Search" query designed as below:

(TS = ("Artificial Intelligence" OR "machine learning" OR "Big Data")) AND (TS = ("design" AND ("sustainability" OR "sustainable development" OR "sustainable consumption*" OR "sustainable production*" OR "cleaner production*" OR "circular economy")) OR TS = ("ecodesign" OR "eco-design" OR "sustainable design" OR "green design") OR TS = ("ICT4D" OR "HCI4D" OR "Positive Technolog*"))

Such a systematic data collection leads to 236 articles from the WoS database (including SCI-EXPANDED, SSCI, CPCI-S, and CPCI-SSH). Both VOSviewer and Python visualization packages were used for analyzing and mapping results.

3. Research mapping results

Given the scope of collected data, the paper presents the mapping results from the broader trends, productive countries, organizations, publication outlets, and their clusters.

3.1. Annual trends

Figure 1(a) shows that the number of publications has largely risen exponentially, especially since around 2013. Also, Figure 1(b) shows that the number of citations takes off after around 2017, indicating the increasing significance of such topics in research. In addition, based on the slope value of regression lines shown in Figure 1(c), the growth rate after 2013 appears to be much faster than before.

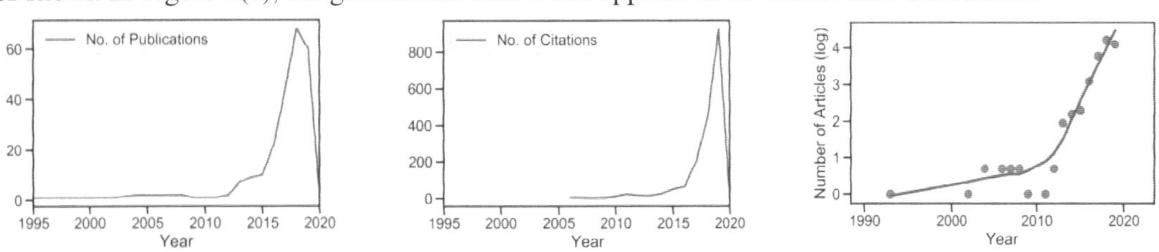

(a) Annual Publications (b) Annual Citations (c) Annual Publications (log)

Figure 1. Annual Trends

3.2. Keyword co-occurrence network analysis

Based on the interconnections among author keywords, or keyword co-occurrence network, we identify 4 clusters, as shown in Figure 2. The first cluster (in red color) features the central concepts of sustainability and big data, along with terms such as data mining, internet of things, circular economy, industry 4.0 and sustainable manufacturing. The second cluster (in green color) features artificial intelligence, sustainable development, case-based reasoning, and decision support system. The third cluster (in blue color) has keywords such as energy efficiency, smart cities and big data analytics. The final cluster (in yellow color) features more technical terms such as machine learning and optimization.

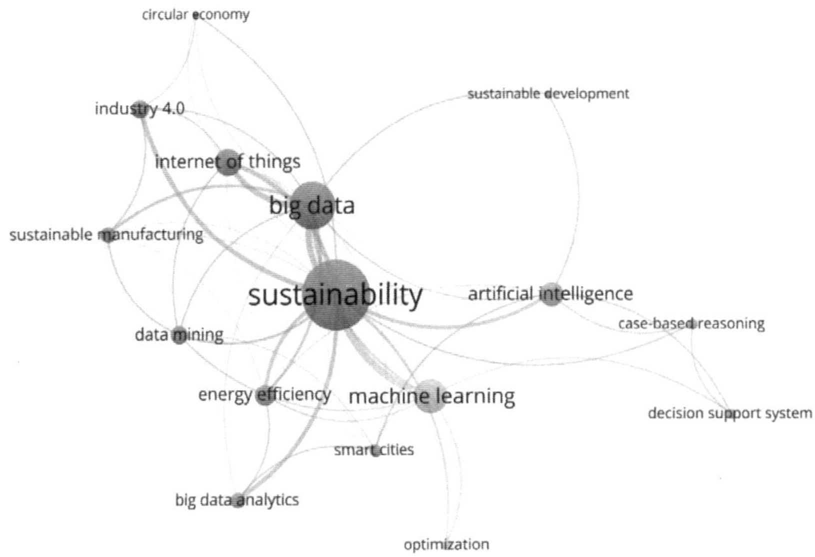

Figure 2. Keyword co-occurrence network

3.3. Top productive countries and organizations

Figure 3 shows the outcomes of the co-authorship network analysis. While the United States, China and England appear to dominate the country network, as shown in Figure 3(a), the organization networks reveal a more nuanced picture of cross-country and -institution collaboration, as shown in Figure 3(b).

(a) Country (b) Organization (c) Organization (detailed)

Figure 3. Co-authorship network analysis

In addition, Figure 3(b) also reveals the loosely-coupled co-citation network at the level of organization, where Chinese institutions such as Zhejiang University and Chinese Academy of Science stand out at the core of the network.

3.4. Two top multidisciplinary publication outlets

Two publication outlets, *Sustainability* and *Journal of Cleaner Production*, clearly dominate the publication landscape of the topic, each with 19 and 18 articles in the dataset. The distant third, *Sustainable Cities and Society*, only has 5 articles. These journals all belong to the Web of Science category of "Green & Sustainable Science & Technology", with the top two also belonging to those categories of "Environmental Sciences" and "Environmental Studies", and the last one belonging to "Construction & Building Technology".

Altogether, the top three journals appear to be more receptive to eco-design research outputs, with topics ranging from circular economy, sustainable consumption, green manufacturing to smart cities. However, it is noted that while the authorship in the top two is distributed more evenly across a wider range of authors, one author has contributed to 4 out of 5 articles published in *Sustainable Cities and Society*, and 3 of them focus on the notion of smart sustainable city [10,15–17].

The prominent role of the top two journals can also be shown in Figure 4(a).

(a) Bibliographic coupling networks (b) Co-citation networks

Figure 4. Intellectual base and research fronts of main publication outlets

3.5. Bibliographic coupling network and clusters

To identify research fronts and intellectual bases [18], both bibliographic coupling and co-citation networks of publication outlets are constructed and visualized, as shown in Figure 4.

Four clusters have been identified on the bibliographic coupling network, as shown in Figure 4(a). The first cluster (in red color) includes *Journal of Cleaner Production*, *Supply Chain Management*, *Management Decision*, *International Journal of Logistics Management*, etc. The second cluster (in green color) contains *Sustainability, Sustainable Cities and Society, Journal of Urban Technology*, etc. The third cluster includes *Computers in Industry*, *International Journal of Production Research,* and *Process Safety and Environmental Protection*, and *CRIP Annals- Manufacturing Technology*

Five clusters have been identified on the co-citation network, as shown in Figure 4(b). The first cluster has *Journal of Cleaner Production, International Journal of Production Research, Supply Chain Management*, etc. as its prominent nodes. The second energy-building cluster has *Energy Buildings*, *Renewable and Sustainable Energy Reviews*, etc. as its central and prominent nodes. The third sustainable city cluster has *Sustainable Cities Society, Sustainability Basel, Cities,* etc. The fourth cluster includes *Waste Management, Resources, Conservation & Recycling*, etc. The last cluster has two often-cited journals such as *MIS Quarterly* and *Nature*.

4. Conclusion

The paper has reviewed AI- and Big Data-related articles on the topic of eco-design, which sees tremendous growth in publications since 2013, and in citations since around 2017, indicating strong-growing attention on the topic. Also, the findings about the productive countries and organizations show a healthy development in research across developing and developed countries, with co-citation networks indicating some level of cross-country collaboration. Further findings on both bibliographic coupling and co-citation networks at the level of publications reveal that *Sustainability*, *Journal of Cleaner Production*, and *Sustainable Cities and Society* are more receptive to eco-design research on the topic of AI and Big Data. In addition, Journal of Cleaner Production has dominant position in the co-citation network. Although the data is limited to the WoS database, the findings nevertheless provide necessary groundwork for exploring the eco-design application of AI and Big Data.

As the notion of eco-design of AI and Big Data is expected to be pertinent and relevant for designing greener strategies, products and services for greener digital transformation, we argue that more concerted efforts among researchers, funding bodies, policy-makers and industry professionals must be made to advance both the theoretical and empirical research on the nascent topic. Several questions can be asked as follows: How can we apply the most relevant design thinking and system thinking knowledge to the advancement of AI and Big Data with Eco-design? Considering the impact of ecological, material, energetic, and societal limits on both interaction and computing resources, what are the new methods to facilitate the innovative design of Big Data, Artificial Intelligence, and their cloud services and applications? In short, how can we conduct the eco-design of digital services for green digital transformation? To begin answering the questions above, researchers, funding agencies, policy-makers and industry practitioners may find the paper helpful in mapping out the fruitful topics and publication outlets.

Acknowledgments
The research is funded by the Application-oriented Curriculum Development Projects of "API, Machine Learning and Artificial Intelligence" (NFU 02-40248) and "Big Data" (NFU 02-40249), under the Guangdong Province Department of Education 2018 "Innovation-Strengthening Higher Education Program" Grants, and partly by a project of HCI for Education Development (2018WQNCX283), under the 2018 Major Projects of Guangdong Province Department of Education "Young Innovative Talents" Grants. Han-Teng Liao and Zijia Wang have made equal contributions to the paper, including the data collection and analysis work and have given the approval of the current version for publication.

References

[1] JRC (2018) Sustainable Product Policy. (Seville, SPAIN: European Commission, Joint Research Centre (JRC))
[2] González, J. (2013) Eco-design and the impact in ICT assets End of Life. 3rd ITU Green Standards Week (Madrid, Spain: ITU)
[3] AGIT (2019) Ecodesign of digital services. (AGIT (Alliance Green IT))
[4] Deloitte (2019) D3.4 Recommendations towards a Policy Action Plan on "green ICT". (ICTFOOTPRINT.eu)
[5] ICT for Sustainable Growth Unit. Energy Efficiency of the ICT Sector: ICT for Sustainable Growth. (Brussels: European Commission)
[6] PAI (2019) The Partnership on AI. Partnersh. AI PAI
[7] Butler, D. (2017) AI summit aims to help world's poorest. Nat. News 546: 196.
[8] Patterson, D. (2016) United Nations CITO: Artificial intelligence will be humanity's final innovation. TechRepublic
[9] UN (2017) Big Data for Sustainable Development. U. N. UN
[10] Bibri, S.E., Krogstie, J. (2017) Smart sustainable cities of the future: An extensive interdisciplinary literature review. Sustain. Cities Soc. 31: 183–212.
[11] Zou, P.X.W., Xu, X., Sanjayan, J., Wang, J. (2018) Review of 10 years research on building energy performance gap: Life-cycle and stakeholder perspectives. Energy Build. 178: 165–81.
[12] Wong, J.K.W., Zhou, J. (2015) Enhancing environmental sustainability over building life cycles through green BIM: A review. Autom. Constr. 57: 156–65.
[13] Kamble, S.S., Gunasekaran, A., Gawankar, S.A. (2018) Sustainable Industry 4.0 framework: A systematic literature review identifying the current trends and future perspectives. Process Saf. Environ. Prot. 117: 408–25.
[14] Lei, G., Zhu, J., Guo, Y., Liu, C., Ma, B. (2017) A Review of Design Optimization Methods for Electrical Machines. Energies 10: 1962.
[15] Bibri, S.E. (2018) A foundational framework for smart sustainable city development: Theoretical, disciplinary, and discursive dimensions and their synergies. Sustain. Cities Soc. 38: 758–94.
[16] Bibri, S.E. (2018) The IoT for smart sustainable cities of the future: An analytical framework for sensor-based big data applications for environmental sustainability. Sustain. Cities Soc. 38: 230–53.
[17] Bibri, S.E., Krogstie, J. (2017) ICT of the new wave of computing for sustainable urban forms: Their big data and context-aware augmented typologies and design concepts. Sustain. Cities Soc. 32: 449–74.
[18] Persson, O. (1994) The intellectual base and research fronts of JASIS 1986–1990. J. Am. Soc. Inf. Sci. 45: 31–8.

IOP Conf. Series: Materials Science and Engineering **806** (2020) 012040 doi:10.1088/1757-899X/806/1/012040

External Sorting Algorithm: State-of-the-Art and Future Directions

Wenhan Chen[1], Yang Liu[1], Zhiguang Chen[2*], Fang Liu[2], Nong Xiao[1,2]

[1]College of Computer, National University of Defense Technology, Changsha, Hunan, 410073, China

[2]School of Data and Computer Science, Sun Yat-sen University, Guangzhou, Guangdong, 510000, China

[*]Zhiguang Chen's e-mail: zhiguang.chen@nscc-gz.cn

Abstract. The advent of the era of big data provides new opportunities and more challenges to sorting algorithms. The traditional internal sorting algorithm cannot adapt to the explosive growth of data, and the memory cannot accommodate all the data for sorting, so the external sorting algorithm arises at the historic moment. Because of the different application scenarios, storage devices and improvement strategies, there are many kinds of external sorting algorithms. Traditional main memory architecture based on DRAM faces the problems of capacity, energy consumption and reliability. Emerging nonvolatile memory technologies are non-volatile, high-density, byte-addressable, low-power, so they can replace persistent storage, main memory or storage class memory. Though NVM devices provide new choices to the revolution of traditional memory and storage system, traditional external sorting algorithms cannot achieve its performance. This paper first sorts out the development of external sorting algorithm, and summarizes it into four kinds of external sorting algorithm based on HDD, embedded device, SSD and NVM. In addition, the classical external sorting algorithms based on different storage devices are listed, and our opinions are put forward. Finally, this paper proposes three problems that need to be solved urgently in the future development of external

1. Introduction

With the large-scale application of sensor devices and the blowout growth of mobile Internet devices, more and more data are generated nowadays. Field data plays an important role in guiding industrial applications, and many related technologies, such as edge computing, cloud computing and so on, emerge as the times require. The rapid processing of huge amount of data has gradually become the bottleneck of data management. Sorting algorithm is one of the most basic algorithms in data management system. The speed of sorting algorithm directly affects the efficiency of data management system. After more than ten years of development of sorting algorithm, the development of internal sorting is more mature. However, with the increase of the amount of data, internal sorting has been unable to meet the current sorting needs of the amount of data. Thus, an external sorting algorithm is proposed. External sorting is mainly used in the scenarios where the amount of data is too large to be loaded into limited memory at one time. When the amount of data is particularly large, due to the limited memory capacity, it is impossible to sort all the data in memory at once, so it is necessary to use external sorting. External sorting is an I/O-intensive algorithm that produces a large number of I/O operations on external memory devices, thus affecting the performance of data processing. Nowadays, the amount of data is exploding, and processing data at the TB and PB levels has become a reality [1].

By 2025, the total amount of data is estimated to reach 185ZB [2]. The capacity of memory has been unable to keep pace with the growth of the data explosion. For data-intensive applications, it is an urgent problem to process a large amount of data quickly. However, at present, no scholar has summarized the research on the external sorting algorithm. In order to facilitate the relevant researchers to do further research quickly, and to point out the direction for the researchers in this field for the first time, this paper collates and summarizes the research on external sorting algorithms in recent years and summarize our direction and content. This is also the first summary of the external sorting algorithm in the past 30 years, which is of great significance to the data management system.

2. Definition of external sort

When the amount of data is particularly large, due to the limited memory capacity, it is impossible to sort all the data in memory at one time, so it is necessary to use external sorting. In 1973, Knuth et al. [3] provided a lot of basic knowledge of sorting algorithm, including the replacement selection and merge sort. Merge sort is the algorithm most commonly used for external sorting. External merge sorting is generally divided into two phases: run formation phase (see figure 1) and run merge phase (see figure 2). Run formation phase means the formation of initial runs. Replacement selection is a popular run formation method. Run merge phase means the merge of the runs into a sorted file [4].

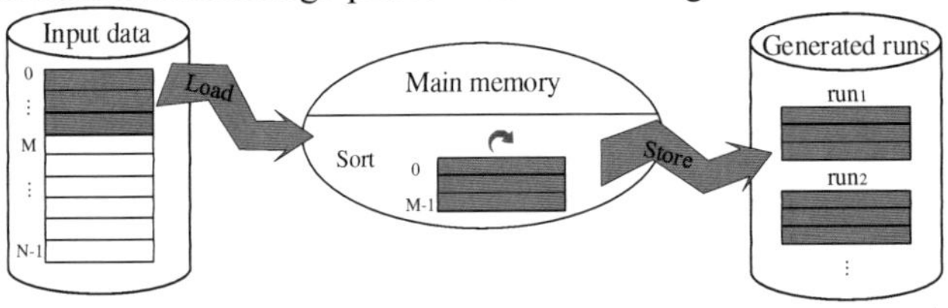

Figure 1. Illustration of the run formation phase.

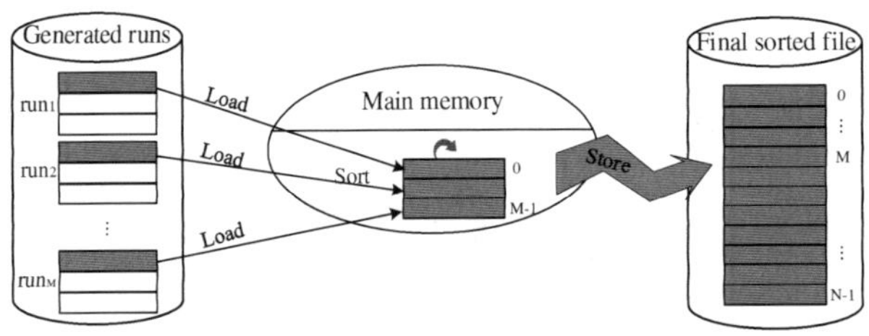

Figure 2. Illustration of the merge phase.

Run formation phase generally has two modes: load-sort-store and replacement selection. In load-sort-store mode, the input data is loaded into memory and sorted, and the sorted data is written to the external memory device to form a run. The length of the run generated in this way is generally the same, except for the last run.

Replacement selection has the following advantages over load-sort-store algorithm. First, it can generate longer runs, averagely generating runs twice the size of memory, thereby reducing the number of runs generated and the number of merge passes [5]. Moreover, replacement selection can take advantage of the partial order of the data to generate a longer run than when the input data is randomly distributed. Besides, replacement selection can overlap CPU processing and I/O, which reduce elapse time, but there exists frequent cache miss problem.

I/O time normally determines the elapsed time for external merge sort. So many external sorting research focus on optimizing algorithms to reduce I/O operations.

3. HDD-based external sorting

The traditional research of external merge sorting is mainly based on hard disk drive (HDD). The performance of read and write operations are symmetric, but sequential I/O operations are more efficient than random ones. The main research direction of disk-based external sorting is to reduce I/O operations to reduce the random read access to external memory devices, reduce seek time and rotational latency.

In 1996, Luoquan Zheng et al. [6] proposed two strategies: interleaved layout and a new reading strategy, to speed up the execution of external merge sorting and improve the performance of external merging sorting based on disk. Interleaved layout reduces the seek time of the disk during the merge phase by placing the blocks of different run in a continuous position on disk. A new reading strategy calculates block consumption sequence, and then allocates extra buffer in memory to prefetch data in that order, thereby reducing seek overhead and read time. Block consumption sequence is the order in which blocks are processed during the merge phase, which can be formed by sorting the maximum key values for each block during the run formation phase. A new reading strategy reduces seek overhead and read time by calculating block consumption sequence and prefetching data to memory buffer in that order.

Traditional merge sorting in the merge phase can only load one block at a time, and cannot prefetch the data, so it cannot overlap CPU process and I/O operations. Forecasting can prefetch the data during the merge phase. In run formation phase, keep the last key of each block, to determine which block will be emptied first. During the merge phase, an extra buffer is set aside in memory to prefetch next block from the run which block was emptied first. Only one input buffer is available for reading, so the disks cannot work in parallel.

Double buffering [7] can better overlap CPU process and I/O operation, which is a space-for-time strategy. During the merge phase, half of the buffer in memory is used as input buffer, and the other half of the buffer is used for prefetching the next block from each run on disk. However, double buffering does not know in advance which block in the input buffer will be emptied next. As a result, the eager block needs to wait for the previous block to finish processing before it can be loaded into memory for processing, so the algorithm is not very efficient.

AlphaSort [8] is a fast, cache-conscious disk-to-disk sort. It uses QuickSort to generate runs and uses replacement-selection to merge the runs, which has good cache performance and overlaps CPU process and I/O operations. Per-Ake Larson has proposed a cache-conscious version of replacement selection [9], called batch replacement selection, which is similar to AlphaSort. Traditional replacement selection can only handle fixed-length records, but batch replacement selection can handle variable-length records through better memory management. Batch replacement selection creates small run in memory and merges them to generate output runs, and both input buffer and output buffer are processed in batches. Overall, batch replacement selection can solve the problem of cache miss very well, reduce the number of comparisons, and run faster than the traditional replacement selection algorithm.

Md Rafiqul Islam et al. [10] proposed a new external sorting algorithm, which does not need to make use of additional disk space and generate no intermediate data. The algorithm makes use of fast sorting and special merging technology, which can reduce the number of comparisons in the sorting process and has a good performance.

4. External sorting for flash memory

Compared to traditional storage media, flash is non-volatile memory introducing many advantages: shock-resistance, fast read access, low production cost and power efficiency. There are two types of flash memory, NOR and NAND [11].

Four critical issues have a significant impact on the efficiency of external sorting in flash memory: (1) write once with bulk erase, (2) the endurance issue, (3) asymmetrical read and write operations, and (4) no mechanical latency. Flash memory has no disk head and disk platter, no seek or rotational latency will be needed. Therefore, random access and sequential access to flash memory will not cause mechanical latency problems [12].

4.1. external sorting on flash-based embedded devices

In recent years, with the rapid development of flash technology, the capacity of flash memory has increased and the price has decreased a lot. Flash memory is widely used in mobile devices, embedded devices and sensing devices. Many embedded devices need to store and calculate huge amount of data. Embedded devices process data internally more efficiently and faster than sending data over the network for analysis. However, the biggest challenge of external sorting in embedded devices is its limited memory space and very few CPU resources.

Fsort [13] is an external sorting algorithm on embedded flash-based sensor devices with small memory capacity. Flash memory has the characteristics of low access delay, power efficiency and shock-resistance [14]. However, the cost of writing access (about 800us) on flash memory is higher than that of reading (about 60us). And the number of times a block can be written is limited (between 10000 to 100000times). This algorithm can minimize the write-delete operation on flash memory. Fsort uses a top-down replacement selection algorithm in run formation phase to generate a longer run, with an average size of twice the memory size. Generating a longer run makes it more likely to decrease the number of runs generated and the number of merge passes, thus reducing the number of I/O operations. The paper compares the performance of online sorting and offline sorting in time and energy consumption, and the effect of online sorting is 2-3 orders of magnitude worse than that of offline sorting. The algorithm is improved based on offline sorting.

FAST [15] is an external sorting algorithm applied to NAND-flash based mobile database system. Because the write cost of flash memory is much higher than the read cost, the algorithm reduces write operations to flash memory by increasing additional read operations, thus improving the execution time and response time, as well as prolonging the lifetime of flash memory. In the merging phase, FAST scans the input data several times and constructs the maximum tree to get the first m minimum values, and the values are sorted and written directly to the output file. Until all the input data is processed, the algorithm terminates. The algorithm also needs to keep the max key value written out and its record number, to skip the data that has been processed in the previous iteration. To our opinion, this algorithm may work better in NVM. By taking profit of byte addressability, we only need to index the last processed data for each run to avoid reloading the processed data into memory again.

Similarly, Flash Minsort is also external sorting algorithms specifically for embedded devices based on flash storage system [16]. Flash Minsort combine run generation and merge phase into a single phase. The main idea of the algorithm is to retrieve only the tuples required by random reads instead of sequential scans. The input data is statically divided into different regions, and the smallest value in each region is stored to form a minimum index. The algorithm reads only the region containing the smallest value, and output the record, and then update minimum index. When all values are output to the external memory device in order, the algorithm is terminated. The performance of the algorithm depends on the distribution of the input data. If the number of different values in the input data is particularly large, the effect is not ideal. When the key values of the input data are clustered, the performance is better. However, the base version of flash Minsort does not adapt to the input relation size. In 2013, Ramon Lawrance et al proposed a modified version of Flash Minsort which are adaptive to memory limits [17]. The number of regions can be adjusted dynamically when the actual size of the input data is unknown. Once the memory limit is reached, the algorithm dynamically constructs the minimum index by merging adjacent regions.

Standard external merge sorting, merging in limited memory requires at least three memory buffers, but the algorithm proposed by Riley Jackson [18] can use only two memory buffers in memory to complete the merge operation, at the expense of more data comparison operations and data movement operations. This is very effective in embedded devices with minimal memory capacity.

4.2. external sorting on flash-based SSD

With the development of flash memory technology, flash-based solid-state drives (SSDs) has been widely used in various fields, as an alternative to HDDs. Compared with HDD, SSD has high I/O bandwidth and short access latency. Moreover, flash-based SSD has the characteristics of good internal parallelism. The performance of sequential read and random read of are symmetrical, but not the performance of read and write operations are [19].

YANG LIU et al. [20] put forward an external sorting algorithm on flash memory via natural page run generation. The algorithm speeds up the running generation phase by taking advantage of the partial sorting nature of the data. The algorithm finds pages without overlapping value intervals, which form naturally occurring page runs. Figure 3 shows an example of a naturally occurring page run of 3 pages in partially sorted data(P_1, P_3, P_4). The algorithm writes the index entries for the pages of naturally occurring page run, instead of sorting it and writing it back to memory, which reduces writes to flash memory. In this paper, a fast heuristic method is proposed to find a high percentage of naturally occurring page runs using a minimum of computational overhead. Firstly, separation distance loading of pages increases the chance of finding naturally occurring page run. Secondly, find non-overlapping intervals using a minimum interval tree (Min tree) implemented by red and black tree. Thirdly, if there are no enough non-overlapping pages can be found to form a naturally occurring page run, remove long intervals using an interval length tree (Length Tree) to form a normal run. In the merge phase, pages of naturally occurring runs are loaded into RAM buffer and sorted before participating in the merging process. However, the performance of the algorithm depends on the distribution of the input data. If the input data is partially ordered, the performance of the algorithm is better. Moreover, the algorithm assumes that it merges in a single pass, and the case of multiple merge passes is not discussed.

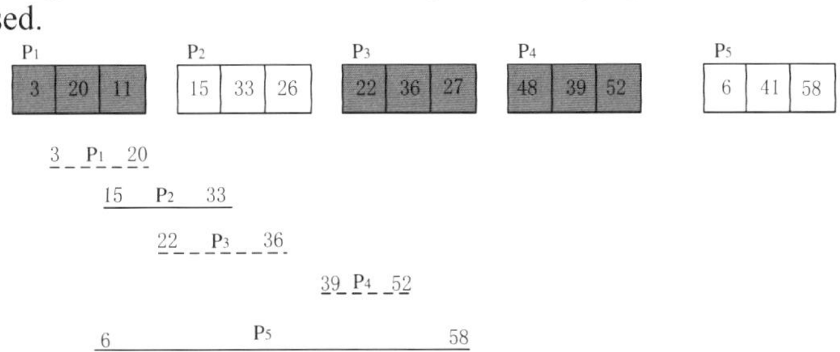

Figure 3. Example of a naturally occurring page run of 3 pages.

MONTRES [21] takes advantage of SSD performance to accelerate the sorting process. MONTRES mainly to solve the problem that other external sorting algorithms do not work well when the amount of input data is too large and the memory capacity is too small. During run formation phase, ascending block selection and continuous run expansion to generate larger runs are used to optimize the algorithm. Besides, by introducing a merge on-the fly mechanism in run formation phase, the algorithm continuously writes some small values directly to the final output file, thus reducing the generation of temporary data and reducing I/O amount. During run merge phase,MONTRES proceeds in a single pass by indexing the minimum value of each block and using a min-max heap to retrieves the small values.The performance of the algorithm is closely related to the distribution of input data. If the input data is partially ordered, the performance will be much better.

FMsort [22] takes advantage of the short access delay and high random I/O bandwidth of flash-based SSD. Flash-based SSD has the characteristics of good internal parallelism [23]. The algorithm can overlap CPU process and I/O operations very well. In run formation phase, FMsort calculates a block read order by sorting the minimum value of each block for each run, which is the order of blocks needed in the merge phase. In the merge phase, the next L blocks are prefetched according to the block read order, which overlaps CPU process and I/O operations, ans speeds up the merge phase.

Yaron Kanza et al. [24] proposed an external sorting algorithm for flash-based SSD. The algorithm can reduce cell wearing, improve the efficiency and prolong lifetime of the flash memory, by reducing the intermediate write operations. The algorithm is mainly used in the case that the input data is only a little larger than the memory, and the input data is partially ordered.

The ActiveSort [25] proposed by Young-Sik Lee makes use of Active SSD [26][27]. Active SSD is a special kind of SSD. Active sort offloads the merge operation to the SSD, which can reduce the number of I/O operations to preserve SSD lifetime. In run formation phase, the input data is loaded into memory and sorted to form a run, and written to SSD. When the host sends a standard read request to read the final sorted data, Active Sort creates the sorted opuput on-the-fly within the active SSD and sends the sorted results to the host side. The application of SSD into Hadoop MapReduce, can take advantage of the high network bandwidth of SSD and increase parallel storage access, thus speeding up the execution of Hadoop MapReduce [28]. Integrating ActiveSort into MapReduce framework can improve the performance of MapReduce framework and prolong lifetime of SSD. Active sort can speed up the data processing process of data-intensive applications based on Hadoop framework. However, this paper does not discuss how to optimize merge phase when the number of runs is greater than the number of chunks, nor does it discuss how to better place data on the active SSD, so as to reduce the time extension caused by channel congestion.

5. External sorting on NVM

Emerging memory technologies has developed rapidly in recent years. Emerging nonvolatile memory (NVM) technologies such as Phase-Change Memory (PCM), Spin Torque Transfer Magnetic RAM (STT-RAM), and Memristor-based Resistive RAM (ReRAM) offer the promise of significantly lower energy and higher density (bits per area) than DRAM [29]. Non-volatile memory (NVM) has the following advantages: high-density, high-speed, low-power, random-accessibility, non-volatility [30]. However, unlike DRAM, there exists endurance problems in NVMs (a cell wears out after 10^8-10^{12} writes) [31]. There is a significant asymmetry between read and write costs [32], which requires NVM-based algorithms to minimize write operations.

Phase change memory (PCM) [33] is a byte-addressable memory that has higher density and better endurance than NAND flash memory. However, compared with DRAM, there is a higher access delay and write operation is more expensive. In 2018, Arezki Laga et al proposed MONTRES-NVM [34], an external sorting algorithm based on hybrid memories that consist of PCM and DRAM. MONTRES-NVM has made optimization based on MONTRES. MONTRES-NVM uses a small amount of DRAM to speed up the sorting process and minimize writes to NVM. This paper mainly makes use of byte addressability of PCM to detect already sorted sequences in the input data and treat the sorted sequences as runs, so as to reduce the write operations to PCM.

The algorithm uses the merge on-the-fly mechanism of MONTRES to write the small values directly to the final sorted file, so as to reduce the generation of intermediate data. MONTRES can merge in a single pass. In the merge phase, the minimum value is retrieved by constructing a minimum heap and some small values are written to the final output file. The performance of the algorithm is closely related to the distribution of the input data. When the input data is partially ordered, it can play a very good performance.

6. External sorting with other technologies

6.1. External sorting with compression

Data compression technology is also often used in external sorting. Data compression can increase the information density of transferred data, and can relieve the I/O bottleneck of high performance database management system [35]. John Yiannis et al. [36] apply data compression to external sorting algorithms, through better management of memory space, which can accelerate the execution of external sorting, reduce the cost of data transmission, and generate a smaller number of runs. During run formation phase, first load the key values of the records into memory and sort them, then compress the records, and write the generated run to the external memory device. In the merge phase, the data is loaded into memory from each run for merging, and the data is decompressed and output to the external memory device. The performance of three data compression techniques applied to external sorting is also compared in the experiment: Canonical Huffman coding of bigrams [37], Bytewise bigram coding, Bytewise common-quadgram coding. In this paper, it is mentioned that for small-scale data, the speed of data compression and data decompression is the main factor affecting external sorting; however, the amount of data compression is the key factor to determine the sorting time for large-scale data.

6.2. Distributed external sorting

Multiway merge sort is a good algorithm for parallel external sorting. Two algorithms based on the multiway merging algorithm, are proposed for sorting a large amount of data on massively parallel machines by Mirko Rahn et al [38]. Merge sort with global striping minimizes I/O operations and works with just two passes over large-scale input data. Canonical Merge sort algorithm has close to minimal communication overhead and uses more conventional specifications of the result.

DEMSort, is a distributed external memory sort proposed in 2009 [39]. The algorithm consists of three phases: run formation, distribution, local merging. It supports at the same time distributed-memory parallelism and shared-memory parallelism, which work well in sorting huge volumes of data.

The distributed external sorting algorithm is also modified based on bucket sorting (an internal sorting) [40]. The input data is divided into different sets with non-overlapping intervals, and the interval of each bucket is defined. The data in each set is then distributed to different buckets for internal sorting. Finally, the ordered data of each bucket form the final ordered file together.

7. Conclusion

To conclusion, traditional external sorting research is mainly based on disk. The main optimization direction is to place the data properly, and to reduce the random read access to external storage, thus reducing seek time and rotational latency. The read and write overhead of flash memory is asymmetric. The cost of writing is much higher than that of reading, and the number of times a block can be written is limited .The research of external sorting based on flash memory is mainly to reduce the write operations to flash memory by increasing additional read operations, and even to avoid the generation of intermediate data. Some methods manage to optimize external sorting based on SSD. The algorithm mainly makes use of the high random I/O bandwidth and internal parallelism of SSD to overlap CPU process and I/O operation, so as to accelerate external sorting. Nowadays, DRAM technology is approaching its scaling limit, and the cost of using traditional technology to increase the capacity of DRAM and improve its energy efficiency and reliability is also greatly increased [41]. NVM has the characteristics of non-volatile, byte addressing, high storage density, low energy consumption, read and write performance close to DRAM, asymmetric read and write overhead, limited lifetime and so on. NVM-based external sorting algorithms mainly utilize the performance asymmetry between reads and writes and the benefits of byte addressing, decreasing write operations and extending lifetime of NVM. Our team mainly studies the external sorting algorithm based on NVM, which makes use of the characteristics of NVM byte addressing to optimize the external sorting algorithm, so as to reduce write amount and random access to NVM. There are three problems that need to be solved urgently in the future. Firstly, the main research direction in the future is to solve the fragmentation problem caused by NVM byte addressing. Secondly, when NVM and DRAM are combined as a hybrid memory, hard-to-find pointers may occur because of non-volatile feature of NVM. Finally, because of endurance of NVM device, so it is necessary to put forward solutions about wear-leveling, reducing write amount and the write operation to enhance lifetime of NVM device.

Acknowledgments

Supported by National Key R&D Program of China under Grant NO.2018YFB0203904 and National Natural Science Foundation of China under Grant NO.61832020, NO.61872392, NO.U1611261 and NO.U1811461, the Pearl River S & T Nova Program of Guangzhou under Grant NO. 201906010008, and Natural Science Foundation of Guangdong Province, China under Grant 2018B030312002.

References

[1] Thusoo, A., Shao, Z., Anthony, S., Borthakur, D., Jain, N., Sarma, J. S., ... & Liu, H. (2010). Data warehousing and analytics infrastructure at facebook. *international conference on management of data.*

[2] J. Boukhobza, and P. Olivier. (2017). Flash Memory Integration: Performance and Energy Issues. 1st. *UK: ISTE Press - Elsevier.*
[3] Wilkes, M. V. (1974). The Art of Computer Programming, Volume 3, Sorting and Searching. *The Computer Journal*, 17(4), 324-324.
[4] Dobosiewicz, W. (1984). Replacement selection in 3-level memories. *The Computer Journal*, 27(4), 334-339.
[5] Estivillcastro, V., & Wood, D. (1992). A survey of adaptive sorting algorithms. *ACM Computing Surveys*, 24(4), 441-476.
[6] Zheng, L., & Larson, P. (1996). Speeding up external mergesort. *IEEE Transactions on Knowledge and Data Engineering*, 8(2), 322-332.
[7] Anderson, T. A., Breitbart, Y., Korth, H. F., & Wool, A. (1998). Replication, consistency, and practicality: are these mutually exclusive?. *international conference on management of data*, 27(2), 484-495.
[8] Nyberg, C., Barclay, T., Cvetanovic, Z., Gray, J., & Lomet, D. (1994). AlphaSort: a RISC machine sort. *international conference on management of data*, 23(2), 233-242.
[9] Larson, P. A. (2003). External sorting: run formation revisited. *IEEE Transactions on Knowledge and Data Engineering*, 15(4), 961-972.
[10] Islam, R., Adnan, N., Islam, N., & Hossen, S. (2003). A new external sorting algorithm with no additional disk space. *Information Processing Letters*, 86(5), 229-233.
[11] Okazawa, T. (1996). Electrically erasable programmable non-volatile semiconductor memory device and method for manufacturing the same.
[12] Wu, C. (2012). Data sorting in flash memory. acm symposium on applied computing, 1847-1849.
[13] Andreou, P., Spanos, O., Zeinalipouryazti, D., Samaras, G., & Chrysanthis, P. K. (2009). FSort: external sorting on flash-based sensor devices. *data management for sensor networks.*
[14] Lee, S., Moon, B., Park, C., Kim, J., & Kim, S. (2008). A case for flash memory ssd in enterprise database applications. *international conference on management of data.*
[15] Park, H., & Shim, K. (2009). FAST: Flash-aware external sorting for mobile database systems. *Journal of Systems and Software*, 82(8), 1298-1312.
[16] Cossentine, T., & Lawrence, R. (2010). Fast sorting on flash memory sensor nodes. *international database engineering and applications symposium*, 105-113.
[17] Cossentine, T., & Lawrence, R. (2013). EFFICIENT EXTERNAL SORTING ON FLASH MEMORY EMBEDDED DEVICES. *International Journal of Database Management Systems*, 5(1), 1-20.
[18] Jackson, Riley & Lawrence, Ramon. (2019). Faster Sorting for Flash Memory Embedded Devices, 1-5.
[19] Boukhobza, Jalil. (2013). Flashing in the Cloud: Shedding some Light on NAND Flash Memory Storage Systems.
[20] Liu, Y., He, Z., Chen, Y. P., & Nguyen, T. (2011). External Sorting on Flash Memory Via Natural Page Run Generation. *The Computer Journal*, 54(11), 1882-1990.
[21] Boukhobza, J., Rubini, S., Chen, R., & Shao, Z. (2017). Emerging NVM: A Survey on Architectural Integration and Research Challenges. *ACM Transactions on Design Automation of Electronic Systems*, 23(2).
[22] Lee, J., Roh, H., & Park, S. (2016). External Mergesort for Flash-Based Solid State Drives. *IEEE Transactions on Computers*, 65(5), 1518-1527.
[23] Chen, F., Lee, R., & Zhang, X. (2011). Essential roles of exploiting internal parallelism of flash memory based solid state drives in high-speed data processing. *high-performance computer architecture*. pp. 266–277
[24] Kanza, Y., & Yaari, H. (2016). External sorting on flash storage: reducing cell wearing and increasing efficiency by avoiding intermediate writes. *very large data bases*, 25(4), 495-518.

[25] Lee, Y., Quero, L. C., Kim, S., Kim, J., & Maeng, S. (2016). ActiveSort: Efficient external sorting using active SSDs in the MapReduce framework. *Future Generation Computer Systems*, 76-89.
[26] Cho, S., Park, C., Oh, H., Kim, S., Yi, Y., & Ganger, G. R. (2013). Active disk meets flash: a case for intelligent SSDs. *international conference on supercomputing*. 91-102.
[27] Tiwari, D., Boboila, S., Vazhkudai, S. S., Kim, Y., Ma, X., Desnoyers, P., & Solihin, Y. (2013). Active flash: towards energy-efficient, in-situ data analytics on extreme-scale machines. *file and storage technologies*. 119-132.
[28] Moon, S., Lee, J., & Kee, Y. S. (2014). Introducing SSDs to the Hadoop MapReduce Framework. *international conference on cloud computing*. 272-279.
[29] Blelloch, G. E., Fineman, J. T., Gibbons, P. B., Gu, Y., & Shun, J. (2015). Sorting with Asymmetric Read and Write Costs. *acm symposium on parallel algorithms and architectures*,, 1-12.
[30] Li, H., & Chen, Y. (2010). Emerging non-volatile memory technologies: From materials, to device, circuit, and architecture. *international midwest symposium on circuits and systems*,, 1-4.
[31] Meena, J. S., Sze, S. M., Chand, U., & Tseng, T. (2014). Overview of emerging nonvolatile memory technologies. Nanoscale Research Letters, 9(1), 526-526.
[32] Kim, H. J., Seshadri, S., Dickey, C. L., & Chiu, L. (2014). Evaluating Phase Change Memory for Enterprise Storage Systems: A Study of Caching and Tiering Approaches. file and storage technologies, 10(4), 33-45.
[33] Lee, B. C., Zhou, P., Yang, J., Zhang, Y., Zhao, B., Ipek, E., ... & Burger, D. (2010). Phase-Change Technology and the Future of Main Memory. international symposium on microarchitecture, 30(1), 143-143.
[34] Bey Ahmed Khernache, Mohammed & Laga, Arezki & Boukhobza, Jalil. (2018). MONTRES-NVM: an External Sorting Algorithm for Hybrid Memory. *IEEE Non-Volatile Memory Systems and Applications Symposium*.
[35] Graefe, G. (1993). Query evaluation techniques for large databases. *ACM Computing Surveys*, 25(2), 73-169.
[36] Yiannis, J., & Zobel, J. (2003). External sorting with on-the-fly compression. *british national conference on databases*,, 115-130.
[37] Alistair Moffat, & Andrew Turpin. (2002). Compression and Coding Algorithms. Kluwer Academic Pub, Holland.
[38] Rahn, M., Sanders, P., & Singler, J. (2010). Scalable distributed-memory external sorting. *international conference on data engineering*,, 685-688.
[39] Rahn, Mirko & Sanders, Peter & Singler, Johannes & Kieritz, Tim. (2010). DEMSort — Distributed External Memory Sort.
[40] Dobosiewicz, W. (1984). Replacement selection in 3-level memories. *The Computer Journal*, 27(4), 334-339.
[41] Mutlu, Onur. (2015). Main Memory Scaling: Challenges and Solution Directions. 127-153.

IOP Conf. Series: Materials Science and Engineering **806** (2020) 012041 doi:10.1088/1757-899X/806/1/012041

Parameter adaptive LS-SVR based on multi-stages division and cluster sampling for remaining useful life prediction

Jian-wen Yan[1,2], Xiao-hu Zhong[1,2], San-min Guo[1,2], Yu Fan[2*] and Qiang Zhang[1]

[1]School of Management, Hefei University of Technology, Hefei 230009, China

[2]Anhui Province Key Lab of Aerospace Structural Parts Forming Technology and Equipment, Institute of Industry & Equipment Technology, Hefei University of Technology, Hefei 230009, China

[*]Corresponding author's e-mail: yufan@hfut.edu.cn

Abstract. Remaining useful life (RUL) prediction is an advanced technology to manage life cycles of equipment and to reduce maintenance cost. Support vector regressive (SVR) is one of the frequently-used data driven method in the field of prognostics, and is much suitable for analyzing small samples and multi-dimensional data. However, there is a limitation of traditional SVR model that the linearity will increase as the training data increase. In order to overcome it, this paper proposes the parameter adaptive least square support vector regressive (LS-SVR) based on multi-stages division and cluster sampling. It divides the degradation process of the equipment into several stages according to the value of degradation data and selects the model parameters stage by stage. The optimal model parameters of each stage are calculated through the novel testing data, which are obtained through appending the cluster sampling data to the original testing data. Therefore, the optimal model parameters can be selected adaptively online as the degradation evolves. Finally, the proposed method is verified systematically by a simulated dataset of fatigue crack growth and a real-world degradation dataset of GaAs-based lasers. It is shown that the proposed method is of better effectiveness, reliability and robustness.

1. Introduction

Equipment degradation and unexpected failures affect the safety, cost and quality in production. With the increasing automation and complexity, performance and health of the modern equipment will inevitably follow degradation before unplanned shutdown or even catastrophic failures. When the degradation reaches a certain extent, the equipment will not be able to complete normal tasks and functions. Prognostics and health management (PHM) as a key technology to ensure equipment safety and reliability, can effectively monitor equipment operation state and predict the remaining useful life (RUL), i.e., the time remaining before it can no longer perform its function once a potential fault occurs [1,2].

Generally, prognostics can be implemented using either physics-based or data-driven methods [3,4]. Physics-based methods need to have a broad understanding of physical mechanism, however there usually involves many physical parameters, and it is difficult to obtain an accurate physical model [5,6]. In addition, the identification of model parameters requires a large amount of experimental and failure history data, which is much too expensive and time consuming. In contrast, data-driven methods don't require to work out a physics model to represent the physical failure mechanism, but do mine the hidden degradation information from the observation data or sensor measured data via some data analysis tools and algorithms [3,7-9]. There are some classical data-driven methods for RUL prediction, such as auto-regressive (AR) models [10,11], fuzzy logic methods [12], hidden Markov models [13], Bayesian networks [8], artificial neural networks (ANN) [14-16], support vector regressive (SVR) methods [17,18], similarity-based RUL prediction (SbRP) methods [19-21] and stochastic process models [22-24], etc. SVR is the common application form of support vector machine (SVM) in the field of prognostics, and is much suitable for analyzing small samples and multi-dimensional data. The core idea is to use condition monitoring data to train a SVR model, and identify the model parameters (non-sensitive coefficients, penalty factors and kernel functions, etc.), accordingly the future degradation trend of equipment can be predicted based on the trained SVR model, therefore the RUL is obtained by comparing the predicted degradation level with the preset failure threshold. In comparison to other data-driven methods, SVR is of better preferable generalization performance in engineering practice due to that the real-world failure historical data are always limitation. For example, Nuhic et al. (2013) [25] employed SVR method to predict the RUL of lithium-ion batteries, the validation shows that it can reliably learn the degradation behavior of Li-ion cells. Loutas et al. (2013) [17] proposed a methodology for rolling bearing online RUL predictions based on probabilistic SVR, the results show adequate agreement with the corresponding actual RUL. Benkedjouh et al. (2015) [26] estimated and predicted successfully the amount of wear and calculate the RUL of the cutting tool by SVR method. Furthermore, Researchers usually combine the SVR model with other methods to predict the RUL of the equipment, in order to achieve better prediction results. Chen et al. (2018) [27] used SVR method to predict the RUL of the reference samples with right censored data, these processed samples are satisfactorily used for the similarity-based method to predict the RUL of the operating equipment, their combination has been successfully used for the RUL prediction of aircraft engines. Ordonez et al. (2019) [28] proposed a hybrid model incorporated of auto-regressive integrated moving average (ARIMA) model and SVR model to predict the RUL of aircraft engines. Wei et al. (2018) [29] employed a particle filter algorithm to suppress the measurement noises of current and voltage, the RUL of lithium-ion batteries was predicted based on the SVR model. In addition, the least square support vector regressive (LS-SVR) as the most widely used SVR model performs well in the field of RUL prediction. It transforms the problem of solving quadratic programming into the counterpart of solving linear least squares in high-dimensional space, and then simplifies the calculation process and improves the training speed.

From the above, the effectiveness, reliability and robustness of SVR methods are well verified in engineering practice. Even so, there is an inevitable limitation that with the increase of sample set, linearity will increase which will lead to the increase of overfitting [30]. From a practical point of view, however, the degradation processes usually show two or even more distinct stages of different degradation rates during the whole life period [31], thus a single SVR model often place too much emphasis on the generalization capability which then lead to that the prediction accuracy isn't satisfactory in some cases. To overcome this limitation, this paper proposes a parameter adaptive LS-SVR based on multi-stages division and cluster sampling for RUL prediction. The novelty is that it divides the degradation process into several stages, and through cluster sampling of the test data of each stage, as well as model construction stage by stage. Therefore, the optimal model parameters can be selected adaptively according to the degradation state, so that each stage of the degradation process can be fitted at the best extent, and the trained SVR model linearization is avoided.

The rest of this paper is organized as below. Section 2 introduces the SVR model and proposes the parameter adaptive LS-SVR based on multi-stages cluster sampling. Section 3 analyzes systematically the proposed method by a simulated dataset of fatigue crack growth and a real-world degradation dataset of GaAs-based semiconductor lasers. Finally, Section 4 concludes the whole paper.

2. Methodology

2.1. Introduction to SVR model

SVR is mainly used for prediction, which aims to find an optimal classification surface in order to minimize the error of the missed training samples set from the optimal classification surface. It's the most common application of support vector machine (SVM). Given a training set $Z = \{x_i, y_i\}_{i=1,2,...,n}$ with input data $x_i \in R^N$ and output data $y_i \in R$. Finding a function $f(x)$ on R^N to infer the output y corresponding to any x, is a regression problem.

Assume that the linear regression function for $f(x)$ constructed in the high-dimensional feature space is shown in (1):

$$f(x) = w * \phi(x) + b \qquad (1)$$

where $\phi(x)$ denotes a nonlinear mapping function.

Defining ε linear insensitive loss function:

$$L(f(x), y, \varepsilon) = \begin{cases} 0, & |y - f(x)| \le \varepsilon \\ |y - f(x) - \varepsilon|, & |y - f(x)| > \varepsilon \end{cases} \qquad (2)$$

The slack variables ξ_i and ξ_i^* are introduced, and the problem of calculating w, b can be expressed as:

$$\begin{cases} \min\left[\frac{1}{2}\|w\|^2 + C\sum_{i=1}^{n}(\xi_i + \xi_i^*)\right] \\ s.t \begin{cases} y_i - w * \phi(x_i) - b \le \varepsilon + \xi_i \\ -y_i + w * \phi(x_i) + b \le \varepsilon + \xi_i^* \\ \xi_i \ge 0,\ \xi_i^* \ge 0 \end{cases} \quad i = 1,2,\dots,n \end{cases} \qquad (3)$$

According to eq. (3), if the penalty factor C is larger, it is proved that the training error is large and the sample penalty of ε is larger, otherwise the opposite is true; if ε is smaller, the error of the regression function is smaller.

Introduce Lagrange function, convert to dual form, as follow:

$$\begin{cases} \max\left[-\frac{1}{2}\sum_{i=1}^{n}\sum_{j=1}^{n}(\alpha_i - \alpha_i^*)(\alpha_j - \alpha_j^*)K(x_i, x_j) - \sum_{i=1}^{n}(\alpha_i + \alpha_i^*)\varepsilon + \sum_{i=1}^{n}(\alpha_i - \alpha_i^*)y_i\right] \\ s.t \begin{cases} \sum_{i=1}^{n}(\alpha_i - \alpha_i^*) = 0 \\ 0 \le \alpha_i \le C \\ 0 \le \alpha_i^* \le C \end{cases} \end{cases} \qquad (4)$$

where $K(x_i, x_j) = \phi(x_i)\phi(x_j)$ is the kernel function.

Assume that the optimal solution of eq. (4) is $\alpha = (\alpha_1, \alpha_2, \dots, \alpha_n), (\alpha_1^*, \alpha_2^*, \dots, \alpha_n^*)$, and then:

$$w^* = \sum_{i=1}^{n}(\alpha_i - \alpha_i^*) * \phi(x) \qquad (5)$$

$$b^* = \frac{1}{N_{ssv}}\left\{\sum_{0<\alpha_i<C}\left[y_i - \sum_j^m(\alpha_j - \alpha_j^*)K(x_i, x_j) - \varepsilon\right] + \sum_{0<\alpha_i^*<C}\left[y_i - \sum_j^m(\alpha_j - \alpha_j^*)K(x_i, x_j) + \varepsilon\right]\right\} \qquad (6)$$

where N_{ssv} is the number of standard support vectors and m is the number of support vectors. Then the approximate regression function is expressed as:

$$f(x) = \omega^* * \phi(x) + b^* = \sum_i^m(\alpha_i - \alpha_i^*)\, K(x_i, x) + b^* \qquad (7)$$

In addition, the key of SVR is choosing a proper kernel function, which directly affects the nonlinear mapping of samples. In order to obtain better generalization and nonlinear regression ability, the RBF kernel function is selected for SVR model construction in this paper.

2.2. The proposed method for RUL prediction

In engineering practice, even the same batch of products operate under the same operating conditions, their degradation rates are often different with each other, even these differences become more and more obvious over time. Such issue leads that the real-world failure historical data present varied degradation curves. Therefore, with the increase of training sample set the linearity will increase in the trained SVR model, thus the prediction error of the trained SVR model for an operating product will increase. However, the degradation process of a product can usually be divided into two or more stages, each stage present different degradation patterns. For example, Hu et al. (2016) [32] divided the degradation processes of generator bearings into four stages based on the change points of confidence levels. Kimotho et al. (2013) [33] divided the degradation processes of bearings into five stages by analyzing the changes of frequency amplitudes in the power spectral density. Their studies show that subdivide the degradation process is beneficial to improve the prediction accuracy. Therefore, based on this view point, this paper proposes a parameter adaptive LS-SVR method to overcome the limitation of traditional SVR model, it's based on multi-stages division and cluster sampling, and can be effectively used for RUL prediction. The flowchart of the proposed method is shown in Fig. 1.

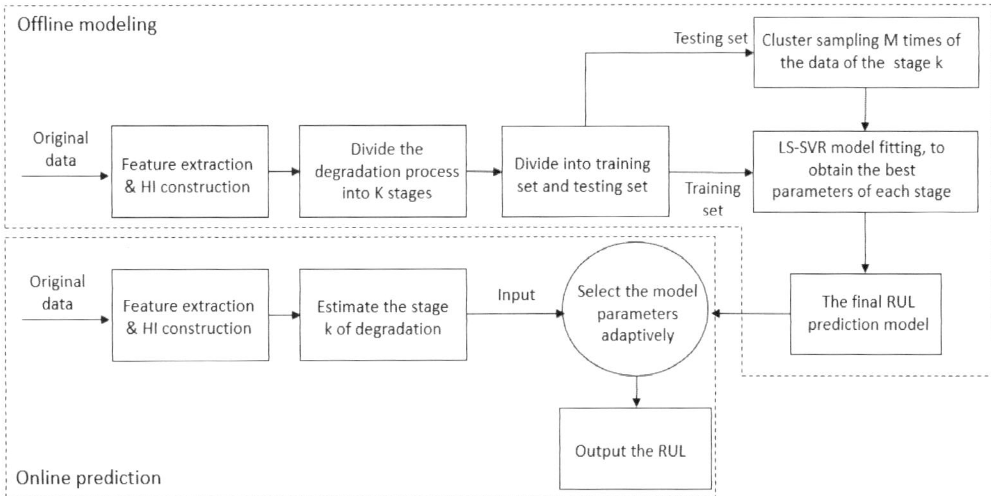

Fig.1 The flowchart of the proposed method.

In sampling theory, the coherent sample set composed of several connected basic units is called a cluster. For cluster sampling, the population is first divided into different sub-clusters according to some criteria, and then each sub-cluster is taken as the sampling unit, finally several sub-clusters are extracted from them by random principle for analyzation. The shortcoming of cluster sampling is that the sampling error is often larger than that of simple random sampling, due to that the difference between different sub-clusters is often remarkable. However, the aim of employing cluster sampling in this paper is to increase the weight of corresponding sub-clusters in SVR model construction, so that such shortcoming can be ignored. The RUL prediction model is constructed at offline modeling part. Reference to Fig. 1, when a suitable health indicator (HI) is obtained to represent the whole degradation process, it can be divided subjectively into stages according to the its value, with the difference between each stage is as large as possible. Suppose that, a process has K stages, of which 0 stage is the initial state; 1, ···, K-1 stages are the degradation state of acceptable degradation level; K stage is the failure state. Then through M times cluster sampling of the test data of each stage, as well as model construction stage by stage. Therefore, the optimal model parameters can be selected adaptively according to the degradation state, so that each stage of the degradation process can be fitted at the best extent. Additionally, the test error of the SVR model decreases as the value of M increases, and finally converges at some level, which is verified in the later sections. Finally, based on the optimal number of cluster sampling, the multi-stages prediction model is constructed with the optimal parameters of each stage. Therefore, for an online operating product, the prediction accuracy of the proposed method is better than the traditional SVR model.

3. Case studies

In this section, we first use simulation study to demonstrate the reliability and robustness of the proposed model. Then, it is applied to a real case study of GaAs-based semiconductor lasers.

3.1. Simulation study

In this subsection, the prediction is illustrated and the performance is evaluated through numerical simulations. The degradation trajectories dataset of 30 fatigue-crack-growth units is obtained by the following polynomial-type Wiener process model [23]:

$$s(t) = s_0 + a \int_0^t \tau b^{\tau-1} d\tau + \int_0^t \sigma_B dB(\tau) \qquad (8)$$

The initial crack length s_0 is 0 inch, and the failure threshold is set to be 50 inches. The parameter a follows a Gamma distribution G(α, β). The preset values of the model parameters are given in Table 1. And the degradation trajectories of 30 units are shown in Fig. 2, which present the growth of the fatigue crack along with the stress cycles. The numerical simulations are all terminated at 25 million of cycles. These degradation trajectories are full of time variability and nonlinearity, can be considered as the real data for analysis. We choose 20 units randomly as the training units, and the remaining 10 units as the testing ones.

Table 1 Preset parameters of the Wiener process model

Model parameters	α	β	b	σ_B^2
Preset values	10	0.2	1.5	4

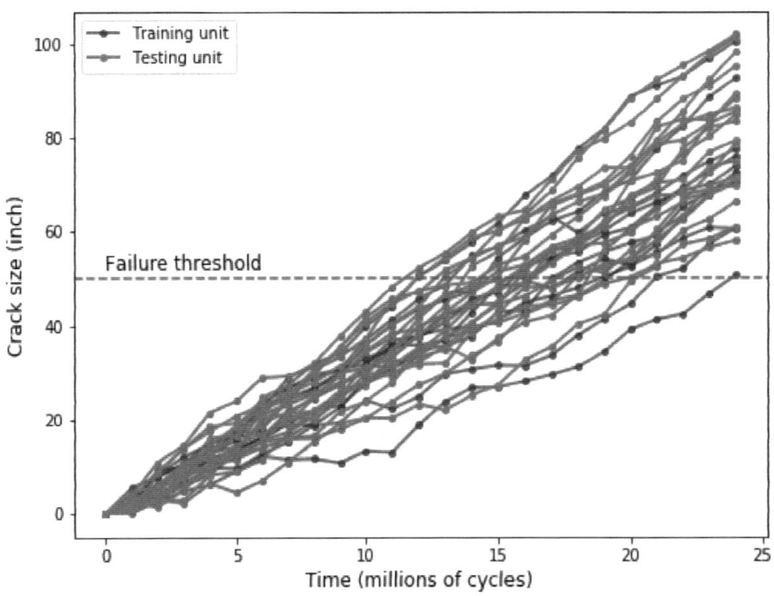

Fig. 2 Simulated degradation trajectories of 30 fatigue-crack-growth units

In this experiment, we divide the degradation process into stages by every 10 inches:

- Stage 0 – initial state, HI=0;
- Stage 1– degradation state, 0<HI<10;
- Stage 2– degradation state, 10≤HI<20;
- Stage 3– degradation state, 20≤HI<30;
- Stage 4– degradation state, 30≤HI<40;
- Stage 5– degradation state, 40≤HI<50;
- Stage 6– failure state, HI≥50.

Clearly, the RUL prediction is performed at stage 1-5. In order to demonstrate the performance of the proposed method, we employ mean absolute error (MAE) as the performance metric, and the lower the MAE, the better the performance. Fig. 3 illustrates the performance changes with the increasing number of cluster sampling for one of the testing unit, while others are much the same as it and don't be provided here. As the number of cluster sampling increases, the performance of the proposed method converges gradually at a constant level which is much better than that of the traditional SVR model without cluster sampling (i.e. the number of cluster sampling M=0). Therefore, in compare with the traditional SVR model, the proposed method can effectively improve the accuracy of online prediction. Although the time consumption of model construction increases as the number of cluster sampling increases, these works are carried out offline.

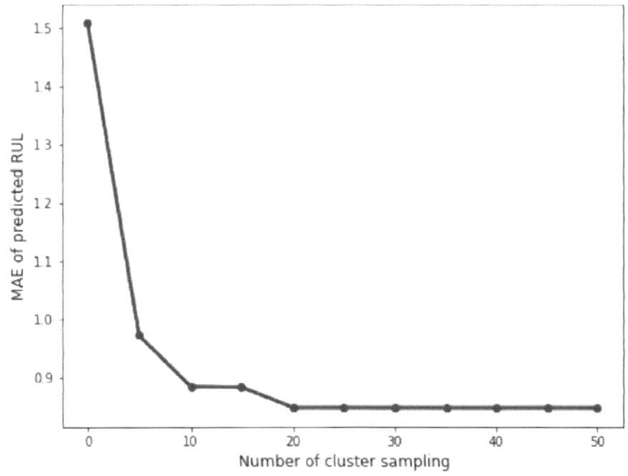

Fig. 3 The performance changes with the increasing number of cluster sampling for one of the testing unit

Due to the limited space, the test results of the 10 testing units are not provided here. As a replacement, we provide the performance metrics of them with the RUL prediction starts at 4 million of cycles, summarized in Fig. 4. The performance of the proposed method is compared with the traditional SVR method and the SbRP method. The MAE of the predicted results are shown in Table 2. Comparing these three methods, it shows clearly that the performance of the proposed method is the best one, and its reliability and robustness are all better than the other two methods.

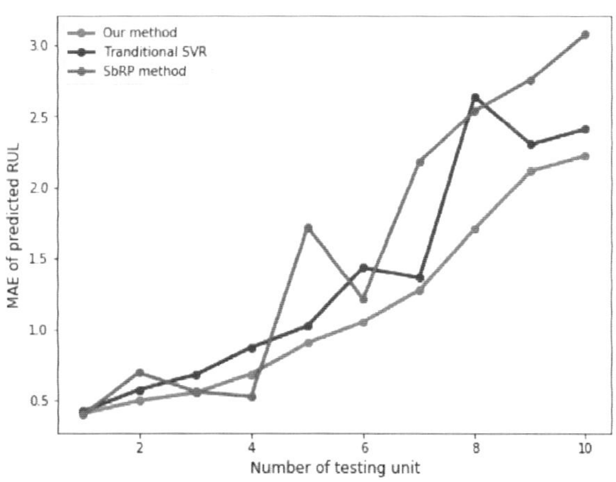

Fig. 4 The comparison of the predictive metrics between the proposed method and the traditional SVR method as well as the SbRP method

Table 2 The comparison of the predicted results for the simulated data

Testing unit	MAE of the proposed method	MAE of the traditional SVR method	MAE of the SbRP method
1	0.407	0.423	0.400
2	0.497	0.574	0.694
3	0.553	0.679	0.560
4	0.684	0.87	0.525
5	0.903	1.022	1.720
6	1.05	1.434	1.211
7	1.275	1.363	2.183
8	1.709	2.641	2.54
9	2.115	2.302	2.761
10	2.222	2.411	3.078
Mean	**1.142**	**1.372**	**1.567**

3.2. Application to GaAs-based semiconductor lasers

In this subsection, the proposed method is applied to a degradation dataset of GaAs laser devices, which is taken from Meeker and Escobar (1998) [34]. Degradation of the laser device is measured as the operation current, which increases over time. Degradation data from 15 testing units are collected over time with measure frequency 250 h and terminal time 4000 h, they are shown in Table 3. This dataset has been extensively studied in degradation modeling [35-37] and RUL prediction [38]. In this paper, a laser is assumed to have failed if the percentage of its operating current exceeds a predefined threshold level FT=6%. Note that the observation time points are recorded at every 250 h, this leads to the inability to know the true life, i.e. the time arriving the failure threshold. Due to the reasonable degradation process of laser is an approximate linear degradation [35-37], the most possible failure time considered as the real one can be calculated by local linear regression method.

Table 3 GaAs-based laser performance degradation data under thermal stress of 80°C (percentage increase of working current)

t/h	No.														
	1	2	3	4	5	6	7	8	9	10	11	12	13	14	15
250	0.47	0.71	0.71	0.36	0.27	0.36	0.36	0.46	0.51	0.41	0.44	0.39	0.3	0.44	0.51
500	0.93	1.22	1.17	0.62	0.61	1.39	0.92	1.07	0.93	1.49	1	0.8	0.74	0.7	0.83
750	2.11	1.9	1.73	1.36	1.11	1.95	1.21	1.42	1.57	2.38	1.57	1.35	1.52	1.05	1.29
1000	2.72	2.3	1.99	1.95	1.77	2.86	1.46	1.77	1.96	3	1.96	1.74	1.85	1.35	1.52
1250	3.51	2.87	2.53	2.3	2.06	3.46	1.93	2.1	2.59	3.84	2.51	2.98	2.39	1.8	1.91
1500	4.34	3.75	2.97	2.95	2.58	3.81	2.39	2.4	3.29	4.5	2.84	3.59	2.95	2.55	2.27
1750	4.91	4.42	3.3	3.39	2.99	4.53	2.68	2.78	3.61	5.25	3.47	4.03	3.51	2.83	2.78
2000	5.48	4.99	3.94	3.79	3.38	5.35	2.94	3.02	4.11	6.26	4.01	4.44	3.92	3.39	3.42
2250	5.99	5.51	4.16	4.11	4.05	5.92	3.42	3.29	4.6	7.05	4.51	4.79	5.03	3.72	3.78
2500	6.72	6.07	4.45	4.5	4.63	6.71	4.09	3.75	4.91	7.8	4.8	5.22	5.47	4.09	4.1
2750	7.13	6.64	4.89	4.72	5.24	7.7	4.58	4.16	5.34	8.32	5.2	5.48	5.84	4.83	4.38
3000	8	7.16	5.27	4.98	5.62	8.61	4.84	4.76	5.84	8.93	5.66	5.96	6.5	5.41	4.63
3250	8.92	7.78	5.69	5.28	6.04	9.15	5.1	5.16	6.4	9.55	6.2	6.23	6.94	5.76	5.38
3500	9.49	8.42	6.02	5.61	6.32	9.95	5.57	5.46	6.84	10.45	6.54	6.99	7.39	6.14	5.84
3750	9.87	8.91	6.45	5.95	7.1	10.49	6.1	5.81	7.2	11.28	6.96	7.37	7.85	6.51	6.16
4000	10.94	9.28	6.88	6.14	7.59	11.01	7.17	6.24	7.88	12.21	7.42	7.88	8.09	6.88	6.62

In order to demonstrate the effectiveness of the proposed method, we choose unit 6-15 as the training units while others as the testing ones. RUL prediction starts at 750 h, using separately the proposed method, the traditional SVR method and the SbRP method. For the proposed method, we divide the degradation process into stages by every 1% increase of working current:

- Stage 0 – initial state, HI=0;
- Stage 1– degradation state, 0<HI<1%;
- Stage 2– degradation state, 1%≤HI<2%;
- Stage 3– degradation state, 2%≤HI<3%;

- Stage 4– degradation state, 3%≤HI<4%;
- Stage 5– degradation state, 4%≤HI<5%;
- Stage 6– degradation state, 5%≤HI<6%;
- Stage 7– failure state, HI≥6%.

Take the unit 3 for example, the predicted results are plotted in Fig. 5, the results by the proposed method is relatively closer to the actual value. In addition, Fig. 6 provides the predicted results of unit 1, unit 2, unit 4 and unit 5. For unit 1 and unit 2, we can see that the predicted results by traditional SVR method don't converge near the real value over time, even the gap expands when the degradation state closes to failure, which is unacceptable in engineering practice. This issue is mainly caused by the limitation of SVR model that the linearity will increase as the training data increase, so that the traditional SVR model may get poor performance for some nonlinear units. Moreover, the predicted results for unit 1 and unit 2 by SbRP method are also of large gap with the real value during the whole lifecycle. Therefore, the proposed method has better reliability and robustness than the other two methods.

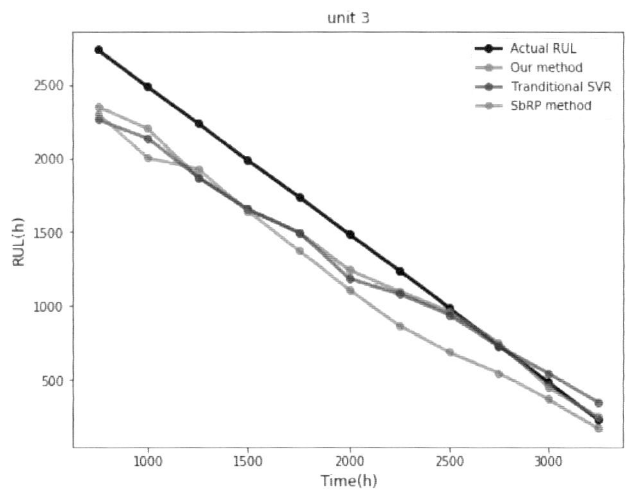

Fig. 5 The comparison of the predicted results of testing unit 3

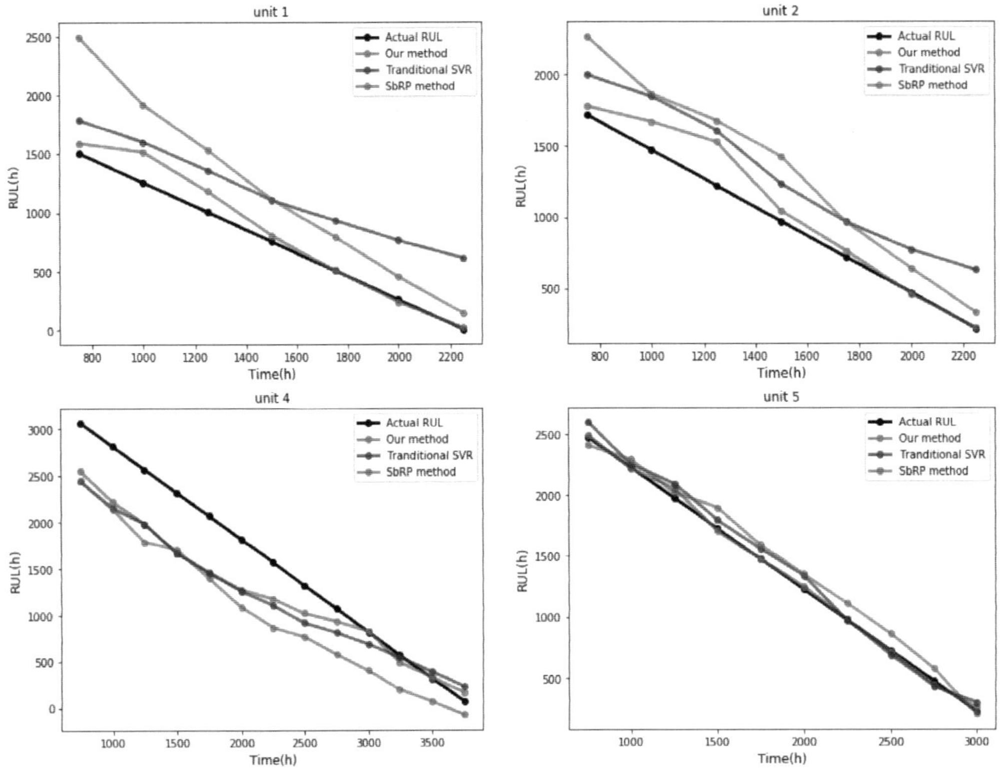

Fig. 6 The comparison of the predicted results of testing unit 1, unit 2, unit 4 and unit 5

Table 4 The comparison of the predicted results for the GaAs-based lasers data

Testing unit	MAE of the proposed method	MAE of the traditional SVR method	MAE of the SbRP method
1	80.765	411.257	451.276
2	99.799	323.853	340.039
3	189.368	222.83	303.851
4	347.734	400.756	540.246
5	22.060	68.630	99.000
Mean	**149.145**	**285.465**	**346.882**

To demonstrate further the effectiveness of the proposed method, Table 4 provides the MAE of the predicted results for the testing units. The proposed method is of the best performance for all the five testing units. Therefore, the proposed method can be used effectively for RUL prediction, and has better reliability and robustness than the traditional SVR method and the SbRP method.

4. Conclusions

This paper proposes a novel LS-SVR method based on multi-stages division and cluster sampling. It divides the degradation process into several stages, and through cluster sampling of the test data of each stage, as well as model construction stage by stage. Therefore, the optimal model parameters can be selected adaptively as the degradation evolves, so that each stage of the degradation process can be fitted at the best extent. This method overcomes the limitation of traditional SVR model that the linearity will increase as the training data increase.

To demonstrate the advantages of the proposed method, a simulated dataset of fatigue crack growth and a real-world degradation dataset of GaAs-based lasers are applied to study. And the predicted results are compared with the counterparts by the traditional SVR method and the SbRP method.

Results show that the proposed method is of better reliability and robustness than the traditional SVR method and the SbRP method. The proposed method extends the studies of SVR methods for RUL prediction, and is of effectiveness in engineering practice.

References

[1] Vogl, G. W., Weiss, B. A., Helu, M. (2019). A review of diagnostic and prognostic capabilities and best practices for manufacturing. *Journal of Intelligent Manufacturing,* 30(1), 79–95.
[2] Javed, K., Gouriveau, R., Zerhouni, N., Nectoux, P. (2015). Enabling health monitoring approach based on vibration data for accurate prognostics. *IEEE Transactions on Industrial Electronics*, 62(1), 647-656.
[3] Si, X. S., Wang, W., Hu, C. H., Zhou, D. H. (2011). Remaining useful life estimation – a review on the statistical data driven approaches. *European Journal of Operational Research*, 213, 1-14.
[4] Hu, Y. W., Liu, S. J., Lu, H. T., Zhang, H. C. (2019). Remaining useful life model and assessment of mechanical products: a brief review and a note on the state space model method. *Chinese Journal of Mechanical Engineering*, 32(1), 15.
[5] An, D., Kim, N. H., Choi, J. H. (2015). Practical options for selecting data-driven or physics-based prognostics algorithms with reviews. *Reliability Engineering & System Safety*, 133, 223-236.
[6] Li, C. J., Lee, H. (2005). Gear fatigue crack prognosis using embedded model, gear dynamic model and fracture mechanics. *Mechanical Systems and Signal Processing,* 19(4), 836-846.
[7] Kontar, R., Zhou, S. Y., Sankavaram, C., Du, X. Y., Zhang, Y. L. (2018). Nonparametric-condition-based remaining useful life prediction incorporating external factors. *IEEE Transactions on Reliability,* 67(1), 41-52.
[8] Mosallam, A., Medjaher, K., Zerhouni, N. (2016). Data-driven prognostic method based on Bayesian approaches for direct remaining useful life prediction. *Journal of Intelligent Manufacturing,* 27(5), 1037-1048.
[9] Fan, Z. L., Liu, G. B., Si, X. S., Zhang, Q., Zhang, Q. H. (2013). Degradation data-driven approach for remaining useful life estimation. *Journal of Systems Engineering and Electronics, 24*(1), 173-182.
[10] Long, B., Xian, W. M., Jiang, L., Liu, Z. (2013). An improved autoregressive model by particle swarm optimization for prognostics of lithium-ion batteries. *Microelectronics Reliability, 53*(6), 821-831.
[11] Barraza-barraza, D., Tercero-Gomez, V. G., Beruvides, M. G., Limon-Robles, J. (2017). An adaptive ARX model to estimate the RUL of aluminum plates based on its crack growth. *Mechanical Systems & Signal Processing,* 82, 519-536.
[12] Fleischer, C., Waag, W., Bai, Z., Sauer, D. U. (2013). On-line self-learning time forward voltage prognosis for lithium-ion batteries using adaptive neuro-fuzzy inference system. *Journal of Power Sources,* 243, 728-749.
[13] Dong, M., He, D. (2007). A segmental hidden semi-Markov model (HSMM)-based diagnostics and prognostics framework and methodology. *Mechanical Systems and Signal Processing,* 21(5), 2248-2266.
[14] Gebraeel, N. Z., Lawley, M. A. (2008). A neural network degradation model for computing and updating residual life distributions. *IEEE Transactions on Automation Science and Engineering,* 5(1), 154-163.
[15] Tian, Z. G. (2012). An artificial neural network approach for remaining useful life prediction of equipment subject to condition monitoring. *Journal of Intelligent Manufacturing,* 23(2), 227-237.
[16] Mahamad, A. K., Saon, S., Hiyama, T. (2010). Predicting remaining useful life of rotating machinery based artificial neural network. *Computers & Mathematics with Applications,* 60(4), 1078-1087.

[17] Loutas, T. H., Roulias, D., Georgoulas, G. (2013). Remaining useful life estimation in rolling bearings utilizing data-driven probabilistic e-support vectors regression. *IEEE Transactions on Reliability, 62*(4), 821-832.

[18] Khelif, R., Chebel-Morello, B., Malinowski, S., Laajili, E., Zerhouni, N. (2017). Direct remaining useful life estimation based on support vector regression. *IEEE Transactions on Industrial Electronics,* 64(3), 2276-2285.

[19] Zio, E., Maio, F. D. (2010). A data-driven fuzzy approach for predicting the remaining useful life in dynamic failure scenarios of a nuclear system. *Reliability Engineering & System Safety, 95*(1), 49-57.

[20] You, M. Y., Meng, G. (2013). A framework of similarity-based residual life prediction approaches using degradation histories with failure, preventive maintenance, and suspension events. *IEEE Transactions on Reliability, 62*(1), 127-135.

[21] Gu M. Y., Chen Y. L. (2019). Two improvements of similarity-based residual life prediction methods. *Journal of Intelligent Manufacturing*, 30(1), 303-315.

[22] Noortwijk, J. M. V. (2009). A survey of the application of gamma processes in maintenance. *Reliability Engineering & System Safety*, *94*(1), 2-21.

[23] Li, N. P., Lei, Y. G., Yan, T., Li, N. B., Han, T. (2019). A Wiener-process- model-based method for remaining useful life prediction considering unit-to-unit variability. *IEEE Transactions on Industrial Electronics*, 66(3), 2092-2101.

[24] Son, K. L., Fouladirad, M., Barros, A., Levrat, E., Lung, B. (2013). Remaining useful life estimation based on stochastic deterioration models: a comparative study. *Reliability Engineering and System Safety*, 112, 165-175.

[25] Nuhic, A., Terzimehic, T., Soczka-Guth, T., Buchholz, M., Dietmayer, K. (2013). Health diagnosis and remaining useful life prognostics of lithium-ion batteries using data-driven methods. *Journal of Power Sources, 239*, 680–688.

[26] Benkedjouh, T., Medjaher, K., Zerhouni, N., Rechak, S. (2013). Health assessment and life prediction of cutting tools based on support vector regression. *Journal of Intelligent Manufacturing, 26*(2), 213-223.

[27] Chen, Z. Z., Cao, S. C., Mao Z. J. (2018). Remaining useful life estimation of aircraft engines using a modified similarity and supporting vector machine (SVM) approach. *Energies*, 11(1), 1-14.

[28] Ordonez, C., Lasheras, F. S., Roca-Pardinas, J., Juez, F. J. D. (2019). A hybrid ARIMA–SVM model for the study of the remaining useful life of aircraft engines. *Journal of Computational and Applied Mathematics*, 346, 184-191.

[29] Wei, J. W., Dong, G. Z., Chen, Z. H. (2018). Remaining useful life prediction and state of health diagnosis for lithium-ion batteries using particle filter and support vector regression. *IEEE Transactions on Industrial Electronics*, 65(7), 5634-5643.

[30] Pei, H., Hu, C. H., Si, X. S., Zhang, J. X., Pang, Z. N., Zhang, P. (2019). Review of machine learning based remaining useful life prediction methods for equipment. *Journal of Mechanical Engineering*, 55(8), 1-13.

[31] Wen, Y. X., Wu, J. G., Das, D., Tseng, T. L. (2018). Degradation modeling and RUL prediction using wiener process subject to multiple change points and unit heterogeneity. *Reliability Engineering & System Safety*, 176, 113-124.

[32] Hu, Y. G., Li, H., Liao, X. L., Song, E. B., Liu, H. T., Chen, Z. (2016). A probability evaluation method of early deterioration condition for the critical components of wind turbine generator systems. Mechanical Systems and Signal Processing, 76–77, 729–741.

[33] Kimotho, J. K., Sondermann-Woelke, C., Meyer, T., Sextro, W. (2013). Machinery prognostic method based on multi-class support vector machines and hybrid differential evolution–particle swarm optimization. *Chemical Engineering Transactions*, 33, 619–624.

[34] Meeker, W., Escobar, L. A. (1998). Statistical methods for reliability data. *Probability and Statistics*, New York: Wiley.

[35] Wang, X., Xu, D. H. (2010). An inverse gaussian process model for degradation data. *Technometrics*, 52(2), 188-197.

[36] Ye, Z. S., Chen, N. (2014). The inverse gaussian process as a degradation model. *Technometrics*, 56(3), 302-311.

[37] Pan, D. H., Liu, J. B., Cao, J. D. (2016). Remaining useful life estimation using an inverse gaussian degradation model. *Neurocomputing*, 185, 64-72.

[38] Liu, Z., Wang, Q., Song, C. L., Cheng, Y. H. (2017). Similarity-based difference analysis approach for remaining useful life prediction of GaAs-based semiconductor lasers. *IEEE Access*, 5, 21508-21523.

Business Model, Lock-in Effect and Network Effect of Virtual Community

Shih-Feng Chang[1], Guo-An Wu[2], Jia Liu[1*]

[1] Nanfang College of Sun Yat-Sen University, Guangzhou, 510000, China

[2] School of Big Data Management of Soochow University, Taipei, 11102, China

*Corresponding author's e-mail: 2228575971@qq.com

Abstract. As progress of era and technology and development of internet, many firms make big change which occurring on communication and interaction among people differing from before. Especially in recent years, virtual community which has characteristics of social networking services appears and grows up, and it make the related topics of network effect more and more attraction between academics and industry. In other words, social network which has multiple functions can fully utilize the powers of these new and popular virtual community to prototype and develop rapidly on internet, even more, to make general interest users to expand human network, interact and exchange information through these internet tools. In view of this, we take Facebook, MySpace, YouTube, Flickr, Twitter, Plurk, Wretch and PTT as research subjects, through qualitative analysis, and finally we discover significant relationships among business model, lock-in effect and network effect.

1. Introduction

In recent years, the vigorous development of virtual community, both at home and abroad, has attracted many netizens' participation, and it has been highly valued by many enterprises. So, firms are willing to invest part of the company's resources to operate virtual community, and they also try to interact with netizens. After all, in virtual communities, knowledge generated through member interaction is the most important asset of virtual community and the main reason for attracting other new members to join.

However, due to the lack of research on these popular virtual community websites, both at home and abroad, and the lack of case comparison, this study is more interested in discussing the above.

In view of this, we attempt to understand the business model of these popular virtual community websites by observing and analyzing the characteristics of Facebook, MySpace, YouTube, Flickr, Twitter, Plurk, Wretch and PTT, and we try to test the effectiveness by using locking effect or network effect.

2. Literature review

2.1. Virtual community

Rheingold (1993) first proposed the concept of virtual community, which means that in the Internet space, there are a group of enough people who have long-term public discussion and information exchange, and the interpersonal network formed by full emotional communication [1]. Baym (1994) further pointed out that virtual community is generated in the following four communication processes:

(1) creating different forms of opinion expression and communication, (2) personal identity, (3) relationships among members, (4) establishing norms of behavior [2].

2.2. Business model

Hamel (2000) believed that a business model must cover the following four elements: (1) core strategy, (2) strategic resources, (3) customer interface, (4) value network [3]; in addition, he also mentioned that the four elements should be connected through three bridges, namely customer interest, activity structure and company boundary; in addition, the following four factors determine profit potential to support this business model factors of force: (1) efficiency, (2) uniqueness, (3) collocation, (4) profit propeller. As for Chesbrough & Rosenbloom (2002) said, the business model should be a middle way between technological development and economic value [4], that is, the characteristics and possibilities of technology are investment. Through the transformation of business model, it will become an economic output acceptable to customers and the market.

2.3. Lock-in effect

Liebowitz & Margolis (1994) further defined lock-in effect as that the user is unwilling or unable to switch to other new products or services due to certain factors [5]. Because the conversion cost is the main factor that causes the locking effect, different lock-in effect types usually have different conversion costs. Therefore, Shapiro & Varian (1998) proposed the following seven types of lock-in effects and explained their related conversion costs: (1) contract lock-in, (2) durable equipment, (3) personnel training, (4) information and database, (5) professional suppliers, (6) search cost, (7) loyalty program [6].

2.4. Network effect

Katz & Shapiro (1994) found that when other users choose to join the network and cause the network to expand day by day, the increase in the number of users will have a positive impact on other users, and this situation has a network effect [7]. Shankar & Bayus (2003) pointed out that the increase of network effect is mainly due to the increase in the number of users of a certain product or service. According to this, network effect can be divided into the following two types: (1) network scale; (2) network strength [8].

3. Research method

We mainly take case study method and comparative method, which will be very helpful for the verification of theory. In addition, in order to further understand the characteristics and operation mode of these virtual communities, we joined Facebook, MySpace, YouTube, Flickr, Twitter, Plurk, Wretch and PTT respectively, and we became members of these virtual community websites, so as to observe various behaviors and phenomena of their members in these virtual community websites.

4. Analysis and discussion

4.1. Business model and network effect

The research objects of this study, such as Facebook, MySpace, YouTube, Flickr, Twitter, Plurk, Wretch and PTT, are not only limited in the number of users of the website, but also unable to improve the traffic of the website when they have not yet imported the social function (see Table 1). For example, before July 2007, users of Facebook could only post text messages, but since then, it has started to open a large number of social interaction functions such as pictures, images, psychological tests and small games, and can share them with friends. Its increased sense of fun really attracted a lot of users. In addition, after May 2007, Facebook began to provide the marketplace service, which enables both parties to make transactions by publishing information and responding to the posted content. Similarly, since its opening in July 2003, MySpace has been mainly focusing on making friends and initially calling on music, but it is unable to attract a large number of users to register on the site. However, it attracted a large number of users in early 2005 after the launch of event invitation, viral promotion of instant messaging software and alumni list. At the same time, its movie leaderboards are the features of its main social platform, and it wants to use them to differentiate itself from YouTube. Flickr, which belongs to video sharing YouTube and photo sharing, sets community operation options on its home page. Therefore, we infer that proposition 1-1 (a): when virtual community website introduces the community function in its business model customer interface, it can rapidly increase network scale; proposition 1-1 (b): when virtual community website introduces the community function in its business model customer interface, it can quickly increase network strength.

In Taiwan, Wretch adopts open platform approach. In addition to its original BBS function, which is intended to compete with PTT, it also provides an online album and even a whole set of basic blog services. Even though it later closed the BBS platform, it established the first brand blog website in Taiwan with the most users and the largest traffic. On the contrary, PTT is limited by the closed BBS architecture. Although PTT enjoys the reputation of the largest BBS station in China, it is unable to break through the limitation of this framework. Therefore, we infer proposition 1-2: in the resource allocation of its business model, virtual community website with open platform can increase network scale more quickly than virtual community website with closed platform.

According to Table 1, the number of members of relational virtual community websites such as Facebook or MySpace, far exceeds the number of members of interested virtual community websites such as YouTube, Flickr, Twitter or Plurk. As for Wretch and PTT, although they are also interest oriented virtual community websites, they are not considered because most of their users are Internet users in Taiwan and they have their own prior development restrictions. Therefore, we infer proposition 1-3: in the knowledge application of its business model, relational virtual community website can increase network scale in a shorter time than interested virtual community website.

4.2. Business model and lock-in effect

Since Facebook and MySpace are providing more dynamic information in each member's personalized page, even if the owner of the personalized page does not update the information, other viewers can still see something different each time. These dynamic information is the result of the website record extraction. In other words, the two websites themselves record all the actions of everyone on the website, and then play relevant updates on other people's pages. Of course, such a move is likely to cause privacy disputes. However, Facebook has caused protests from netizens because of this function, it has only protested for half a year, but the original protest waves have gradually disappeared. It can be seen that members have relaxed the scale of how much their own information can be disclosed. Even because these information is too funny, the traffic of the whole website has been successfully surging up. As for YouTube that shares video and audio or Flickr that shares photos, because of the strong rendering power of videos and photos, plus the community power of free sharing, as well as the unique functions of embedding and semantic tags, the cost of conversion has also been greatly increased. Therefore, we infer proposition 2-1: when the virtual community website introduces the community function into its business model customer interface, it will produce the lock-in effect of high conversion cost.

Whether it's Twitter or Plurk, it can be made into a plug-in, incarnate itself as a small frame, and then embed it in personal websites, forums or blogs at will, and update it anytime and anywhere, so this kind of intimate little action can accidentally trigger the super strong network effect of "one to ten, ten to 100". As for wretch, which is loved by Taiwanese netizens, although it competes with PTT in the first place in the service, it failed, but because it adopted the open platform strategy, skillfully combined the BBS function with the network album function, and even further integrated into the blog service later, finally let it gradually surpass and replace the PTT which always adheres to the closed architecture in the number of users and traffic of the website, or in the influence of the website. Therefore, we infer proposition 2-2: in the resource allocation of its business model, virtual community website with open platform can produce the lock-in effect with high conversion cost more quickly than virtual community website with closed platform.

According to Table 1, the number of members of relational virtual community websites such as Facebook or MySpace is far more than that of interested virtual community websites such as YouTube, Flickr, Twitter or Plurk, and it has reached such an amazing number in a relatively short time. As for Wretch and PTT, although they are relational virtual community websites, they are not considered due to their innate development or postnatal special use. Therefore, we infer proposition 2-3: in the knowledge application of its business model, relational virtual community website can produce lock-in effect of improving the conversion cost in a shorter time than interest virtual community website.

4.3. Lock in effect and network effect

Both Facebook and MySpace allow users to have their own personalized home page, and they can also place or change the information and style which they want to present at will, but they don't have to stick to the traditional standardized forms. Since then, young netizens can have a face that they can dress meticulously on the Internet, and use it to communicate and share with others. The embedding of YouTube not only enables the videos and videos shared on its website to be viewed anytime and anywhere, but also because the semantic tags provided by Flickr are deeply attractive to powerful users. Therefore, we infer proposition 3-1 (a): when virtual community website has a high conversion to native lock-in effect, it can rapidly increase the network scale; proposition 3-1 (b): when virtual community website has a high conversion to native lock-in effect; when the lock-in effect of high conversion cost is achieved, network strength can be increased rapidly.

However, although Twitter or Plurk are only micro blogs, their functions are easy to understand and easy to use, because they are designed as small plug-ins, they can achieve similar embedded functions with YouTube, so they can achieve the same powerful effect. As for Wretch, because of its open platform architecture and integration of related services and functions, it has formed the blog effect of ant hero. Even through it continue to innovate and provide updated services according to the needs of netizens, it can attract the use and love of netizens in Taiwan. On the contrary, PTT adopts a closed BBS architecture. Although it can publish and receive information in real time and attract a large number of college students, it is still limited in growth speed and expansion scale. Therefore, we infer proposition 3-2: when virtual community website has the lock-in effect of high conversion cost, open platform virtual community website can increase network scale more quickly than closed platform virtual community website.

According to Table 1, the number of members of relational virtual community websites such as Facebook or MySpace will exceed the number of members of interested virtual community websites such as YouTube, Flickr, Twitter or Plurk. In addition to the reputation effect of loyal users, it also includes Xinjia The time and energy invested by the entrant himself has transformed him into another loyal user. Under such a chain effect fermentation, the entrant can reach a certain number of members in a relatively short time, or even surpass the original competitors. As for Wretch and PTT, although they are relational virtual community websites, they are not considered due to their innate development or postnatal special use. Therefore, we infer proposition 3-3: when virtual community website has lock-in effect of high conversion cost, relational virtual community website can increase network scale in a shorter time than interest virtual community website.

Table 1. Comparison of virtual community websites.

Website	Founded	Number of Users	User Type	Purpose
Facebook	2004.2	300 million	from the initial group of students to all levels	building and maintaining relationships and promoting public goods or music
MySpace	2003.7	114 million	musicians and teenagers	building and maintaining relationships and promoting public goods or music
YouTube	2005.2	100 million person times per month	general Internet users	sharing videos
Flickr	2004.2	3.3 million items	general Internet users	sharing photos
Twitter	2006.3	4.1 million	heavy users	share mood and information
Plurk	2007.11	2.73 million	heavy users	share mood and information
Wretch	1999.7	5 million	mainly Taiwan netizens	self-expression
PTT	1995.9	1.5 million	mainly Taiwan netizens	instant messaging

5. Conclusions and suggestions

5.1. Conclusions
Based on the above discussion, this study analyzes and explores Facebook, MySpace, YouTube, Flickr, Twitter, Plurk, Wretch and PTT, and draws the following three important conclusions:

1. In the business model, if the community function is introduced into the customer interface, it will help to directly form network effect.

2. In the business model, if the community adopts open platform architecture in resource allocation, it will help to form lock-in effect and lead to network effect.

3. For business model, lock-in effect and network effect, they should adapt to each other and support each other in design and implementation, so as to play the best role.

5.2. Suggestions
First, cross industry research of related industries can be done to expand the applicability and value of this study by gradually spreading and extrapolating the research results to other fields. Second, the research propositions deduced in this study can be transformed into research variables after considering and supplementing the relevant theories to carry out quantitative research, so as to obtain more widely applicable results. Third, it is suggested that firms should try to adopt open platform architecture when they design business models, so as to have the chance to achieve the desired lock-in effect, or even strong network effect.

References
[1] Rheingold, H. (1993) *Virtual community: Homesteading on the electronic frontier*. Addison-Wesley, Cambridge.
[2] Baym, K. N. (1994) The emergence of community. In Jone, S. G. (ed.), *Cybersociety: Computer-mediated communication and community*. SAGE, London.
[3] Hamel, G. (2000) *Leading the revolution*. Harvard Business School Press, Boston.
[4] Chesbrough, H. & Rosenbloom, R. S. (2002) The role of the business model in capturing value from innovation: Evidence from Xerox corporation's technology spin-off companies. *Industrial and Corporate Change*, 11(3): 529-555.
[5] Liebowitz, S. J. & Margolis, S. E. (1994) Network externalities: An uncommon tragedy. *Journal of Economics Perspectives*, 8(2): 133-150.
[6] Shapiro, C. & Varian, H. (1998) Information rules: A strategic guide to the network economy. Harvard Business School Press, Boston.

[7] Katz, M. & Shapiro, C. (1994) System competition and network effect. *Journal of Economics Perspectives*, 8(2): 93-115.
[8] Shankar, V. & Bayus, B. L. (2003) Network effects and competition: An empirical analysis of the home video game industry. *Strategic Management Journal*, 24(4): 375-419.

IOP Conf. Series: Materials Science and Engineering **806** (2020) 012043 doi:10.1088/1757-899X/806/1/012043

Palm-vein verification based on U-Net

Peng Wang[1*] and Huafeng Qin[2]

[1] National Research Base of Intelligent Manufacturing Service, Chongqing Technology and Business University, Chongqing, 400067, China

[2] Chongqing Engineering Laboratory of Detection Control and Integrated System, Chongqing Technology and Business University, Chongqing, 400067, China

*Corresponding author's e-mail: 2017658007@email.ctbu.edu.cn

Abstract: As one of the biometric characteristics, palm-vein have received more and more attention in recent years. However, in practical applications, the vein image capturing is affected by various factors, so many low quality images are produced in recognition system. Generally, the vein networks extracted from such low quality images contains many noises, which degrades the recognition accuracy. To address these problems, this paper proposes an end-to-end convolutional neural network to extract vein feature for verification. Firstly, we label the palm-vein pixel by the combination of some handcraft-based palm-vein image segmentation methods and build a training dataset. Secondly, a U-Net network is trained based on the resulting dataset and its outputs are the probability of pixels to belong to vein pattern. Thirdly, we propose a scheme to encode the outputs of U-Net to obtain the vein network patterns. The experiment results on the public CASIA palm-vein dataset implies the effectiveness of our proposed method.

1. Introduction

With the development of the economy and wide applications of internet technology, the demand for information and property security is increasing[1]. As one of personal identification, biometrics characteristics have developed rapidly in past years. Biometric recognition technologies are divided into two categories: (1) Extrinsic characteristics: finger print[2], iris[3], gait[4], face[5]. (2) Intrinsic characteristics: finger-vein[6], palm-vein[7]. Unlike the extrinsic biometric features, vein verification technology has many advantages such as intrinsic features, living identification, uniqueness, not easy to replicate, etc. In addition, compared with the finger-vein, the palm vein has more complex lines and structures, which are beneficial to improve the verification rate of palm-vein. Based on these excellent characteristics, the palm-vein verification technology has received more and more attention from researchers and industry in recent years.

2. Related Works

Despite recent advances in palm-vein recognition, palm-vein still faces many challenges in real life. In practical application, imaging process is affected by many factors such as temperature, equipment, user habits, illumination, etc. so the palm-vein images contains ambiguous regions where the separability between vein and background is poor. In general, matching such regions degrades the recognition accuracy. To solve these problems, many palm-vein segmentation methods are proposed to extract vein network. In general, they can be divided into two categories as follows. Handcraft-based segmentation approaches. These methods mainly build mathematical models to extract vein based on the assumption that the vein pattern shows in valleys or straight-line, etc. Han[7] proposed an innovative and robust adaptive Gabor filtering method, which encodes palm vein features into bit string representations. Based on the Scale-invariant feature transform (SIFT) method, Kang[8] proposed the ROOTSIFT method, a more stable local invariant feature extraction. Zhang[9] extracted the palm vein by multi-scale filtering. Deep learning-based segmentation methods. Convolutional neural network have outperformed the state of the art in computer vision field[10-12]. Different with handcraft-based segmentation methods, deep learning segmentation methods is an end-to-end architecture without the manual attribute distribution assumption. Some researchers have made some explorations. For example, Qin[13] proposed an iterative deep neural network for hand-vein verification. These existing methods do not always perform well in practical application. For example, handcraft-based segmentation methods are based on the distribution assumption. However, many vein patterns show more complex shape instead of valley or straight-line, so these assumption may not be always effective.. Based on the good performance of U-Net[14] network in the field of image segmentation, we employ it for palm vein segmentation.

3. The Proposed Method

In this paper, we design a convolutional neural network structure to extract palm vein features. Firstly, we assign the vein pixel label by some handcrafted segmentation methods in the training datasets. Secondly, we build a U-Net and train it. Finally, we match the vein features for verification. The overview of the proposed method is shown in Figure1.

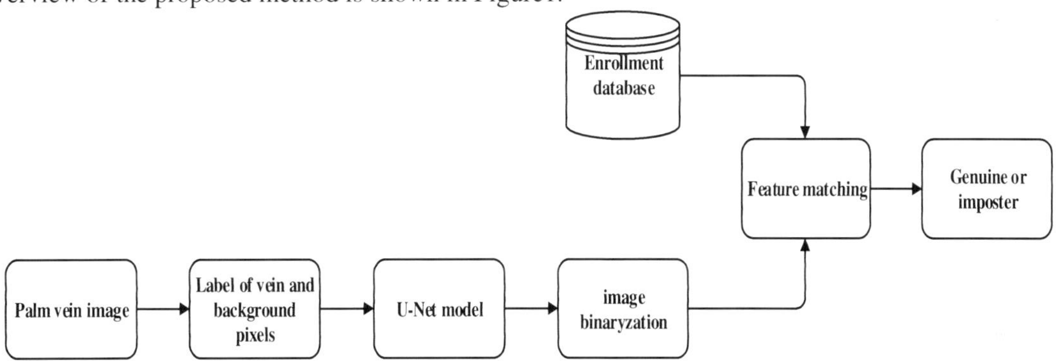

Figure 1. The overview of the proposed system

3.1. Label vein pixel

In this Section, we will use four baselines e.g. repeated line tracking[15], mean curvature[16], maximum principle curvature[17], and Gabor filters[18] to segment vein pattern for each vein images. For each image, four binary vein images is extracted based on four approaches. Finally, we average four binary images to obtain a map which is subject to binarization by a threshold of 0.5 to compute a labeling image (vein pixel and background are labeled as 1 and 0). The label approach is detail in [13].

3.2. U-Net model

Unlike general convolutional neural network, this network consists of encoder layers and decoder layers. Encoder layers gradually reduce the spatial dimension of the layers, and decoders gradually repair the details and spatial dimension of objects. There is usually a fast connection between the encoder layers and the decoder layers, so it can help the decoder layers to repair the details of the target better. In the final layer, the outputs are the probability of each pixel which belongs to palm vein or the background. The network architecture is illustrated in Figure. 2.

IOP Conf. Series: Materials Science and Engineering **806** (2020) 012043 doi:10.1088/1757-899X/806/1/012043

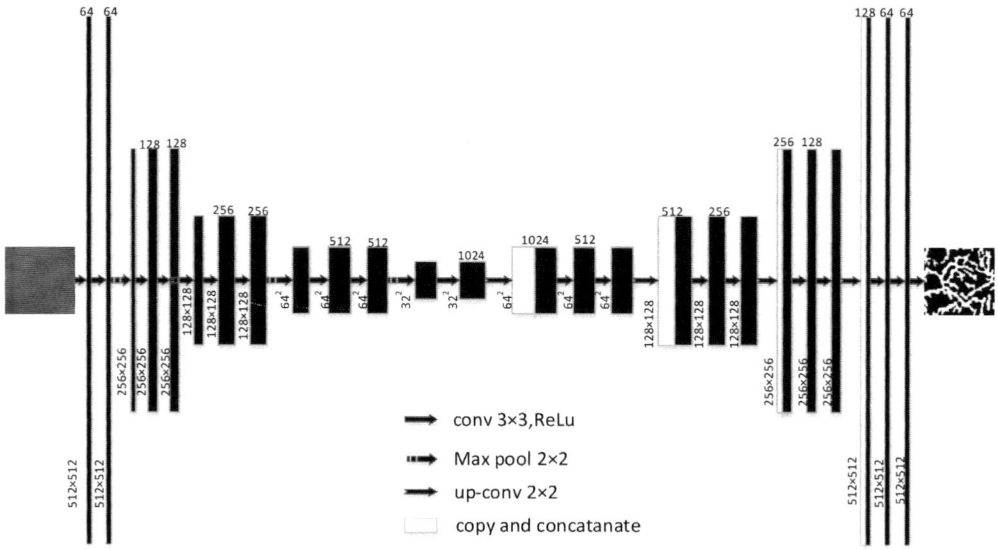

Figure.2.The architecture of U-Net

As shown in the Figure.2, the U-Net architecture consists of a contracting path and an expansive path. In the contracting path, down-sampling is a classical convolutional neural network, which consists of five layers. In each layer, there are two 3× 3 convolution layers, followed by a rectified linear unit (ReLU) and a 2× 2 max pooling layer with stride 2. And the number of convolution kernels in the next layer is twice times than the number of convolution kernels in the previous layer, and finally the number of convolution kernels in the fifth layer is 1024. In the expansive path, each layer have two 2× 2 up-convolution layers, followed by two 3× 3 convolution layers and a rectified linear unit (ReLU). With the number of layers increasing, we halve the number of convolution kernels. After training， the U-Net outputs a map in which the values of each pixel is its probability being to vein pattern. Then, the map is further subject to binarization for matching.

3.3. Supervised Feature Encoding

In some approach, the output map of U-Net is encoded by a threshold of 0.5 for matching. To achieve robust matching, we encode the feature map to a threshold which is related to matching scores. In our encoding scheme, a threshold is obtained by maximizing the distance between intra-class and inter-class, and employed to encode feature map. We assume that there are m samples, and each sample have n classes. Then there are m × n images. We generate intra-class scores by matching binary images from the same classes using Hamming distance in [19], and generate inter-class scores by matching binary images from the different classes. Finally, we obtain $a_1 = m \times C_n^2$ genuine matching scores $U_1 = \{d_1(T),\ d_2(T), \dots, d_{a_1}\}$ and $a_2 = \mathrm{n} \times \mathrm{m} \times C_m^2$ imposter matching scores $U_2 = \{d_1'(T),\ d_2'(T), \dots, d_{a_2}'\}$. To get a robust threshold, we formulate the following optimization objective function:

$$maxJ(T) = \frac{|u_1(T) - u_2(T)|}{D_1(T) + D_2(T)} \quad (1)$$

Where T denote the threshold, $u_1(T)$ and $u_2(T)$ are the means of the scores in the sets U_1 and U_2, and $D_1(T)$ and $D_2(T)$ are the variance of the scores in the sets U_1 and U_2. The parameter T is assigned from 0 to 255, so 256 different values *J(T)* are computed using Eq. (1)

3.4. Enhanced hamming distance

Hamming distance [19] is used to calculate the difference between two vectors. The vein network are stored in binary image, so we employ Hamming distance to compute the similarity between two palm-vein images. Because there are translations among palm-vein images, we propose an enhanced hamming distance to match them. We assume E and F are two binary feature images with size of $w \times h$. The length and width of image E are extended to $w + 2i$, $h + 2j$. Its expanded images $\bar{E}$ are expressed as:

$$\bar{E}(w,h) = \begin{cases} E(w-i,h-j) & \text{if } 1+i \le w \le g+i, \\ & \quad 1+j \le h \le y+j \\ -1, & \text{otherwise} \end{cases} \tag{2}$$

The parameters w and h that control the translation distance in horizontal and vertical directions are set to 100 and 100, and the matching distance between E and F is obtained by

$$d(E, F) = \min_{0 \le m \le 2i, 0 \le n \le 2j} \frac{\sum_{w=1}^{x} \sum_{h=1}^{y} hamdis(\bar{E}(w+m,h+n),F(w,h))}{\sum_{w=1}^{x} \sum_{h=1}^{y} \theta(\bar{E}(w+m,h+n),-1)} \tag{3}$$

hamdis denotes the hamming distance between two binary images and

$$\theta(U,V) = \begin{cases} 1 & if\ U \neq V \\ 0 & otherwise \end{cases} \tag{4}$$

4. Experiment and results

4.1. CASIA Datasets

There are 1200 images in the CASIA Multi-Spectral Palm-print Image Database. In our experiments, a pre-processing method[13] is employed to extract the region of interest (ROI) image. we split the database into two datasets. There are 100 hands associated with 600 images in the training dataset, and 100 hands associated with 600 images in the test dataset.

4.2. Verification results

In this experiments, we visually assess the performance of various approach. We employ the proposed to extract the vein network and segmentation results is shown in Figure.3. In addition, the vein patterns extracted by four baselines are illustrated in Figure. 3 for comparison. From Figure. 3, we can see that the proposed model can extract more continuous and smoothness vein patterns compared to four handcrafted approaches. This can be explained by the following fact. the proposed model can directly learn rich prior knowledge acquired by training the U-Nets on a huge train set from different images without depending any assumptions.

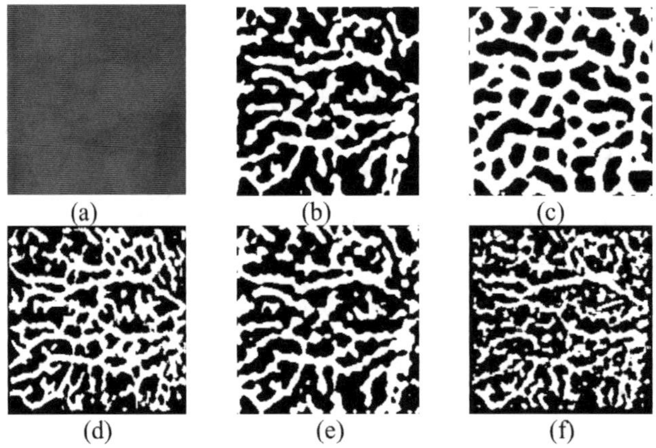

Figure 3. Segmentation results from varoius apporaches. (a) Original palm-vein image, (b) vein feature using proposed approach, (c) vein feature using Gabor filtes, (d) vein feature using Maximum principle curvature, (e) vein feature using Mean curvature, (f) vein feature using Repeated line tracking

In this section, the experiments are carried out to estimate the verification performance of our approach. Based on Eq (1), we encode the output of U-Nets and obtain the binary images. In test dataset, there are .600 (100 × 6) genuine scores are generated by matching samples from same hand and (100×99×6/2) impostor scores are generated by matching samples from different hands. The FAR and FAR are computed using genuine sores and imposter scores. The EER is the error rate when FAR is equal to FRR. The receiver operating characteristics of various approaches is depicted in Figure.4 and the corresponding EER is listed in Table 1. The experimental results from Figure.4 and Table 1 implies that the proposed approach outperform handcrafted approaches and achieve lowest EER, e.g. 0.47%.

Table 1. Experiment results

Methods	EER(%)
Gabor filters	0.93
Maximum principle curvature	1.85
Mean curvature	0.93
Repeated line tracking	3.59
Proposed method	0.47

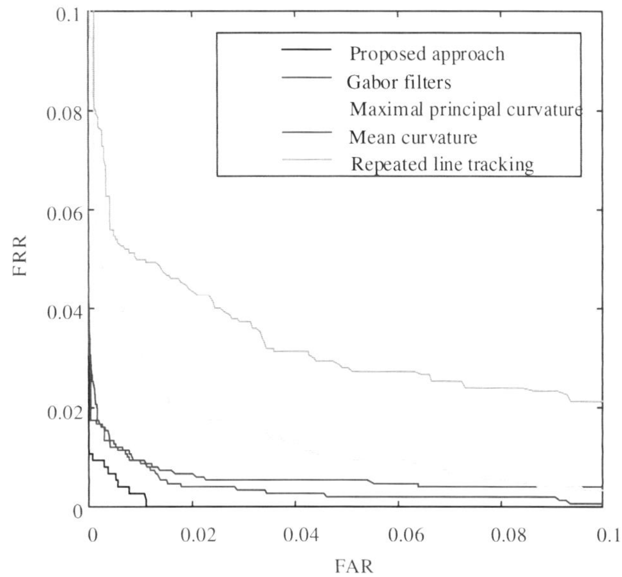

Figure.4. EER of various approaches.

5. Conclusion

In this paper, we presents a U-Net based palm segmentation approach to extract palm vein pattern. Firstly, we label the vein pixel by the combination of several handcraft-based segmentation methods. Secondly, the U-Net model is employed to predict the probability of pixel to belong to vein pattern. Thirdly, we proposed a robust scheme to encode the vein patterns. Finally, experimental results on the public dataset demonstrate the proposed approach's effectiveness.

References

[1] Su J.F, Yang Y, Yang. (2018) Measuring knowledge diffusion efficiency in R&D networks. Knowledge management research and practice. 16(2):208-219.

[2] Kasban H. (2016) Fingerprints verification based on their spectrum J. Neuro computing, 171: 910--920.

[3] Wildes, R. P. (1997). Iris recognition: an emerging biometric technology. Proceedings of the IEEE, 85(9): 1348-1363.

[4] Isaac E R H P, Elias S, Rajagopalan S, et al. (2019) Gait Verification System Through Multiperson Signature Matching for Unobtrusive Biometric Authentication J. Journal of Signal Processing Systems, 91(2): 147-161.

[5] Wang F, Cheng J, Liu W, et al. (2018) Additive margin softmax for face verification J. IEEE Signal Processing Letters, 25(7): 926-930

[6] Qin H, El-Yacoubi M A. (2017) Deep representation-based feature extraction and recovering for finger-vein verification J. IEEE Transactions on Information Forensics and Security, 12(8): 1816-1829.

[7] Han W Y, Lee J C. (2012) Palm vein recognition using adaptive Gabor filter J. Expert Systems with Applications, 39(18): 13225-13234.

[8] Kang W, Liu Y, Wu Q, et al. (2014) Contact-free palm-vein recognition based on local invariant features J. PloS one , 9(5): e97548.

[9] Zhang Y B, Li Q, You J, et al. (2007) Palm vein extraction and matching for personal authentication. In: International Conference on Advances in Visual Information Systems, pp.154-164.

[10] Krizhevsky A, Sutskever I, Hinton G E. (2012) Imagenet classification with deep convolutional neural networks. In: Advances in neural information processing systems, pp. 1097-1105.

[11] Oquab M, Bottou L, Laptev I, et al. (2014) Learning and transferring mid-level image representations using convolutional neural networks. In: Proceedings of the IEEE conference on computer vision and pattern recognition (CVPR), pp. 1717-1724.

[12] Gatys L A, Ecker A S, Bethge M. (2016) Image style transfer using convolutional neural networks. In: Proceedings of the IEEE conference on computer vision and pattern recognition (CVPR), pp. 2414-2423.

[13] Qin H, El Yacoubi M A, Lin J, et al. (2019). An iterative deep neural network for hand-vein verification J. IEEE Access, 7: 34823-34837.

[14] Ronneberger O, Fischer P, Brox T. (2015) U-net: Convolutional networks for biomedical image segmentation. In: International Conference on Medical image computing and computer-assisted intervention, pp.234-241.

[15] Miura N, Nagasaka A, Miyatake T. (2004) Feature extraction of finger-vein patterns based on repeated line tracking and its application to personal identification J. Machine vision and applications, 15(4): 194-203

[16] Song W, Kim T, Kim H C, et al. (2011) A finger-vein verification system using mean curvature J. Pattern Recognition Letters, 32(11): 1541-1547

[17] Miura N, Nagasaka A, Miyatake T. (2007) Extraction of finger-vein patterns using maximum curvature points in image profiles J. IEICE TRANSACTIONS on Information and Systems, 90(8): 1185-1194

[18] Han, W. Y., & Lee, J. C. (2012). Palm vein recognition using adaptive Gabor filter. Expert Systems with Applications, 39(18): 13225-13234.

[19] Hamming, R.W. (1950) Error detecting and error correcting codes. Bell System technical journal, 29: 147–160.

Thinking on the construction of big data service for public security in megacities -- a case study of Chongqing

LIU Ge[1]

[1]Department of Information Security, Chongqing Police College, Chongqing 401331, PRC

Abstract. According to the national outline of development of Big Data, combined with the current situation of big data development in Chongqing and the action plan of urban development in Chongqing, this paper, from the perspective of Chongqing public security, analyses the social risks, subjective and objective conditions related on big data, puts forward the Big Data construction goal of Chongqing Municipality public security , and gives the construction and development strategies and suggestions for the implementation of the Big Data project.

1. Introduction

Science and technology are rapidly changing human beings and human civilization. There's no era in history while people can make such great progress at such an unexpected speed as today. With the progress of science and technology in the era represented by information technology, this era is clearly defined as the information society. The main sign of the information society is the explosive growth of information and information dissemination through the network. Massive information is transmitted and calculated in the form of data, forming "big data". When the wheel of history drives into the era of big data, the realization of information service and social public security not only needs the close cooperation of related departments at different levels and in different fields, but also needs to focus on integrating the massive data scattered in different departments and mining high-quality information from them, especially when the information service needs span multiple management authorities, supported by big data Collaborative integration is very important.

The so-called "big data era" refers to the era when all kinds of organizations and individuals apply big data thinking or big data technology to realize or create value.

Big data brings some unique characteristics to the society. **First, everything can be quantified**. Quantifying everything is the core of data and the cornerstone of the era of big data[1]. In the book "when we become a bunch of numbers", Stephen Baker put forward the idea of "everything can be quantified". He pointed out that every day people click on the web, switch TV channels, swipe cards for shopping, drive through the automatic toll station, answer and call mobile phones and other behaviors - these previously ignored information, are now stored by different ways of data recording. Through GPS positioning system, direction and position can be digitized; access to sensors, air quality can be digitized; analysis of BLOG status, mood can also be digitized. Things that have never been quantified in the world seem to be presented to the world overnight in a variety of data forms. **Second, data is wealth**. Andreas Weigend, former Chief Scientist of Amazon, once pointed out that "data is new oil"[2]. Moreover, the value of data will not be reduced due to the improvement of its utilization rate. Data can be repeatedly processed and constantly generate new value. According to the search results of Internet users, the Centers for Disease Control and prevention in the United States studied and judged the spread of global influenza and other diseases. Alibaba uses the more than 100PB data accumulated by users and businesses to make accurate decisions on whether to lend or not and how much to lend. **Third, the future can be predicted.** American Professor Albert Laszlo Barabasi pointed out that human behavior conforms to the law of power law and can be predicted in advance. Prediction is the core usage of big data. On the basis that everything can be quantified, we can predict the basic development tendencies of things in a short time by mining and analyzing the data of the past and the present. In this regard, it has been proved by practice that Obama's big data team and Google's influenza prediction are both powerful manifestations of big data prediction ability.

2. Risks in the digital age

With the rapid development of digital technology, human society is undergoing profound changes, showing a new social form. In a congratulatory letter to commemorate the December 10, 2018 Universal Declaration of human rights, President Xi Jinping pointed out in his congratulatory letter to the 70th anniversary World Conference on human rights: "people's happy life is the greatest human right". These words shows that what a a basic attitude in this new era Human Beings should hold.

Information technology represented by Internet, Big Data, Cloud Computing, Artificial Intelligence, etc. has become the symbol of this era[3].The extensive use of information technology has become an indispensable part of people's life, survival and development. Information society is an intelligent digital society. Information society is a sharing society under the digital economy. Information society is also a risk society under the digital network environment.

Under the promotion of greedy capital, while greatly facilitating their lives, science and technology will gradually deprive people of the opportunities and rights to work and create wealth. More importantly, digital technology is rapidly invading people's private space. It can be predicted that in the near future, citizens' privacy rights and personal dignity will become an optional setting option. And this option will be the consideration that people must give up in order to integrate into the intelligent life. In addition, the Artificial Neural Network closed, hidden and selective nature of artificial intelligence algorithm will create extremely convenient conditions for the systematization and generalization of the discriminatory consequences against fairness and justice, public order and good custom. More attention should be paid to the spread of various risks with the development of digital information and communication technology. Digital technology is a double-edged sword. What digital technology brings is not only the gospel of rights, but also the crisis of rights. When people enjoy the personal liberation, life convenience, economic growth, democratic progress, cultural diversity, social security and other benefits brought by digital technologies such as big data technology and artificial intelligence technology, they are also facing the risks of privacy disclosure, unequal treatment (discrimination), illegal supervision and other risks caused by excessive collection and improper use of personal data. [4]

3. Advantages and disadvantages on the implementation of big data project

As for the construction of big data, at present, 1 pilot comprehensive Experimental Zone has been built in Guiyang, and 2 cross regional comprehensive Experimental Zone are Beijing Tianjin Hebei and the Pearl River Delta. In 2016, the state established 4 regional demonstration comprehensive Experimental Zone including Chongqing national big data comprehensive Experimental Zone and 1 Experimental Zone for overall infrastructure development. Chongqing big data Experimental Zone was approved in October 2016. It is positioned to lead the development of "four plates" as the East, central, Western and Northeast China, pay attention to the overall planning of data resources, strengthen the agglomeration of big data industry, play the role of radiation driving, promote the coordinated development of regions, and realize the quality and efficiency improvement of economic.[5]。

“Big Data, Cloud Computing”, when we see these words, the first thing we think about may be Silicon Valley, Zhongguancun Beijing and other places, or Amazon, Google and other enterprises, but it is difficult to connect them with Guizhou, a backward province in the southwest corner. However, at the 4th China Data Center Industry Development Conference and IDC product exhibition and resource negotiation and Trade Conference in 2014, China Data Center Industry Development Alliance awarded Guiyang, the capital city of Guizhou "The Most Suitable City for Data Center Investment".[6]

What are the advantages of Guizhou? First, the province and the city work together to advance the layout; second, the unique favorable nature environment; third, to seize the historical opportunity. Compared with the big data development of Chongqing Municipality, it's "Got up early and had a late arrival." In 2017, the national development and Reform Commission officially announced the National Engineering Laboratory of "big data and Internet +", to determine the list of commitment units. In total of 19 national engineering laboratories were selected, including 11 in the big data area and 8 in the "Internet +" field[7]. In additional the first National Engineering Laboratory of big data application technology for improving government governance capacity which located in Guiyang, there are 20 altogether. Through the online research, all of these are not located in Chongqing. The situation is critical for Chongqing. It should be said that the development of Chongqing big data is not late. As early as August 2013, Chongqing municipal government issued "a big data action plan", proposing to form a big data hub and industrial base with international influence by 2017. As time went into 2018, Chongqing's big data industry has not splashed much water. As mentioned in the work report of Chongqing municipal government in 2018, Chongqing would focus on building "Eight Action Plans". The first is "Big Data intelligent development strategic action plan", and the keyword "Big Data" has appeared 12 times in the report. It is generally believed that this is the "second propulsion" of Chongqing's big data industry. In fact, in addition to the strategic position of the national strategic rear and pillar of the third line construction, the latest central government also gives Chongqing more core and important roles. President Xi Jinping clearly defined as the "two points" status (the important strategic fulcrum point of the western development, the connection point of "Belt and Road" and the Yangtze River Economic Belt); the goal of "two places" (inland open Highlands place, green hills and clear waters-picturesque scenery place); and "two high" trends (to promote high-quality development and create high quality of life). Chongqing, as the "fifth economic city" in China, as one of the important central cities in China, is a famous historical and cultural city in China, an economic center in the upper reaches of the Yangtze River, an important modern manufacturing base in China, and also a comprehensive transportation hub in the southwest region. It has met with developing bottlenecks in the economic growth, industrial upgrading, and innovation developing in recent two years. No matter the opportunity and tuyere brought by the development of big data industry, or the grasp of urban development and transformation, Chongqing must seize the opportunity.

The following figure shows the number of big data information of Beijing, Shanghai and Chongqing municipality by searching Baidu in Chongqing on December 5, 2019.

Figure 1 Baidu big data search results of Beijing, Shanghai and Chongqing

From the picture, it can be seen that Chongqing are far from the first-tier cities as Beijing and Shanghai. Guiyang, Chengdu, Xi'an, Chongqing, the four big digital economy west cities taken for statistical reference and research objectives, Chongqing's advantages and disadvantages are obvious.

3.1. First, each mega city has its' own unique natural environment.
Chongqing is geographically mountainous and watery with a hot surface climate, which facilitates the construction of air defense data center and underwater data center. In fact, Chongqing itself has a considerable scale of air raid shelters that have not been developed and utilized. Yet it has not developed a large area of underwater data center either. This is mainly due to the fact that Chongqing's power production is not rich enough and the power supply cost required by big data is on the high side. But depending on the advantages of Chongqing's national central city resources gathering and using the advantages of tourism market position and geographical and cultural fields, we can build various kinds of data centers, landscapes, simulation and cultural integrated projects suitable for livable work, tourism and visiting experience, through the earning to appropriately subsidize the cost of electric power.

3.2. Second, Big one facing big difficulties
Chongqing, as a megacity, has a strong economic overall strength and a large number of transfer payments from the central government. However, Chongqing is also responsible for the provincial population area jurisdiction and the burden of becoming rich. The land soil available is enough, but the funds are not enough. In this regard, we should develop various favorable policies for cooperation and investment.

3.3. Third, Talent predicament
Chongqing has enough attraction and talent pool, but the talent absorption, training and reserve urgently needed for the development of big data have obvious shortcomings. Compared with Chengdu and Xi'an, the universities and other scientific and technological units on which Chongqing's data talents trained are obviously not thick enough. There is not only the lack of leading role of various universities in Chongqing, but also the great lack of national scientific research units (This conclusion comes from the online survey results analyzing on the distribution of the Chinese Academy of Sciences, National Engineering Laboratory and National Key Laboratory). On one hand, at present, the municipal government had made great efforts to introduce the Chinese Academy of Sciences and other units into Chongqing, Chongqing University has built the first Internet Of Things(IOT) big data laboratory in China cooperated with enterprises, Chongqing University of Posts and telecommunications has a Key Laboratory of the Ministry of Education for industrial Internet Of Things(IOT) and networked control. But these achieved results, the introduction efforts are not enough compared with other 3 provinces and/or cities(Compared with Guiyang, 3 national engineering laboratories not under its own name were introduced into the construction, Xi'an was approved 3 national engineering laboratories, and Chengdu made many efforts to introduce the secondary unit layout of engineering laboratories, such as Huawei R & D center, Tsinghua University Energy Internet Innovation Research Institute, etc.). On the other hand, under the situation of fierce competition in talent policies all over the country, we should also be based on self-reliance and talent policies. Think about the big data talent problem as a whole with the idea of a game of chess, whether we can integrate the intensive efforts of talents (such as setting up an experts working committee), whether we can better build a nest to attract talents, and whether we can grow and cultivate our existing talents (related matching of policies and treatment), and so on. There are many problems worthy pondering.

4. The goal and vision of Big Data construction
Based on various subjective and objective conditions, this paper proposes to establish a Comprehensive Analysis and Layered Service Project of Big Data footing on the existing, focusing on the long term, with police-civilian integration oriented. There are some reasons for this:

4.1. The existing limitations
Due to the limitation of funds, geography, nature and other resources, Chongqing has missed the opportunity to take the lead in the construction. It was plagued by the fundamental problem of electric

and capital, so it is not suitable to put forward the concept and goal of absolute leading. Moreover, there are different understandings about the concept and goal of leading.

4.2. Fast shifting era

The modern society shifting rapidly on technology development and the trend of the times. In order to avoid the decision-making being too ahead of time or backward, as well as deviated from the change of technical standards and the direction, it is necessary to consider the keep-up-withing and win-win strategy and avoid the strategy of rat race or championship.

4.3. Integrated progressing thinking

The Big Data project itself needs the coordination of the level of science and technology, the overall development strategy of the whole city and the investment of funds and energy. It should strive for progress in stability and avoid slogans. In order to actively respond to the pursuit of the central government's strategic positioning of Chongqing, it is not appropriate to put forward any bold, breakthrough, cutting-edge and other considerations. After all, the public security has the unique responsibility of collecting, managing and protecting all kinds of public and social information in the country, the goal puts forward another implicit meaning, which is to actively integrate the construction of big data of public security into the construction of big data of the whole city, so as to draw support from resources and save the whole city's expenditure.

4.4. Development vision on results

This project will help the government guarantee the governance environment and public demand in real time, accurately and efficiently, and effectively provide the public with integrated, coordinated and convenient services. Specifically, the development vision lies bellow:

4.4.1. An efficient and cooperative agile government. It realizes the information sharing among government departments, and the department boundary is gradually blurred. At the same time, it guarantee the governance environment and public demand in real time, accurately and efficiently with the help of big data technology, and effectively provides the public with integrated, coordinated and convenient services.

4.4.2. A humanized and self-service government. Through the development of a series of data sharing facilities such as government cloud platform and data sharing and exchange platform, the government's services resources are effectively integrated to provide one-stop, integrated services for social justice. The advent of the era of big data also forces the government to open more data related to environmental protection, medical and health transportation, etc. it also requires more and more individuals and social organizations to participate in the government's public services, carry out data value-added applications, create more service applications, and realize the social value-added government data.

4.4.3. A precise and networked public-participated government. With the powerful collection and analysis abilities of big data technology, data governance is realized and decision-making is more accurate. Because of the development of big data, the data network covers the whole city. In the field of public governance, the public is no longer a demand role, individuals can participate in the government management through the network.

Whether it is to build an agile government, a service-oriented government or a public-participated government, in the final analysis, it is to build a government that the people are satisfied with. It's hopeful that big data can make people's lives better. [8]

To turn the new vision into a better reality, we need to have a keen insight, be good at seizing opportunities and courageous to meet challenges. Big data has become an extremely important wing of national strategy. How to seize and make good use of the strategic opportunity of national big data development, accelerate data aggregation, fusion and application, and deeply explore the economic and social value of big data has become a new proposition in front of us.

In order to turn the new vision into a better reality, we must build a unified sharing and exchange platform for government data, and deeply promote the combination of regional development and national strategy[9]. In this regard, the key is to make efforts to build a platform, on the basis of scientific planning of the spatial layout of big data industry, and focus on building a number of big data industry platforms. Take advantage of Chongqing's positioning as a national center city to become the propositional person of big data major scientific infrastructure. For example, there is no Internet of Things (IOT) big data engineering laboratory among the 20 national engineering laboratories at present. With the help of the "Butterfly Effect" generated by Chongqing industrial foundation status combined with the permanent address status of China International Intelligent Industry Expo, its' available raise funds to establish a national Internet of Things (IOT) big data engineering laboratory. Thus, it can indirectly drive Chongqing to develop big data and artificial intelligence industries, and form a promising prospect with obvious characteristics, complementary advantages.

Correspondingly, Chongqing public security takes advantage of the construction of Internet Of Things(IOT) big data engineering laboratory, builds and implements the public security Internet Of Things(IOT) big data project on the basis of video and Skynet projects in the past few years, which also provides reliable boosting resources for the development of smart cities.

Looking at the whole country, local incentive policies and measures are practical, powerful and high in gold, setting off a new wave of big data construction. Chongqing must seize the opportunity and take advantage of the times.

5. Big data construction and development strategy

The conveniences brought by science and technology are pervasive. Science and technology make society better and life more exciting. With the development of modern information society, the masses of people put forward new expectations and requirements for public security work. What the trends are: From more solving cases to less accidents, the public security work mode is required to change from emergency response to risk management and control;

From good attitude to excellent service, public security work is required to change from window service to online service;

From small safety to wide security, the government is required to provide all-inclusive security needs such as ecological environment, food and drug, personal privacy, etc.

5.1. How to adapt to the current rapid and dynamic change

5.1.1. Change the concept and set the important position of big data engineering. In the era of big data, departments at all levels must attach great importance to information management and raise it to a strategic level that affects survival and development. Making use of modern scientific and technological means to build an intelligent security system and precise police mode with ubiquitous perception, multi-dimensional research and judgment, flat command and efficient disposal[10], it will draw the public security work to deepen reform, promote police process reengineering and team management, upgrade and share information resources.

5.1.2. Reform and improve the organizational structure of the information team. In order to ensure the full play of big data technology, a unified data analysis and processing department and data police team must be settled on the shift from the original division of science and technology, technical investigation, network security and their respective duties.

According to the principle of "centralized management, division of labor with responsibility, system integrated, high efficiency and accuracy, data sharing", the construction of unit big data is planned in a unified way to form a complete big data service system and realize data sharing.

Strengthen the construction of decision support system, use system engineering, value engineering, operational research, statistical analysis and prediction, big data, artificial intelligence technology methods to make decision analysis on major issues, and provide scientific basis for decision-making departments.

5.1.3. Revise the institution to realize the standardization and scientization of the index system. On the basis of scientific formulation of index system and improvement of information network, as the report automatically generated by the information center processing system, the report period can be shortened. At present, each unit has developed various information systems, and has realized internal networking. However, there are many problems such as imperfect index system, low level of system intelligence, poor connection of subsystems, and weak real-time data processing ability. In order to adapt to the new requirements of information processing, the original information system must be upgraded and transformed from the center operation oriented to cloud computing and blockchain oriented, using modern network technology to fully realize information sharing.

5.1.4. Establishing platform to realize the modern mode of information sharing management. Relying on the blockchain technology, through the establishment of a comprehensive, real-time, efficient and perfect cloud platform and data center, you can master all kinds of relevant data and external information to be pushed at any time.

Meanwhile, because the data center has a wealth of information resources, it can use big data and artificial intelligence technology to conduct comprehensive analysis and thematic research on various problems and development trends of social and government departments, so as to provide implementation opinions and suggestions for scientific decision-making and management of up-class organs.

According to the current situation, its proposed to build the public security information filtering and desensitization processing system at all levels (categories) and external information pushed processing system, and to share the information value safely and conditionally (pushed to society or other departments) to serve the society.

5.2. Strategies on construction and development under the current situation

5.2.1. Security situation on big data and the countermeasures

Big data technology increases both the risk of information disclosure and the difficulty of network security protection, which puts forward new challenges to the security of infrastructure and the maintenance of national sovereignty. Data leakage, hacker attacks and other phenomena occur from time to time. China has certain achievements in the development and application of big data, with market advantages and development potential. However, there are also some problems, such as the lack of open and sharing of government data, the weak industrial foundation, the lack of top-level design and overall planning, the lag in the construction of laws and regulations, and the lack of innovation and application fields, which need to be solved urgently. [11]

For public security, there are five major pain points: inaccurate awareness of public security situation; insufficient matching of results; poor cloud-to-end link; insufficient middleware support services; unbalanced development of regional differences[12].

Therefore, it is necessary to:

- formulate relevant laws and regulations as soon as possible, clarify the responsibilities and rights of all sides;
- strengthen the system construction, strengthen the safety supervision of key data in key fields and industries;
- strengthen the management of big data related products and services, establish an independent and controllable information technology ecosystem;
- accelerate the development of big data related technologies, establish a network security defense system in depth;

- make full use of blockchain technology to promote large data value realization, ensure the information security [13];
- adhere to active prevention, and build a big data defense in depth protection architecture based on level protection[14].

First, we need to mobilize the prevention awareness of all; Second, we need to develop an effective security system; Third, we need to build a security system for the whole life cycle of big data; Fourth, we need to do a good job in the supervision of sensitive data; Fifth, we need to use information security defense means flexibly; The last but the most, we need to establish and improve big data policies and regulations in time.

In August 2015, the State Council issued the action plan for promoting the development of big data, pointed out that China should pay attention to the safety and security issues in the process of using big data while vigorously developing the big data industry.

On March 17, 2016, the 13th five-year plan adopted at the fourth session of the 12th National People's Congress requires the implementation of the national big data strategy and the strengthening of information security.

In the "network security law" implemented on June 1, 2017, it puts forward the responsibilities and obligations of network operators such as "maintaining the integrity, confidentiality and availability of network data", "preventing network data from leaking or being stolen or tampered with", as well as the requirements for exit security assessment and network information security protection of personal information and important data.

On the morning of April 14, 2018, the National Information Security Standardization Technical Committee officially released the big data security standardization white paper (2018 version) and other standardized technical documents. This paper focuses on introducing the domestic and foreign big data security laws and regulations, the status quo of standardization, analyzing the risks and challenges faced by big data security, giving the framework of big data security standardization system, planning the key work of big data security standards, and putting forward suggestions for carrying out the work of big data security standardization.

On June 13, 2018, the National Information Security Standardization Technical Committee issued the notice on Soliciting Opinions on the draft of the National Standard Guide for the assessment of the impact of information security technology on personal information security. The standard specifies the basic concepts, framework, methods and processes of personal information security impact assessment, and proposes specific methods for assessment in specific scenarios.

From the standards and regulations in the field of big data security every year, it can be seen that the information security problem in the big data environment has been promoted to the national strategic level.

Big data security standard is an important support to ensure big data security and promote the development of big data. It is urgent to speed up.

In addition to improving the relevant system institution, hierarchy and standards, strengthening the research on the network security issues and technology in the big data, and implementing, such as information security level protection and risk assessment, are also the key to solve the information security problems.

At present, Chongqing has made great achievements in big data standardization[15], which is the result of active efforts.

5.2.2. Current strategy for government big data construction and development

5.2.2.1. Using system engineering. Big data engineering is a strategic system engineering project, which needs intensive use of the city's resources. It should use the methods of system engineering to analyze, plan and build from the whole.

In view of the huge resource consumption and strategic role of big data project, the construction must contain a certain degree of resource intensive and technology frontier characteristics, and that means it should be footed on existing resources and focus on long-term development.

The strategic project requires that the construction of big data project of public security can not be restrained in the scope of public security. It also requires to provide support not only for the public security work, but also to provide data support and services for the comprehensive management and development of the city.

The so-called comprehensive analysis and layered services of big data for police-civilian integration orientated emphasizes the integration of police and people.

It is not only to strengthen interactive sharing and solve the chimneys problem under the system of belt-block division within the public security, but also to comprehensively consider and solve the chimneys problem of various information systems of public security and society.

The so-called comprehensive analysis and hierarchical services is to not only solve the needs of different levels and different divisions of public security within the public security, but also provide services for the external social needs of wide security and other data.

In addition, it should also be ensured the data interface consistency and the desensitization processing and transmission standardized and legitimacy.

5.2.2.2. From data sharing to data product construction. Although the government is advocating and calling for big data sharing and development, from the perspective of the masses or users, it has not deeply felt the value of data openness, Why? Because there is no visible and touchable data product.

The data products solidify the data, data model and analysis decision logic into a software system as much as possible, and play the decision value of data in a more automatic, accurate and intelligent way.

The value of data products is more extensive than deep, and the wider the customer base, the better reflection of the work effect.

This requires a high degree of sensitivity to the business of the target customer group in product design, to meet the urgent needs first, and to abstract the diversified data requirements into a Point, Line and Plane layered demand model.

For example, the application of Alipay, the demonstration effect of one point "paying", drives the online payment chain and even the development of Internet finance.

5.2.2.3. From big data to intelligence. Intelligence is the next stage of big data development. Not only Chongqing, but also Tianjin, Guangzhou, Hefei and Beijing, etc., have issued the development of intelligence strategy. The policies of talent, tax, land others are very attractive, which are good incentives for entrepreneurs.

The development of intelligence is far more complicated than that of data. The development competition of intelligence is a competition of basic science and technology ecology.

Xi'an, Beijing, Shanghai, and Wuhan have great advantages on these. Chongqing needs to pay more attention to the development of education and make farsighted layouts before it can occupy a place.

Shanghai public security, with the attitude of actively integrating into "Shanghai Urban brain", broke through the shackles, carried out institutional mechanism innovation, and firstly took the form of "one center, one platform, and the Internet" to try to integrate the public security business into the construction of smart city in China. It also made great achievements in the 1st China International Import Expo and was fully affirmed by the state leaders.

5.2.2.4. From internal interconnection to all things' interconnection.

At present, the opening of data is more based on the opening of organizations, governments and industries. Through the network penetration in R & D, production, transaction, circulation and financing and other links, we can improve efficiency and optimize resource allocation. It is to get through the upstream and downstream to help the organization rebuild its core competitiveness. And then, the next is the interconnection of all things.

Therefore, the circulation value sharing between data industries becomes an important content in the next stage. Thus, it becomes "infrastructure construction" of "information society", full of "data" mineral resources and "flow" resources.

5.2.2.5. From data benefiting people to data managing. Data resources is first and foremost open to serve the public. It's easy to benefit people with data, but it's hard for data "managing" the city.

Now let data directly manage the business, which needs to reshape the business process. The transformation of government governance mode in the era of big data is the transformation of government functions. The State Administration of Taxation of Shaanxi Province has a good exploration of data "tariffing"[16], which is worth learning. If Chongqing can explore some cases to provide data governance solutions, it will be a great contribution.

5.2.2.6. From business innovation oriented to technology innovation oriented. It is necessary to use various algorithm models to drive business transformation and lay a foundation for future competition. The change of thinking and opinion is very challenging. It is also the top priority of big data development in the next stage, so that the value of data technology can be truly released into the business.

5.2.2.7. Giving full play to their own advantages. It should vigorously promote the application of emerging technologies and the implementation of projects in the advantageous industries of the region. The development of economy has its own law. The development of each region have its own connotation. On the one hand, we need endogenous growth. On the other hand, we need denotation development. But the fundamental thing is to make use of emerging technology to ensure the original local advantage industries.

6. Implementation of big data projects

There is no essential difference between the implementation process of big data projects and general IT projects. The difference is that the technology used in big data project is not as simple as the traditional database SQL development one. It requires higher abilities for programming and troubleshooting in case of problems. Because of the distributed operation, the troubleshooting becomes very complex.

6.1. Problems to pay attention to

In the process of organizing and implementing big data projects, the following six mistakes should be avoided to keep away from the pitfalls and ensure the smooth implementation:

- *Negative attitude.* Such as passive waiting, has nothing to do with oneself, conservative wait-and-see, awe inspiring, etc.;
- *Improper big data plan, or blind pursuit big*;
- *Bypassing security for the project*;
- *The user participation fit is not good.* One is the user experience is not friendly, another is pursuing endless user participation projects;
- *Data is not prepared or cleaned correctly, and lack of checking its accuracy;*
- *Limiting innovation.* Either the application mechanism of big data does not match, or continuing to develop and manage big data project as an ordinary business one.

6.2. Implementation method and suggestions of the project

6.2.1. Layering projects. Divide the project into: Edge, Perception Layer → Cognition, Reflection Layer → Intelligent Reasoning, Prediction, Decision Layer → Wisdom Discovery Layer - Artificial Intelligence Layer driven by big data. Layering is a method of system engineering for any difficult question. If the layered or divided question is still hard, then continuously split them in small questions till all the divided questions feasible before assemble them. On the others, it should use different methods of human resource in construction and Implementation for different levels. For example, before analyzing data, data needs to be cleaned and prepared, which should not be carried out by senior data engineers. Else while, once algorithms and applications using big data are developed, they should be tested and phased before deployment. The best way to achieve these goals is to add a skilled project manager to the data science team, or use the project management skills and human resources of the IT department.

6.2.2. Jointed with traditional IT development team. Big data engineering, as well as artificial intelligence and machine learning applications, must be integrated with other IT applications and systems to get the most value. This is the only way to truly integrate big data analysis with other systems and applications across the organization.

6.2.3. Setting up big data maintenance and monitoring team. Whether the network and hardware infrastructure, or ensuring that big data is executed correctly within or independent of the application, once big data are deployed in production, the process must be monitored and maintained continuously.

6.2.4. Using agile development method of software engineering. In other words, it's to say "Running in small steps and iterating quickly". With a small run, fast iteration can quickly find a balance between demand and supply. Furthermore, since data change and algorithm modification are iterative and continuous processes, data scientists need to understand the concept of algorithm iterative modification. In this case, both from IT and data science teams, they must learn management on how to combine traditional IT project. Data algorithms emerge as the change of application requirements.

6.2.5. Business demand oriented. It was well known that designing and R & D based on business requirements. So, go back to user requirements and analyze what the market really needs. Otherwise, you'll lose the goal of big data projects. As Information needs to share, data needs to flow. Thus, big data needs application and service scenarios to generate real value.

6.2.6. Personnel skills and technical team training is the key. What the most lack is not the concept, methodology, market and policy promotion, but the talents and the environment to nurture talents. So many provinces and cities are working on big data with a lot of emphasizing. There may be others. Big data talents still have been in short supply. Big data doesn't work without talents. The responsibility of colleges and universities lies here. At present, the education of talents and the training of skills that the market really needs in Colleges and universities have been lagging behind. Talent project is a long-term strategic project of every country. Both managers and ordinary engineers need to have certain big data capabilities.

6.2.7. Bottom-up and top-down running parallel
It should not only care about the concepts and requirements of the upper level, but also pay attention to the business details of the lower level and the real usage scenarios. This is a very important thinking when doing big data projects.

6.2.8. Not pursuing new. when the hustle and bustle past, is the beginning of maturity. New technology and new platform emerge in endlessly, don't blindly pursue new. Case is forever in the race after a vicious circle, continuously adjusted in confusion. No matter how the outside world jollificatly, it should keep rational analysis, to ensure the security and stability of the solution. If a new technology hasn't yet been fully understood thoroughly, it is best not to apply. Otherwise it will cause a lot of solution risks and uncertainty factors.

7. conclusion

Beginning from analyzing the risks of big data, this paper analyzes the advantages and disadvantages of Chongqing's big data development. It proposes to establish a Comprehensive Analysis and Layered Service Project of Big Data footing on the existing, focusing on the long term, with police-civilian integration oriented. This paper also describes the future vision of Chongqing's big data construction. Then it puts forward feasible project implementation suggestions and opinion based on the analysis of big data construction and development strategies.

References

[1] [UK] Viktor M.S, Kenneth C. (2013) Big Data: A revolution that will transform how we live, work and think. Zhejiang people's publishing house. Hangzhou. pp. 61-62.
[2] [USA] Albert, L.B. (2012). Explosion: a new way of thinking about the future in the age of big data. China Renmin University Press, Beijing. p109.
[3] [PRC] Zhangwenxian. (2019) Seven propositions of human rights jurisprudence in the new era. Human Rights. 2019/03. pp12-27.
[4] [PRC] Zhangwenxian. (2019) No Data, No Human Rights. http://www.aisixiang.com/data/11 6468.html. d. 20191205
[5] [PRC] Big data standard working group of National Information Technology Standardizatio n Technical Committee. (2018) Big data standardization white paper (2018). China In stitute of electronic technology standardization (CESI). Beijing. pp10-12
[6] [PRC] Yangjun. (2014) Development concept of a western city: Guiyang wants to transfor m high-tech by big data. http://www.rmlt.com.cn/2014/0422/260668.shtml. d. 20191205
[7] [PRC] Fynlch. (2017) "Internet +" and National Engineering Laboratory of big data area t o confirm the list of commitment units. http://www.cbdio.com/BigData/2017-01/04/cont ent_5422405.htm. d. 20191205
[8] [PRC] Huangzhenjie. (2016) Research on the development system of government big data. http://www.echinagov.com/viewpoint/48509.htm. d. 20191205
[9] [PRC] Anonymous Commentator of Guiyang Daily. (2015) Lightening the new vision of b ig data development. http://www.gysdj.gov.cn/djxx/llyj/56939.shtml. d. 20191205
[10] [PRC] Hanyong, Xumin, Wenghaiguang, Cuilei. (2018) Exploration and practice of data p olice training. Public Security Education. 2018/09. pp7-12
[11] [PRC] The State Council. (2015) White paper of action platform for big data developme nt. Bejing
[12] [PRC] Chengdu Tianma Technology Co., Ltd. (2018) Information security risks and strate gies faced by big data technology development and innovation in the new situation. http://www.sohu.com/a/257119857_506171. d. 20191205.
[13] [PRC] Liuquan. (2019) Big data confronted situation and Countermeasures. http://master.in fosws.cn/20190401/20002.html. d.20191205.
[14] [PRC] Shenchangxiang. (2016) Strengthening information security protection in the era of big data. http://it.people.com.cn/n1/2016/0527/c404362-28383933.html. d. 20191205
[15] [PRC] Luoxuemei, Chenyi, Zhangwei. (2019) Chongqing starts the construction of nationa l big data comprehensive standardization system. http://cq.people.com.cn/n2/2019/0331/ c365402-32795327.html. d. 20191205
[16] [PRC] Shubangke. (2019) Six leading strategies for Guiyang big data. http://www.sohu.co m/a/244618153_468714. d. 20191205

IOP Conf. Series: Materials Science and Engineering **806** (2020) 012045 doi:10.1088/1757-899X/806/1/012045

The possibility of evolution from non-strategic risk to strategic risk: The role of strategic assets

Daoping Chen*

School of Economics and Management, Chongqing Normal University, Chongqing, 401331, China

*E-mail: 983119486@qq.com

Abstract. It is of great importance for the survival and development of an enterprise to effectively manage the risks based on the strategy. On the basis of combing the relevant literature, this paper redefines strategic risk and non-strategic risk, discusses the measurement of the possibility of non-strategic risk evolving into strategic risk by introducing strategic assets, and puts forward the opportune moment of management for strategic risk. The study suggests that strategic assets play an important role in the evolution of non-strategic risk into strategic risk. In order to improve the effectiveness of management, enterprises should start strategic risk management when the risk is still non-strategic rather than has evolved into strategic risk. Big data analysis method and artificial intelligence technology can help enterprises calculate the probability of non-strategic risk evolving into strategic risk and grasp the opportunity of strategic risk management.

1. Introduction

Some small events may turn into a big crisis, which will bring huge losses to the enterprise, and even make the enterprise lose the opportunity to survive. When people analyze these event, they tend to be attributed to such factors as weak supervision, weak sense of responsibility, etc., but the most important reason may be that enterprises do not carry out strategic management or effective strategic management on the risks they face.

Strategic risk management is to manage the risks of the enterprise based on the strategic perspective, and mobilize the resources and capabilities of the enterprise to eliminate those factors that lead to the non-strategic risks evolve into strategic risks[1-2]. Effective management for strategic risk must solve two problems. The first problem is how to choose non-strategic risks. The non-strategic risks faced by enterprises constitute a big data set. How to select the risk that is likely to change from non-strategic risk to strategic risk? The second problem is how to determine the strategic risk. Which risks can be called strategic risks? This paper studies the evolution of non-strategic risk to strategic risk and the determination of strategic risk from the perspective of strategic assets, and proposes the timing of intervention of strategic risk management.

2. Literature review

It is generally believed that the concept of strategic risk was first proposed in the field of management from the decision theory, and its initial meaning is the risk brought by strategic decision[3]. Scholars' research on strategic risk is mainly carried out from the following aspects: the concept and characteristics of strategic risk, the composition and measurement of strategic risk.

Although there are many literatures on the concept of strategic risk, the definition of the concept of strategic risk and the main research methods of strategic risk have not been unified. Of course, this also provides impetus for the increase of strategic risk knowledge. Baird points out that a complete concept of strategic risk should include all reasonable definitions of risk after analyzing the definition of risk in the fields of finance, market, psychology, crisis management and previous strategic management[4]. At present, the differences in the definition of strategic risk mainly focus on whether strategic risk is a strategic risk or a risk of strategy[5]. The author tends to the former. In fact, if the implementation of strategy is regarded as an enterprise's behavior, the risk of implementing strategy will be strategic and unified under the strategic risk.

The risk has the characteristics of loss, uncertainty and dynamic, which are often used to define risk. As a kind of risk, strategic risk must have these characteristics. In addition, it also has the basic characteristics of subjectivity and manageability. The manageability of strategic risk seems to excite managers, but in reality, there are often cases of crisis or even bankruptcy caused by the lack of effective management of strategic risk. It is not necessarily correct to think that the ability of large-scale enterprises to resist risks is stronger than that of small enterprises and infer that the objective risks are smaller in the eyes of relatively large enterprises[6]. For enterprises, the most important thing is to confirm what the risk is and whether the risk is effectively managed from the strategy.

Strategic risk has different components in different researchers' research. Budd thinks that strategic risk consists of entrepreneurial risk, operational risk and competitive risk in the multidimensional model of strategic risk system[7]. Therefore, the strategic management of risk must solve the problems of entrepreneurship, engineering technology and market. Slywotzky classifies strategic risk into technology risk, new product risk, customer risk, brand risk, competition risk, industrial economic risk and market stagnation risk[8]. These studies on the composition of strategic risk can help managers understand the source of strategic risk, but they can't explain why a small operational risk turns into strategic risk when a product or business process makes a serious mistake. In fact, products and processes based on resources and capabilities are strategic assets of an enterprise. The author believes that when a small risk is highly related to the strategic assets of an enterprise, it is more likely to evolve into strategic risk. Therefore, in the study of the evolution from non-strategic risk to strategic risk, the recognition of strategic assets is the premise of effective management of strategic risk.

The measurement technology of strategic risk can help managers to make quantitative analysis of the strategic risk faced by enterprises. Compared with other management fields, the research on risk in financial field is relatively early, and there are many methods to measure risk[9]. Capital asset pricing model (CAPM) is often used to measure the strategic risk of enterprises. In fact, the risks in the financial field are generally not classified according to the strategic and non-strategic risks, but classified into the systematic and non-systematic risks. The reason is that the main consideration in the financial field is the decentralization of risks rather than the strategic problems of risks. Of course, there are also researchers who consider the strategic risks in the financial field and classify them as systematic risks and non-systematic risks[10]. In addition, a series of methods based on value at risk (VaR) in the financial field are often used to measure the risk of enterprises[11]. It is obvious that the method of measuring the risk in the financial field applied to the measurement of the strategic risk of enterprises is insufficient, because the strategic risk of enterprises is very different from the financial risk.

In the measurement of strategic risk, a method called change is widely used because of its simplicity[5]. Its essence is to use the change characteristics of specific variables (such as variance or standard deviation) to estimate the strategic risk level of an enterprise. The estimation object can be either the whole enterprise or the strategic business unit of an enterprise, and the most commonly used variable is revenue type index. This method will derive many specific methods along with difference in change characteristics and preferences of indicators selected by researchers. Of course, the diversity of specific methods brings benefits to researchers as well as disadvantages, because different specific methods do not have the same assessment results on the strategic risk of the same object. Moreover, the results of using this method to evaluate strategic risk may exaggerate the risk or not find the risk, because predictable changes and the use of historical data to evaluate risk do not focus on the uncertainty as the essential characteristics of risk.

Collins and Ruefli developed an ordinal method to measure strategic risk, and used it to measure the strategic risk of American aviation industry[12]. Later, they developed the technology into a state determination method to measure strategic risk. This method takes the competitive position of an enterprise as a state. The change of the state is described by the change of the competitive position. The decline of competitive position and the extent of decline reflect the strategic risk of enterprises. The calculation of strategic risk is based on the state transition probability matrix. In practice, the challenge of this method is to measure the competitive position and the availability of data, because it needs to use the relevant data of other enterprises in the industry when measuring the strategic risk of an enterprise.

Many scholars have also carried out in-depth research on the relationship between strategy and risk, the relationship between strategy risk and the overall income of the enterprise, and obtained many meaningful but not completely unified conclusions. Bowman's study points out that in most industries, there is a negative correlation between risk and return, which means that high risk does not necessarily have high return[13]. However, research by Aaker and Jacobson shows that there is a strong positive correlation between earnings and system risk, while the positive correlation between earnings and non-system risk is weak, which means that high risk brings high earnings[14]. Later, Maurer reexamined Bowman's return paradox by using the ordinal method to measure strategic risk. His conclusion did not simply agree with or oppose the paradox, but reinterpreted the paradox. Finally, he came to the conclusion that when reducing the strategic risk of the company, shareholders' wealth was maximized, which seemed to be a new paradox[15].

In recent years, how to effectively manage the strategic risk of enterprises has become the focus of strategic risk research. Woodard introduces a new profit and loss model of strategic risk management, which comprehensively discusses risk management and emphasizes the importance of risk identification and measurement in the process of strategic risk management[16]. Hopkins Studies the organizational problems in strategic risk management, and believes that using different institutions with different risk management capabilities to manage different types of risks can change the distribution of risks[17]. Other studies examine strategic risk management from the perspective of the responsibilities of top managers and the coordination among various departments[18].

Although the study of strategic risk has a history of more than 30 years, it is far from the formation of a generally accepted understanding. It is believed that various views are still in a jungle state, and the systematic management of strategic risk has not been effectively solved. This paper will study the evolution from non-strategic risk to strategic risk from the perspective of strategic assets.

3. The possibility of evolution from non-strategic risk to strategic risk

3.1. Risk, strategic risk and non-strategic risk

The most basic definition of risk is the uncertainty of future loss, but this uncertainty refers to the uncertainty that future loss may or may not occur, or the uncertainty of its loss size if future loss occurs, or both of them have meanings. The author thinks that it is not accurate to define risk by the uncertainty of future loss, because the loss is caused by other uncertain events. For example, a car manufacturer faces the risk of rising steel price, which means that if the steel price rises, it will bring relative losses to the enterprise. The relative loss here refers to the uncertainty of the price rise, which is not reflected in the uncertainty of the loss, but in the uncertainty of the steel price rise. The rise of steel price is a more direct event that may or may not happen. It is uncertain. The uncertainty of loss is caused by it. Therefore, risk should be defined by uncertain events.

In this paper, a risk is defined as a random event which will bring loss to the enterprise if it occurs. Obviously, a risk must be a random event, but a random event is not necessarily a risk, and the risk is a random event with a certain nature. For example, the rise of steel price is a risk for one enterprise and not a risk for another, even though it is a random event. Similarly, strategic risk can be defined as a random event which will bring great or overall loss to the enterprise if it occurs. Further, non-strategic risk is the risk which is not strategic risk, or can be defined as a random event which will only bring small or local losses to the enterprise if it occurs.

Non-strategic risk may be transformed into strategic risk, or it may be further transformed into crisis. Crisis can be regarded as the end of strategic risk, and the relatively early non-strategic risk can be called the front of strategic risk. When an enterprise is facing a crisis, it must carry out crisis management. Before the crisis, risk management is the management of all kinds of random events that may bring losses to the enterprise. The effective management for strategic risk is to keep the enterprise in a small risk state, so that the strategic risk of the enterprise does not occur.

3.2. Strategic assets

Strategic assets will play an important role in examining the transition from non-strategic risk to strategic risk. Strategic assets refer to the combination of unique resources and capabilities of an enterprise. Owning strategic assets is the precondition for an enterprise to gain future competitive advantage. In order to succeed in the competition, what kind of key resources and capabilities are necessary for the enterprise? This is not only an important issue in the strategic practice of enterprise managers, but also the focus of strategists' strategic research. In order to explain this problem, scholars put forward many ideas based on company resources, such as core competitiveness, key success factors and so on. However, most of these ideas or their application focus on the internal resources and capabilities of the enterprise, or on identifying the important capabilities of the enterprise in the past rather than the capabilities needed by the enterprise in the future. Obviously, in a rapidly changing global competitive environment, it is very important for an enterprise to recognize the capabilities it needs in the future and focus on cultivating these capabilities for its sustainable competitive advantage. Strategic assets is a concept put forward by combining internal resources and capabilities with external competitive environment and key success factors.

Enterprises must choose strategic assets. An enterprise's choice of strategic assets mainly includes two main stages. The first stage is to analyze the strategic situation from the outside to the inside, and analyze the strategic business areas that enterprises may participate in the competition in the future, so as to determine the strategic industry elements that enterprises need in different competition situations. The second stage is to analyze the competitive relationship between the participating enterprises, and at the same time considers the situation of the first stage to select the strategic assets needed by the enterprises.

3.3. The evolution from non-strategic risk to strategic risk and strategic risk management

According to the above definition of risk, the evolution of risk refers to the process from one risk or a series of risks to another, or it can be expressed as the process of initiating or evolving from one or a series of events to another or a series of events. The evolution from non-strategic risk to strategic risk is a process of initiating or evolving from a non-strategic risk or a series of non-strategic risks to one or a series of strategic risks, or it can be expressed as a process from one or a series of events that bring small or partial losses to the enterprise to one or a series of events that bring large or overall losses to the enterprise.

How to measure the possibility of evolution from non-strategic risk to strategic risk? Now F_A and F_B are used to represent the set of all non-strategic risks and strategic risks faced by an enterprise at T_0. Let non-strategic risk A_1, A_2 ,$\cdots$,A_n be the elements of F_A, and let strategic risk B_1, B_2 ,$\cdots$,B_m be the elements of F_B. Then all possible evolution from non-strategic risk to strategic risk in the period from T_0 to T_1 is shown in Table 1. In order to simplify the problem, table 1 only considers the situation that one risk evolves into another rather than a series of risks. For example, $A_i \to B_j (i=1,2,\cdots,n, j=1,2,\cdots,m)$ means that a non-strategic risk A_i evolves into a strategic risk B_j.

Table 1. Evolution from non-strategic risk to strategic risk.

non-strategic risk	strategic risk			
	B_1	B_2	$\cdots$	B_m
A_1	$A_1 \to B_1$	$A_1 \to B_2$	$\cdots$	$A_1 \to B_m$
A_2	$A_2 \to B_1$	$A_2 \to B_2$	$\cdots$	$A_2 \to B_m$
$\vdots$	$\vdots$	$\vdots$	$\vdots$	$\vdots$
A_n	$A_n \to B_1$	$A_n \to B_2$	$\cdots$	$A_n \to B_m$

The probability of $A_i \to B_j$ is measured by the probability $P(B_j|A_i)$ of B_j under the condition of A_i. The total probability formula and Bayes formula can be used to calculate $P(B_j|A_i)$. But when $P(B_j|A_i)$ is calculated, it will face the problem of non-strategic risk of big data, which makes a lot of conditional probability $P(B_j|A_i)$ need to be calculated. The solution is to first examine the strategic assets of an enterprise, and then examine whether the risk A_i in the non-strategic risk set F_A is highly related to the strategic assets. If it is highly correlated, the risk A_i is called a strategic event. Obviously, strategic events are only a few events in the non-strategic risk big data set F_A. When $A_{i_k}\ (k \le n)$ is used to represent all strategic events, calculating $P(B_j|A_i)$ becomes calculating $P(B_j|A_{i_k})$. In this way, the amount of calculation will be much less, and it is possible to calculate the probability of evolution from non-strategic risk to strategic risk.

Because the non-strategic risk may evolve to the strategic risk, the enterprise must carry on the risk management based on the strategy to each link which may evolve. The key of management is to determine the degree and the timing of strategic management intervention according to $P(B_j|A_{i_k})$. Figure 1 shows the timing of interventions of strategic risk management.

Figure 1. Strategic risk management opportunity.

As shown in Figure 1, strategic management should not only intervene when strategic risks appear, but also move forward to the front of risk evolution. Strategic management should be carried out in case of non-strategic risks. As there is a possibility that non-strategic risk may evolve into strategic risk and then further evolve into crisis, moving forward strategic risk management to non-strategic risk will help to increase the effectiveness of strategic management. In the specific risk management, big data analysis and artificial intelligence technology can help enterprises calculate the probability of non-strategic risk to strategic risk evolution and accurately grasp the opportunity of strategic risk management.

4. Conclusion

Non-strategic risks do not always evolve into strategic risks, but non-strategic risks highly related to strategic assets are likely to evolve into strategic risks. Therefore, it is of great significance to examine the possibility of non-strategic risk evolving into strategic risk by examining strategic assets.

In order to improve the effectiveness of strategic risk management, strategic risk management should start from the front of risk evolution. First of all, the enterprise should analyze the general events to help select the non-strategic risks, then determine the strategic events based on the strategic assets of the enterprise, then calculate the probability that the strategic events evolve into strategic risks, and finally, propose strategies on risk management and carry out strategic risk management. The most important thing is that strategic risk management does not only appear at the end of risk evolution from non-strategic risk to strategic risk, but should start to intervene from the beginning when non-strategic risk appears. Strategic management should run through the whole process of risk evolution rather than only aiming at strategic risk. In this process, strategic assets play an important role. Big data analysis and artificial intelligence technology can be applied to help enterprises manage strategic risk.

References

[1] Namkyung Oh. (2017) Dimensions of strategic intervention for risk reduction and mitigation: A case study of the MV Sewol incident. J. of Risk Research, 20:1516–1533.
[2] Desislava V. P. (2017) Strategic risk management. Economic archive, 3:60-76.
[3] Andrews K. (1971) The concept of corporate strategy. Homewood, IL: Dow Jones-Irwin.
[4] Baird, I. S. (1986) Defining and predicting corporate strategy risk: An application in the telecommunications industry. https://www.ideals.illinois.edu/handle/2142/70461.
[5] Collins J. M., Ruefli T. W. (1996) Strategic risk: A state-defined approach. Norwell, MA: Kluwer Academic Publishers.
[6] Boulton W. R., Lindsay W. M., Franklin, S. G. (1982) Strategic planning: Determining the impact of environmental characteristics and uncertainty. Academy of Management Journal, 1982, 25: 500-509.
[7] Budd J. L. (1994) Characterizing risk from the strategic management perspective. Kate state: Kate State University.
[8] Slywotzky A. (2004) Exploring the strategic risk frontier. Strategy & Leadership, 32: 11-19.
[9] Chatterjee S. T. (1999) A strategic theory of risk premium: moving beyond CAPM. Academy of Management Review, 24: 556-567.
[10] Janos A. (1985) A comparison of models for strategic planning, risk analysis and risk. Theory and Decision, 19: 205-248.
[11] Harris, Richard D. F., Shen J. (2004) Estimation of VaR with bias-corrected forecasts of conditional volatility. Journal of Derivatives, 11: 10-20.
[12] Collins J. M., Ruefli T. W. (1992) Strategic risk: An ordinal approach. Management Science, 38: 1707-1731.
[13] Bowman E. H. (1980) A risk/return paradox for strategic management. Sloan Management Review, 21: 31-37.
[14] Aaker D. A., Jacobson R. (1989) The role of risk in explaining differences in profitability. Academy of Management Journal, 30: 277-296.
[15] Maurer F. (2004) Ordinal strategic risk and return: A fresh look at Bowman's paradox. Academy of Management Proceedings, B1-B6.
[16] Woodard M. (2005) Measuring the payoffs of strategic risk management. CMA Management, 79: 30-35.
[17] Hopkins M. M. (2006) Strategic risk management using complementary assets: Organizational capabilities and the commercialization of human genetic testing in the UK. Research Policy, 35: 355-374.
[18] Aabo T. F., John R. S., Simkins B. J. (2005) The rise and evolution of the chief risk officer: Enterprise risk management at Hydro One. Journal of Applied Corporate Finance, 17: 62-75.

Deep Feature Attention and Aggregation for Real-Time Head Semantic Segmentation

Zhiqin Zhang[1], **Bo Zhang**[2,*]

School of Computer Science, Wuhan Donghu University, Wuhan 430212, China

*Corresponding author's e-mail: bob.cheung@ovspark.com

Abstract. This paper proposes a mobile CNN(Convolutional Neural Networks) architecture named FAANet(Feature Attention Aggregation Net) , which focused on self-attention and feature aggregation for real time head semantic segmentation using limited computation resource. The proposed network is based on a single lightweight backbone cnn network and aggregates discriminative features through multiple different scale feature maps, and the different feature maps are fused using self-attention sub-block. Using the proposed multi-scale feature communication and propagation, FAANet substantially reduces the size of model, but still achieves big receptive field and strengthen the generalization ability, which achieves a balance between the computation cost and performance. Experiments on head segmentation dataset demonstrate the improved performance of FAANet with 10 times less FLOPs and 3times faster than the existing state-of-the-art fast semantic segmentation methods and the proposed method can achieve comparable accuracy. Specifically, when computed on one NVIDIA Titan X card , it can achieve 98.3% Mean IOU on our mixed test dataset with only 1.2 GFLOPs and 100 FPS while inferring on a 1024*1024 input image.

1.Introduction

Semantic segmentation is a pixel-wise classification task, and it is a basic and important task in computer vision. A lot of practical applications has been developed in the fields of video effects, robot sensing, video surveillance, autonomous driving and so on. For most such applications, especially for video effects on mobile phone, how to keep real time inference speed and achieve good performance is important.

Some real-time semantic segmentation methods [1-5] have already been proposed and obtained good performances on various benchmarks [6-7]. However, it is time consuming if most of the operations are calculated on the high-resolution feature maps when using the U-shape backbone. Some researches decrease the computation complexity by reducing the input image size, distilling learning or pruning the feature map channels to suppress the computation cost. Though it is possible effective, they always damage the spatial details and can not achieve good segmentation results on the object boundary and small objects. Also, if a network is shallow, the feature discriminative ability will be weakened. In order to improve the segmentation results, it is important to selectively combine the spatial details and context information[4,8]. Nevertheless, it is not the best choice to add more layers on the high-resolution image which will cost more computation resource, and the limited connection between different scale layers will decrease the model learning and generalization ability.

Commonly, pretrained model will be used for semantic segmentation task, such as ResNet, googlenet, inception, DenseNet and so on. Aiming to save inference time, we adopt a lightweight modified model and research for improving the segmentation performance even with small model.

Progressively increasing the feature maps is a traditional method from the several last outputs of a single path architecture, in this kind of design, it is difficult to achieve fine segmentation results because the high-level context is not fused with the spatial level detail features. In order to improve the model learning capacity and expand the receptive field contemporary even in restrictive mobile device, feature reuse is an intuitional thought. So, it is necessary to find a lightweight method as well as incorporating multi-level context into encoded features by using self-attention fusion mechanism named SA-block.

In this paper, we proposed two methods to improve cross-scale feature fusion results. First, for fusing global semantic information and local structure details, we reuse high-level features, meanwhile, different feature maps from different stages was combined for enhanced feature representation ability. Second, we use SA-block selectively fuse different level features. Our whole network structure is visualized in Figure 1.

The proposed FAANet was tested on one standard benchmark dataset Cityscapes and our practical portrait videos dataset which include 500 short video clips. With a 1024×1024 input, FAANet achieves 98.3% Mean IOU with 1.2G FLOPs and speed of 100 FPS on a NVIDIA Titan X card. While implemented on a smaller input size such as 512*512, we achieved the 96.3% Mean IOU with only 0.5G FLOPs which is suitable for mobile application and comparable with most of the state-of-the-art real-time segmentation model.

2. Related Work

2.1Time Limited Segmentation

Real-time semantic segmentation algorithms are focused on generating the SOTA comparative segmentation mask under limited calculation. ENet achieved good performance but sacrificed miou by reducing the number of downsampling times. Because of the erasing for the last stages of the model, the receptive field is too small and it is difficult to segment larger objects subtly. ICNet uses different scale image inputs and a cascade network was used to decrease computation cost. BiSeNet proposes to prune channels by using spatial path and semantic path to reduce calculation. In the two methods, there is only one branch is used for deep feature extraction, and the other branches are designed to extract different scale details. On the contrary, we obtained distinguished feature by multiple level feature aggregation and self-attention feature fusion.

2.2Multiple Level Feature Aggregation

In traditional approaches, the encoder-decoder structure was used to solve pixel-to-pixel prediction. Relative to deep network, how to aggregate different scale features between blocks deserves more attention. RefineNet proposed a complex refine module in each magnifying stage before concatenating multi-scale features which is too time-consuming. DenseNet uses the dense connection to aggregate different feature for easy training.

3.Our Method

We start with our observation and research of calculation cost when applying traditional semantic segmentation strategy in our head segmentation task on mobile device. This motivates our multiple scale feature aggregation strategy based on self-attention to combine detail and spatial information in different level feature maps of the network to achieve comparable accuracy even in tiny model size. The whole architecture of Deep Feature Attention and Aggregation Network (FAANet) is illustrated in Figure 1. We focus on making the selectively fusion of different scale features in networks, it is important to aggregate different receptive field features for semantic segmentation task. Our aggregation strategy is composed of sub-network aggregate and using attention sub-block selectively fuse different level features.

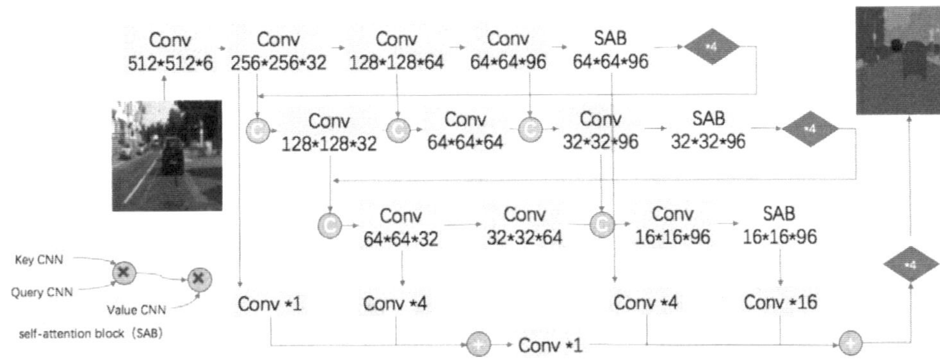

Figure 1.Overview of our Network Architecture

3.1 How To Aggregate Feature At Tiny Backbone

The structure is composed with a coarse to fine encoder-decoder network which can refine different scale feature. Features aggregation implements mixture of different scale features at the network level. In general, we implement our architecture as a stack of feature refining sub-network by selectively feeding the output of the previous sub-network to the next. A sub-network process can be defined as y = Φ(x), the output of sub-network Φn is the input of sub-network Φn+1, so the whole network aggregate thought can be formulated as: Y = Φn(Φn−1(...Φ1(X))). Sub-network aggregation allows these high-level features to be reprocessed on account of low-level spatial feature, it is beneficial for further evaluating and reassessing higher order spatial relationships.

We also use multiple stage feature aggregation strategy, for a single sub-network $\emptyset_{n(x)}$, a stage process can be defined as $\emptyset_n^i$. The stage in the previous sub-network is $\emptyset_{n-1}^i$, i means the index of the stage. Sub-stage aggregation method can be formulated as:

$$x_n^i = \begin{cases} x_n^{i-1} + \phi_n^i(x_n^{i-1}) & \text{if } n = 1, \\ [x_n^{i-1}, x_{n-1}^i] + \phi_n^i([x_n^{i-1}, x_{n-1}^i]) & \text{otherwise,} \end{cases} \tag{1}$$

While, x_{n-1}^i is coming from:

$$x_{n-1}^i = x_{n-1}^{i-1} + \phi_{n-1}^i(x_{n-1}^{i-1}) \tag{2}$$

In our strategy, high-resolution features block will flow into the low-resolution block. Our formulation learns a new mapping of nth feature maps and fuses (n−1)th features and receptive field. So, the full connection of feature Information flow can greatly improve the feature discrimination. Details was showed in Figure 1.

3.2 Self-Attention Block

First, the object context estimation will be calculated, p is the object context for each pixel and is defined as the probability if a set of pixels belong to the same category as p. The object context map shows the probability if the pixel p and each other pixel belong to the same object. The computation of object context map is given as follows,

$$w_{pi} = \frac{1}{Z_p} \exp(f_q(\mathbf{x}_p)^\top f_k(\mathbf{x}_i)), \tag{3}$$

where xp and xi are the representation vectors of the pixels p and i, Zp is the normalization number: $Z_p = \sum_{i=1}^{N} \exp(f_q(\mathbf{x}_p)^\top f_k(\mathbf{x}_i))$N = W × H. fq(·) and fk(·) are the query transform function and the key transform function which can be formed by convolution, matrix multiply and reshape operation. Second, object context aggregation is calculated, object context representation of the pixel p can be constructed by aggregating the representations of the pixels as follows,

$$\mathbf{c}_p = \sum_{i=1}^{N} w_{pi}\phi(\mathbf{x}_i), \quad (4)$$

where $\emptyset(\cdot)$ is the value transform function following the self-attention[9].
Our self-attention block was shown in Figure.2.

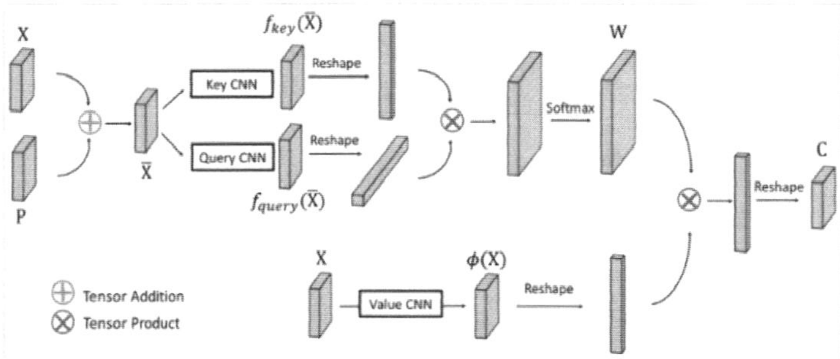

Figure 2. self-attention block(SA-Block)

4.Experiments

we evaluate the proposed method on two challenging benchmarks: Cityscapes and our head test dataset, and it is a big challenge for real-time semantic segmentation. We first compare the accuracy and speed results on Cityscapes with the existing fast segmentation algorithms. After that, a practical experiment on head dataset was made to prove that the architecture is fit for real mobile application. We used the following training strategy: mini-batch stochastic gradient descent (SGD) with batch size 64, momentum 0.8 and weight decay 1e − 6. For fair comparation, we used the traditional cross-entropy loss at each pixel. Data augmentation includes mean subtraction, resizing, cropping and random horizontal flip for training. Some segmentation results are shown in Figure.3.

Figure 3. Segmentation results mask of the proposed FAANet

In Table 1, Different structure was evaluated on Cityscapes val dataset.'Scale' means scaling ratio of input image. the results are shown, as can be observed, the miou metrics is kept comparable, while the inference speed of the proposed method is superior to state-of-the-art methods. The baseline of the proposed method achieves 98.3% Mean IOU on the our test dataset with only 1.2 GFLOPs and a speed of 100 FPS on one NVIDIA Titan X card while inferring on a 1024*1024 resolution image.

Table.1 Different structure was evaluated on Cityscapes

Model	Scale	FLOPS	mIOU(%)
ResNet-50	0.25	9.3G	64.5
Xception A	1.0	1.6G	59.2
Xception B	1.0	0.83G	55.4
Our	1.0	1.2G	66.6

5.Conclusion

In this paper, we propose multiple level aggregation and self-attention fusion strategy to tackle real-time semantic segmentation on high resolution image. Aggregation strategy connects different receptive field convolution layers to refine different level features from coarse to fine, which can vastly reduce calculation time. Quantitative experimental results on Cityscapes and our own head dataset are detailed to show the effectiveness of our strategy.

Acknowledgement

This paper was co-supported by the Foundation for Young Scholars in Wuhan Donghu University under grant 2018dhzk006. Thanks for the support of Wuhan Donghu University.

References

[1] Vijay Badrinarayanan, Alex Kendall, and Roberto Cipolla. Segnet: A deep convolutional encoder-decoder architecture for image segmentation. IEEE transactions on pattern analysis and machine intelligence, 39(12):2481–2495, 2017.

[2] Michael Treml, José Arjona-Medina, Thomas Unterthiner, Rupesh Durgesh, Felix Friedmann, Peter Schuberth, Andreas Mayr, Martin Heusel, Markus Hofmarcher, Michael Widrich, et al. Speeding up semantic segmentation for autonomous driving. In MLITS, NIPS Workshop, 2016.

[3] Zifeng Wu, Chunhua Shen, and Anton van den Hengel. Real- time semantic image segmentation via spatial sparsity. arXiv preprint arXiv:1712.00213, 2017.

[4] Changqian Yu, Jingbo Wang, Chao Peng, Changxin Gao, Gang Yu, and Nong Sang. Bisenet: Bilateral segmentation network for real-time semantic segmentation. arXiv preprint arXiv:1808.00897, 2018.

[5] Hang Zhang, Kristin Dana, Jianping Shi, Zhongyue Zhang, Xiaogang Wang, Ambrish Tyagi, and Amit Agrawal. Context encoding for semantic segmentation. arXiv preprint arXiv:1803.08904, 2018.

[6] Bolei Zhou, Hang Zhao, Xavier Puig, Sanja Fidler, Adela Barriuso, and Antonio Torralba. Semantic understand- ing of scenes through the ade20k dataset. arXiv preprint arXiv:1608.05442, 2016.

[7] Tsung-Yi Lin, Michael Maire, Serge Belongie, James Hays, Pietro Perona, Deva Ramanan, Piotr Dollár, and C Lawrence Zitnick. Microsoft coco: Common objects in context. In European conference on computer vision, pages 740–755. Springer, 2014.

[8] Hengshuang Zhao, Xiaojuan Qi, Xiaoyong Shen, Jian- ping Shi, and Jiaya Jia. Icnet for real-time semantic segmentation on high-resolution images. arXiv preprint arXiv:1704.08545, 2017.

[9] A. Vaswani, N. Shazeer, N. Parmar, J. Uszkoreit, L. Jones, A. N. Gomez, Ł. Kaiser, and I. Polosukhin. Attention is all you need. In NIPS, pages 5998–6008, 2017.

IOP Conf. Series: Materials Science and Engineering **806** (2020) 012047 doi:10.1088/1757-899X/806/1/012047

Rule Extraction of Generalized Incomplete Variable Multi-granularity Rough Set

MO Jing-lan

Mathematics Teaching Department, Lushan College of Guangxi University of Science and Technology, Liuzhou 545616, Guangxi, China

e-mail: jinglan022@163.com

Abstract—According to the advantages and disadvantages of feature relation and multigranularity decision rough set in generalized incomplete information system, a multi-granularity rough set model based on feature relation is proposed. Firstly, a variable multi-granularity rough set model is proposed, and an optimistic and pessimistic variable multi-granularity decision rough model based on feature relation is constructed. By combining variable multi-granularity and approximate distribution reduction, the variable multi-granularity approximate distribution reduction is proposed, and all the decision rules and possible decision rules are obtained. The results show that the method of variable multi-granularity approximate distribution reduction and rule acquisition is more reasonable and effective than the original method of single granularity approximate distribution reduction and decision acquisition.

1. Introduction

The rough set theory proposed by the famous polish scholar z.pawlak [1-2] is a mathematical tool to deal with fuzzy and uncertain data, mainly dealing with complete information system. However, due to various reasons such as limited means of obtaining information and loss of recording information, the information system is incomplete. Kryszkiewicz presented olerance relation [3] and his binary relation is able to deal with incomplete information system. In order to be able to use the rough set theory to deal with more complex generalized incomplete information systems, J.G rzymala. Busse's feature relation [4] are put forward. Granular computing theory [5-7] is a core part of the study of the current field of artificial intelligence, multi-granularity is a core concept of granular computing theory, said in two or more different granularity constitute a multiple granularity spatially analyzed problem solving research, based on this perspective, Qian Yuhua et al. [8], puts forward the concept of multi-granularity rough set are given more granularity and pessimistic optimistic granularity rough set model, in 2010, Qian Yuhua et al. [9] and presented based on tolerance relation in incomplete multi-granularity rough set model. In the literature [10], wang lijuan et al proposed an incomplete multi-degree rough set model based on similarity relation. Literature [11-12] studied the granularity reduction of pessimistic multi-granularity rough sets and proposed different granularity reduction algorithms. Literature [13] studies the granularity reduction of variable multigranularity rough sets, and literature [14] proposes a multigranularity rough decision analysis method based on the possibility degree tolerance relationship in incomplete information system. In the incomplete multi-granularity rough set model proposed in literature [14], any known attribute value can be compared with any unknown attribute value, and it is considered that they are equal. However, in reality, many unknown attribute values cannot be compared with known attribute values.

The incomplete multi-granularity rough set model proposed in literature [14] is based on the same assumption: the unknown attribute values in the information system are of omission type, and the incomplete information system cannot process noisy data. There is another interpretation of unknown attribute value in incomplete information system, that is, missing unknown attribute values are not allowed to be compared. In this paper we discuss omissions and lost unknown attribute values at the same time, the existence of generalized incomplete information system, introducing the ideas of multi-granularity generalized incomplete information system, put forward based on the characteristics of the relationship of multi-granularity rough set model and variable multiple granularity rough set model, and puts forward the variable multiple granularity (under) the concept of approximate distribution reduction, thus the most Jane determine the decision rules and to obtain all the minimalist may decision rules.

2. Ease of Use

2.1. generalized incomplete information system

Definition 1[4] Let $S=(U,AT,V,f)$ is an information system, Where U is a non-empty finite set of objects, called on the field; AT is the set of non-empty finite attributes; $V=\bigcup_{a\in AT} V_a$ is the set of property values, and V_a is the range of the attribute $a\in AT$; $f:U\times AT\rightarrow V$ is an information function. For A given object x, assign a value to the object x under the attribute a .If $AT=C\bigcup D, C\bigcap D=\varnothing$, C and D are condition attribute set and decision attribute set respectively, and V is attribute value, $V_C=\{V_q \mid q\in C\}$ and $V_D=\{V_d \mid d\in D\}$ are conditional attribute value set and decision attribute value set respectively, then S is called decision system. Unless otherwise stipulated in the S, $\exists x\in U, a\in C, a(x)=*$ or $a(x)=?$ ("$*$" represents the missing unknown value, "?" stands for missing type unknown value), If two types of unknown attribute values exist simultaneously, S is called generalized incomplete information system.

2.2. Characteristics of the relationship

Definition 2[4] Let $S=(U,AT,V,f)$ is a generalized incomplete information system, for $\forall A\subseteq AT$, the characteristic relationship determined by a is defined as:

$K(A)=\{(x,y)\in U\times U \mid \forall a\in A\wedge f(x,a)\neq ?, f(x,a)=f(y,a)\vee f(x,a)=*\vee f(y,a)=*\}$.

Feature relation $K(A)$ is regarded as A mixed manifestation of tolerance relation and asymmetric similarity relation, which retains both the relevant properties of tolerance relation and asymmetric similarity relation. Feature relation can deal with incomplete information system with missing type or missing type unknown attribute value. It inherits tolerances and nought.

The advantage of symmetric similarity relation, has more reasonable classification ability, but also has some problems.

3. Variable multigranularity rough set based on feature relation

3.1. Multi-granularity rough set based on feature relation

This section introduced the ideas of multi-granularity generalized incomplete decision system, based on the characteristics of the relationship between the granularity of the rough set is from the perspective of granular computing, there is a binary relation from the export of a single grain structure extended to multiple granularity structure composed of multiple binary relations export, so as to build more granularity rough set model.

Definition 3 $IIS=(U,AT\bigcup\{d\},V,f)$ is a generalized incomplete decision system, let $A_1,A_2,\cdots,A_m$ is subset of m properties of AT, for $X\subseteq U$, the upper and lower approximations of optimistic multi-granularity rough sets under feature relation family $K(A_1),K(A_2),\cdots,K(A_m)$ are as follows:

$$\underline{\sum_{i=1}^{m}A_{i_K}^{O}}(X)=\{x\in U:K_{A_1}(x)\subseteq X\vee K_{A_2}(x)\subseteq X\vee\cdots\vee K_{A_m}(x)\subseteq X\};$$

$$\overline{\sum_{i=1}^{m}A_{i_K}^{O}}(X)=\sim\underline{\sum_{i=1}^{m}A_{i_K}^{O}}(\sim X).$$

Theorem 1 $IIS=(U,AT\bigcup\{d\},V,f)$ is a generalized incomplete decision system, let $A_1,A_2,\cdots,A_m$ is subset of m properties of AT, for $X\subseteq U$, under the characteristic relation, the following equation is true:

$$\overline{\sum_{i=1}^{m}A_{i_K}^{O}}(X)=\{x\in U:K_{A_1}(x)\bigcap X\neq\varnothing\wedge K_{A_2}(x)\bigcap X\neq\varnothing\wedge\cdots\wedge K_{A_m}(x)\bigcap X\neq\varnothing\}.$$

Proof: we know from definition 3,

$$x\in\overline{\sum_{i=1}^{m}A_{i_K}^{O}}(X)\Leftrightarrow x\notin\underline{\sum_{i=1}^{m}A_{i_K}^{O}}(\sim X),$$

$$\Leftrightarrow K_{A_1}(x)\nsubseteq(\sim X)\wedge K_{A_2}(x)\nsubseteq(\sim X)\wedge\cdots\wedge K_{A_m}(x)\nsubseteq(\sim X),$$

$$\Leftrightarrow K_{A_1}(x)\bigcap X\neq\varnothing\wedge K_{A_2}(x)\bigcap X\neq\varnothing\wedge\cdots\wedge K_{A_m}(x)\bigcap X\neq\varnothing.$$

Definition 4 $IIS=(U,AT\bigcup\{d\},V,f)$ is a generalized incomplete decision system, let $A_1,A_2,\cdots,A_m$ is subset of m properties of AT, for $X\subseteq U$, the upper and lower approximations of pessimistic multi-granularity rough sets under feature relation family $K(A_1),K(A_2),\cdots,K(A_m)$ are as follows:

$$\underline{\sum_{i=1}^{m}A_{i_K}^{P}}(X)=\{x\in U:K_{A_1}(x)\subseteq X\wedge K_{A_2}(x)\subseteq X\wedge\cdots\wedge K_{A_m}(x)\subseteq X\},$$

$$\overline{\sum_{i=1}^{m}A_{i_K}^{P}}(X)=\sim\underline{\sum_{i=1}^{m}A_{i_K}^{P}}(\sim X).$$

Theorem 2 $IIS=(U,AT\bigcup\{d\},V,f)$ is a generalized incomplete decision system, let $A_1,A_2,\cdots,A_m$ is subset of m properties of AT, for $X\subseteq U$, the following equation is true:

$$\overline{\sum_{i=1}^{m}A_{i_K}^{P}}(X)=\{x\in U:K_{A_1}(x)\bigcap X\neq\varnothing\vee K_{A_2}(x)\bigcap X\neq\varnothing\vee\cdots\vee K_{A_m}(x)\bigcap X\neq\varnothing\}.$$

Proof: we know from definition 4,

$$x\in\overline{\sum_{i=1}^{m}A_{i_K}^{P}}(X)\Leftrightarrow x\notin\underline{\sum_{i=1}^{m}A_{i_K}^{P}}(\sim X)\Leftrightarrow K_{A_1}(x)\nsubseteq(\sim X)\vee K_{A_2}(x)\nsubseteq(\sim X)\vee\cdots\vee K_{A_m}(x)\nsubseteq(\sim X),$$

$$\Leftrightarrow K_{A_1}(x)\bigcap X\neq\varnothing\vee K_{A_2}(x)\bigcap X\neq\varnothing\vee\cdots\vee K_{A_m}(x)\bigcap X\neq\varnothing.$$

3.2. γ variable multi - granularity rough set based on feature relation

Definition 5 $IIS=(U,AT\bigcup\{d\},V,f)$ is a generalized incomplete decision system, let $A_1,A_2,\cdots,A_m$ is subset of m properties of AT, and $A_i\subseteq AT,i=\{1,2,\cdots,m\}$, let $A\in\{A_1,A_2,\cdots,A_m\}=\{a_1,a_2,\cdots,a_n\}$, $0\prec\gamma\leq 1$, for $X\subseteq U$, The upper and lower approximations of R variable multi-granularity rough set of attribute A are defined as follows:

$$\underline{\sum_{i=1}^{m}A_{i_{\gamma K}}^{o}}(X)=\{x\in U:(K_{A_1}(x)\subseteq X\vee K_{A_2}(x)\subseteq X\vee\cdots\vee K_{A_m}(x)\subseteq X)\wedge\frac{|N_{AT}(X)|}{m}\geq\gamma\},$$

$$\overline{\sum_{i=1}^{m}A_{i_{\gamma K}}^{o}}(X)=\sim\underline{\sum_{i=1}^{m}A_{i_{\gamma K}}^{o}}(\sim X).$$

By defining 5, when $0 \prec \frac{1}{m} \le \gamma \le 1$, γ variable multiple granularity for multi-granularity rough set, rough set degradation when $\gamma = 1$,γ variable multiple granularity for the pessimistic multi-granularity rough set, rough set degradation accordingly under the characteristic relations of optimism multi-granularity rough set and rough set is γ variable granularity more pessimistic multi-granularity two special forms of rough set.

Theorem 3 $IIS = (U, AT \cup \{d\}, V, f)$ is a generalized incomplete decision system, let $A_1, A_2, \cdots, A_m$ is subset of m properties of AT, and $A_i \subseteq AT, i = \{1, 2, \cdots, m\}$, let $A \in \{A_1, A_2, \cdots, A_m\} = \{a_1, a_2, \cdots, a_n\}$, $0 \prec \gamma \le 1$, for $X \subseteq U$, acquire

$$\overline{\sum_{i=1}^{m} A_{i_{\gamma K}}^{o}}(X) = \{x \in U : (K_{A_1}(x) \cap X \ne \varnothing \wedge K_{A_2}(x) \cap X \ne \varnothing \wedge \cdots \wedge K_{A_m}(x) \cap X \ne \varnothing) \wedge \frac{|N_{AT}(X)|}{m} \ge \gamma\}.$$

Proof: we know from definition 5

$$x \in \overline{\sum_{i=1}^{m} A_{i_{\gamma K}}^{o}}(X) \Longleftrightarrow x \notin \underline{\sum_{i=1}^{m} A_{iK}^{\gamma}}(\sim X)$$

$$\Leftrightarrow \{K_{A_1}(x) \not\subseteq (\sim X) \wedge K_{A_2}(x) \not\subseteq (\sim X) \wedge \cdots \wedge K_{A_m}(x) \not\subseteq (\sim X)\} \wedge \frac{|N_{AT}(X)|}{m} \ge \gamma$$

$$\Leftrightarrow \{K_{A_1}(x) \cap X \ne \varnothing \wedge K_{A_2}(x) \cap X \ne \varnothing \wedge \cdots \wedge K_{A_m}(x) \cap X \ne \varnothing\} \wedge \frac{|N_{AT}(X)|}{m} \ge \gamma.$$

Theorem 4 $IIS = (U, AT \cup \{d\}, V, f)$ is a generalized incomplete decision system, let $A_1, A_2, \cdots, A_m$ is subset of m properties of AT, and $A_i \subseteq AT, i = \{1, 2, \cdots, m\}$, let $A \in \{A_1, A_2, \cdots, A_m\} = \{a_1, a_2, \cdots, a_n\}$, $0 \prec \gamma \le 1$, for $X \subseteq U$, Under the characteristic relation, the γ variable optimistic multi-granularity rough set has the following properties:

(1) $\underline{\sum_{i=1}^{m} A_{i_{\gamma K}}^{o}}(X) \subseteq X \subseteq \overline{\sum_{i=1}^{m} A_{i_{\gamma K}}^{o}}(X)$;

(2) $\underline{\sum_{i=1}^{m} A_{i_{\gamma K}}^{o}}(U) = U = \overline{\sum_{i=1}^{m} A_{i_{\gamma K}}^{o}}(U)$,

$\underline{\sum_{i=1}^{m} A_{i_{\gamma K}}^{o}}(\varnothing) = \varnothing = \overline{\sum_{i=1}^{m} A_{i_{\gamma K}}^{o}}(\varnothing)$;

(3) $\underline{\sum_{i=1}^{m} A_{i_{\gamma K}}^{o}}(\sim X) = \sim \overline{\sum_{i=1}^{m} A_{i_{\gamma K}}^{o}}(X)$,

$\overline{\sum_{i=1}^{m} A_{i_{\gamma K}}^{o}}(\sim X) = \sim \underline{\sum_{i=1}^{m} A_{i_{\gamma K}}^{o}}(X)$;

(4) $\gamma_1 \le \gamma_2 \Rightarrow \underline{\sum_{i=1}^{m} A_{i_{\gamma_1 K}}^{o}}(X) \supseteq \underline{\sum_{i=1}^{m} A_{i_{\gamma_2 K}}^{o}}(X)$,

$\Rightarrow \overline{\sum_{i=1}^{m} A_{i_{\gamma_1 K}}^{o}}(X) \subseteq \overline{\sum_{i=1}^{m} A_{i_{\gamma_2 K}}^{o}}(X)$.

Proof :(1), (2) can be obtained by definition 3.

(3) We know $\underline{\sum_{i=1}^{m} A_{i_{\gamma K}}^{o}}(\sim X) = \sim \overline{\sum_{i=1}^{m} A_{i_{\gamma K}}^{o}}(X)$ from definition 3, for $X = \sim X$,There are $\underline{\sum_{i=1}^{m} A_{i_{\gamma K}}^{o}}(X) = \sim \overline{\sum_{i=1}^{m} A_{i_{\gamma K}}^{o}}(\sim X)$, so, to prove to get $\overline{\sum_{i=1}^{m} A_{i_{\gamma K}}^{o}}(\sim X) = \sim \underline{\sum_{i=1}^{m} A_{i_{\gamma K}}^{o}}(X)$.

(4) For any $x \in \underline{\sum_{i=1}^{m} A_{i_{\gamma_2 K}}^{o}(X)}$, We know from definition 3 that there is $A_i \subseteq AT, i=\{1,2,\cdots,m\}$, and $A_i \in \{A_1, A_2, \cdots, A_m\}$, make $\frac{|N_{AT}(X)|}{m} \geq \gamma_2$, Because we know what we have $\gamma_1 \leq \gamma_2$, So there are $\frac{|N_{AT}(X)|}{m} \geq \gamma_1$, That is $x \in \underline{\sum_{i=1}^{m} A_{i_{\gamma_1 K}}^{o}(X)}$, Therefore, it can be proved $\underline{\sum_{i=1}^{m} A_{i_{\gamma_1 K}}^{o}(X)} \supseteq \underline{\sum_{i=1}^{m} A_{i_{\gamma_2 K}}^{o}(X)}$. Something similar can be done $\overline{\sum_{i=1}^{m} A_{i_{\gamma_1 K}}^{o}(X)} \subseteq \overline{\sum_{i=1}^{m} A_{i_{\gamma_2 K}}^{o}(X)}$.

3.3. *γ variable multi-granularity rough set reduction and rule acquisition based on feature relation*

Reduction and rule extraction are the core contents of rough set theory. Reduction is to remove unnecessary attributes and attribute values without changing the classification ability of information system, so as to extract rules for decision makers.This paper introduces the theory of multi-granularity and puts forward the concept of γ variable multi-granularity upper and lower approximate distribution reduction based on feature relation, so as to obtain the rules of generalized incomplete information system.

Definition 6 $IIS=(U, AT \cup \{d\}, V, f)$ is a generalized incomplete decision system, let $A_1, A_2, \cdots, A_m$ is subset of m properties of AT, and $A_i \subseteq AT, i=\{1,2,\cdots,m\}$, let $A \in \{A_1, A_2, \cdots, A_m\} = \{a_1, a_2, \cdots, a_n\}$, $0 \prec \gamma \leq 1$, for $X \subseteq U$, $U / ind(D) = \{D_1, D_2, \cdots, D_n\}$, remember to

$$\underline{AT_{\gamma K}^{o}}(D) = \left\{ \underline{\sum_{i=1}^{m} A_{i_{\gamma K}}^{o}(D_1)}, \underline{\sum_{i=1}^{m} A_{i_{\gamma K}}^{o}(D_2)}, \cdots, \underline{\sum_{i=1}^{m} A_{i_{\gamma K}}^{o}(D_n)} \right\}$$

$$\overline{AT_{\gamma K}^{o}}(D) = \left\{ \overline{\sum_{i=1}^{m} A_{i_{\gamma K}}^{o}(D_1)}, \overline{\sum_{i=1}^{m} A_{i_{\gamma K}}^{o}(D_2)}, \cdots, \overline{\sum_{i=1}^{m} A_{i_{\gamma K}}^{o}(D_n)} \right\}.$$

(1) If $\underline{A_{\gamma K}^{o}}(D) = \underline{AT_{\gamma K}^{o}}(D)$, then, it is said that attribute set A is the under approximate distributed coordination set of conditional attribute set AT γ variable multi-granularity of decision attribute set D, if $\underline{A_{\gamma K}^{o}}(D) = \underline{AT_{\gamma K}^{o}}(D)$ and only if $\forall B \subset A$, $\underline{B_{\gamma K}^{o}}(D) \neq \underline{A_{\gamma K}^{o}}(D)$, then attribute set A is the under approximate distribution reduction γ variable multi-granularity of the conditional attribute set AT about the decision attribute set D.

(2) If $\overline{A_{\gamma K}^{o}}(D) = \overline{AT_{\gamma K}^{o}}(D)$, then, it is said that attribute set A is the upper approximate distributed coordination set of conditional attribute set AT γ variable multi-granularity of decision attribute set D, if $\overline{A_{\gamma K}^{o}}(D) = \overline{AT_{\gamma K}^{o}}(D)$ and only if $\forall B \subset A$, $\overline{B_{\gamma K}^{o}}(D) \neq \overline{A_{\gamma K}^{o}}(D)$, then attribute set A is the upper approximate distribution reduction γ variable multi-granularity of the conditional attribute set AT about the decision attribute set D.

Theorem 5 $IIS=(U, AT \cup \{d\}, V, f)$ is a generalized incomplete decision system, let $A_1, A_2, \cdots, A_m$ is subset of m properties of AT, and $A_i \subseteq AT, i=\{1,2,\cdots,m\}$, let $A \in \{A_1, A_2, \cdots, A_m\} = \{a_1, a_2, \cdots, a_n\}$, $0 \prec \gamma \leq 1$, for $X \subseteq U$, if

$$\underline{AT_{\gamma K}^{o}}(x) = \{X_l \in U / IND(\{d\}) : x \in \underline{\sum_{i=1}^{m} A_{i_{\gamma K}}^{o}(X_l)} \},$$

$$\overline{AT_{\gamma K}^{o}}(x) = \{X_l \in U / IND(\{d\}) : x \in \overline{\sum_{i=1}^{m} A_{i_{\gamma K}}^{o}(X_l)} \}.$$

Then

(1) $\underline{A_{\gamma K}^{o}}(\{d\}) = \underline{AT_{\gamma K}^{o}}(\{d\}) \Leftrightarrow \forall x \in U, \underline{A_{\gamma K}^{o}}(x) = \underline{AT_{\gamma K}^{o}}(x)$;

(2) $\overline{A_{\gamma K}^{o}}(\{d\}) = \overline{AT_{\gamma K}^{o}}(\{d\}) \Leftrightarrow \forall x \in U, \overline{A_{\gamma K}^{o}}(x) = \overline{AT_{\gamma K}^{o}}(x)$.

Theorem 6 $IIS=(U,AT\cup\{d\},V,f)$ is a generalized incomplete decision system, let $A_1,A_2,\cdots,A_m$ is subset of m properties of AT, and $A_i\subseteq AT,i=\{1,2,\cdots,m\}$, let $A\in\{A_1,A_2,\cdots,A_m\}=\{a_1,a_2,\cdots,a_n\}$, $0\prec\gamma\le 1$, then

(1) $\forall x\in U,\underline{A^o_{\gamma K}}(x)=\underline{AT^o_{\gamma K}}(x)\Leftrightarrow\exists A_i\in A$, when $\underline{AT^o_{\gamma K}}(x)\not\subseteq\underline{AT^o_{\gamma K}}(y)$, $K_{A_i}(y)\not\subseteq K_{A_i}(x)$ set up;

(2) $\forall x\in U,\overline{A^o_{\gamma K}}(x)=\overline{AT^o_{\gamma K}}(x)\Leftrightarrow\exists A_i\in A$, when $\overline{AT^o_{\gamma K}}(y)\not\subseteq\overline{AT^o_{\gamma K}}(x)$, $K_{A_i}(y)\not\subseteq K_{A_i}(x)$ set up .

Definition 7 $IIS=(U,AT\cup\{d\},V,f)$ is a generalized incomplete decision system, $0\prec\gamma\le 1$, if

$$D^{oL}_{\gamma K}=\left\{(x,y)\in U^2:\underline{AT^o_{\gamma K}}(x)\not\subseteq\underline{AT^o_{\gamma K}}(y)\right\},$$

$$D^{oU}_{\gamma K}=\left\{(x,y)\in U^2:\overline{AT^o_{\gamma K}}(y)\not\subseteq\overline{AT^o_{\gamma K}}(x)\right\},$$

We can be defined

$$D^{oL}_{\gamma K}(x,y)=\begin{cases}a\in AT:(x,y)\notin K(a), & (x,y)\in D^{oL}_{\gamma K}\\ \varnothing & otherwise\end{cases},$$

$$D^{oU}_{\gamma K}(x,y)=\begin{cases}a\in AT:(x,y)\notin K(a), & (x,y)\in D^{oU}_{\gamma K}\\ \varnothing & otherwise\end{cases},$$

$$M^{oL}_{\gamma K}=\left\{D^{oL}_{\gamma K}(x,y):x,y\in U\right\},M^{oU}_{\gamma K}=\left\{D^{oU}_{\gamma K}(x,y):x,y\in U\right\},$$

then $M^{oL}_{\gamma K}$ and $M^{oU}_{\gamma K}$ are respectively called the approximate distribution identification matrix under γ variable multi-granularity for generalized incomplete decision system and the approximate distribution identification matrix under γ variable multi-granularity for generalized incomplete decision system.

Theorem 7 $IIS=(U,AT\cup\{d\},V,f)$ is a generalized incomplete decision system, let $A_1,A_2,\cdots,A_m$ is subset of m properties of AT, and $A_i\subseteq AT,i=\{1,2,\cdots,m\}$, let $A\in\{A_1,A_2,\cdots,A_m\}=\{a_1,a_2,\cdots,a_n\}$, $0\prec\gamma\le 1$, then

(1) $\underline{A^o_{\gamma K}}(\{d\})=\underline{AT^o_{\gamma K}}(\{d\})\Leftrightarrow\forall D^{oL}_{\gamma K}(x,y)\neq\varnothing,A\cap D^{oL}_{\gamma K}(x,y)\neq\varnothing$;

(2) $\overline{A^o_{\gamma K}}(\{d\})=\overline{AT^o_{\gamma K}}(\{d\})\Leftrightarrow\forall D^{oU}_{\gamma K}(x,y)\neq\varnothing,A\cap D^{oU}_{\gamma K}(x,y)\neq\varnothing$.

Definition 8 $IIS=(U,AT\cup\{d\},V,f)$ is a generalized incomplete decision system, $0\prec\gamma\le 1$, if

$$\Delta^{oL}_{\gamma K}=\underset{(x,y)\in D^{oL}_{\gamma K}}{\wedge}\vee D^{oL}_{\gamma K}(x,y),\quad \Delta^{oU}_{\gamma K}=\underset{(x,y)\in D^{oU}_{\gamma K}}{\wedge}\vee D^{oU}_{\gamma K}(x,y),$$

then it is said that $\Delta^{oL}_{\gamma K}$ and $\Delta^{oU}_{\gamma K}$ are the approximate distribution identification function of γ variable multi-granularity and the approximate distribution identification function of γ variable multi-granularity.Where $\wedge$ and $\vee$ represent conjunctive and disjunctive operations respectively.

Theorem 8 $IIS=(U,AT\cup\{d\},V,f)$ is a generalized incomplete decision system, let $A_1,A_2,\cdots,A_m$ is subset of m properties of AT, and $A_i\subseteq AT,i=\{1,2,\cdots,m\}$, let $A\in\{A_1,A_2,\cdots,A_m\}=\{a_1,a_2,\cdots,a_n\}$, $0\prec\gamma\le 1$, then

(1) If A is an under approximate reduction for γ variable multi-granularity,then $\forall x\in U,a_i\in A,j\in V_{\{d\}}$, $r_x:\wedge(a_i,v_{a_i})\to\vee(d,j)$ is the simplest deterministic decision rule;

(2) If A is an upper approximate reduction for γ variable multi-granularity,then $\forall x\in U,a_i\in A,j\in V_{\{d\}}$, $r_x:\wedge(a_i,v_{a_i})\to\vee(d,j)$ is the simplest possible decision rule.

Proof: the definition of under (upper) approximate reduction under γ variable multi-granularity is obviously obtained.

4. Case analysis

$IIS=(U,AT\cup\{d\},V,f)$ is a generalized incomplete decision system,as shown in table 1, Where, $U=(x_1,x_2,x_3,x_4,x_5,x_6)$ is the object set, $AT=(a_1,a_2,a_3,a_4)$ is the conditional attribute set, $\gamma=0.4$ is the threshold value, d is decision attribute set, the unknown value can be compared with any other known attribute value. “*”represents the missing unknown value, which cannot be compared with any other known attribute value. “?” represents a lost type unknown value that cannot be compared with any other known attribute value.

Table 1. Generalized incomplete decision system

U\A	a1	a2	a3	a4	d
x1	1	1	1	1	1
x2	2	*	1	1	2
x3	*	?	*	2	2
x4	1	1	1	2	2
x5	?	*	1	2	3
x6	2	1	1	*	2

From table 1,we get $U/IND\{d\}=(D_1,D_2,D_3)$, $D_1=\{x_1\},D_2=\{x_1,x_3,x_4,x_6\},D_3=\{x_5\}$. Suppose the attribute subset family of generalized incomplete decision system is as follows: $A=\{A_1,A_2,A_3,A_4\}=\{\{a_1,a_2\},\{a_2,a_3\},\{a_2,a_4\},\{a_1,a_4\}\}$, take the threshold $\gamma=0.4$.

(1) In the case of single particle size, the attribute reduction method of generalized incomplete decision system based on feature relation is proposed based on the general rough set theory, and an approximate reduction $red_L=\{a_1,a_4\}$ at single particle size and an approximate reduction $red_U=\{a_1,a_4\}$ at single particle size are obtained. According to the lower approximation reduction of the generalized incomplete decision system, four decision rules of the system can be obtained, and the same is true for the upper approximation reduction of the generalized incomplete decision system. Therefore, in the case of single granularity, the upper and lower approximate sets in table 1 are completely consistent, which cannot reflect the advantages of upper and lower approximate distribution reduction.

(2) In the case of variable multi-granularity of threshold $\gamma=0.4$, the generalized incomplete decision system can obtain approximate reduction $red1_{\gamma L}=\{a_1,a_4\}$, $red2_{\gamma L}=\{a_3,a_4\}$ and $red3_{\gamma L}=\{a_2,a_4\}$ under three variable multi-granularity. It can be seen that in the case of variable multi-granularity of threshold $\gamma=0.4$, the generalized incomplete decision system not only obtains a lower approximation reduction under 3 single granularity, but also obtains another two approximate reductions $red2_{\gamma L}=\{a_3,a_4\}$ and $red3_{\gamma L}=\{a_2,a_4\}$ under variable multi-granularity.

In the same way, three approximate reduction $red1_{\gamma U}=\{a_4\}$, $red2_{\gamma U}=\{a_1,a_4\}$ and $red3_{\gamma U}=\{a_3,a_4\}$ on variable multi-granularity can be obtained. Compared with the upper approximation reduction of single particle size, it is found that $red1_{\gamma U}=\{a_4\}$ is simpler than $red_U=\{a_1,a_4\}$, and another upper approximation reduction $red3_{\gamma U}=\{a_3,a_4\}$ can be obtained under variable multi-particle size.

Taking $red1_{\gamma U}=\{a_4\}$ as an example, the following 5 decision rules with variable multi-granularity can be obtained:

(1) $(a_4,1)\to(d,1)$;

(2) $(a_4,1)\to(d,2)$;

(3) $(a_4,2)\to(d,2)$;

(4) $(a_4,2)\to(d,3)$;

(5) $(a_4,*)\to(d,2)$。

By the case 1, the multi-granularity approximate distribution reduction method can not only **obtain** approximate single particle size distribution reduction method of approximate distribution reduction, and can get more approximate distribution reduction, and the reduction to further eliminate the redundant information, get more qualified reduction, thus can get more Jane's decision rules. The upper and lower approximate distributions obtained under the theory of multi-particle size are inconsistent, which can be approximated better and meet the needs of real life.

5. Conclusion

This article employed multi-granularity theory to deal with the decision rules of generalized incomplete decision system and proposed multiple granularity rough set model and Variable multi-granularity rough set model based on feature relationships. Based on that, combining Variable multi-granularity theory with the approximate distribution reduction methods, a decision theorem for upper and lower approximate distribution reduction of generalized incomplete variable multi-granularity, Variable multi-granularity upper and lower approximate distribution resolution matrix and Resolution function and reduction formula were put forward. As the simplest rule of acquisition tactic is given, variable multi-granularity minimization rules and variable multi-granularity minimization rules can be obtained. Combined with case studies, it is found that the variable multi-granularity approximate distribution reduction method can obtain more reasonable upper and lower approximate distribution reduction than the original approximate distribution reduction method in the case of single granularity, so as to obtain more effective decision rules. It can be seen that the variable multi-granularity approximate distribution reduction and decision acquisition method based on feature relation proposed in this paper is not only an extension of multi-granularity rough set, but also provides a new theoretical method and technical means for the processing of generalized incomplete information.

Foundation item:
Supported by the basic ability improvement project of middle and young teachers in colleges and universities of Guangxi (2018KY0869).

Biography:
Mo Jing-lan was born in 1984. She is a associate professor. Her research interests include Rough Sets and Data Mining.

References:

[1] Pawlak Z.Rough sets[J].International Journal of Computer and Science, 1982,11(5):341-356.
[2] Zhang wenxi, Wu weizhi. Rough set theory and method [M].Beijing: Science Press,2001
[3] Kryszkiewicz M. Rough set approach to incomplete information system [J]. Information Sciences, 1998,11(2) :39-49.
[4] Grzymala -Busse J W. Data w ith missing attribute values: generalization of indiscernibility relation and rule induction[C]// James F Peters.Transactions on Rough Sets I. Berlin: Springer-Verlag, 2004: 78-95.
[5] Thiele H. On semantic models for investigating computing with words [A]. Proceedings of the Second International Conference on Knowledge Based Intelligent Electronic Systems [C]. USA: IEEE,1998:32 -98.
[6] Lin T Y.Granular computing on binary relations I: data mining and neighborhood systems[A]. Rough Sets and Knowledge Discovery[C] . Heidelberg, Germany: Physica-Verlag, 1998:107 -121.
[7] Lin T Y. Granular computing on binary relations II: Rough set representations and belief functions [A]. Rough Sets and Knowledge Discovery [C]. Heidelberg, Germany: Physica-Verlag,1998:107-121.
[8] Qian Yuhua, Liang Jiye,Yao Yiyu, et al. MGRS: A multi-granulation rough set [J]. Information Sciences, 2010, 180(6) :949-970.
[9] Qian Yuhua, Liang Jiye, Dang Chuangyin. Incomplete multi-granulation rough set[J] . IEEE Transactions on Systems, Man and Cybernetics, Part A,2010, 40(2) : 420-431.
[10] Wang lijuan, Yang xibei, Yang jingyu, et al. Incomplete decision rule acquisition based on multigranulation theory[J]. Journal of Nanjing University of Science and Technology,2013,37(1):12-18.
[11] Qian Yuhua, Li Shunyong, Liang Jiye, et al. Pessimistic rough set based decisions: A multigranulation fusion strategy[J]. Information Sciences,2014(264) :196-210.

[12] Meng huili, Ma yuanyuan, Xu jiucheng. The granularity reduction of pessimistic multi-granulation rough set based on the information quantity[J]. Journal of Nanjing university(nature sciences),2015, 51(2) :343-348.
[13] Wang Xiaoyan, Guo Yunting, Shen Yuanxia. Research on granularity reduction of variable multi-granulation rough set[J].Natural sciences journal of haebin normal university,2019,35(1):24-30.
[14] Luo Gongzhi, Xu Xinxin. Rough analysis method of multi-granularity decision rough set based on possible degree tolerance relation[J].Application Research of Computers,2018,36(12):2912-2020

IOP Conf. Series: Materials Science and Engineering **806** (2020) 012048 doi:10.1088/1757-899X/806/1/012048

Analysis of Tourism City Image Based on Big Data of Network Text——Take Guangzhou as an Example

Yang Sen[1*], Zhang Ling[1] and Xu Liping[2]

[1]School of Geography and Tourism, Guangdong University of Finance and Economics, Guangzhou, Guangdong, 510320, China

[2]Qianshan Yuantan High School, Anqing, Anhui, 246300, China

[*]Corresponding author's e-mail: 1028943403@qq.com

Abstract. Based on 16979 tourist reviews of six scenic spots in Guangzhou by ctrip.com, using the triple process of open coding, spindle coding and selective coding based on grounded theory, it is found that the tourist image of Guangzhou is mainly composed of cognitive image, emotional image and conative image; the tourists are satisfied with a high degree of willingness to revisit and recommend to their relatives and friends; the tourists are emotional towards Guangzhou tourism High evaluation and strong willingness of tourists to travel.

1. Introduction

In recent years, with the rise and popularization of mass tourism, the tourist demand is more and more personalized and diversified, and the competition among tourist destinations is more and more fierce. Destinations with distinctive destination characteristics and typical images are more attractive to tourists and are favored by more tourists. It is found that a good image of a tourist destination can attract tourists and affect their willingness to revisit[1-2]. The traditional tourism image is perceived by the public through TV, advertising and other media. Now with the arrival of the era of big data, the network has become the backbone of tourism destination image communication, an important medium for tourists to obtain tourism information, as well as an important channel for tourists to share their tourism experience and express their emotions. Because of the experience of tourists, online reviews are believed and recognized by more tourists, and can also affect more tourists' travel decisions.

At present, there are many scholars who use Internet text to analyze the image of tourism destination, but most of them use the Internet travel notes of tourism websites to explore the image of tourism destination, few of them use Internet comments to study the image of tourism destination. This paper attempts to analyze the tourist image of Guangzhou perceived by tourists by mining 16977 comments from six of the most popular scenic spots in Guangzhou in the past two years.

2. literature review

Tourism destination image research began in the 1970s. After Dr. Hunt of the United States discussed the tourism destination image in 1971, the tourism destination image has become a hot spot of tourism research[3]. Domestic research on tourism destination image started late. In 1998, Li Lei published a paper on urban tourism image[4], and in 1999, Li Lei published the first monograph on tourism destination image planning: Theory and practice[5], and then domestic scholars began to enter this research field.

At present, the research on tourism destination image mainly focuses on the definition, composition, promotion, function and formation mechanism of tourism destination image. There is not a universally accepted definition of tourism destination topography. Crompton believes that tourism destination image is the overall perception of tourism destination formed by tourists and potential tourists through processing information from various channels, and a series of impressions, views and emotional expressions of tourism destination[6]. Tourism destination image is the sum of tourists' beliefs, views and impressions of tourism destination[7]. Xu Cuiping defines the destination image from the perspective of supply and demand: from the perspective of supply, the destination image is the impression that tourists hope to perceive and form through publicity; from the perspective of demand, the destination image is the specific internal psychological cognition that tourists form through various information sources and personal experience to the tourist destination[8]. Most studies believe that the destination image is a three-dimensional structure. Gunn divides the formation process of the target terrain image into the organic image and the induced image. The organic image refers to the image formed through newspapers, magazines, relatives and friends and other information sources, while the induced image refers to the image formed through commercial information sources such as advertisements and promotions[9]. Fakeye and Crompton expand Gunn's theory, and divide the target terrain image into organic image, induced image and complex image. The complex image refers to the image formed by tourists visiting the destination through personal experience[10]. Gartner's research optimizes the theoretical basis of destination image structure, and puts forward the "three-dimensional structure" of destination image: cognitive image, emotional image and conative image[11]. Baloglu and McCleary proposed a "new three-dimensional structure" of destination image: cognitive image, emotional image and overall image[12]. Now, the "new three-dimensional structure" model of the destination image has been basically recognized by scholars, and has carried out more in-depth and extensive research. The research of Liu Li[7], Zhang Hongmei et al.[13] also confirmed the three-component structure of the destination image and its applicability in the context of Chinese culture.

The existing research shows that the tourism destination terrain image affects the decision-making and behavior tendency of tourists[14-15], has important value and significance for the satisfaction and loyalty of tourists[16-17], and becomes an important way of tourism destination marketing and promotion[18]. Zhang Hongmei et al.[13] took Chinese potential consumers as samples and Bai Kai et al.[19] took South Korean potential tourists as samples, which also confirmed the positive correlation between the perception of destination terrain image and tourism intention. Liu Li found that the cognitive dimension of destination image significantly affects the emotional dimension, which together determine the overall image of the destination, and the overall image perception of potential tourists' destination significantly affects their tourism intention[7]. Secondly, the destination image will affect the tourists' tourism experience and experience, and the tourists' evaluation of their own tourism experience will affect the behavior after tourism, such as revisiting intention, recommendation intention and positive word-of-mouth[20-21].

A large number of studies in the field of consumer behavior and tourism show that word-of-mouth communication is a far more reliable information source than corporate advertising, which has a significant impact on tourists' purchase decision-making and destination selection[22-23]. Wang Jinsong thinks that the network is one of the important modes of tourism destination image communication, and discusses the network communication mode[24]. Wu Xuefei proposed the influence of online word-of-mouth marketing and customer loyalty on the promotion of tourism destination image [25]. Ruzzier discusses the construction of tourism destination image, and proposes that the survey and evaluation of tourism destination image from the perspective of tourists should be applied to tourism brand building[26]. Lee develops and shapes the marketing strategy of shopping tourism image based on the perspective of tourist attraction, which is a new way of tourism destination image publicity[27].

Tourism city image is one of the most important categories of tourism destination image, but there is a lack of research on tourism city image at present, which is mostly extended from tourism destination image to tourism city image.

3. Research design

3.1. Case selection

As the capital of Guangdong Province, Guangzhou is one of the first tier cities in China and one of the important cities in Guangdong Hong Kong Macao Bay area. It is not only the transfer station of Guangdong tourism, but also the important distribution center of international tourism. Its tourism destination development level can directly affect the impression of International tourists on China. Based on the ranking of online tourism operator Ctrip on Guangzhou scenic spots, the top 6 scenic spots are selected, which are Guangzhou tower, Chimelong wildlife world, Shamian, Chimelong happy world, Chenjiaci and Baiyunshan. All of the six scenic spots scored above 4.5 points (full score 5 points), and were able to accommodate 60% of tourists to Guangzhou.

3.2. Data acquisition and analysis tools

In this study, we use the crawler software octopus to capture the tourist comments of six scenic spots on ctrip.com from December 1, 2017 to November 30, 2019. There are 16979 tourist comments, including 2985 from Guangzhou tower, 2960 from Changlong wildlife world, 3001 from Shamian, 2941 from Changlong happy world, 2100 from Chenjiaci and 2992 from Baiyun Mountain. In this study, the text analysis software RST CM6 and NVivo11.0 are used to assist the analysis, and the grounded theory and content analysis method are used.

3.3. Construction of tourism city image

Open coding is a process of classifying and summarizing the acquired preliminary data to form a conceptualized category, from being simple and disorderly to being conceptualized and categorized. After sorting out and coding 16979 comments of tourists, refining the spoken words of tourists, and refining and summarizing for many times, more than 1000 original sentences and corresponding initial codes are finally extracted. Because the number of initial codes is large and there is a certain cross, the repeated initial codes are eliminated and refined again. At last, more than 1000 initial codes eventually form 30 initial categories, which are festivals, sightseeing, pension, sports, shopping, natural attractions, human culture primers, public leisure places, public services and ancillary facilities, guide services, price level , popularity, external transportation, internal transportation, catering, accommodation, tourism information consulting service, safety, social customs, residents' attitude, landscape style, climate, urbanization level, biology, folk culture, education, entertainment, tourist behavior, staff quality and creative goods. See Table 1 for specific process examples.

Table 1. The example of open coding.

Some Original Data	Code	Initial Categories
The overall environment of Shamian is very good; the main street is relatively broad, which is a garden style pedestrian street with good greening; it is quiet and leisurely.	Environment, greening, quiet and leisure	Natural attraction
In the rainy days of Guangzhou, the northerners are very cold; it's really worth tasting the morning tea in the White Swan Hotel.	Rainy, cold	Climate
	Morning tea	Catering
Baiyun Mountain tickets 5 yuan, half price for students, tickets are very cheap! Like to climb the mountain can climb to the main peak of Moxing ridge, do not want to climb can take cableway!	Admission ticket	Price level
	Cableway	Internal traffic
Indoor steles, outdoor steles, cliff inscriptions and other forms are integrated into the garden and natural scenery, showing the splendid culture of the Chinese nation.	Inscriptions, culture	Human attraction
Guangzhou tower Ferris wheel 460 meters high experience, the night scene is particularly beautiful.	Night scene	Urbanization level
If you find the right angle to take photos, you can take photos with Xiaomanyao to create countless interesting shapes.	Photograph	Entertainment
All kinds of animal performances, animal science popularization, catering, toilet, baby cart rental, luggage storage are all done very well.	Polular Science	Education
Need to explain, but the service desk didn't have an interpreter, just met another 6 people, so we rented an interpreter together, about an hour's explanation, it felt very good.	Interpreter	Tour Guide Service
The stunt theater is worth watching. Although it's hot, the staff are dedicated.	Employee dedication	Staff quality
I hope it can be improved very well. I can't buy any anti mosquito or medicine. It's a bit troublesome.	Mosquito	Security

Spindle coding is to cluster the initial categories extracted from open coding and merge the interrelated initial categories. Through refining and clustering 30 initial categories, five main categories are obtained, namely, urban soft environment, tourism services, public places and services, tourism attractions and tourism activities. See Table 2 for details.

Table 2. Formatting sections, subsections and subsubsections.

Open categories	Axial categories	Selective categories
Security, Social customs, Resident attitude, Price level, Urbanization level, Tourist behaviour, Staff quality	Urban soft environment	
External traffic, Internal traffic, Catering, Accommodation, Tour Guide, Service	Tourism service	Cognitive image,
Tourism information consulting service Public leisure places, Public services and ancillary facilities	Public places and services	
Natural attraction, Human attraction, Landscape style and features, Biology, Climate Folk culture, Popularity, Creative Goods	Tourist attractions	Emotional image
Festival, Sightseeing, Pension, Sports, Shopping, Entertainment, Education	Tourism activities	Conative image

Selective coding is the process of selecting a core category and systematically connecting it with other categories after systematic analysis of all discovered categories. Through the systematic and comprehensive analysis of the five main categories, we can get three choice categories, including cognitive image, emotional image and emotional image. See Table 2 for details.

3.4. Emotional analysis of tourist destination

The six scenic spots in this study can be divided into three categories according to natural landscape, historical sites and man-made landscape, among which Baiyun Mountain and Changlong wildlife world belong to natural landscape, Chenjiaci and Shamian belong to historical sites, Guangzhou tower and Changlong happy world belong to man-made landscape. The three types of scenic spot visitors' comments were analyzed with the rose CM6, and the analysis results are shown in Table 3.

Table 3. An analysis of tourists' feelings towards the scenic spot.

Scenic type	Positive emotion (%)	Neutral emotion (%)	Negative emotion (%)
Natural landscape	66.42	24.13	9.45
Historical site	73.28	19.14	7.58
Man-made Tourist Site	62.32	26.44	11.24
Whole	67.34	23.24	9.42

4. Research conclusion and Prospect

4.1. Research conclusion

First, this study concludes that the tourism image of Guangzhou is mainly composed of cognitive image, emotional image and emotional image. Among them, tourists have the strongest perception of the cognitive image. Tourists are mainly interested in Guangzhou tourist attractions, catering, accommodation, Lingnan architecture and unique street landscape, especially the image of "eating in Guangzhou" is deeply rooted in the hearts of the people. Tourists' voluntary image is the actual tourism activities that make tourists make decisions on tourism behavior, which is the most important for tourists, and tourists can make decisions on the purpose What kind of tourism activities does the city carry out. Secondly, most of the tourists are satisfied with the trip and express their willingness to revisit and recommend to their relatives and friends. A few tourists are not satisfied with the long queue time, insufficient parking space and high price of some tourist commodities. Third, the six scenic spots are divided into three categories. Through the emotional analysis of the tourists' comments, it is found that the positive emotions of the tourists in the three categories are all greater than 62%, and the positive emotions and neutral emotions are all more than 88%, which indicates that the overall tourism image given to tourists by Guangzhou is relatively good. The tourists' attitude towards the destination directly affects the tourists' willingness to travel, and the tourists' willingness to travel is also correspondingly high Enhance.

4.2. Research limitations and Prospects

First, this study only grabs the tourist comments of one website, and the data source is monotonous. In the later stage, we can try to grab the research data from more tourism related websites, even the comments and travel notes of foreign tourism websites about Guangzhou. Second, this study only analyzes the tourists' comments, not the data of travel notes, photos and videos. In the future, it can analyze the pictures and videos taken by tourists, so as to obtain the most accurate image of the tourist city perceived by tourists. Third, the young and middle-aged tourists are the majority who comment and publish the travel notes on the Internet. The perception of the young children and the elderly tourists to the destination image is not analyzed. In the future, we can interview or questionnaire the two groups to obtain more accurate tourism city image.

References

[1] Fang M.C., Qu Y. (2018) Research on the attention of online homestay reputation evaluation index based on text review. Regional research and development, 37:123-127.
[2] Zhang H.Z., Huang K.J., Wu M.Y. (2015) Sustainable development of coastal cultural tourism based on Game Theory—Taking Putuo Mountain as an example. Economic geography,35: 202-208.
[3] Tapachai N.， Waryszak R. (2000) An Examination of the Role of Beneficial Image in Tourist Destination Selection． Journal of Travel Research， 3:37-44.
[4] Li L. (1998) Discussion on urban tourism image design. Tourism Tribune, 1:46-48
[5] Li L.(1999)Tourism image planning: Theory and practice, Guangdong Tourism Press, Guangzhou.
[6] Crompton J. L. (1979) An assessment of Mexico as a Vacation Destination and the Influence of Geographical Location upon that Image． Journal of Travel Research,4:18-23.
[7] Liu L. (2013) Screen-induced Tourism: Perceived Destination Image and Intention to Visit. Tourism Tribune,28:61-72.
[8] Xu C. P. (2019) Multidimensional Analysis of the Representation and Reception of Luoyang’ s City Image. Journal of Henan University of science and Technology, 37:5-11
[9] Gunn C. A. (1972) Vacationscape: Designing Tourist Regions. Austin: Bureau of Business Research， University of Texas,114.
[10] Fakeye P. C., Crompton J. L. (1991) Image differences between prospective， first time， and repeat visitors to the Lower Rio Grande Valley． Journal of Travel Research,30:10-16.
[11] Gartner W.C.(1993)Image formation process.Journal of Travel and Tourism Marketing,2:191-215.
[12] Baloglu S., McCleary K. W. (1999) A model of destination image formation. Annals of Tourism Research,26:868-897.
[13] Zhang H. M., Lu L.， Cai L. P.， et al. (2011) Tourism destination image structural model and visitors’behavioral intentions: Based on a confirmatory study of localization of potential consumers. Tourism Science, 25:35-45.
[14] Jeong G. C., Tamara T., Shomir S. (2011) On the Destination Image of Korea by Russian Tourist. Tourism Management,1:193-194.
[15] Nikolaos S., Victoria B., Andreas A., et al. (2017) Linking the Dots Among Destination Images， Place Attachment, and Revisit Intentions: A Study Among British and Russian Tourists. Tourism Management,15-29.

[16] Assaker G., Vivzi V. E., Connor P. O. (2011) Examining the Effect of Novelty Seeking, Satisfaction, and Destination Image on Tourists'Return Pattern: A Two Factor, on Linear Latent Growth Model. Tourism Management, 4):890-901.

[17] Zhang H. M., Fu X. X., Al L. P., et al. (2014) Destination Image and Tourist Loyalty: A Meta-Analysis. Tourism Management, 213-223.

[18] Castro C. B., Armario E. M., Ruiz D. M. (2007) The Influence of Market Heterogeneity on the Relationship between a Destination's Image and Tourists´ Future Behavior. Tourism Management, 1:175-187.

[19] Bai K., Chen N., Zhao A. Z. (2012) Potential Korean tourists' cognition of Chinese destination image and their behavioral intentions. Tourism Science, 26:82-94.

[20] Chen C. F., Tsai D. (2007) How destination image and evaluative factors affect behavioral intentions? Tourism Management,28:1115-1122.

[21] Wang C., Hsu M. K. (2010) The relationships of destination image, satisfaction, and behavioral intentions: An integrated model. Journal of Travel & Tourism Marketing,27:829-843.

[22] Gitleson R. J., Crompton J. L. (1982) Insights into the repeat vacation phenomenon. Annals of Tourism Research,11:199-217.

[23] Gremler D. D., Gwinner K. P., Brown S. W. (2001) Generating positive word-of-mouth communication through customer-employee relationships. International Journal of Service Industry Management, 12:44-59.

[24] Wang J. S. (2008) Study on the network communication mode of tourism destination image. Journal of Central South Forestry University of science and Technology, 4:48-50.

[25] Wu X. F. (2010) The relationship between tourism destination image, online reputation and customer loyalty [J]. Journal of Shenyang Normal University, 4:37-40.

[26] Ruzzier M. K. (2013) Extending the tourism destination image concept into customer-based brand equity for a tourism destination. Social Science Electronic Publishing, 23:24-42.

[27] Lee Y. (2013) Effect on the tourism motivation and tourism destination image affected to shopping tourism destination choice. International Journal of Digital Content Technology & Its Applic, 7:416-421.

Vehicle-specific retrieval based on aerial images

Guoliang Yang, Zhigang Tang* and Jianchen Huang

Department of Electrical Engineering and Automation, Jiangxi University of Science and Technology, Ganzhou, Jiangxi, 341000, China

*Corresponding author's e-mail: 1397414695@qq.com

Abstract. UAVs(Unmanned Aerial Vehicle) are currently widely used in civil and military fields. Collecting ground information through UAVs is their main application direction.This paper proposes a vehicle retrieval system based on aerial images, which can retrieve the corresponding target vehicle according to the set conditions.For the retrieval algorithm, we optimized its feature extraction network based on YOLO v3, and optimized it as a two-way residual based on the original residual module to improve its ability to detect small targets. Considering that the position deviation of the prediction frame in the actual detection will greatly affect the detection result, the GIoU position calculation index is adopted to improve the network's sensitivity to the position information.At the same time, retrieval conditions such as vehicle type and color are added on the basis of detection, so that the model can complete the retrieval of specific vehicle targets. The test results of the finally optimized network on the VEDAI dataset reached 91.2%mAP.

1. Introduction

UAVs have been popular in our lives from the initial military use to the present. In recent years, the rapid development of drones has made their application areas not limited to military operations, terrain surveys, and high-altitude gas detection. Drones rely on their high-altitude advantages to obtain some perspectives that we usually cannot obtain. Aerial images obtained by drones can contain more ground information.Although the remote sensing images taken by satellites are similar to high-altitude aerial photos taken by drones, they contain rich ground information, but the image quality of remote sensing images is not ideal due to the influence of external factors such as clouds and light. The addition of additional noise makes convolution. The neural network has problems such as missed detection and false detection.The drone can improve the overall imaging quality by changing the position and aerial photography height, and it is flexible and easy to use. Therefore, more and more high-altitude image acquisition methods are obtained by aerial photography of mobile devices such as drones. Aerial target detection is mainly aimed at aerial images, and detects targets of interest, such as vehicles, ships, roads, and natural landforms. The aerial image can clearly express the geographical form, and it is less affected by the terrain and ground architecture, and can hold more information. However, because of the small proportion of the target in the image due to high-altitude shooting, it is difficult to extract target features, resulting in poor detection results.

Although the concept of convolutional neural networks was proposed as early as the 1980s, it was limited by the hardware equipment at the time and could not complete the calculation of large-scale convolutional neural networks, making it not a research hotspot. In recent years, the rapid development of GPU hardware technology can rely on its high-bandwidth, multi-threaded parallel computing capabilities to easily complete inference calculations on a complex network. Especially in the 2012 ILSVRC competition, Alex Krizhevsky won the championship with the AlexNet model, and brought the field of deep learning back to everyone's perspective. The automatic learning ability of convolutional neural network for target features is more robust and the detection effect is better than that of the traditional target detection method through artificially designed features. Therefore, we use YOLO v3 convolutional neural network to detect vehicles in aerial images according to the retrieval conditions. Compared with other vehicles, we can further narrow the detection range by setting the rigid characteristics of the vehicle, and the detection results obtained will have more Targeted.

2. Small target detection and problem analysis based on YOLO v3

The YOLO series of algorithms rely on the end-to-end network structure, and can complete the regression of target locations and category predictions in a network. The overall model has a small amount of calculations and can quickly complete inference, but it sacrifices detection accuracy. Based on the actual needs, we hope to be able to detect the video returned by the drone in real time, and require the algorithm to reason quickly. And different from the RCNN series of algorithms based on sliding window target detection, YOLO's target detection is for a complete feature map, so it can reduce the false detection due to the background of the image, especially the small target area in aerial images Most of them are background images. In summary, in the end, we chose YOLO v3 algorithm as the basic algorithm for this design, and optimized its network structure to improve the detection ability for small targets.

2.1. Principle of YOLO v3

The YOLO algorithm inputs a complete picture into an end-to-end network. Because the image is not segmented, the detection is performed based on a complete feature map. The feature map input to the YOLO detector is first divided into S * S grids. When the center point of a potential target is in a certain grid, the grid will generate B bounding boxes to predict them, and each bounding box has its own confidence. At the same time, each bounding box is composed of five parts: x, y, w, h, and c. (x, y) represents the center position of the bounding box relative to the grid. And height; c indicates the ratio of the area of the boundary prediction box to the actual boundary box. In addition to the prediction of position information, the boundary prediction box also predicts the category of the target in the box. The final confidence score of the boundary prediction box is composed of the prediction results of position and category. The calculation formula is as follows:

$$C = \Pr(class_i \mid object) * \Pr(object) * IOU_{pred}^{truth} = \Pr(class_i) * IOU_{pred}^{truth}$$

In the above formula, $\Pr(class_i)$ Indicates the category prediction score for the target in the box; IOU_{pred}^{truth} The ratio of the predicted bounding box to the actual bounding box area is the location prediction score.

Compared with YOLO v1 and YOLO v2, YOLO v3 has greatly improved the feature extraction network structure[1]. First of all, the idea of residual network is borrowed, and the residual module is constructed through the fast link layer (Figure 1), so that the total number of layers in the network reaches 106 layers[2]. The more network layers, the stronger the ability to extract features, but blindly increasing the number of network layers will cause problems such as disappearance of gradients during training, and the residual module solves the network degradation during training by means of identity mapping. This problem also makes YOLO v3 with more network layers have stronger feature extraction capabilities.

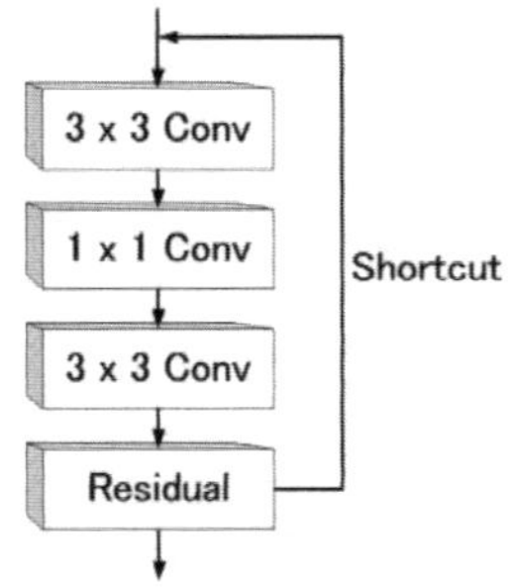

Figure 1. YOLO v3 residual module

In addition, in order to improve the detection ability of small targets, YOLO v3 uses a multi-scale detection mechanism, which is set up with 13 * 13, 26 * 26, and 52 * 52 three grid division scales. Among them, the 52 * 52 scale is responsible for detecting small targets. The addition of fine-grained scale detection effectively reduces the occurrence of missed detection of small targets and improves the overall network robustness

2.2. Test and analysis of YOLO v3 algorithm based on aerial image

In order to determine YOLO v3's ability to detect small targets in aerial images, we use the VEDAI data set as the training and testing set for this experiment, and at the same time train the YOLO v3-tiny network model as a reference. The test results are shown in Table1.

Table1.Test results on the VEDAI datase

Model	mAP
YOLO v3	76.5%
YOLO v3-tiny	71.7%

Refer to the test results of YOLO v3 and YOLO v3-tiny on the COCO dataset.The test results are shown in Table2.

Table2.Test results on the COCO dataset

Model	mAP
YOLO v3	55.3%
YOLO v3-tiny	33.1%

Compare the test results of the two data sets. In the comparison of the test results of the VEDAI data set, the detection accuracy of YOLO v3 is not much higher than that of YOLO v3-tiny.

Because YOLO v3 uses a multi-scale detection mechanism, the object detected based on the actual application scene is a small target. In order to verify the ability of three scales to detect small targets in aerial images, we split YOLO v3 into three single scales for testing. The test results are shown in Table3.

Table3.Single-scale test results

Model	mAP
YOLO v3-13*13	33.4%
YOLO v3-26*26	48.6%
YOLO v3-52*52	70.2%

It can be seen from the above table that the detection accuracy of 13 * 13 and 26 * 26 scales is much lower than 52 * 52 scales.It shows that the detection of small targets is mainly performed on the 52 * 52 scale, and the contribution rate of the 13 * 13 and 26 * 26 scale detection is not large.

3. Optimization based on YOLO v3 algorithm

According to the test results and analysis in section 1.2, we mainly optimize the YOLO v3 algorithm from the following aspects. First, streamline the overall network structure to avoid loss of features caused by excessively deep network layers;Second, add more fine-grained YOLO detection to further improve the detection ability for small targets;Third, replace the original IOU calculation method with the GIOU calculation method to optimize the accuracy of the target position prediction.

3.1. Network Structure Optimimization

The reason why the original YOLO v3's backbone network is deeper is that the residual module used by it is smaller. Each scale feature detection has multiple residual modules for stacking. First, we perform the convolution structure in the residual module. In a large residual module, there are residual modules of two receptive fields. The structure of the optimized residual module is shown in Figure2.

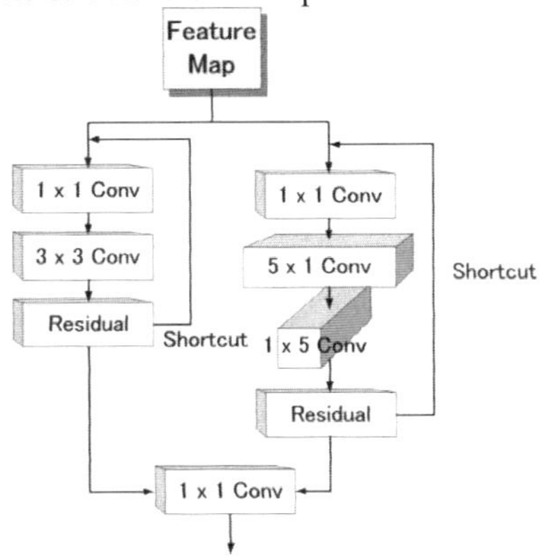

Figure 2. Two-way residual module

It can be seen from the above figure that when a feature map is input into a residual module, two small residual modules will perform feature extraction operations on it simultaneously.In the residual module on the left, the receptive field size of its convolution kernel is 3*3, which mainly extracts detailed features.The size of the convolution kernel receptive field in the residual module on the right is 5*5, which is responsible for extracting the overall features of the target.Finally, the 1*1 convolution layer is used to extract the two channels to the features for fusion, so as to obtain more comprehensive features[3].At the same time, in order to reduce the amount of calculation caused by the increase of the module structure, we perform a low-rank decomposition on the 5*5 convolution kernel and decompose it into two serial convolution kernels of 5*1 and 1*5.The low-rank decomposition method can greatly reduce the calculation amount, and the effect of feature extraction is the same as the effect before decomposition[4].The test results can reduce the calculation amount by using the low-rank decomposition method, and the effect of feature extraction is the same as that before decomposition. The test results are shown in Table4.

Table4.Residual block replacement test results

Model	mAP
1 residual module per scale	78.6%
2residual module per scale	88.5%
3 residual module per scale	82.4%
4 residual module per scale	72.9%

From the data analysis results in the table above, it can be concluded that the four scales 104 * 104, 52 * 52, 26 * 26, and 13 * 13 are better for feature extraction by two optimized residual modules.

According to the test results of detection accuracy of different scales in Chapter 1.2, in the new network structure, we have removed the detection of two dimensions of 13 * 13 and 26 * 26, and increased the scale of 104 * 104 to achieve the purpose of more fine-grained detection.There are two main improvements in this step. The first is to add an inverted pyramid structure after the 13 * 13 scale residual module, and use the deconvolution layer to re-expand the feature map to 52 * 52 and 104 * 104, and then perform YOLO. Detection.After the expansion of the feature map is completed, the shallow-scale features of the same scale are merged into the deep layer, and the shallow and deep feature information is synthesized.For shallow neural networks, because the feature map has not been down-sampled many times, more target position information can be extracted;In deep networks, feature maps contain rich semantic information after multiple samplings.Before the YOLO layer is input, the features extracted from the shallow layer and the deep layer are fused to improve the richness of the feature information and thereby enhance the detection capability of the YOLO layer.We call the optimized YOLO v3 YOLO-E, and its network structure is shown in Figure3.

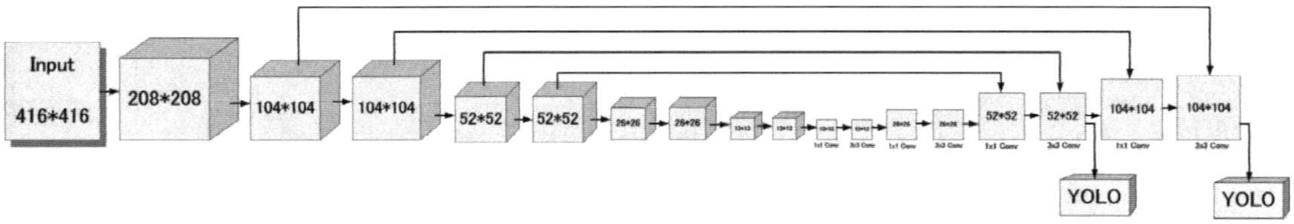

Figure 3. YOLO-E network structure

3.2. Use of GIOU indicators

The accuracy of object detection in this application scenario for the prediction of the bounding box position is more accurate than that in ordinary scenes. This is because the detected target occupies a small pixel area in the picture, and even if the bounding box appears slightly offset, There will also be significant visual errors in the display on the prediction map. In order to enhance the sensitivity of the network to the location, we replaced the original IoU with a GIoU indicator and improved its loss function[5].

In object detection algorithms, IoU is widely used as an indicator of the accuracy of position prediction. Its advantages are mainly that the calculation method is simple to meet most calculation scenarios, but its disadvantage is also obvious, that is, it is not sensitive to position deviation. GioU adds the calculation of the non-intersecting area on the basis of IoU, which is more sensitive to the offset between the prediction box and the actual bounding box. The calculation formula and schematic diagram are shown in Figure4:

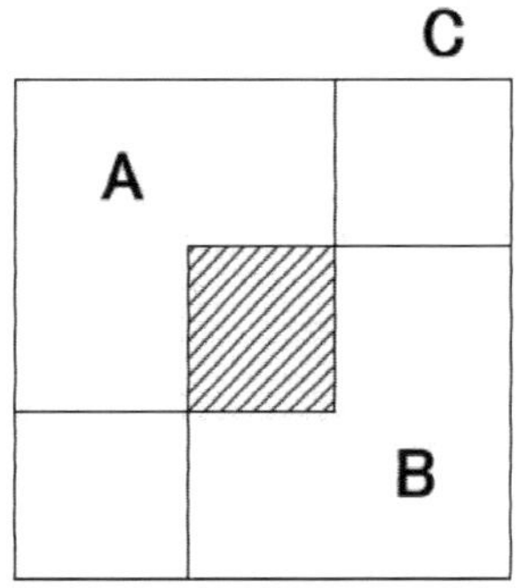

Figure 4. GIOU schematic

In the above formula, C represents the minimum rectangular area including the predicted bounding box A and the actual bounding box B, $|C\backslash(A\cup B)|$ is represents the area area of C that does not include A and B,Finally, it is compared with the area of C, and then subtracted from IoU to obtain the final GioU value.The value of IoU ranges from [0,1]. When IoU is 0, the gradient value is also 0 during the back propagation calculation. The model cannot be further optimized.The value range of GioU is [-1,0) U (0,1], which avoids the problem of a gradient value of 0. At the same time, the introduction of C makes the non-overlapping regions get the attention of the network.We added GioU loss calculation to the original YOLO v3 loss function calculation. The calculation formula of the loss function is as follows.

$$L_{GIoU} = 1 - GIoU$$

3.3. Fine search process design

In addition to requiring vehicles to be detected in aerial images, we also need to further narrow down the detection range based on search conditions.Analyze the characteristics of the samples on the aerial image, and filter the detected vehicle targets based on the model and color. The vehicles in the VEDAI dataset are divided into the following three categories:Cars, pickup trucks and SUVs.First, we cut out the vehicles in VEDAI through the coordinate information in the label file, and divided these vehicles into three types: cars, pickup trucks, and SUVs.Then, based on the trained YOLO-E network model, transfer learning is performed to complete the training of the classification network.The color selection of vehicles is relatively simple. When YOLO-E detects the vehicle, it extracts nine pixels centered on the center point of the predicted bounding box and sends them to the opencv color classifier for color determination.We set the upper and lower RGB thresholds for the six common colors of red, green, blue, white, gray, and black. The color of the vehicle can be determined when the color of the extracted pixel block belongs to a certain color interval.

4. Experimental testing and analysis

We make clear sample from VEDAI the data sets, picking out 2000 pictures which contain the vehicle 8: 2,dividing the data set into training. The first is the comprehensive test of YOLO-E network with multiple indicators to improve the overall robustness of the network after a more intuitive response. We have added YOLO v3 to YOLO v3-tiny for comparison, and the comparison test results of the three network models are shown in Table5:

Table5.Comparative test results of network models

	precision	recall	F1-score	mAP	FPS
YOLO v3	74%	67%	70%	76.5%	45
YOLO v3-tiny	65%	59%	61%	71.7%	69
YOLO-E	88%	81%	84%	91.2%	42

The calculation formulas for the above five indicators are as follows.

$$precision = \frac{TP}{TP+FP}$$

$$recall = \frac{TP}{TP+FN}$$

$$F1-score = \frac{2*precision*recall}{precision+recall}$$

Precision reflects the accuracy and recall of the model. F1-score indicator is a comprehensive indicator based on precision and recall ,reflecting the robustness of the network, mAP is the average accuracy, which is also the detection accuracy of the vehicle in this experiment. We use the FPS frame rate indicator to reflect the model's inference speed, the values vary depending on the hardware equipment test. This test was performed on a NVIDIA 2080Ti graphics card.

From the table above, it can be directly seen that the overall performance of YOLO-E is better than YOLO v3 and YOLO v3-tiny. Although we have reduced the network layer, the network structure in the new residual module is more complex, which makes it less obvious for the optimization of inference speed.

In order to verify the improvement of the model of the GIoU indicator and the optimized loss function, we set up two different YOLO layers based on the YOLO-E network structure, one using the IoU indicator and the original loss function, and the other using the GIoU indicator and newly added L_{GIoU} loss function, the comparison test results are shown in Table6.

Table6.IOU and GIOU test comparison

	precision	recall	F1-score	mAP	FPS
YOLO-E(IoU)	87%	77%	81%	88.5%	42
YOLO-E(GIoU)	88%	81%	84%	91.2%	42

According to the data in the above table, it is proved that the optimization of GIoU and the loss function will improve the accuracy of the network.

The actual results of the three networks in the test set are shown in Figure5.

YOLO v3-tiny YOLO v3 YOLO-E

Figure 5.Comparison of several detection results

From the test comparison chart above, you can also see the pros and cons of the three models. The main problem with YOLO v3-tiny and YOLO v3 is the feature extraction network. The simple YOLO v3-tiny backbone network leads to insufficient extraction of features . Some houses will be detected as vehicles by mistake.YOLO v3 has a deep backbone structure , because the detected targets are small, and the size of the feature map of the deep network is also small. It is easy to lose the small target information, causing the final detection result to be leaky check the situation. For the YOLO-E network, the network structure and the YOLO layer have been optimized accordingly, and the above problems of false detection and missing detection have not occurred.

At the same time, we test the accuracy of the network classification on the transfer learning. The test results are shown in Table7:

Table7.Classification accuracy test

	car	pickup trucks	SUV
AP	98.7%	96.2%	95.9%
mAP		96.8%	

The classification accuracy of the three categories is as high as 93% or more. Pickup trucks and SUVs have certain similarities in the vehicle models, resulting in slightly lower accuracy than the cars, but the overall average accuracy also reaches 96.8%, which can be effectively classified. Then we tested the exact target retrieval, and set the retrieval targets to red cars and white pickup trucks respectively. The detection results are shown in Figure6.

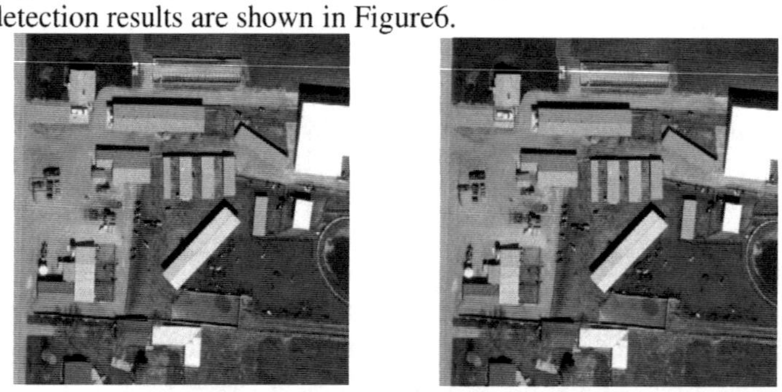

Figure 6.Accurate target detection results

5. Conclusion

This paper is based on the YOLO v3 algorithm to optimize the design of its feature extraction network and YOLO layer, which improves the performance of its backbone network for small target detection and the sensitivity of the YOLO layer to target locations. At the same time, based on the optimized YOLO-E network model, a small target vehicle classification network was migrated and trained. Later, the detected vehicle targets can be further detected according to the model and vehicle color, and reduce the target detected by YOLO-E network to the specified interval. Compared with ordinary vehicle detection, this experiment is aimed at vehicle detection of satellite images, which can detect multiple targets simultaneously in a large range. It has some practical value in real life. The optimized neural network has good robustness for the detection of such small targets. However, due to the richer structure of the improved residual module, the overall computing speed of the network has not been greatly improved. This is also the direction to continue to optimize the network. Compressing the network and improving the speed of network inference will enable it to be carried in the aerial cameras running on embedded devices, which will make the network more practical.

References

[1] Redmon J , Farhadi A . YOLOv3: An Incremental Improvement[J]. 2018.
[2] He K , Zhang X , Ren S , et al. Deep Residual Learning for Image Recognition[J]. 2015.
[3] Zhu,Yousong,zhao,Chaoyang,Wang,Jinqiao,et al.CoupleNet:Coupling Global Structure with Local Parts for Object Detection[J].
[4] Shafiee,Mohammad Javad,Chywl,Brendan,Li,Francis,et al.Fast YOLO:A Fast You Only Look Once System for Real-time Embedded Object Detection In Video[J].
[5] Rezatofighi H , Tsoi N , Gwak J Y , et al. Generalized Intersection over Union: A Metric and A Loss for Bounding Box Regression[J]. 2019.

Classification of Diabetic Retinopathy Using Adversarial Training

Dawen Wu[1*], Shishi Liu[2] and Jian Ban[3]

[1]School of Software and Microelectronics, Peking University, City, Province, 100871, China

[2]Central University of Finance and Economics, Beijing, 100081, China

[3]Chinese Academy of Sciences, Beijing, 100049, China

[*]Corresponding author's e-mail: wudawen@pku.edu.cn

Abstract. In recent years, the classification of medical images has become more and more important. With the help of the deep learning approach, the classification accuracy of diabetic retinopathy has been greatly improved, and it has brought great benefits to the residents living in the suburbs. However, the researchers found that these neural network models became extremely vulnerable when confronted with an adversarial example. Specifically, some minor changes to the samples can fail the model immediately. In this paper, we use adversarial training methods instead of traditional training methods to improve the model robustness against adversarial example. Experiments showed that using this method in APTOS data set the adversarial accuracy increased from 43% to 83%.

1. Introduction

Diabetic Retinopathy is a progressive condition with microvascular alterations that lead to retinal ischemia, retinal permeability, retinal neovascularization and macular edema. If a patient is left untreated for a long time, they can develop severe visual problems. Diabetic retinopathy is one of the main causes of blindness, especially in the working-age population. With proper management, visual loss from diabetic retinopathy can be reduced by up to 90 percent, so it is extremely important to detect the severity of diabetic retinopathy before the disease gets serious. However, this kind of detection requires a large number of doctor resources, and large-scale detection can lead to inaccuracy due to various factors. From the perspective of computer vision, detecting the severity of diabetic retinas becomes a simple image classification problem.

As machine learning methods have been greatly improved in recent years, many research has been put into how to improve the generalization ability of models, i.e. its performance on unseen test scenarios. However, most studies overlook the reliability of the model. As more and more machine learning models are applied to the real world, the reliability of the models has become critical. In this paper, we focus on improving the reliability or robustness of the model instead of accuracy.

We now give an example of the importance of model robustness. One pair of example images is shown in the figure 1. The picture on the left is the original picture, which the model predicts severe illness. The right one is just similar to the original picture, but with some pixels little changed. However, the model wrongly predicts the right one to no illness at all, and we call this sample an adversarial example. It can cause a severe consequence and how can we trust a model like this. So, in this paper, we are trying to improve the model robustness against the adversarial example, and in the above example, we are trying to make the model predict the right one correctly.

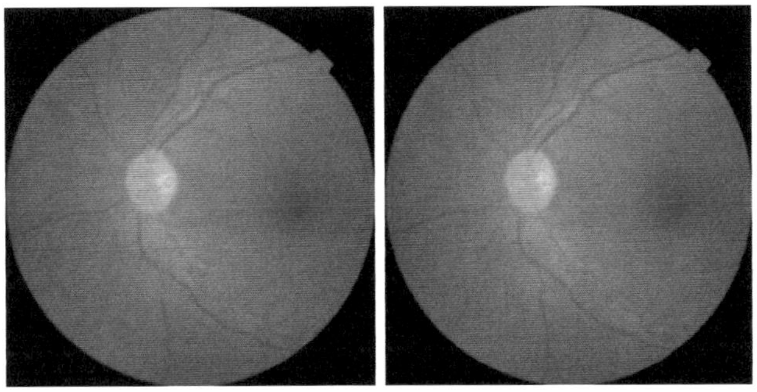

Figure 1: An adversarial example in Diabetic Retinopathy

Our contribution is two-fold: Firstly, we demonstrate the Diabetic retina classification tasks that adopted state-of-the-art deep learning models are not immune to adversarial examples. Secondly, we utilize the adversarial training method to increase model robustness, and the effectiveness of the method is verified by experiments.

2. Releated Work

As neural networks have been shown to be expressive both theoretically[5,4,1] and empirically[13], however, Szegedy et al.[12] found that very small changes to the input image can fool a high accuracy neural network, and name these sample as adversarial examples.

Thereafter, many approaches have been proposed in order to increase model robustness against adversarial example, namely defense method. Along with these defense methods, there are also many studies on how to craft more complicated adversarial example to fool the model, namely attack method.

2.1. Defense method

Madry[8] proposed an iterative defense strategy namely PGD, to improve the robustness of neural network to adversarial examples.

Similar to the most machine learning task, Kannan[6] add a regularization term to the adversarial training loss function, in order to shorten the difference between the original sample and the adversarial sample predicted values.

Based on the calibrated loss analysis given by Jordan[2], Zhang[14] introduces the TRADES method. This method is similar to the ALP in form, but it provides a solid theoretical analysis.

For alleviating the high computation cost of adversarial training, Shafahi et al.[11] present a "free" version of adversarial training with cost nearly equal to natural training. Free training can be further combined with other defenses to produce robust models without a slowdown.

2.2. Attack method

Many algorithms for crafting adversarial example has been proposed with the objective of generating a smallest modified sample x that can be misclassified by the classifier.

Ian has introduced a fast method for generating adversarial examples FGSM [7]. The algorithm uses only one step gradient descent to generate the adversarial example, and it shows the high efficiency of this algorithm through the experiment on mnist data set.

Mohsen et al.[9], proposed an algorithm, DeepFool, to compute adversarial examples. It is based on an iterative linearization of the classifier to generate minimal perturbations that are sufficient to change classification labels.

Papernot[10] introduced a new class of algorithms to craft adversarial samples based on computing forward derivatives. This technique allows an adversary with knowledge of the network architecture to construct adversarial saliency maps that identify features of the input that most significantly impact output classification.

In addition to a large number of white-box attacks mentioned above, Guo et al.[3] proposed SimBA, a simple black-box adversarial attack that takes small steps iteratively guided by continuous-valued model output.

3. Methodology

3.1. Adversarial example generating

Among many adversarial examples generating method mentioned above, the projected gradient descent(PGD) method is the most popular one.

This method generates adversarial samples iteratively using a small step size.

We now give a brief description to the elements in Algorithm 1.

• Input x: A sample from the data set, also called the original sample.

• Epsilon : The infinite norm bound, is the maximum allowing distance between adversarial example and the original one.

• Step size α: The adversarial example generating step size, similar to the usual gradient descent based training method.

• Output x_0: The adversarial example of the original sample.

The method of generating adversarial examples is similar to the traditional neural network training procedure. The essential difference is that we calculate and utilize the gradient of the data sample x instead of the model parameter θ.

Algorithm 1: PGD Adversarial Example Generating

input : Traing sample: x
output: Adversarial sample: x'
$x_0 = x$
for $i = 0, 1, ..., n-1$ **do**
  $x_{i+1} = Clip_{x,\epsilon}\{x_i + \alpha_1 * sign(\nabla_x \ell(x_i, y_{true}; \theta))\}$
end
$x' = x_n$

3.2. Adversarial training

with the arm of adversarial example generating method, we can present the complete adversarial training method that we will use in the next section.

As usual, the algorithm starts with extracting a batch from the data set. Then, based on the current network model, generate an adversarial example batch. We call this the original batch in compare to the adversarial batch. One can also generate adversarial examples without knowledge of the neural network model, which is called the black-box attack. Finally, use the adversarial example batch instead of the original batch.

From the perspective of optimization, adversarial training can be viewed as optimizing the following:

$$\min_\theta \mathbb{E}_{(X,Y)\sim\mathcal{D}}\left[\max_{x' \in N_\epsilon(X)} \ell(x', Y)\right]$$

So this optimization problem involves an inner maximization problem and an outer minimization problem. The inner maximization problem aims to find a modified version of a given data point x that achieves a high loss. This is precisely the problem of attacking a given neural network. On the other hand, the goal of the outer minimization problem is to find model parameters so that the adversarial loss given by the adversarial attack problem is minimized. This is precisely the problem of training a robust classifier using adversarial training techniques.

Algorithm 2: Adversarial training

```
input  : Data set X
output: A trained neural network
for each epoch do
    for each batch x ⊂ X do
        x_0 = x
        for i = 0, 1, ..., n − 1 do
            x_{i+1} = Clip_{x,ε} {x_i + α_1 * sign(∇_x ℓ(x_i, y_true; θ))}
        end
        x' = x_n
        θ ← α_2 * ∇_θ ℓ(x', y_true; θ)
    end
end
```

4. Experiment

4.1. Dataset

We now give a brief introduction to the data set used in this paper. An Indian organization has provided a high quality data set on diabetic retinopathy. It contains 3662 training samples and The sample label contains five class ranging from low risk to high risk. We split it into a training set and a validation set. Training sets are used for training models, and the validation set are used for testing.

4.2. Result

The experiment is conducted using a single GPU on a node equipped with an NVIDIA GeForce GTX 2070 and i7-9700 processors, and the whole experiments are implemented in PyTorch.

We use a ResNet-32 as our deep neural network model, and choose cross-entropy loss as the loss function.

For optimization of the empirical training loss, we run standard minibatch SGD with a momentum term with parameter 0.9 and weight decay parameter 0.0002. We use an initial learning rate of 0.05 which is divided by 10 after half and three and three-quarters of the training steps.

We now defined the metric to show the effectiveness of our method.

Adversarial accuracy :

$$ACC_{adv} = \frac{1}{m}\sum_{i=1}^{m} 1(\arg\max model(X_{adv}^{i}) == Y^{i})$$

In Table 1, we report the result of the adversarial accuracy value after 100 epochs training. The results clearly show that after using adversarial training, the adversarial accuracy remarkably increases compared to the traditional training.

Figure 2 has shown more information about the adversarial accuracy with respect to time.

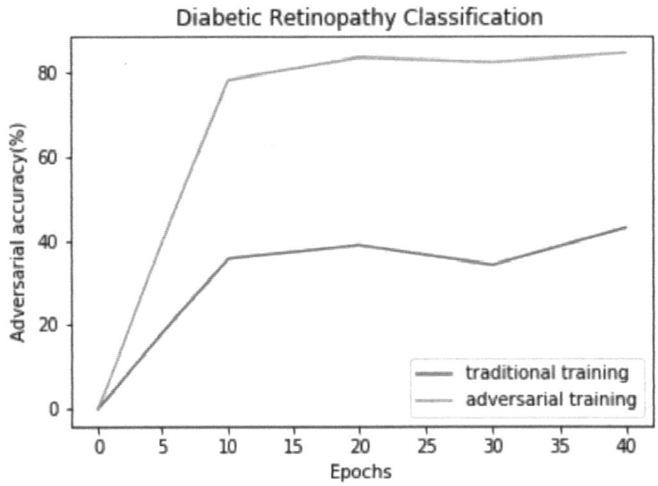

Figure 2: Comparation between traditional training and adversarial training

5. Conclusion

In this paper, we first demonstrate the lack and importance of robust research in the classification of diabetic retinopathy. Among the various studies on adversarial training, we chose a method suitable for our case. We give a full explanation of the method we used in the experiment.

And finally, in experiments, we have verified that the robustness of the model has been significantly improved through adversarial training.

References

[1] Andrew R Barron. "Universal approximation bounds for superpositions of a sigmoidal function". In: IEEE Transactions on Information theory 39.3 (1993), pp. 930–945.

[2] Peter L Bartlett, Michael I Jordan, and Jon D McAuliffffe. "Convexity, classification, and risk bounds". In: Journal of the American Statistical Association 101.473 (2006), pp. 138–156.

[3] Chuan Guo et al. "Simple Black-box Adversarial Attacks". In: CoRR abs/1905.07121 (2019). arXiv: 1905.07121. url: http://arxiv.org/abs/1905.07121.

[4] Boris Hanin. "Universal function approximation by deep neural nets with bounded width and relu activations". In: arXiv preprint arXiv:1708.02691(2017).

[5] Kurt Hornik, Maxwell Stinchcombe, and Halbert White. "Multilayer feed forward networks are universal approximators". In: Neural networks 2.5(1989), pp. 359–366.

[6] Harini Kannan, Alexey Kurakin, and Ian Goodfellow. "Adversarial logit pairing". In: arXiv preprint arXiv:1803.06373 (2018).

[7] Alexey Kurakin, Ian Goodfellow, and Samy Bengio. "Adversarial machine learning at scale". In: arXiv preprint arXiv:1611.01236 (2016).

[8] Aleksander Madry et al. "Towards deep learning models resistant to adversarial attacks". In: arXiv preprint arXiv:1706.06083 (2017).

[9] Seyed-Mohsen Moosavi-Dezfooli, Alhussein Fawzi, and Pascal Frossard. "Deepfool: a simple and accurate method to fool deep neural networks". In: Proceedings of the IEEE conference on computer vision and pattern recognition. 2016, pp. 2574–2582.

[10] Nicolas Papernot et al. "The limitations of deep learning in adversarial settings". In: 2016 IEEE European Symposium on Security and Privacy (EuroS&P). IEEE. 2016, pp. 372–387.

[11] Ali Shafahi et al. "Adversarial Training for Free!" In: arXiv preprint arXiv:1904.12843 (2019).

[12] Christian Szegedy et al. "Intriguing properties of neural networks". In: arXiv preprint arXiv:1312.6199 (2013).
[13] Chiyuan Zhang et al. "Understanding deep learning requires rethinking generalization". In: arXiv preprint arXiv:1611.03530 (2016).
[14] Hongyang Zhang et al. "Theoretically principled trade-offff between robustness and accuracy". In: arXiv preprint arXiv:1901.08573 (2019).

IOP Conf. Series: Materials Science and Engineering **806** (2020) 012051 doi:10.1088/1757-899X/806/1/012051

Implementation of Distribution Network Integrated Blackout Management Platform Based on Business Central Platform

Zhang Jiang-feng[1,2], Zhao Fu-quan[3], Su Ye[1,2], Zeng Liang-liang[3], Wang Zi-xiang[1,2]

[1] State Gird Zhejiang Electric Power Research Institute, Hangzhou, Zhejiang, 310014, China

[2]Hangzhou Yineng Electric Power Technology Co.,Ltd,Hangzhou, Zhejiang, 310014, China

[3]NARI Group Corporation/State Grid Electric Power Research Institute, NanJing, Jiangsu, 211106, China

*Corresponding author's e-mail: zhangjiangfeng0725@163.com

Abstract. With the rapid development of information technology in power grid enterprises, there are many information systems in relevant business departments. It is difficult to avoid the dilemma of duplication of business functions and waste of data resources because of the "chimney" function development model oriented to business lines. Aiming at the decentralized management status of data related to blackout business such as smart grid dispatching technology support system, production management system, distribution automation system and marketing system, based on the system architecture of business center, this paper takes integrated blackout intelligent optimization as the construction goal, and optimizes blackout inspection from the standardisation of data. Repair plan, improve the coordination efficiency of relevant business departments, avoid repeated blackouts, improve the reliability of power supply, and provide decision support for power grid enterprises in blackout management.

1. Introduction

With the continuous improvement of social industrialization, the demand of power supply reliability is also increasing. In order to reduce the impact of power outage caused by users, scientific and effective power outage management is a necessary technical ways. In order to ensure the quality of power supply and the security of power grid, considering the operation mode of power grid, the allocation of operation, maintenance resources and other related factors. It is inevitable to involve the data sharing and business collaboration of multiple departments for power grid enterprises, such as dispatching, equipment, marketing and so on. Literature [1] introduces the research and judgment scheme of distribution network fault based on dispatching management, literature [2] introduces the distribution network production emergency repair technology based on operation and distribution connection, literature [3,4] introduces the on-site power failure diagnosis technology based on power consumption information collection system, literature [5] introduces the equipment fault location technology based on intelligent algorithm, and literature [6,7] introduces the distribution network fault location technology based on big data, It can be seen that at present, the power outage business management of power grid enterprises mostly relies on the internal single management, which mainly has the following problems:

(1) repeated construction leads to waste of resources: independent construction of similar systems by multiple departments, chimney construction mode is easy to cause waste of construction capital;

(2) the separation of services leads to low management efficiency: the artificial separation of the analysis chain of power outage services, especially the centralized management of master distribution network diagram and model data, is easy to lead to low efficiency of power outage services management;

(3) data duplication leads to untrustworthy data: power outage related business data exist in the business system independently constructed by each department. Due to the inconsistency of data collection channels and basic data, the indicators of power outage statistical analysis are contradictory;

(4) data sunk: resulting in the data value can not be mined: in the process of information system construction, different business information is digitized, but with the advancement of time, most of the data only become 0 and 1, the value behind it can not be deeply mined, and it is difficult to realize the data commercialization.

Referring to the construction experience of Internet enterprises, taking Alibaba and didi as examples, table[1] shows the optimization scheme of realizing common business through business Cnteral Pataform.

Table 1. Internet application in Business Cnteral Pataform.

Business line	Present situation	Solution
Alibaba: Taobao, tmall, Alibaba, Ali's mother	Each business has its own system and data source, and has done a lot of services on its own system. These systems are not connected	The data center platform completes the storage, calculation and productization process of massive data. The business in the middle platform builds a "big middle platform, small front-end" through "service reuse" to meet the market demand of rapid business distribution
Didi car: Express, taxi, special car, hitchhiker, Valet	The essence of all businesses is travel, which has synergistic effect. However, in the case of independent development, there is no synergy between businesses	In the process of building business in Taiwan, we should gradually build up the synergy through service, asynchronization, configuration, plug-in and data.

In this paper, the integrated power failure management platform of distribution network is set up from the equipment management model, the power grid control model and the power grid marketing model. The unified panoramic model of the power grid is established. Through the key data of the EMS system, the distribution automation system, the power information collection system and so on, based on the topology of the electrical network equipment and the real-time equipment status information, the power failure of distribution network is comprehensively analyzed. At present, the power grid equipment maintenance plan, distribution network outage management and other businesses have changed from passive acceptance to active perception, gradually improving the closed-loop management of outage business, providing power grid enterprises with outage management decision support, and improving the refined management level of power grid enterprises..

2. Business platform overview

The core requirement of business platform is to store the data associated with power grid business in the shared platform, and provide data calculation, business calculation and other data services, so as to solve the problems of data isolation and data inconsistency caused by "chimney" architecture. Through the common design of data storage, data tagging service and business calculation of public model, the basic data service ability of power grid enterprises is improved.

The basic implementation of the business platform depends on the data middle platform. The technical architecture of the data middle platform is shown in Figure 1.

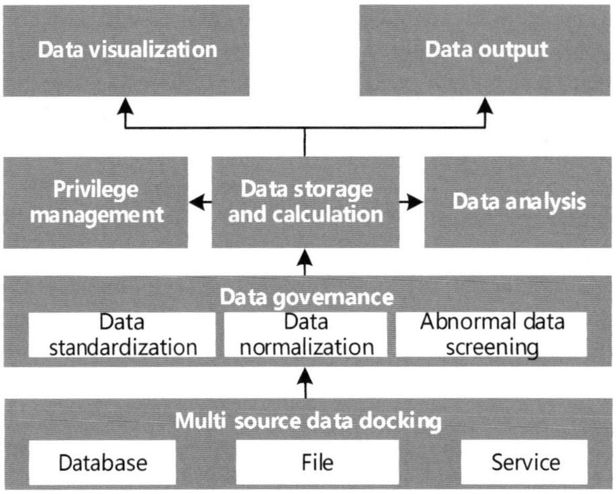

Figure 1. Technical architecture of data center

The process of data processing is described as follows:

① access target data from business services, file services, database services and other diversified data sources, including structured data sources and unstructured data sources;

② after obtaining different data sources, standardize and standardize the data, and screen the abnormal data according to the business rules;

③ after the completion of data governance, it is put into the database for management and operation;

The data need to be labeled;

⑤ in order to support the data security in different business scenarios, it is necessary to manage the permissions of different users and scenarios at different levels;

⑥ connect different business systems and drive business scenarios through data.

It can be seen from the above description that the data center realizes cross domain data integration through data tagging processing of primary equipment, secondary equipment, transformer load, switch status, etc., and data encapsulation and development through data services. Based on the data middle platform, the business middle platform realizes the decoupling of data and business in the way of business rule processing interface, and meets the personalized requirements of different business systems to quickly build new business scenarios. Figure 2 shows the business middle platform for comprehensive intelligent power outage analysis.

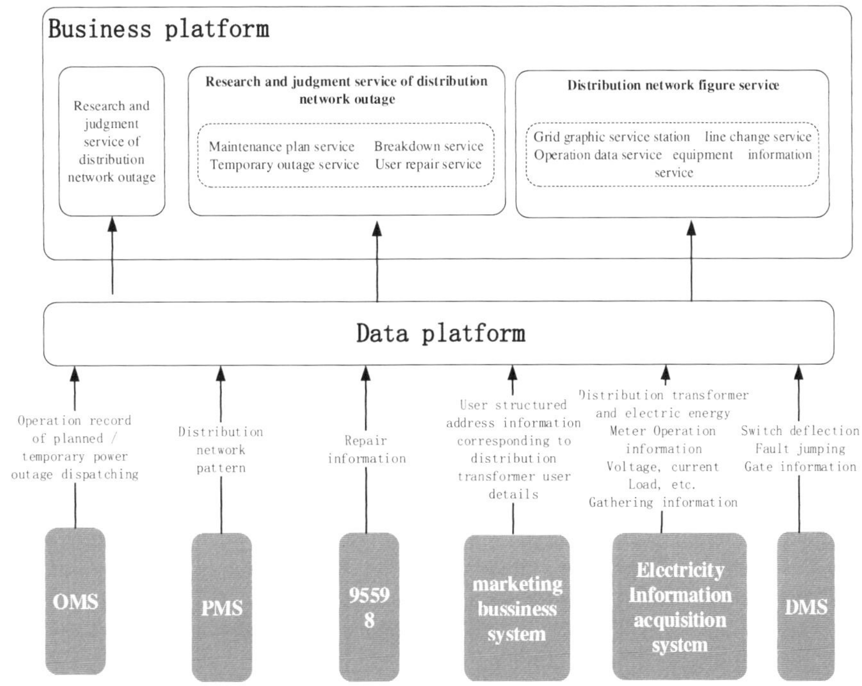

Figure 2. Business Center for integrated intelligent outage analysis

Taking the topological structure service of power grid as an example, through the results of operation, distribution and commissioning, the emergency repair resource information of PMS system is supplemented, and finally a node service provider of "station line transformer (emergency repair) - Household" is formed to realize the relationship between power grid equipment and topology. The construction of panoramic model of marketing users. Based on the panoramic model, combined with the real-time operation data of switch state, energy meter (metering point), distribution transformer, maintenance plan, switching operation and other grid operation management data, consolidate the data base of intelligent analysis and optimization of power outage.

3. Overall system architecture

Based on the dispatching automation system, distribution automation system, power consumption information collection system, production management system and other systems, the power outage business management of distribution network comprehensively analyzes the real-time information of power grid operation, equipment maintenance information and so on, in order to find out the final optimal power outage scheme.

Figure 3 shows the overall architecture of the integrated power outage management platform of the distribution network, in which the ODS layer of the data center platform realizes the unprocessed storage of data from other systems, the CDM layer is subdivided into the detailed wide surface layer (DWD layer) and the public summary data layer (DWS layer), and the ads layer stores the personalized statistical index data of data products; the data map is a unified query of the data in the whole data center platform Query and management map; data model management, mainly to solve the inconsistency between architecture design and data development; data quality around integrity, accuracy, consistency, timeliness monitoring and analysis of data quality problems, improve the quality of enterprise data [8].

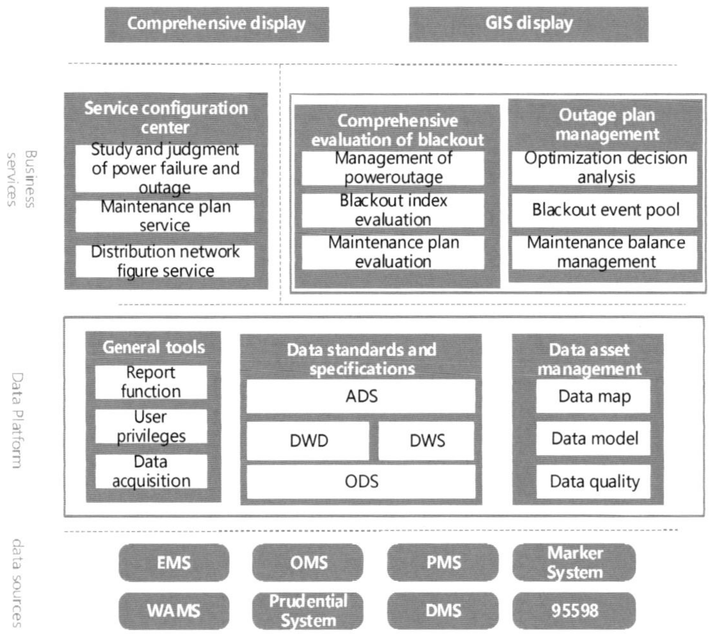

Fig.3 System functional architecture

It can be seen from Figure 3 that there are many data sources in the integrated power outage management platform of the power grid. The power outage business should be interfaced with energy management system (EMS), power grid operation management system (OMS), production management information system (PMS), distribution automation system (DAS), power consumption information collection system, Power Grid Geographic Information System (GIS) and other systems to realize data collection and status analysis of power outage business And the intelligent management of fault study and judgment, and later provide data support for equipment maintenance plan optimization through big data analysis technology.

Through the application of data center and business center, the system can meet the data source requirements of power outage business model, realize the decision optimization of power outage plan, support the business requirements of intelligent management of power outage business, and realize the visual management of power outage business through intelligent report and GIS platform.

4. Key business scenarios

4.1. Panoramic model construction

There are many data sources in the integrated outage management platform of power grid, such as the real-time operation control data of power grid from EMS, the operation management data of power grid from OMs, the operation data of power grid equipment from PMS, and the power consumption information data of power supply users from the power consumption information collection system. How to solve the problem of inconsistent description of the same equipment object in each related system is the key to the intelligent analysis of integrated outage Where. In this paper, we build a panoramic model based on the data in the middle of the data, based on the operation and distribution data, to realize the information from the main network equipment in EMS, OMS, PMS, marketing system and GIS system to the power supply user according to the voltage level. The realization mechanism of the panoramic model is shown in Figure 4. The unique identification of equipment based on panoramic model realizes the association with the real-time operation data model of power grid equipment.

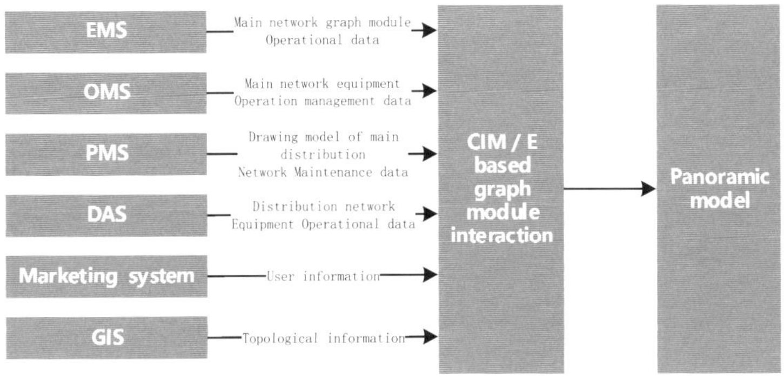

Fig.4 Panoramic model implementation mechanism

Based on the results of operation, distribution and dispatching data penetration, taking the topological relationship construction of "station line transformer (emergency repair) - Household" as an example [9], the realization mechanism of panoramic model is described.

① based on PMS system, acquire station line relationship and line transformer relationship;

Secondly, based on the marketing business system, the relationship between the station and transformer, between the station and user, between the metering point and the station and user is acquired;

③ based on PMS system, obtain emergency repair resource information and equipment account information;

④ based on the results of operation, distribution and dispatching data penetration, the core business data based on "station line transformer (emergency repair) - customer" is formed.

Based on CIM model, a power grid panoramic model for distribution network outage management is systematically constructed with the core data of the relationship between station and line, equipment account, drawing model and emergency repair resource information. On the basis of panoramic model, the system of power consumption information collection is connected to obtain the live state of distribution network equipment in real time, and the live state of distribution transformer and meter can be called to support fault diagnosis.

4.2. Optimization of outage maintenance plan

The outage maintenance plan management includes four stages: declaration, optimization, adjustment and release. In the declaration process, it is required to comprehensively consider distribution transformer, new energy and other factors. The lower level dispatching organization shall declare the maintenance plan according to the annual and monthly cycle. The maintenance system shall summarize and analyze the rationality of the maintenance plan, and the maintenance plan shall be balanced and optimized manually by personnel [10 ~ 12].

Distribution network maintenance plan optimization is a multi-objective and multi constraint optimization process with maintenance time range, maintenance outage range and maintenance resources as control variables. Maintenance time optimization includes operation time balance, maintenance duration balance and other processes. Maintenance outage range includes restoration outage balance, maintenance correlation balance, load transfer and other processes. Maintenance resource optimization includes maintenance Window balance, maintenance density balance and other processes. Based on the principle of comprehensive outage, a decision-making system of intelligent outage plan is established to change the passive outage balance into the active outage balance, push forward the gateway of equipment outage plan arrangement, and realize the automatic balance of outage plan in advance.

According to the priority of each outage plan and the requirements of repeated outage, realize the balance of repeated outage; automatically analyze the topological relationship of power grid through the visual window, realize the balance of Maintenance Association; carry out the balance of maintenance density according to the construction unit and operation team involved in the maintenance plan; carry out the balance of maintenance density according to the seasonal operation characteristics of power grid section, load characteristics change, power maintenance situation and important sensitive user time Requirements, etc., take the initiative to prepare maintenance window period; carry out maintenance period and operation time balance according to historical work time of the same type and typical maintenance standard working hours, and finally give optimization suggestions.

The principle of maintenance plan balance optimization is as follows:

① according to the principle of "overall planning of overhaul and technical transformation, rural power grid engineering, municipal engineering, industrial expansion engineering, equipment defects, overhaul and pre test", the system achieves the optimized coordination of main distribution network, primary and secondary equipment, power generation, transmission and transformation;

② give weight to different types of maintenance work (infrastructure construction, industrial expansion, government, rural power grid, etc.) and adjust the maintenance plan time according to the priority;

③ according to the special time periods such as power saving and peak load, no outage maintenance work shall be arranged, and the rationality of maintenance window period shall be judged;

④ judge whether the maintenance period is reasonable according to the rules that the maintenance time of the more difficult power-off equipment is not arranged in the high load period, and the maintenance time of the equipment that cannot be cut off at the same time must be staggered;

⑤ through the analysis of power outage scope, judge whether it meets the requirements of repeated power outage;

⑥ judge whether the maintenance density is balanced through the analysis of maintenance resource allocation.

Based on the analysis process of maintenance plan balance model, it can avoid unreasonable arrangement of maintenance plan, provide intelligent service for the planners, improve the rationality and enforceability of planning, and optimize the scope of maintenance outage and maintenance human resource cost.

4.3. Study and judgment of power failure in distribution network

The research and judgment of distribution network outage fault is based on the business service realization of multi system integration. It is an intelligent management and control of outage work order based on the combination of 95598 system, PMS system, distribution network automation system, dispatching operation management system, marketing system, etc.

Based on the panoramic model of power grid, the system realizes the intelligent management of the outage event pool, and analyzes the real-time signals of the power distribution network obtained by combining with the operation, distribution and commissioning through state. The analysis conclusion should include the outage type, outage failure point, outage scope, outage impact on users. On the basis of accurate low-voltage diagram, the low-voltage power failure management and control and fault analysis are based on the intelligent analysis of the collected equipment state operation data, so as to carry out the low-voltage power failure management and control and fault analysis; the medium-voltage fault analysis of the power-off time of distribution transformer is based on the battery shutdown, combined with the intelligent analysis of the supply circuit path, to achieve the research and judgment of the distribution transformer fault source, so as to form a complete distribution network Failure analysis scheme.

4.3.1. Blackout event pool.
Power failure event pool is the data base of power failure analysis of distribution network. The data source of power failure intelligent analysis mainly comes from EMS system, distribution automation system, power consumption information collection system, 95598 and other systems. Power failure of distribution network mainly includes planned power failure and fault power failure. In the outage intelligent analysis platform, the outage event pool is divided according to the outage location.

① main line fault: through the study and judgment of switch displacement and protection action of EMS system, feeder switch trip and main network fault cause feeder voltage loss event, which is synchronized to power failure event pool;

② branch line fault: comprehensively judge the tripping event of feeder automation switch through EMS system and distribution automation system, and synchronize it to power failure event pool;

③ distribution transformer fault: according to the power failure event of the power consumption information collection system, it is judged as distribution transformer fault according to the corresponding relationship between the power failure event and distribution transformer, and it is synchronized to the power failure event pool;

④ low voltage line fault: the real-time operation data of distribution transformer and electric energy meter can be provided by the power consumption information collection system to determine that the low-voltage line fault is synchronized to the power failure event pool;.

4.3.2. Study and judgment of power failure
The power supply path represents the topological connection between the power supply point and the load [1], which is one of the core businesses of integrated power outage management. Based on the service of power supply path, the system can identify the relevant equipment and areas affected by power failure, and can count the number of users and repeat power failure households and other key indicators. In order to realize the real-time perception and accurate positioning of faults and improve the efficiency of distribution network fault disposal, the service of power supply path is used to accurately locate the fault area and provide auxiliary decision-making of fault disposal.

According to the 95598 power supply user's repair information, the user's account number and address information are used as the source of information. Based on the power supply path service, the distribution transformer information and power supply points of the power supply user are analyzed, and then the intelligent matching of the equipment repair plan and power failure information of the power failure event pool is realized. The real-time operation of key equipment such as distribution transformer and power meter in the power consumption information collection system and distribution automation system is combined Call test function of information, analyze and locate suspected fault equipment and influence range [13].

The flow of study and judgment of distribution network outage is shown in Figure 5.

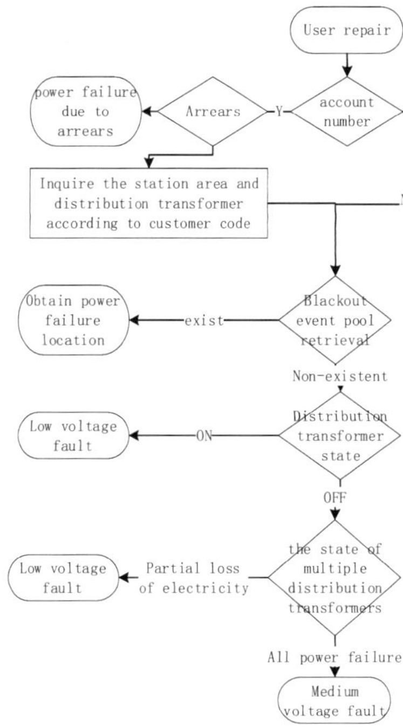

Fig.5 Research and judgment process of distribution network outage

According to the repair information of the power supply user, determine the user's ownership, analyze the matching relationship between the fault work order and the power failure event pool through the fault judgment of the distribution network, if there is a corresponding relationship between the fault work order and the power failure event pool, the upper level active repair service has the function of not re dispatching the order, if there is no corresponding relationship, the system has the function of adding power failure events in the power failure event pool according to the power failure location Function.

5. Appendices

Based on the power grid integrated outage management platform of business middle platform, the multi department data barrier is solved through data middle platform, and the optimal utilization of computer resources is realized. Then, the business functions of power grid topology analysis, power supply path analysis, outage fault analysis and other multi-system reuse are realized by using business middle platform, which meets the requirements of power outage business management of dispatch and control center, equipment department and marketing department Under the premise of management, the ability of business sharing and collaboration between systems is improved.

Through the business application of the power grid integrated outage management platform and the outage event pool mechanism, the possibility of repeated dispatch is minimized. The fault point range is basically determined by using the medium and low voltage fault research and judgment to improve the efficiency of fault emergency repair, which provides a strong support for the development of active emergency repair service.

Acknowledgments

Authors wishing to acknowledge assistance or encouragement from colleagues, special work by technical staff or financial support from organizations should do so in an unnumbered Acknowledgments section immediately following the last numbered section of the paper.

IOP Conf. Series: Materials Science and Engineering **806** (2020) 012051 doi:10.1088/1757-899X/806/1/012051

References

[1] Zhang Yuan-lai, Yi Wen-tao, Fan Qi-jun, etc. Distribution Network Fault Analysis Scheme Based on Dispatching Operation Management System[J]. Automation of Electric Power Systems, 2015, 39(1): 220-225

[2] Zhou Wei-feng, Zhuang Xiao-dan. The Distribution Network Production Repairing Command Platform Based on CIS and PMS[J]. Power System and Clean Energy, 2015, 31(4): 79-84

[3] Ye Aliao-sha, Liu Xuan, Zheng Guo-quan, etc. Optimization method of power outage site diagnosis in advanced metering infrastructure[J] . Electrical Measurement&Instrumentation, 2017, 54(18): 125-128

[4] Liu Liping, Cai Guangiin, Chen Xu, etc. Electricity information management system considering power utilization reliabilitye[J]. Electric Power Automation Equipment, 2018, 38(4): 204-209

[5] Feng Jun. Research on power equipment fault detection technology based on image analysis[J]. Modern Electronic Technique,2015,38(24):7-11

[6] Yao Ying, Xi Xiao-guang, Gao Shi-wei, etc. Fault Diagnosis of Distribution Network Based on Multiple Data Sources[J]. Proceedings of the CSU-EPSA, 2017, 29(2): 50-55

[7] Chai Qian-yi, Zheng Wen-bin, Pan Jie-kai, etc. Research on smart distribution network condition monitoring and fault processing method based on big data analysis[J]. Modern Electronic Technique, 2018, 41(4): 105-108

[8] Zhong Hua. The way of enterprise IT architecture transformation: Alibaba's strategic thinking and architecture practice in Data Cnteral Pataform[M]. Beijing: China Machine Press, 2017

[9] Hao Sipeng, Chu Chenbiao, Fang Quan, etc. Research on marketing - distribution - dispatching integration platform and its key technologies[J]. Electrical Measurement & Instrumentation, 2014, 51(24): 101-105

[10] LI Gang, LI Shan-zong, CHENG Chun-tian, etc.Intelligent visualization method and application of making maintenance scheduling for provincial power grid[J]. Power System Protection and Control, 2012, 40(13): 145-150

[11] Yu Qun, Fan Chang-jian, Cao Na. Assessment of Power Failure Risk in Interconnected Power Grid under Multi-scenario Considering New Energy Uncertainty[J]. Science Technology and Engineering, 2019, 19（13）: 98-103

[12] He Min, Xu Ming-ming, Xu Heng-bo, etc. Annual Maintenance Plan Optimization for Distribution Transformers Considering the Reliability Performance[J] . Science Technology and Engineering, 2017, 17(16): 213-219

[13] You Da-ning, Qu Han-bing, Huo Jian, etc. Research on fault analysis and identification strategy for distribution network[J]. Power System Protection and Control, 2018, 46(13): 84-91

An approach for monitoring sand mining based on sound feature

Qingmei Zhou[1] and Xiping He[1,2,3*]

[1] National Research Base of Intelligent Manufacturing Service, Chongqing Technology and Business University, Chongqing 400067, China

[2]School of Artificial Intelligence, Chongqing Technology and Business University, Chongqing 400067, China

[3]Chongqing Engineering Laboratory for Detection, Control and Integrated System, Chongqing Technology and Business University, Chongqing, 400067, China

[*]Corresponding author's e-mail: jsjhxp@ctbu.edu.cn

Abstract. In order to strengthen the management of sand excavation in river courses and prevent the occurrence of illegal sand mining activities, this paper proposes an approach for monitoring sand mining based on sound. Firstly, Mel Frequency Cepstral Coefficients (MFCCs) abstractor and Autoencoder are combined to extract features of every frame of a sound sequence, and then each frame of the sound sequence is classified by a specific classifier. Finally a voting strategy is used among the frames to determine the final category of the sound sequence. Experiments show that whether the classifier is SVM, KNN, or BP neural network, the result of combined features is better than the result of features extracted by the MFCC abstractor. Therefore, it is feasible to use the artificial intelligence method based on sound features extracted through MFCC abstractor and Autoencoder to monitor sand mining.

1. Introduction

River (lake) sand is a very good building material, and its market demand is quite huge, resulting in frequent illegal sand mining activities, so the effective control of sand mining activity has become an urgent problem to be solved by river management departments. Considering that sound contains a lot of useful information, and the process of collecting sound is simple and the cost of collecting is low, a method of monitoring sand mining based on sound is proposed in this paper.

In today's sound analysis, the most common acoustic features are MFCCs [1]. For example, MFCCs and two parallel feed forward neural networks are used to classify aircraft [2]; MFCCs and minimum mean-square error (MMSE) are used to evaluate speech features of the underlying noise-free speech signal for automatic speech recognition (ASR) [3]; MFCCs and extreme learning machine are used to classify excavation equipment [4]. These papers first use the MFCC abstractor to extract features of sound, and then perform further sound analysis. However, MFCCs are hand-crafted feature representation [5]. Studies have shown that there are always different types of noise mixed in the actual sound environment. Even if noise is limited within a narrow frequency band, the noise effect will still spread over all MFCCs [6]. So hand-crafted features may not always be able to adapt to complex application scenarios in practice [5].

Autoencoder can automatically capture the most important factor that represents the input data and find the main component that can represent the original information [7], and be robust to the noise, especially non-stationary noise [8]. Therefore, this paper considers using MFCC abstractor to extract the initial features, and then uses Autoencoder as the second sound features learning model in order to enhance the performance of feature learning.

The rest of this paper is organized as follows. This paper firstly introduces the technology roadmap of monitoring sand mining based on sound in Section 2, of which 2.1 introduces the data preprocessing before extracting the sound features, 2.2 introduces how to extract the features of sound, and 2.3 describes the process of classification. Section 3 shows the experimental setup and results. Section 4 concludes the paper.

2. Technology roadmap

The technology roadmap of monitoring sand mining based on sound features includes three steps: pretreatment, feature extraction, and classification.

2.1. Pretreatment

In order to extract audio features smoothly, the audio signal must be preprocessed firstly. The process is as follows:

(1) Data preparation: First of all, the acquired audio files are converted into audio signal data. Then, multi-channel is converted to mono. All audio signals are resampled to 16KHZ. Every audio file is split into many more audio files, each with the length of 2s.

(2) Pre-emphasis: Pass the sound signal through a first-order high-pass filter:

$$y[n] = x[n] - a * x[n-1] \tag{1}$$

Where a is between 0.9 and 1.0, and this paper takes 0.97. The purpose of pre-emphasis is to boost the energy in the high-frequency part so that the spectrum can be obtained with the same signal-to-noise ratio over the entire frequency band.

(3) Framing: In order to facilitate sound analysis, one sound sequence can be divided into consecutive small segments, that is, N sample points are grouped into a unit of observation called frame. In order to avoid excessive changes in adjacent frames, an overlapping area is left between adjacent frames. In this paper, each frame contains 128 sample points, with 64 sample points overlapping between adjacent frames, and the time length of each frame is 16ms.

2.2. Feature extraction

2.2.1. Using MFCC abstractor to initially extract features

Mel frequency cepstral coefficients, as the name suggests, feature extraction involves two key steps: transforming to Mel frequency, then cepstral analysis. The correspondence between the frequency(Hz) and the Mel scale is

$$mel(f) = 1127 \log(1 + \frac{f}{700}) \tag{2}$$

The flow chart for extracting MFCCs is shown in figure 1, and each step is explained as:

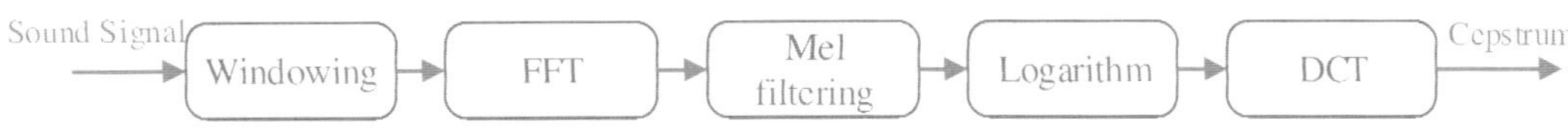

Figure 1. The flow chart for extracting feaures of MFCCs.

(1) Windowing: Sound changes over a wide range, and there is no fixed features operator for processing, so each frame is multiplied by the window function to increase the continuity of the left and right ends of the frame. Assume that the framed signal is $x(n)$, n=0,1...,N-1, N is the number of frames in a sample.

$$x'(n) = x(n)w(n) \tag{3}$$

Where $w(n)$ is the Hamming window:

$$w(n) = 0.54 - 0.46\cos\left[\frac{2\pi n}{N-1}\right], 0 \le n \le N-1 \tag{4}$$

(2) Fast Fourier Transform(FFT): The tool for extracting discrete frequency spectrum information from a discrete signal (sampled signal) is the FFT. After performing FFT on each frame of signal $x'(n)$, the discrete power spectrum $X(k)$ is obtained by modulo.

(3) Mel filtering: We divide the spectrum into several Mel filter banks according to human ear sensitivity. When calculating the MFCCs, the FFT spectrum can be converted into the Mel spectrum by passing through a set of Mel filter banks. The Mel filter bank is generally a set of Mel triangular filter banks. The frequency response of each triangular filter is

$$H_m(k) = \begin{cases} 0 & k < f(m-1) \text{ or } k > f(m+1) \\ \dfrac{k - f(m-1)}{f(m) - f(m-1)} & f(m-1) \le k \le f(m) \\ \dfrac{f(m+1) - k}{f(m+1) - f(m)} & f(m) < k \le f(m+1) \end{cases} \tag{5}$$

Where $f(m)$ is the center frequency and m=1, 2 , . . . , M, M is the number of Mel filters. The center frequency is converted from Mel scale.

(4) Logarithm and Discrete Cosine Transform(DCT): Calculating the logarithmic energy of the output of each filter bank and then doing a discrete cosine transform to obtain a set of cepstral coefficients:

$$m_i = \log[X(k)H_m(k)], i = 1,2,...,M \tag{6}$$

$$c_n = \sqrt{\frac{2}{M}} \sum_{i=1}^{M} m_i \cos\left(\frac{\pi n}{M}(i - 0.5)\right) \tag{7}$$

M is the number of filters, n = 1, 2 , . . . , L, L is the number of cepstral coefficients, and c_n is the MFCCs per frame. The 12 MFCC parameters of each frame are taken as one column of the matrix, and the matrix formed in order is the MFCC feature matrix.

2.2.2. *Further learning of features using Autoencoder*

Autoencoder is essentially a 3-layer BP neural network, except that its output is equal to the input. In other words, the following relationship exists between the output layer and the input layer of Autoencoder:

$$\hat{x} \approx x \tag{8}$$

x is the input vector and $\hat{x}$ is the output vector, called the reconstructed input vector.

Autoencoder is mainly composed of two parts: encoding process and decoding process. The specific process is as follows:

(1) Encoding process:

$$h = f\left(W^{(1)}x + b^{(1)}\right) \tag{9}$$

Where $f(\ )$ is the activation function. $W^{(1)}$ represents the coding weight matrix and $b^{(1)}$ represents a vector of bias parameter.

(2) Decoding process:

$$\hat{x} = f\left(W^{(2)}h + b^{(2)}\right) \tag{10}$$

Where $W^{(2)}$ represents the decoding weight matrix and $b^{(2)}$ represents a vector of bias parameter. The equation of the error at this time is as shown in equation (11).

$$loss = \frac{1}{m}\sum_{i=1}^{m}\left(\hat{x} - x\right)^2 \tag{11}$$

The error is back propagated. By adjusting the weights and biases, the error decreases along the gradient. Repeating learning and training until determining the weight and bias of each layer is corresponding to the minimum error, then stop training. At this point h removes the redundant input information and captures the most important factors that can represent the input data.

2.3 Classification and voting strategy

The features are input into a specific classifier. After classifying each frame of an audio sequence, then using a voting strategy to determine the final category. The idea of voting strategy is: The class ID is determined based on weak results voting, i.e. according to the principle that the minority is subordinate to the majority, which can be used to determine which category the sound sample belongs to. In this way, multiple weak classifiers are combined into one strong classifier, greatly improving the accuracy of classification [9].

3. Simulation experiment and result analysis

3.1. Sound data set

The sound data used in the experiment included sound of sand mining collected at different locations, sound of train running near the sand mining boat, sound of the engine, sound of flowing water, and sound of birds and wind. All the sounds are mixed with noise. These sounds were collected on the banks of the Yangtze River, where the number of no-sand-mining sounds is significantly less than that of sand mining.

3.2. Feature analysis

In order to directly observe the advantages of combined features, an audio file was randomly selected from sand mining sounds. The extracted MFCCs are drawn in figure 2(a), the features extracted by the combination of MFCC abstractor and Autoencoder are shown in figure 2(b). Where the coefficient length of each frame of MFCCs is 12; The number of neurons in the input and output layers of the Autoencoder is 12, the number of neurons in the hidden layer is 8, the activation function of the hidden layer is tanh and the activation function of the output layer is softmax.

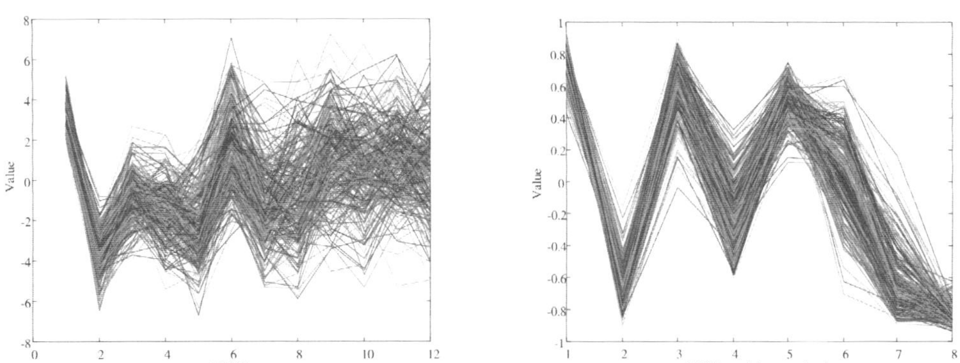

(a) (b)

Figure 2. Features. (a) MFCCs, (b) The combination of MFCCs and Autoencoder

As can be seen from figure 2(a), in the first 6 dimensional features, the variations of each frame of the 249 frames are roughly the same, but in the latter 6 dimensional features, although the variation of the majority of the frames is uniform, the variation of the minority of the frames is very confusing. This shows that there is indeed a small part of noise in the MFCCs. But it can be seen from figure 2(b) that the feature changes of each dimension of the 249 frames are consistent, and the first 6 dimensional changes are also more uniform than those of the first 6 dimensional changes in figure 2(a).

By comparing the two figures, we can see that features extracted using the MFCC abstractor and Autoencoder can indeed filter out most of the noise and better express the features of sand mining sound.

3.3. Perform experiments and results

After learning the features using MFCC abstractor and Autoencoder, the features are input into SVM, KNN and BP neural network algorithms respectively. At the same time, in order to compare the advantages of the combined features, the features extracted only using MFCC abstractor are input into the same classifiers. In this paper, the kernel of SVM is Radial Basis Function, k=1 in KNN, and BP neural network is a 3-layer network structure. The specific results are shown in table 1, table 2, and table 3.

Table 1. Results of multi-classification using SVM.

Sound type	MFCC+ Autoencoder + SVM	MFCC + SVM
Sound of sand mining	100%	95.93%
Sound of train running	97.14%	100%
Sound of the engine	100%	100%
Sound of flowing water	82.98%	46.81%
Sound of birds and wind	87.50%	79.17%

Table 2. Results of multi-classification using KNN.

Sound type	MFCC+ Autoencoder + KNN	MFCC + KNN
Sound of sand mining	100%	100%
Sound of train running	91.43%	91.43%
Sound of the engine	100%	73.91%
Sound of flowing water	93.62%	68.09%
Sound of birds and wind	91.67%	91.67%

Table 3. Results of multi-classification using BP neural network.

Sound type	MFCC+ Autoencoder + BPNN	MFCC + BPNN
Sound of sand mining	100%	100%
Sound of train running	88.57%	80%
Sound of the engine	100%	91.30%
Sound of flowing water	95.74%	85.11%
Sound of birds and wind	95.83%	91.67%

As seen in table 1, table 2 and table 3, the experimental results of combined features are mostly better than the experimental results of features extracted by the MFCC abstractor. The results show that the combined features can greatly improve the accuracy of sound recognition compared with the features extracted only by MFCC.

If only two categories are considered, that is, only the "Sound of sand mining" and "Not sound of sand mining" are judged. The specific results are shown in table 4, table 5, and table 6.

Table 4. Results of binary-classification using SVM.

Sound type	MFCC+ Autoencoder + SVM	MFCC + SVM
Sound of sand mining	100%	100%
Not Sound of sand mining	72.87%	59.69%

Table 5. Results of binary-classification using KNN.

Sound type	MFCC+ Autoencoder + KNN	MFCC + KNN
Sound of sand mining	100%	99.59%
Not Sound of sand mining	92.25%	65.12%

Table 6. Results of binary-classification using BP neural network.

Sound type	MFCC+ Autoencoder + BPNN	MFCC + BPNN
Sound of sand mining	100%	100%
Not Sound of sand mining	99.22%	79.84%

It can be seen from table 4, table 5, and table 6 that among the three algorithms, the results of experiments using combined features to identify each type of sound are best.

It can also be seen from the above six tables that the accuracy of identification of sand mining sounds using the combined features is all 100%, which indicates that the combined features are particularly suitable for the identification of sand mining sounds, and it can also perfectly realize the task of monitoring sand mining.

Figure 3 below shows the average experimental accuracy of multi-classification and binary-classification.

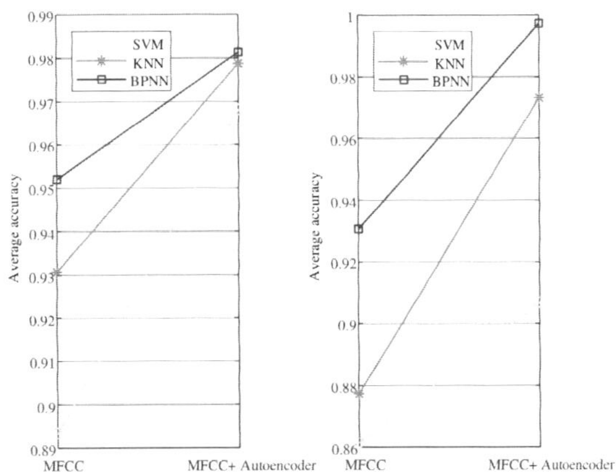

Figure 3. Average experimental accuracy.

It can be seen from figure 3 that the average experimental accuracy of multi-classification and binary-classification have been greatly improved by using the combined features, whether the classifier is SVM, KNN or BP neural network. It shows that this way of extracting features by combination of MFCC abstractor and Autoencoder can indeed improve the ability to express features and better accomplish the task of monitoring sand mining.

At the same time, it can be clearly seen from the results that whether it is binary-classification or multi-classification, regardless of which classifier is adopted, the average experimental accuracy is above 86%. That's because every method of classification uses the idea of voting strategy. It is illustrated that multiple weak classifiers are combined into a strong classifier, in which each weak classifier votes, and the minority is subordinate to the majority, returning the result of the final vote, so the accuracy of classification is indeed very high.

4. Conclusion

In this paper, an approach for monitoring sand mining based on sound is presented. Different from the method of extracting acoustic features by using MFCC abstractor, this paper uses a combination of MFCC abstractor and Autoencoder to extract acoustic features and a voting strategy is used to determine the category of sound, so as to judge whether sand mining activities have occurred. The above experiments proved that the features extracted by combination of MFCC abstractor and Autoencoder can well express the sand mining sounds and the various sounds associated with them, and also proved that the voting strategy based on combination of features can achieve good results in sand mining sound recognition. It is believed that the sand mining monitoring system with this artificial intelligence method can play a very good role in the sand mining management of the river. The next study of us is to apply this combination of features to more types of recognition of environmental sound, so that we can better use the information in natural environment sounds to serve humans.

Acknowledgements

This work is supported by the National Natural Science Foundation of China (No. 51605061), the Fundamental Science and Frontier Technology Research Foundation of Chongqing, Chongqing, China. (cstc2017jcyjA1592 and cstc2017jcyjA1475), and project of Chongqing Huize Technology Development Co., Ltd.(2018500015000110).

References

[1] Zahorian S, Guzewich P, Chen X and Zhang H 2017 The relative importance of static versus spectral change acoustic features for automatic speaker identification, *J. Acoust. Soc.Am.* 141 (5) 3915-3915.

[2] Mrquez-Molina M and Surez-Guerra S 2014 Aircraft take-off noises classification based on human auditory's matched features extraction, *Appl Acoust.* 84 83-90.

[3] Jensen J and Tan Z H 2015 Minimum Mean-Square Error Estimation of Mel-Frequency Cepstral Features–A Theoretically Consistent Approach, *IEEE Trans. Audio Speech Lang. Process.* 23 (1) 186-197.

[4] Cao J, Zhao T, Wang J, Wang R and Chen Y 2017 Excavation equipment classification based on improved MFCC features and ELM, *Neurocomputing* 261 231-241.

[5] Liu J H, Zheng W Q and Zou Y X 2015 A Robust Acoustic Feature Extraction Approach Based on Stacked Denoising Autoencoder, in: *IEEE International Conference on Multimedia Big Data*, pp 124-127.

[6] Nishimura Y, Shinozaki T, Iwano K and Furui S 2004 Noise-robust speech recognition using multi-band spectral features, *J. Acoust. Soc. Am.* 116 (4) 2480-2480.

[7] Feixiang Z, Yongxiang L, Kai H, Shuanghui Z, and Zhongshuai Z (2018). Radar hrrp target recognition based on stacked autoencoder and extreme learning machine. *Sensors* 18(2) 173-187

[8] Zhou C and Paffenroth R C 2017 Anomaly Detection with Robust Deep Autoencoders, in: *ACM SIGKDD International Conference on Knowledge Discovery and Data Mining,* pp 665-674. doi:10.1145/3097983.3098052.

[9] Tamvakis A, Tsirtsis G, Niros A D and Spatharis S 2018 Optimized Classification Predictions with a New Index Combining Machine Learning Algorithms. *Int. J. Artif. Intell. Tools.* 2

IOP Conf. Series: Materials Science and Engineering **806** (2020) 012053 doi:10.1088/1757-899X/806/1/012053

Research on the Construction and Application of Knowledge Graph in Military Domain

XING Meng[1], YANG Chao-hong[1], JIN Li-ya[1], BI Jian-quan[1],

[1]Department of Information and Communication, Army Academy of Armored Forces, Beijing, 100072, China

Address: no.21 yard, dujiakan, fengtai district, Beijing, tel: 15510676806.

[*] XING Meng's e-mail: xingmykx@163.com

Abstract: Construction of military domain knowledge graph is the foundation of military application intelligence development, current research hotspots and development situation of domain knowledge graph at home and abroad are introduced, application scenarios of military domain knowledge graph is given in the state of peacetime and wartime respectively, points out the problems needing attention and corresponding countermeasures in each life cycle.

1. Introduction

Knowledge graph originated from semantic network, proposed by Google to optimize search results. With the rapid development of knowledge graph technology, it has been applied in various vertical fields. Domain-specific Knowledge Graph (DKG) refers to the Knowledge Graph focusing on a specific vertical Domain, the entities and concepts are almost related to the Domain [1].

The purpose of DKG is to give domain knowledge to machines instead of professionals to do simple knowledge work, this technology will further increase machine productivity and free up mental labor resources. Therefore, the introduction of KG technology in military domain and DKG constructing lay the foundation for Smart Q&A, knowledge engineering, reasoning decision-making, personalized recommendation and other artificial intelligence applications in the military domain.This paper introduces the current research status of DKG, including domestic and foreign research hot-spots and existing problems. Gives the application scenario of military DKG, points out the possible problems, countermeasures and evaluation criteria in the life cycle of military DKG construction. At last, the opportunities and challenges of military DKG are discussed.

2. Research status of DKG

At present, the research hotspots at home and abroad mainly include domain brain construction, storage and calculation of large-scale DKG, interaction of KG, application of DKG and so on.

Since the knowledge of GKG comes from data of various structures, while the DKG needs to deal with almost unstructured data when it is constructed. Therefore, in the construction of DKG, it is urgent to enhance the representation ability of domain knowledge, identify domain entities, extract relationships, and discover implicit relationships.

In terms of the storage and calculation of large-scale DKG, in addition to entity relationships, there are also time, space, rules, labels and other important data. Graph database cannot meet the storage requirements of multiple data structures alone, the hybrid storage model based on graph database has become a research hot-spot [2].

Reasoning technology based on GKG can not be transplanted into DKG well. Some scholars began to study knowledge reasoning from the perspective of procedural knowledge. How to transform this human thinking into a machine path in the form of procedural knowledge is a current research focus [3].

The application of DKG mainly includes intelligent search, intelligent Q&A, decision making, explanation and personalized recommendation. At present, DKG has been applied in medical, e-commerce, finance, military industry, electric power, justice, education, public security, oil and other fields, such as the application of credit evaluation, risk control, anti-fraud in the financial field, as well as the application of intelligent consultation in the medical field [4].

3. Application scenarios of military DKG

Application scenarios can be divided into peacetime application scenarios and wartime application scenarios according to the combat status of the troops. In peacetime there are intelligent questions and answers of military knowledge, personalized recommendation, hidden knowledge reasoning, etc. In wartime, it mainly includes combat planning and action control, auxiliary judgment and decision recommendation.

3.1. Peacetime application scenarios

At present, intelligent question-and-answer application based on DKG, application scenarios become complex and diversified, intelligent recommendation and other complex tasks appear. It is applied to the army's ordinary training scenes, such as tank shooting training, including learning and using the battle room, aiming and shooting, and other more than 200 exercises. The number of steps for each exercise varies from three to more than ten steps, and each step has corresponding completion conditions and implementation standards. For example, ask about the content and method of the "check before shooting", the intelligent question answering system based on DKG, practice content should be able to return accurate, such as "fire control computer self-checking, security protection function, hydraulic locking function, working mode function, night-vision units function", etc., and intelligent recommendation for the user to possible use of equipment and auxiliary tool in the practice, in the process of operation can always recommend related Q&A answers and solutions, such as "what preparation work of artillery control system check needed", "suggested goggles check" before the function checks, etc.

The war-fighter's demand for combat data is diversified and personalized, and the result of information retrieval system cannot meet the war-fighter's personalized demand. A promising solution to this problem is personalized recommendation based on DKGs, which studies war-fighters' information needs and points of interest, draws user portraits, and creates interest description files. The traditional statistical learning method uses word sets to describe users' interests. Due to the synonymy and polysemy of words, it cannot accurately and comprehensively summarize and describe users' characteristics. The military domain oriented knowledge graph entity is adopted to describe the user's interest range and reflect the user's concern about the entity range. Moreover, it is verified by the entity-attribute relationship in the knowledge graph to eliminate ambiguous wrong words and combine synonyms repeated terms, so as to describe the user interest more accurately.

3.2. Wartime application scenarios

The wartime application scenario is mainly based on the combat task as the traction, on the basis of the command information system, support the combat planning and operation control, auxiliary judgment and decision recommendation, etc.

Due to the rapid development of communication, computer, sensor and satellite technology, the combat data acquisition technology has made a breakthrough, making it impossible for the war-fighter to distinguish and obtain valuable information in the mass data. In the field of operational planning and control action, knowledge graph can according to the war-fighter application requirements, identification, selection and processing of information needed, based on the simulation of the human cognitive mechanism, establish a commander fighting cognitive framework and action control theory model of thinking, through the knowledge of intelligent organization and visualization analysis technology, fight for the commander of the theory of multi-scale dynamic cognitive to offer help.

In the process of directing, DKG through learning military command combat ordinance, in-depth analysis of operational mechanism of the winning command, by studying the classic case of command decision-making mechanism, provide operational scheme based on similarity of the blues is recommended and decision support, thus forming forces combat experience and real-time situational data based auxiliary judgment and decision recommendation model, can realize the battlefield intelligence system, multi-dimensional and multi-level analysis, to support complex battlefield more comprehensive analysis and decision-making recommendations.

4. Issues and countermeasures of Military DKG life cycle

In the process of constructing the military DKG, there are mainly the following life cycles: knowledge representation, knowledge acquisition, knowledge management and knowledge application. Each link iterates in the construction process [5].

4.1. knowledge application

As an open knowledge management mechanism for man-machine collaboration, KG has specific value and cost, and can only solve limited problems. For different application problems, KG is not always the best solution [6]. The application scenarios of military DKG are much more complex than the GKG. The current domain knowledge system is mostly centered on the basic concepts and entities of domain data when it is built, which has natural defects in understanding user needs in complex application scenarios.

Practical experience shows that application scenarios should be carefully selected for knowledge graph, not only the maturity of technology should be evaluated, but also the adaptability of the technology to the corresponding application. In addition, the balance between cost and effect should be considered, so as to construct knowledge resources to meet the needs of the scenario. Knowledge graph is more likely to succeed in the application scenarios with the following characteristics, and the domain knowledge is relatively closed, less involved in common sense, and simple. The simplicity of the knowledge involved in the application scenario can be determined by setting three types of questions. The first category of questions: whether to use meta-knowledge, whether to use a single model, whether to use long-range reasoning, whether to involve multiple types of knowledge; The second kind of problems: whether knowledge is easy to change, whether it involves complex process description, whether it involves complex reasoning of branches; The third kind of questions: whether there is knowledge resource accumulation, such as domain ontology, thesaurus, domain dictionary, etc.

4.2. knowledge representation

The knowledge graph is not sufficient to express all semantics. For example, in application scenarios such as equipment maintenance and configuration in the military field, it is difficult to express if-then rules. For example, if A and B then C, the knowledge graph is difficult to express complex atomic expressions. Some military application scenarios add expressive elements such as spatial and temporal dimensions, far beyond the simple associative facts expressed by triples of knowledge graphs.

To enhance the representation ability of complex knowledge, knowledge graph can be combined with other representation methods. For example, in the current research on equipment fault diagnosis, production rules are introduced into knowledge graph. Military application scenarios with the addition of expression elements such as spatial and temporal dimensions can be realized by defining quaternions or quaternions, and some entities represented across media, such as those containing sound, pictures and video, can be realized by defining related attribute associations.

4.3. knowledge acquisition

Because the application scenarios in the military field are extremely serious, and the requirements for knowledge acquisition quality are very strict, the knowledge acquisition scheme should take into account the requirements and costs, so as to avoid the problems of missing knowledge base, mistakes and outdated information. In addition, compared with the GKG, the military DKG has the problem of small sample, and the knowledge extraction model construction technology based on small sample should be adopted.

Puts forward a few here on the premise of guarantee quality as far as possible to reduce the cost principle: first, clear the data source, according to the principle of refuge is Jane try to choose better quality and high degree of structured data sources, suggest the preferred command information system of structured data, and military comprehensive information network of semi-structured data, reduce knowledge acquisition cost. Secondly, it is necessary to avoid starting from scratch. In the process of military informatization construction, there are already many knowledge resources, such as ontology and thesaurus, which should be fully utilized. Finally, cross-domain migration, reusing knowledge of similar fields [7].

4.4. knowledge management

The main problem involved is the selection of management tools. Graph database and relational database are the most commonly used ones, and they have different applicable scale and operation complexity. Therefore, reasonable management schemes should be selected according to different scenarios.

From the map scale and complex operation, for millions, thousands of nodes, and the following the size of the knowledge graph can be used in a relational database or figure database, more than million node size chart is recommended for the database, involving complex operations such as calculation, complex global traversal of map figure database, it is necessary to adopt in order to realize efficient access [8]. In addition, the ease of use, popularity and maturity of the system should be taken into account.

References:

[1] He liang. Research on system and application of large-scale knowledge graphping service [D]. Hefei: university of science and technology of China,2018.
[2] hu fanghuai. Research on Chinese knowledge graph construction method based on multiple data sources [D]. Shanghai: east China university of science and technology,2015.
[3] guan saiping, jin xiaolong, jia yantao, et al. Progress in knowledge reasoning oriented to knowledge graphping [J]. Journal of software, 2008,29(10):2966-2994.
[4] Qi guilin, gao huan, wu tianxing. Progress in knowledge graphping research [J]. Information engineering,2017,3(1) : 4-25.
[5] Xu zenglin, sheng yongpan, he lirong, et al. Overview of knowledge graphping technology [J]. Journal of university of electronic science and technology of China,2016,45(4):589-606.
[6] xiao yanghua. Problems and countermeasures in the practice of DKGping [EB/OL].(2018-08-06).
[7] Li baozhen, su jing. DKG construction based on the content generated by experts [J]. Intelligence science, 2008,36(10):13-19.
[8] Li tao, wang qichen, li huakang. Development and construction of knowledge graph [J]. Journal of nanjing university of science and technology (natural science edition),2017,41(1):22-34

Appearance-based Gaze Estimation Using Multi-task Neural Network

Mihan Liu[1], Hui Li[2*], Huanhuan Dai

[1]College of Computer Science, Sichuan University, Chengdu, Sichuan, 610065, China

[2]College of Computer Science, Sichuan University, Chengdu, Sichuan, 610065, China

[3]College of Computer Science, Sichuan University, Chengdu, Sichuan, 610065, China

[*]Corresponding author's e-mail: lihuib@stu.scu.edu.cn

Abstract. In order to improve the gaze estimation accuracy through using the information in the eye image more adequately, an appearance-based gaze estimation neural network EyeNet has been proposed in this article. On the one hand, by decomposing the main task into two subtasks, this network could gain more information during the training stage. On the other hand, EyeNet could extract features in the image more efficiently by adding an auxiliary task which is detecting the iris center and eye ball center of the eye. Experiments show that by using these two tricks, the error decreased from 5.42 degree to 5.16 degree.

1. Introduction

Human's eyes are the main organs of getting information from the outside world. By estimating the gaze direction of eyes, we can develop many applications which can be applied in many scenarios, such as computer-human interaction, VR, psychological research and driver monitor system, so gaze estimation has great research significance.

Gaze estimation methods can be divided into two categories as intrusive and non-intrusive based on the device it used to get the eye images. Non-intrusive gaze estimation methods don't need subjects to wear device like glasses or helmet, thus non-intrusive gaze estimation is friendlier to subjects and has much more potential to be widely used in daily life.

In non-intrusive methods, gaze estimation can be categorized as feature-based, model-based and appearance-based[1]. With the development of deep learning[2] and massive gaze estimation datasets emerging, appearance-based gaze estimation made a huge progress. Researcher have made a lot study on appearance-based gaze estimation using neural networks. For example, Zhang[3] used LeNet-5 and VGG-16 as their base network and added the head pose as auxiliary information into the fully-connected layer to estimate the gaze direction. Zhang[4] argued that by giving different area with different weights, the information can be enhanced or depressed, so they put the whole face into the network and give different area with different weights and get better performance.

All current appearance-based methods regard gaze estimation as a simple regression task. They just send the image into the network and then get two numbers which stands for the pitch and yaw of the gaze direction. But getting the mapping from a raw eye picture to gaze direction is an ill problem due to the head pose variation, illumination changing and image resolution variation. We argue that simply regressing two numbers can't fully utilize the information in eye images, thus may not gain the best performance in inference stage.

In order to extract features in eye images more efficiently and get better accuracy, we proposed an appearance-based gaze estimation network EyeNet. It has two main characteristics:

1. EyeNet divide the main task gaze estimation into two sub-tasks. Each sub-task is designed to predict one angle, yaw or pitch, and in every task we convert the simple regression task into a combination of regression and classification task. In this way, the network can learn more knowledge during the training stage and by adding the classification task the network can learn the relationship among the similar gaze direction pictures to get better performance.

2. By adding an auxiliary task, eye landmark detection, EyeNet can extract features more efficiently. Research in multi-task learning[5] shows that by adding an auxiliary task which is related to the main task, better performance of the main task can be achieved. We add an auxiliary task, eye landmarks localization, to the network and make the network focus on the vital features during the training stage to achieve better performance.

2. Network design

Here we will introduce the tricks we used in the EyeNet, the full picture of EyeNet and the loss function design of EyeNet.

2.1. Main task dividing

Gradient vanishing is a deadly problem for neural networks at training stage. In order to solve this problem many tricks have been proposed. ResNet[6] is one of the most successful way for solving the gradient vanishing problem. ResNet uses short-cuts in their network. The multi-branch way can make more information back propagate to the network to avoid the gradient vanishing problem, and as a result improve the network's performance. In the similar way, we use two separate branches at the end of the network, each one for an angle, to predict gaze angles. The doubled output could bring doubled signal to the networks which could make the network learn more knowledge and get better performance.

Gaze estimation has been regarded as a regression problem for a long time. These days researchers like Ruiz [7] found that the gap between classification and regression might not be so big. They proposed a network to estimate human head pose in a combination way and get an awesome result. Gaze estimation and head pose estimation have some common features. For example, they both are finding a mapping from image to numbers. So, we apply this approach in our EyeNet. For every output branch of EyeNet, the loss of each predicted angle are the combination of regression loss and classification loss. The classification loss could let network to learn the relationships between similar angle pictures and the regression loss could make sure that the output angle of network is accurate.

2.2. Auxiliary task

In Sec. 2.1, we said that gaze estimation is similar to head pose estimation, but they also have many differences. First, eye images are usually smaller than face images. More importantly, gaze direction could be inferred through the features like iris center and eyeball center. For example, Park [8] proposed GazeMap which is a pictorial representation of 3D gaze. Their work proved that by adding some auxiliary information like iris margin and eyeball margin, neural network could concentrate on the important features during training stage and the performance of neural network could be improved in this way.

Rajeev [9] proposed a multi-task neural network, HyperFace. They use one neural network to predict face landmarks, gender, head pose from one image at the same time. Their work showed that every single sub-task's performance has been improved due to that every sub-task cooperates with each other and let the network learning more information in the image than single-task neural network.

Inspired by their excellent works, the second trick that EyeNet used is adding an auxiliary task, eye landmarks detection. Specifically speaking, the auxiliary task is detecting iris center and eyeball center, because of that the ray starts from the eyeball center and ends with iris center can roughly indicates the gaze direction in the 3D world.

There are two types of landmark detection methods, regressing coordinates directly or predicting a heatmap which stands for the landmarks. In our work, we choose the latter. Heatmaps in landmark detection are grayscale images, and the value of pixels which are nearer to the landmarks is greater than farther ones. Figure 1 shows the heatmaps we used in EyeNet.

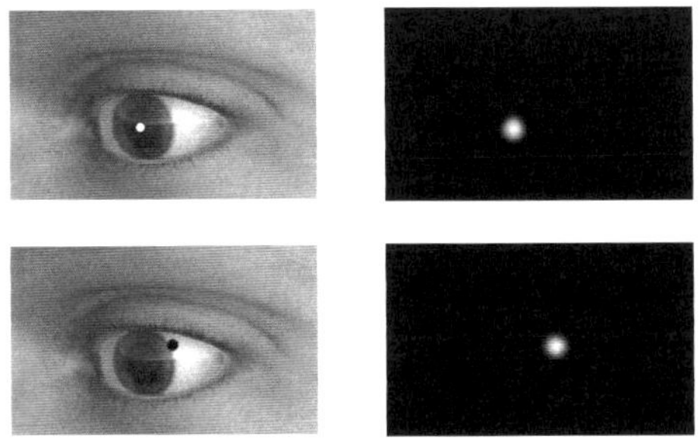

Figure 1. Eye landmarks and corresponding heatmaps.

2.3. Neural network architecture

Figure 2 shows the structure of EyeNet. The back bone of EyeNet is hourglass module that proposed by Newell [10]. They used it in human pose estimation because it can learn the relation among human's body parts. Here we use 3 hourglass modules because the size of eye crop is relatively smaller than human body image.

We can see from figure 2 that there are two ouputs path of EyeNet. The top path is the auxiliary task we mentioned at sec2.2. It outputs two heatmaps, one for iris center and one for eyeball center.

The bottom path is the main task, gaze estimation. It splits into two sub-paths that each path stands for one angle, pitch or yaw. We take the upper path, path of yaw, as an example. The output of hourglass modules will go through a full-connected layers and outputs a probability distribution, and then we caculate the expectation of the probability distribution to get prediction of yaw.

2.4. Loss Functions

As shown in figure 2, there are three outputs in our network and each one has its own loss. The total loss is the combination of all the losses.

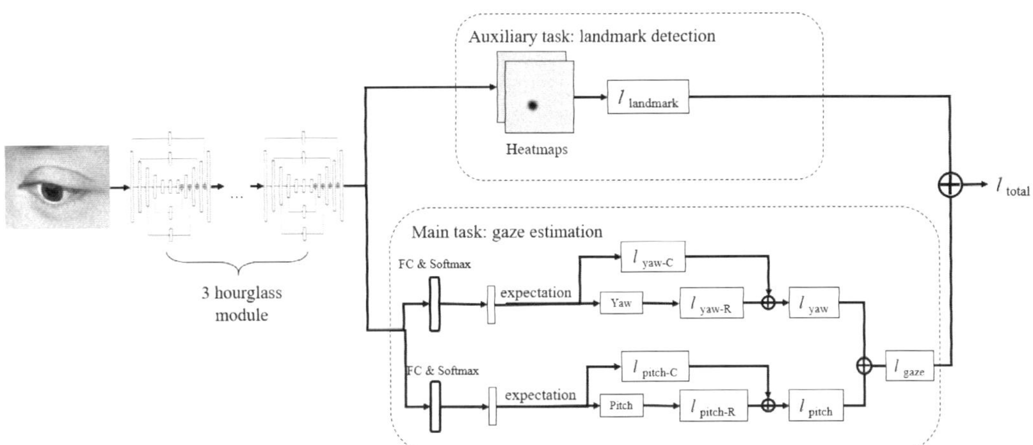

Figure 2. Network architecture

The total loss l_{total} contains two big parts, gaze estimation loss l_{gaze} and landmark detection loss $l_{landmarks}$, as in equation (1). Equation (2) gives the computation of $l_{landmarks}$. In equation (2), h and $\hat{h}$ stand for heatmap outputted from EyeNet and ground truth heatmap respectively, and subscript i and

j stands for horizontal and vertical coordinates respectively. Gaze estimation loss is composed of two parts, l_{pitch} and l_{yaw}. The computation of each angle is demonstrated as equation (4). In equation (4), y and $\hat{y}$ stand for predicted angle and ground truth angle respectively, C and R stand for classification loss function and regression loss function respectively. We used cross-entropy loss for classification and mean square loss for regression in this paper.

$$l_{total} = l_{gaze} + l_{landmarks} \tag{1}$$

$$l_{landmarks} = \left(\sum_{i,j}\left(h_{ij} - \hat{h}_{ij}\right)^2\right)^{1/2} \tag{2}$$

$$l_{gaze} = l_{pitch} + l_{yaw} \tag{3}$$

$$l_{angle} = R(y,\hat{y}) + C(y,\hat{y}) \tag{4}$$

3. Experiments and results

3.1. Training

We use MPIIGaze[3] as our training and testing datasets. The MPIIGaze dataset has 213659 images which are captured among 15 subjects during their daily life. The variation of illumination, resolution and head pose is huge, thus estimating gaze direction on this dataset is a big challenge.

We trained 15 models, one for one subject. We use cross-subject training procedure which is that using one person for testing and using the rest persons for training. Using cross-subject training could show the generalization ability of the proposed network.

All of these models are trained using Adam[11] as optimizer. The hyper-parameters are as follows: learning rate 0.0002, batch size 32, weight decay 0.0001 and epochs 10. The image send into the network are all of size 156 * 96.

3.2. Metrics

In this paper, error is determined as the Euclidean distance between the predicted angles (p, y) and ground truth angles ($\hat{p}$, $\hat{y}$) as given by equation (5). The error of single model is the average of all samples tested on it and the final error is the average error of all 15 models.

$$error = ((p-\hat{p})^2 + (y-\hat{y})^2)^{\frac{1}{2}} \tag{5}$$

3.3. Results

In order to prove that the tricks we applied are effective, we compared the performance among three conditions: hourglass-only, hourglass + main task dividing, hourglass + main task dividing + auxiliary task. From the result showed in table 1, we can see that the two tricks that we proposed can both improve the performance of the network.

Table 1. Effectiveness experiment

Model	Error (degree)
Hourglass only	5.42
Hourglass + main task dividing	5.27
Hourglass + main task dividing + Auxiliary task	5.16

We also compared the performance with some methods that others proposed at table 2.

Table 2. Comparison with other methods

Method	Error (degree)
Wood[12]	9.58
Nie[13]	7.1
Zhang[3]	5.5
Ours	5.16

4. Conclusion

In this paper, we improved the performance of gaze estimation using neural networks in a multi-task way. Our experiments show that both of the tricks we applied are effective. It proved that by adding more signals at the end of neural networks, the network will gain better predicting accuracy at testing stage. And our experiments also show that the positions of iris center and eyeball center are important information in gaze estimation. With good using of these features, great improvement can make in gaze estimation.

References

[1] Park, S., Zhang, X., Bulling, A., Hilliges, O. (2018). Learning to find eye region landmarks for remote gaze estimation in unconstrained settings. In: Proceedings of the 2018 ACM Symposium on Eye Tracking Research & Applications. Warsaw.

[2] LeCun, Y., Bengio, Y., Hinton, G. (2015). Deep learning. Nature, 521: 436-444.

[3] Zhang, X., Sugano, Y., Fritz, M., Bulling, A. (2017). Mpiigaze: Real-world dataset and deep appearance-based gaze estimation. IEEE Transactions on Pattern Analysis and Machine Intelligence, 41: 162-175.

[4] Zhang, X., Sugano, Y., Fritz, M., Bulling, A. (2017). It's written all over your face: Full-face appearance-based gaze estimation. In: Proceedings of the IEEE Conference on Computer Vision and Pattern Recognition Workshops. Hawaii. pp. 51-60.

[5] Zhang Y, Yang Q. (2017) An overview of multi-task learning. National Science Review, 5: 30-43.

[6] He, K., Zhang, X., Ren, S., Sun, J. (2016). Deep residual learning for image recognition. In: Proceedings of the IEEE conference on computer vision and pattern recognition. Las Vega. pp. 770-778.

[7] Ruiz, N., Chong, E., Rehg, J. M. (2018). Fine-grained head pose estimation without keypoints. In: Proceedings of the IEEE Conference on Computer Vision and Pattern Recognition Workshops. Salt Lake City. pp. 2074-2083.

[8] Park, S., Spurr, A., Hilliges, O. (2018). Deep pictorial gaze estimation. In Proceedings of the European Conference on Computer Vision (ECCV). Munich. pp. 721-738.

[9] Ranjan, R., Patel, V. M., Chellappa, R. (2017). Hyperface: A deep multi-task learning framework for face detection, landmark localization, pose estimation, and gender recognition. IEEE Transactions on Pattern Analysis and Machine Intelligence, 41, 121-135.

[10] Newell, A., Yang, K., Deng, J. (2016,). Stacked hourglass networks for human pose estimation. In: European conference on computer vision. Amsterdam. pp. 483-499.

[11] Kingma, D. P., Ba, J. (2014). Adam: A method for stochastic optimization. arXiv preprint arXiv:1412.6980.

[12] Wood, E., Baltrušaitis, T., Morency, L. P., Robinson, P., & Bulling, A. (2016). Learning an appearance-based gaze estimator from one million synthesised images. In: Proceedings of the Ninth Biennial ACM Symposium on Eye Tracking Research & Applications pp. 131-138.

[13] Nie, S., Zheng, M., Ji, Q. (2018). The deep regression bayesian network and its applications: Probabilistic deep learning for computer vision. IEEE Signal Processing Magazine, 35, 101-111.

IOP Conf. Series: Materials Science and Engineering **806** (2020) 012055 doi:10.1088/1757-899X/806/1/012055

The Research Status and Hotspots in the Domain of Smart Learning in China from 2012-2019

Wang HaiBin[1], Ming Zijing[1], Zhang Mengjie[1], He Jiaxiang[1], Cheng Lin[2*], Zhou Qian[3*]

[1] School of Educational Science, Huangshan University, Huangshan, 245041, China

[2] School of Tourism, Huangshan University, Huangshan, 245041, China

[3] Student Office, Huangshan University, Huangshan, 245041, China

*Corresponding author. Cheng Lin, chenglin@hsu.edu.cn or Zhou Qian, zq0214@hsu.edu.cn

Abstract. To explore the research status and hotspots in the domain of Smart learning in China, 118 related articles form the China National Knowledge Infrastructure (CKNI)were analyzed by the software of CiteSpace. The results showed that: (1) the area of smart learning in China began in 2012 and shows an upward trend year by year from 2012 to 2017, and the number of articles reached the historical peak in 2017 with 30 articles, but it showed a downward trend after 2017. Zhiting Zhu, Beijing Normal University and E-education Research topped the lists of contributing authors, organizations and publications respectively. (2) "smart learning", "strategy", "smart learning environment" and "smart education" were the top four largest clusters on the area of smart learning in China from 2012-2019.

1. Introduction

With the development of information technology, the current infiltration of educational technology into the education field faces many problems. How to effectively use educational technology has always been an important issue in the application of educational technology [1]. "Smart Learning" is a new learning paradigm that learners are self-directed, learner-centered, and have a complete learning experience [2-3]. In September 2010, the e-LearningWeek2010 conference held in South Korea used "Smart Learning" as its key issue. Relevant literature has initially shown that South Korea, Malaysia, Australia, Japan, and global enterprises IBM have paid more attention to "smart learning", especially in South Korea [4-5]. Only a few scholars in China payed attention to the "smart learning environment" and "smart classroom", but the research on "smart learning" was rare [6-7]. Until 2012, Chinese researchers first thought that the future classroom needed the smart learning environment [8], and then it showed an upward trend year by year since 2012. Chinese researchers from various disciplines started the study of smart learning, and obtained important achievements, including the Learning analytical techniques and MOOCs, the reform of teaching methods and method under technical support, the applications of mobile and ubiquitous learning, the flipping classroom and micro-course teaching, the gamification learning, and so on [9-10]. As mentioned above, in the past eight years, Chinese researchers have achieved fruitful achievements in the domain of smart education, however, a thorough quantitative analysis to objectively identify influential authors and organizations, as well as the emergent research trends and topics, has not been completed. In order to better promote the development of smart education in China, this paper presents a quantitative review of the research of smart education with the help of the CiteSpace.

2. Methodology

2.1. data sources
118 articles were yielded by the search conducted in the China National Knowledge Infrastructure (CKNI) under the heading of "smart learning". Among them, there were 96 journal articles and 22 Dissertation.

2.2. data processing
The RIS format of the 118 articles was carried out by the bibliometric analysis and visual analysis by the visualization software CiteSpace.5.5. R2 [11].

3. Results

3.1. Bibliometric analysis

3.1.1. Publishing trend. Fig.1 demonstrates the publishing trend of studies on smart learning in China. The earliest research on smart learning in China began in 2012, and then it showed an upward trend year by year from 2012 to 2017, and it appeared at a historical peak in 2017 with 30 articles. But after 2017, smart learning research showed a downward trend.

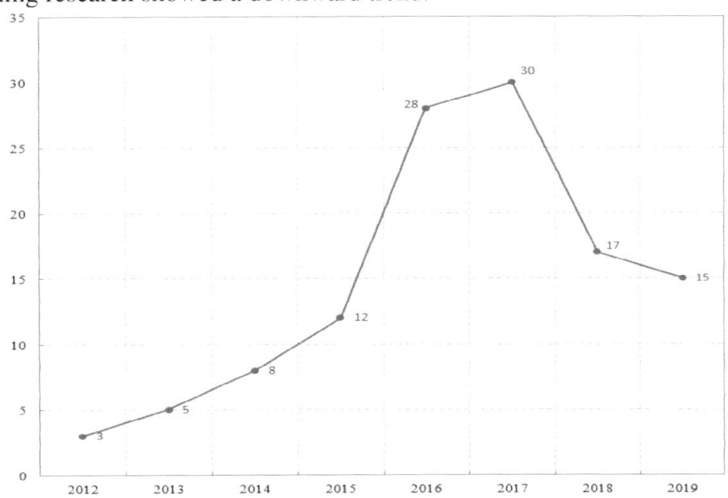

Fig. 1 Publishing trend.

3.1.2. Author statistics. As seen in Tab.1, all of the top 10 contributing authors had published more than two articles. Zhiting Zhu topped the list with eight articles, followed by Shusheng Shen and Lin Chen, each of them published more than five articles.

Tab.1 Top 10 contributing authors

Rank	Authors	Frequency
1	Zhiting Zhu	8
2	Shusheng Shen	5
3	Lin Chen	5
4	Ronghuai Huang	4
5	Youru Xie	3
6	Dongdai Zhou	3
7	Yongbin Hu	2
8	Linke Guo	2
9	Yang Yang	2
10	Zhen Zhao	2

3.1.3. Affiliation statistics. As seen in Tab.2, all of the top 10 organizations had contributed more than two articles on the studies of smart learning in China. Beijing Normal University was the best performer among the list with seventeen articles, followed by Jiangsu Normal University and East China Normal University, each of them contributed more than ten articles in the domain of smart learning in China.

Tab.2 Top 10 contributing organizations

Rank	organizations	Frequency
1	Beijing Normal University	17
2	East China Normal University	11
3	Jiangsu Normal University	11
4	Northeast Normal University	7
5	South China Normal University	6
6	Nanjing Normal University	6
7	Guangdong Polytechnic Normal University	3
8	Southwest University	2
9	Shanghai International Studies University	2
10	Tianjin University of Technology and	2

3.1.4. Source publication statistics. As seen in Tab.3, all of the top 10 source publications contributed more than three articles in the domain of smart learning in China. E-education Research topped the list with nineteen articles, followed by Modern Educational Technology and China Educational Technology, each of them contributed more than fifteen articles in the domain of smart learning in China.

Tab.3 Top 10 contributing source publication

Rank	Source publication	Frequency
1	e-Education Research	19
2	Modern Educational Technology	17
3	China Educational Technology	15
4	Journal of Distance Education	10
5	Open Education Research	8
6	Distance Education in China	4
7	Modern Distance Education Research	4
8	Journal of Teaching and Management	3
9	Journal of Educational Studies	3
10	Adult Education	3

3.2. Knowledge mapping results

3.2.1. High-frequency keywords. The top 10 high-frequency keywords in the domain of smart learning in China, as well as their frequencies on the studies of smart education were shown in Tab.4 and Fig.2. "smart learning" was the most frequent keywords, and "smart learning environment", "smart education", "smart learning space" and "learning analysis" were positioned in the 2nd, 3rd 4th and 5th places on the list. It is indicated that the smart learning environment, smart learning space, and learning analysis attracted researchers' attention.

Tab.4 Top 10 keywords

Rank	Keywords	Frequency
1	smart learning	49
2	smart learning environment	46
3	smart education	8
4	smart learning space	6
5	learning analysis	4
6	strategy	3
7	learning structure	2
8	smart classroom	2
9	flipping classroom	2
10	technology	2

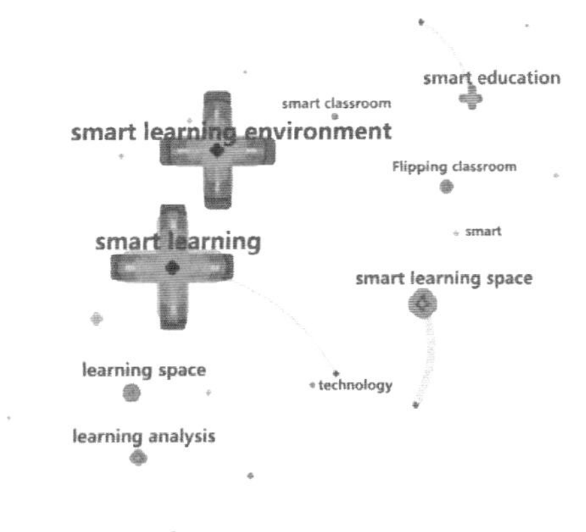

Fig. 2 Top 10 keywords

3.2.2. Popular research topics. Fig 3 outlines the top four largest clusters on the studies of smart learning in China, which were "smart learning", "strategy", "smart learning environment" and "smart education" and the Modularity Q value is 0.66, the Mean Silhouette value is 0.54. As shown, the largest cluster is "smart learning (#0)", which contains 7 articles and its silhouette value is 0.93. The high-frequency keywords of this cluster contain "smart learning", "learning analysis" and "learning space". The second-largest cluster is "strategy (#1)", which contains 7 articles and its silhouette value is 0.85. The high-frequency keywords of this cluster contain "smart learning space", "smart classroom", "strategy" and "technology". The third-largest cluster is "smart learning environment" (#2), which contains 5 articles and its silhouette value is 1.00. The high-frequency keywords of this cluster contain "smart learning environment", "creativity", "electronic bookbag" and "generative teaching". The fourth-largest cluster is "smart education" (#3), which contains 4 articles and its silhouette value is 1.00. The high-frequency keywords of this cluster contain "smart education", "flipping classroom", "ubiquitous learning" and "ecosystem".

IOP Conf. Series: Materials Science and Engineering **806** (2020) 012055 doi:10.1088/1757-899X/806/1/012055

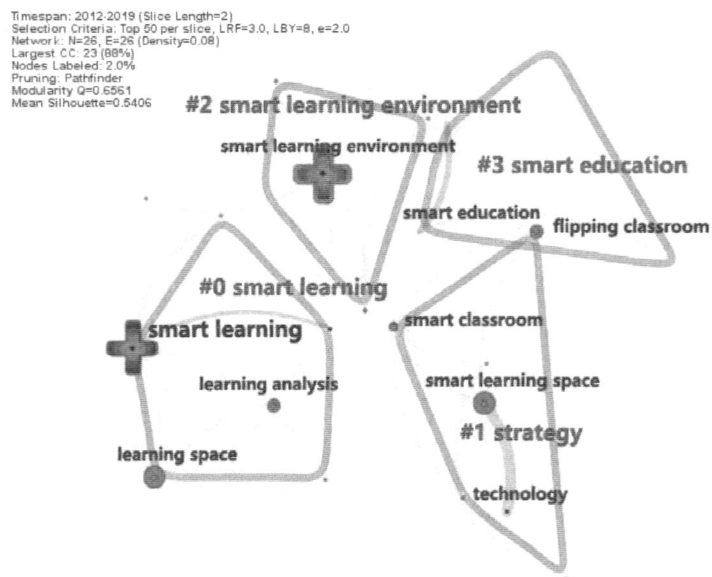

Fig. 3 Popular research topics

4. Conclusions and further research

In order to better promote the development of smart education in China, a quantitative review the research of smart education was conducted with the help of the CiteSpace. The bibliometric analysis results showed that the studies on smart learning in China began in 2012 and it showed an upward trend year by year from 2012 to 2017, and the number of articles reached the historical peak in 2017 with 30 articles, but it showed a downward trend from 2017 to 2019. Zhiting Zhu topped the list of contributing authors, followed by Shusheng Shen and Lin Chen, each of them published more than five articles. Beijing Normal University topped the list of contributing organizations, followed by Jiangsu Normal University and East China Normal University, each of them contributed more than ten articles. E-education Research topped the list of the contributing source publication, followed by Modern Educational Technology and China Educational Technology, each of them contributed more than fifteen articles. Additionally, most of the top 10 contributing organizations were normal universities, indicating that normal universities were the main research institution on the studies of smart learning in China.

In the section of knowledge mapping analysis, the results of high-frequency keywords showed that the "smart learning" was the most frequent keywords, and "smart learning environment", "smart education", "smart learning space" and "learning analysis" were positioned in the 2^{ed}, 3rd 4th and 5th places on the list. It is indicated that the smart learning environment, smart learning space, and learning analysis attracted researchers' attention in China. The clusters analysis results showed that "smart learning", "strategy", "smart learning environment" and "smart education" were the top four largest clusters on the area of smart learning in China from 2012-2019. Among them, "smart learning (#0)" was the largest cluster and the high-frequency keywords of this cluster contained "smart learning", "learning analysis" and "learning space", indicating that smart learning was achieved and completed through learning, smart learning space, etc.

Acknowledgement

This work was financially supported by the Later-stage projects funded by the National Social Science in China (Grant NO. 18FKS007); Key project for university outstanding young talents support program of Anhui province (Grant NO. gxyqZD2019068); Pilot project for smart classroom of Anhui province (Grant NO. 2017zhkt414) and Project for education science research of Anhui province (Grant NO. JK18071).

References

[1] Bin, H. (2013). Smart learning: connotation, evolution and trend - the perspective of learners. *e-Education Research, 34*(11), 24-33.

[2] Hua, L., Jianbin, D., Hannian, X. (2015). Smart Learning supported by intelligent technology——review and review of GCCCE2014. *e-Education Research, 36*(06), 15-20.

[3] Xili, H. (2016). The connotation, essence and characteristics of smart learning ability. *Education Review* (09), 38-41.

[4] Lin, C., Wei, W., Bingbing, L., Ying, Y. (2016). The connotations of wisdom learning and the wisdom learning methods. *China Educational Technology* (12), 31-37.

[5] Xiaxia, Z. (2018). The status quo and development direction of the construction of smart learning in china. *Teaching and management* (30), 1-3.

[6] Hong, Z., Ling, C. (2018). The theory, essence and systematic research of smart learning activities. *Modern Educational Technology, 28*(04), 55-60.

[7] Jizhen, P. (2018). Smart learning: a new path for teacher professional development. *Journal of the Chinese Society of Education* (08), 84-88.

[8] Weidong, C., Xindong, Y., Yafeng, X. (2012). Future classroom: smart learning environment. *Journal of Distance Education, 30*(05), 42-49.

[9] Hanrong, J. (2019). Smart Learning: a new ecology of vocational education and new challenges for teacher professional development. *Journal of Vocational Education* (05), 89-93.

[10] Yinli, L., Yinghao, L. (2019). Smart learning in the age of artificial intelligence: principles, trends and challenges. *China Educational Technology* (02), 16-21.

[11] Chen, C., Leydesdorff, L. (2014). Patterns of connections and movements in dual-map overlays: A new method of publication portfolio analysis. *Journal of the Association for Information Science & Technology, 65*(2), 334-351.

Research on the application of big data in the management and service of College Students

Wenhua Sun[1*], Lifeng Wu[2], Yang Lu[3], Xiaohong Zha[4], Mingyuan Cui[5]

[1,3,4,5]Network Information Center, Nanchang Institute of Technology, Nanchang, China

[2]College of Water Conservancy and Ecological engineering, Nanchang Institute of Technology, Nanchang, China

[*]Corresponding author e-mail: sunwenhua168@163.com

Abstract. The explosive growth of information and the maturity and application of cloud computing, data mining and analysis technology mark that today's society has ushered in the era of big data. College student management and service work are practical, but the traditional way of it is mostly manual and single qualitative management, which has been unable to meet the growing number of students demands, gradually revealing lots of problems like low efficiency, slow speed, and poor quality. The emergence of big data will effectively improve this problem. College students' education management workers will be able to accurately control the ideological and behavioral dynamics of students, properly carry out public opinion analysis, and promote the personalized learning and development of college students.

1. Introduction

With the development of information technology in higher education and the construction of digital campus, colleges and universities have accumulated a large amount of multi-dimensional and multi-level data in daily student management and data service. How to give full play to the value of these data, better to serve the management of colleges and universities, in a real sense, to make the data run on the road, teachers and students run less errands, which requires colleges and universities constantly update and develop the concept of information management of colleges and universities. It also need colleges and universities to establish a talented team to promote its digital management, integrate the data sources in the field of multi business systems in Colleges and universities, and in a real sense, research and improve the theory of big data Only by combining the practice and business of university information management and ensuring the information security, so we can effectively promote the gradual improvement in university information management ability in the era of big data.[1]

The work of student management and service in Colleges and universities lies in the students. The amount of data generated by students in the process of continuous learning and growth is very large. Structural and non structural coexist. Colleges and universities can not only use big data to carry out personalized analysis on students in the enrollment stage, but also carry out personalized analysis on students in every stage after they enter the school. Through the mining and analysis of student data, we can give full play to the value of student data and improve the pertinence and efficiency of management and service.

2. The application of big data in the management and service of College Students

The continuous development of Internet and related industries has further expanded the impact on students. The variety and quantity of data generated by various kinds of data terminals which were used by students have increased significantly, [2]which provides a data base of the application of big data technology in the management and service of college students. Compared with foreign countries, the number of college students in our country is huge, involving a lot of student information data, which to a certain extent increases the workload of big data application. [3] Although there are many colleges and universities actively trying to use big data, after all, the huge amount of data and technical ability have become the factors restricting the application of big data in Colleges and universities. At present, the application of big data in Colleges and universities is mainly reflected in the following aspects.

2.1. Using education information platform to realize shallow big data processing

At present, many colleges and universities have used a variety of management information service platforms, and many colleges and universities have tried to use information data platforms similar to smart campus, which are used in PC and mobile terminals at the same time. Taking the educational administration management information system as an example, after inputting the data information of all students, we can obtain the grade ranking, average to score rankings, grade data analysis, etc. of all students in the class or the same major in a short time. [4]

2.2. Create big data management center

When the process of data mining continues and extends, the amount of data grows explosively, and the simple management information system has been unable to deal with a large number of data. Therefore, many colleges and universities began to try to establish big data analysis center, through the combination of education and modern technology to achieve efficient big data management and application. [5]

2.3. MOOC and cloud education platform to realize multi-dimensional data teaching space

After the emergence of cloud computing, online open courses are increasing day by day. Cross School and cross regional online open courses are gradually entering students' learning life. Using big data technology combined with the storage advantages of cloud computing platforms, MOOC teaching and other multimedia and digital teaching methods have gradually covered the traditional classroom. Teachers' teaching and educational administration management can be realized through the online teaching platform. The teaching data comes from the application of the network in the actual classroom. Students and teachers can use mobile phones and other terminal devices in class to interact on site, sign in, upload homework, answer questions or discuss. [6]

3. Big data in the management and service of College Students

The management and service of college students is a practical work. The traditional way of work is mostly manual and single qualitative management, which can't meet the increasing number of students' needs, and gradually shows the problems of low efficiency, slow speed and poor quality. The emergence of big data will effectively improve this problem. However, as a new technology and concept, there are still some problems in the application of big data in the management and service of college students.

3.1. Lack of data understanding and backward thinking concept

As a new technology and new concept, big data has not been paid enough attention in many colleges and universities, and the innovative role of big data in the management and service of college students has not been recognized. At the same time, university staff lack of understanding of big data, backward concept, traditional thinking mode, and low acceptance of big data.

3.2. Lack of professionals and backward technical facilities

Some colleges and universities set up big data center to try big data application, but most of the staff are transferred from various departments, which makes the application of big data lack of professional technical personnel. At the same time, most colleges and universities do not invest enough in the application of big data, the existing hardware facilities and equipment is relatively backward, unable to support the use of big data technology.

3.3. Difficult data integration and slow progress

In the statistical data information of colleges and universities, the data sources are mostly collected by different management systems. Due to the different data information required by different management departments, the application rules are not uniform, often resulting in a lot of data redundancy and repetition, which will make the work of data integration more difficult, thus reducing the efficiency of data integration.

3.4. Data information involves privacy and security risks increase

The number of college students is increasing year by year, and the amount of big data information is also showing a sharp growth trend. The data information contains a large number of student privacy information, and security risks such as information leakage and hacker attack are correspondingly increasing.

3.5. Single application environment, lack of attempts

At present, the application of big data in Colleges and universities only stays in some simple data extraction and display. The analysis and management of data is at a shallow level, lacking in-depth research and mining of data, single application environment, and lack of attempts on various working environments of university student management and service.

4. The application strategy of big data in the management and service of College Students

4.1. Building a data platform for student education management

The basis of applying big data to college students' education management lies in data. Without data collection and integration, we can't talk about the application of big data and its related technologies. At present, many colleges and universities have initially established a certain scale of education management related information system, but most of them are relatively isolated, limited to the internal management of the Department, and fail to truly realize the sharing and integration of data. At the same time, in the aspect of data collection and collation, there is often a lack of standardization, and the data is complex, which is not conducive to further analysis and processing. Therefore, it is necessary for colleges and universities to set up a special organization or department to coordinate and integrate the information platforms of the departments closely related to students, such as the library, logistics management department, academic affairs office and scientific research office, integrate all the information related to students, build a systematic data collection platform for education management, and form a university wide database for student education management In order to ensure the timely and comprehensive collection, update and sharing of all data of all students, fully tap the value of data, and more comprehensively grasp the situation of students.

4.2. Building a professional management team with big data analysis ability

In the era of big data, in all walks of life, it is not so much big data that creates value as big data thinking that triggers new value growth. The key to the application of big data and its related technologies in the management of College Students' education lies in the people who have big data thinking and big data analysis ability. However, there are few teachers who has the ability of big data analysis and student education management. Therefore, to give full play to the role of big data technology in student education management, it is necessary to establish a professional management team with big data analysis ability. In the team, through the introduction of specific training projects related to big data, help members of the team to establish big data awareness, make them fully realize the important value and significance of big data and its related technologies for students' education management, pay attention to the collection, sorting and mining of students' relevant data from the ideological level, and master the data platform of students' education management from the ability level The theory, method and tool of data analysis.

4.3. Improve big data management and use specifications

The premise of the application of big data and its related technologies in college students' education management lies in the collection of basic data. At present, the amount of basic data related to students is huge, various, and scattered in different institutions and departments. In the process of decentralized collection, due to the disunity of data standards, the final data will inevitably be disordered, which is not conducive to further summary and analysis, and thus unable to maximize the value of data. In addition, the widespread use of big data may also lead to the leakage of students' privacy information, threaten the safety of students' personal information, and make students suffer from different degrees of harm. Therefore, in the application process of big data and its related technologies, colleges and universities should establish and improve the corresponding management and use specifications, unify the data standards, ensure the seamless connection of the data of all departments, so as to facilitate the integration and sharing of data, standardize the process of data collection and the scope and authority of use, and ensure the safety of data collection and use from the system.

5. Conclusion

The management of students in Colleges and universities is a huge, abstract work, which cannot be measured by specific indicators. The information age has given many new methods and means to the management of college students. Using big data to manage students can not only save a lot of human and material resources, but also have a more accurate and scientific evaluation of the effectiveness of management, problems and deficiencies of management, and help managers to make more scientific and correct decisions. In addition, according to the strong analysis and prediction ability of big data technology, giving full play to the role of big data in student management and service, it is bound to make the management and service of students become not only scientific, but also conducive to the development of students, and truly cultivate more excellent talents for the country.

Acknowledgments

This work was supported by the science and technology project of Jiangxi Provincial Department of Education: the application research of big data in university student management (No.GJJ171560).

References

[1] Victor Meier scheenberg. Big data age [M]. Translated by Sheng Yangyan, Zhou Tao, et al. Hangzhou: Zhejiang People's publishing house, 2013:39. (in Chinese)
[2] Wan Hui. The application of big data in the management of college students [J]. Journal of college counselors,2014(04):48-51. (in Chinese)
[3] Fu Xin, Bao Wei, Wang Geng. Research on the application of big data in the management and service of college students [J]. Journal of Guangxi Youth Cadre College,2015(4):19—22. (in Chinese)
[4] Wang Qian. Research on the current situation of information construction of university student management in the era of big data [J]. Journal of Liaoning Agricultural Vocational and technical college,2018(02):55-57. (in Chinese)
[5] Xu Wenxun. The current situation and Countermeasures of information construction of university student management in the era of big data [J]. Journal of Mianyang Normal University,2016(01):53-56. (in Chinese)
[6] Zhang Yue, MI junkui. Application of big data in university management and service [J]. Journal

of Inner Mongolia Normal University,2017(11):56-58. (in Chinese)

IOP Conf. Series: Materials Science and Engineering **806** (2020) 012057 doi:10.1088/1757-899X/806/1/012057

Research on Semantic Segmentation of Portraits Based on Improved Deeplabv3 +

Kaihui Zhang[1*], Xianhui Liu[1] and Yufei Chen[1]

[1]College of Electronic and Information Engineering, Tongji University, Shanghai, 201804, China

[*]Corresponding author's e-mail: 418277390@qq.com

Abstract. At present, semantic segmentation is one of the hottest research directions in the field of computer vision, and the Deeplab series is one of the best neural network models in semantic segmentation. Based on the original Deeplab-v3 + neural network model, this paper makes a series of improvements to make it perform better on semantic segmentation of portraits: 1. Add channel attention module to speed up the convergence speed of model training and improve segmentation accuracy. 2. Improve the original AtrousSpatial Pyramid Pooling (ASPP), adjust the receptive field of the network, and improve the segmentation accuracy. Finally, this paper compares the segmentation performance of different semantic segmentation network models on human-parsing dataset and proves that the proposed neural network structure has better segmentation effect.

1. Introduction

Portrait semantic segmentation is to segment the human body from the image containing the human body and perform detailed segmentation processing on each part of the human body (Figure 1). It has high requirements for the accuracy and real-time of image segmentation,so it is a challenging task. As the basis of human behavior analysis and understanding, the quality of the semantic segmentation of portraits directly determines the effectiveness of subsequent work, such as 3D model of human bodies, motion recognition, detection and tracking. Due to these challenges of portrait semantic segmentation, the research and application of portrait semantic segmentation are still in the exploratory stage[1].

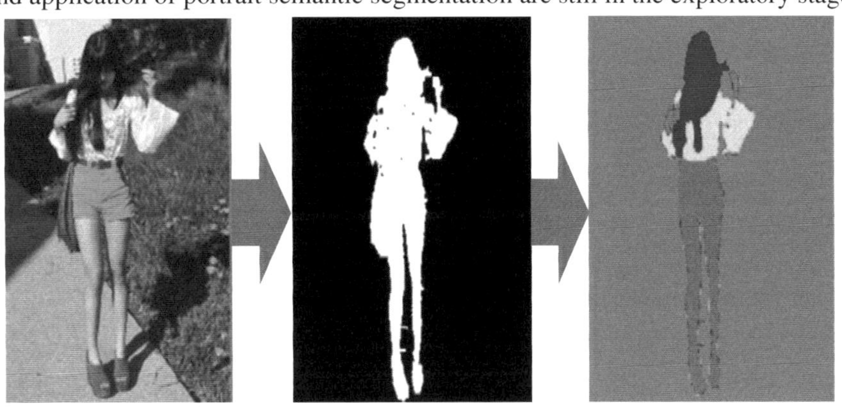

Figure 1.Portrait semantic segmentation

The essence of portrait segmentation is to divide each pixel in the image into a foreground and a background, and continue to subdivide the pixels that belong to the foreground into sections such as the head, upper body, arms, and legs. There are two research directions in the field of semantic segmentation: one is image semantic segmentation based on image graphics, and the other is image semantic segmentation based on deep learning.

The methods based on image graphics generally use the pixel values in the image for analysis. For example, the colour component Red-Green-Blue(RGB) is used as the feature input of the colour feature rough classifier to calculate the colour similarity feature grayscale map. Sobel or Canny operators are used to find target boundaries to achieve image semantic segmentation [2].

The emergence of convolutional neural networks(CNN) provides a new research direction for the study of semantic segmentation. CNN have powerful feature extraction capabilities and have achieved satisfactory segmentation accuracy on human semantic segmentation. With the introduction of fully convolutional networks(FCN)[3], the research on end-to-end semantic segmentation methods has made a major breakthrough, which is a milestone progress. Subsequently, Seg-Net[4], U-Net[5], Deeplab series and other neural networks for semantic segmentation appeared, which further improved the accuracy of segmentation.

The rest of this thesis consists of the following: The section2 introduces the related research and work before the experiment. In section3, the thesis introduces the improved principles and methods of the algorithm in detail. In section4, we carry out comparative experiments and analyse the experimental results. Finally, we conclude this paper in section 5.

2. Related Work

The Deeplab series is one of the most popular CNN architectures in the field of semantic segmentation. Since 2015, v1[6], v2[7], v3[8] and v3 + series have been successively proposed. When traditional CNNs are used for feature extraction, spatial information is often lost during down-sampling and pooling processing, which causes the problem of reduced resolution. In response to this problem, Deeplabv1 proposed atrousconvolution to expand the perceptual field to obtain more contextual information and reduce the loss of location information. Inspired by SPP-Net[9], the v2 network proposed an AtrousSpatial Pyramid Pooling(ASPP) structure, which used multiple atrousconvolution to extract features in parallel and then fused the features. In addition, v2 replaced the backbone network from VGG to resnet-101. v3 is a further improvement of ASPP. It uses deep-level atrousconvolution and adds BN layers, and then the obtained feature map is restored by bilinear interpolation.

Deeplabv3+ is currently the latest neural network structure of the Deeplab series, which is mainly improved based on Deeplabv3. This network mainly borrows the traditional encoder-decoder architecture,expands a simple and effective module for recovering boundary information. Based on the above-mentioned encoder-decoder architecture, inspired by work such as Xception[10], deep separation convolution was applied to ASPP and decoder modules to quickly calculate and maintain the powerful learning ability of the model.

Based on the original Deeplabv3 + architecture, this paper makes the following improvements to the problem of portrait semantic segmentation: introduces the attention module of the channel, improves the accuracy of model segmentation, and accelerates the model's convergence speed. At the same time, the ASPP structure is improved, a smaller convolution kernel is used, and the size of the rate is reduced to improve the accuracy of network segmentation in thin parts.

3. Methods

In this section, we first introduce the working principle of the channel-attention module and its application in this research.Second, we explain in detail the application and improvement of the ASPP structure.Finally, we give the final Improved Deeplabv3 + network structure.

3.1. Channel Attention Module

The essence of channel-attention is to adopt a new "feature recalibration" strategy. Specifically, the importance of each feature channel is automatically obtained through learning, and then according to this importance, useful features are promoted and features that are not useful for the current task are suppressed. This paper introduces the Squeeze-and-Excitation (SE) block in SE-Net for channel-attention learning, as shown in Figure 2:

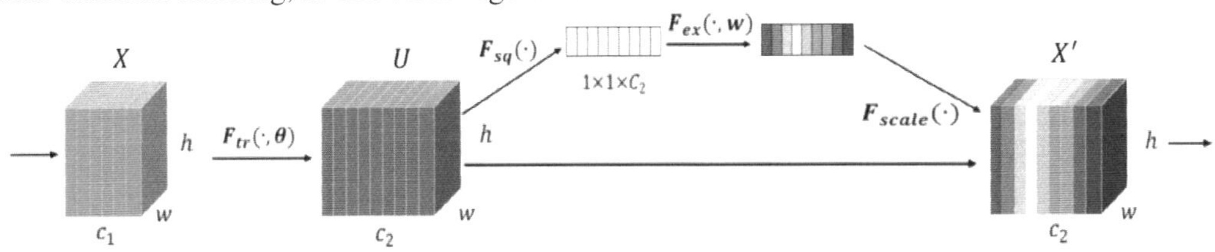

Figure 2.Squeeze-and-Excitation (SE) block

According to figure2, a feature map X is first given, and the number of feature channels is c_1. After a series of general transformations, a feature map U with the number of feature channels c_2 is obtained:

$$F_{tr}: X \rightarrow U, X \in R^{c_1 \times h \times w}, U \in R^{c_2 \times h \times w} (1)$$

The formula of F_{tr} is shown in (3) (convolution operation, v_c represents the c-th convolution kernel, and x^s represents the s-th input):

$$u_c = v_c * X = \sum_{s=1}^{c_2} v_c^s * x^s (2)$$

Next, we need to perform the squeeze operation, that is, perform the global_average_pooling operation: generate a vector with global receptive fields:

$$z_c = F_{sq}(u_c) = \frac{1}{H \times W} \sum_{i=1}^{H} \sum_{j=1}^{W} u_c(i, j) (3)$$

The next step is excitation operation, as shown in (4):

$$s = F_{ex}(z, W) = \sigma(g(z, W)) = \sigma(W_2 \delta(W_1 z)) (4)$$

It is the core of the entire attention mechanism. First multiply z by W_1, which is a fully connected layer operation. The dimension of W_1 is C / r * C. This r is a scaling parameter. In this paper, r is 16. The purpose of this parameter is to reduce the number of channels and thus reduce the amount of calculation. And because the dimension of z is 1 * 1 * C, the result of W_1 z is 1 * 1 * C / r; then, after passing through a relufunction, the output dimension is unchanged; then multiply by W_2 and multiply by W_2. It is also a fully connected layer process, where the dimension of W_2 is C * C / r, so the dimension of the output is 1 * 1 * C; finally, it passes the sigmoid function to obtain s.

After s is obtained, channel-wise multiplication can be performed, as shown in (5). Where u_c is a two-dimensional matrix, and s_c is a number, which is the weight, so it is equivalent to multiplying each value in the s_c matrix by s_c. That corresponds to F_{scale} in figure 2[11].

$$X' = F_{scale}(u_c, s_c) = s_c \cdot u_c (5)$$

3.2. ASPP

Because in the semantic segmentation problem, there are large and small instances that need to be segmented. If the convolution kernels of the same size are used, there may be a problem that the receptive field is not large enough, and the accuracy of segmentation of large objects may decrease. In response to this problem, atrous convolution emerged at the historic moment, that is, the receptive field of the convolution kernel was changed by adjusting the size of the dilation rate.

However, the effect of using atrous convolution on a branch convolutional neural network is not good. For example, in portrait segmentation, we use large atrous convolutions to obtain the upper body information of the portrait, but it is not effective for the feature extraction of smaller hands and feet. If we continue to use smaller atrous convolutions to re-acquire the information of small objects, there will be a lot of redundancy.

ASPP uses the dilation rate of different sizes to capture multi-scale information on different scales on the network decoder. Each scale is an independent branch. It is merged at the end of the network and then a convolutional layer is output to predict the label. Infigure 3. This design effectively avoids the acquisition of redundant information on the encoder and directly focuses on the correlation with the object.

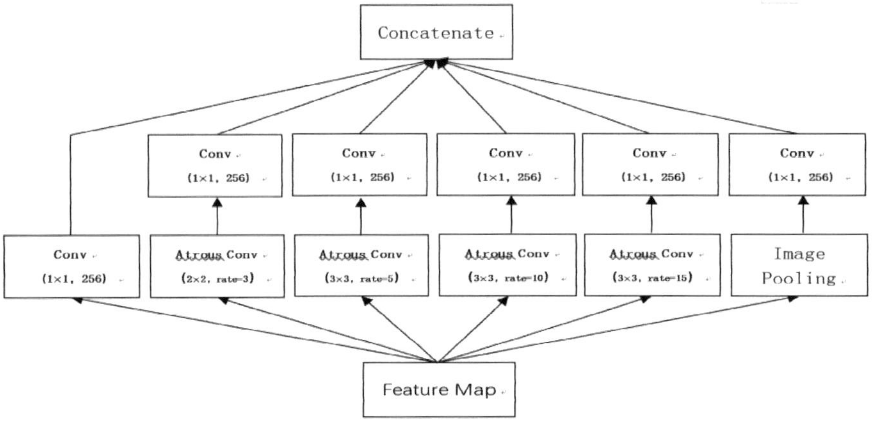

Figure 3.ASPP Structure

Among them, after each convolution, BatchNormalization is used, and relu is used as the activation function. Finally, the results obtained by branch convolution are Concatenated.

3.3. Improved Deeplabv3+

This paper is based on the original Deeplabv3+ network[12], of which BackBone uses an improved Xception network. This paper mainly makes 2 improvements based on the original network:

1. Improve the original ASPP, add 2 × 2 hole convolution, and adjust the dilation rate to better detect the portrait part.
2. The feature map spliced after ASPP branch convolution, because the channel is very large. The paper uses channel attention to process to improve the segmentation accuracy and speed up the model convergence speed.

After these two improvements, the resulting Improved Deeplabv3 + network structure is shown in Figure 4:

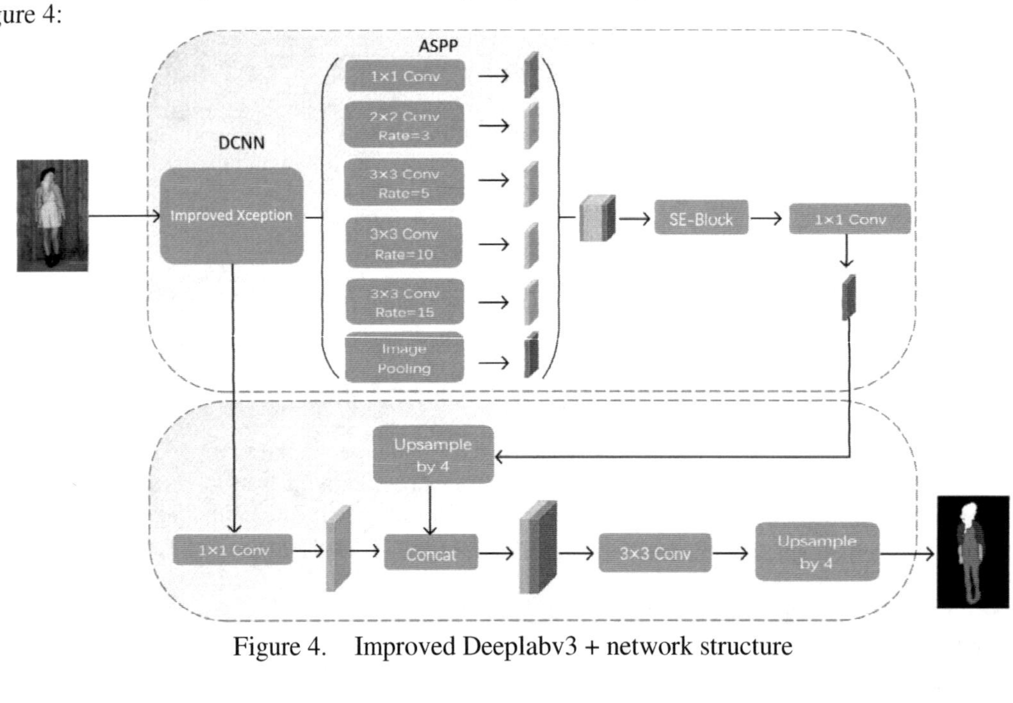

Figure 4. Improved Deeplabv3 + network structure

4. Experiment

4.1. dataset
This paper uses HumanParsing-Dataset data set for experiments. The dataset contains 17,706 images, including a total of 16 + 1 object classes. The dataset is segmented for portrait details, and hats, tops, and pants are segmented. In this experiment, the segmented regions are merged and divided into 4 + 1 categories, which are background, head, upper body, hands, and lower body. In this project, 12,706 pictures were selected. After data enhancement, it was expanded to 63,539 pictures as the training set, 3000 pictures as the verification set, and the rest as the test set.

4.2. loss function
In this experiment, categorical_crossentropy is used as the loss function during training. Suppose there are K label values, and the probability that the i-th sample predicts the k-th label value is$p_{i,k}$, that is, $p_{i,k} = \Pr(t_{i,k} = 1)$. If there are N samples in total, the loss function of the data set is:

$$L_{log}(Y,P) = -logPr(Y|P) = -\frac{1}{N}\sum_{i=0}^{N-1}\sum_{k=0}^{K-1} y_{i,k} logp_{i,k} \quad (6)$$

4.3. results
This experiment uses mIOU (Mean Intersection over Union) as an index to measure the quality of model segmentation, and uses Seg-Net, U-net, Deeplabv3 + and the Improved Deeplabv3+ in this paper for comparison experiments. As shown in Table 1, the Improved Deeplabv3+ has significantly improved the segmentation effect.

Table 1.Performance of each segmented network model on the human-parsing dataset

Method	Head	Upperbody	Both hands	Lower body	Mean
SegNet	0.404	0.261	0.027	0.379	0.268
U-net	0.665	0.486	0.270	0.665	0.522
Deeplabv3+	0.708	0.667	0.499	0.845	0.679
Improved Deeplabv3+	**0.715**	**0.685**	**0.512**	**0.854**	**0.691**

Deeplabv3+has two kinds of Backbone networks: ResNet-101 and Xception. The experimental results are shown in Table 2.

Table 2. Performance of the two Backbone networks on the human-parsing dataset

BackboneNetwork	mIOU
Improved Deeplabv3+(ResNet-101)	0.673
Improved Deeplabv3+(Xception)	**0.691**

In order to prove that the improved ASPP does improve the accuracy of portrait semantic segmentation, this article also did a comparison experiment using the improved ASPP and the original ASPP, as shown in Table 3.

Table 3. Performance of the two ASPP on the human-parsing dataset

BackboneNetwork	mIOU
Improved Deeplabv3+(Original ASPP)	0.657
Improved Deeplabv3+(Improved ASPP)	**0.691**

By observing the training process of the model, we also found that adding the channel-attention module can speed up the model's convergence speed, as shown in Figure 5.

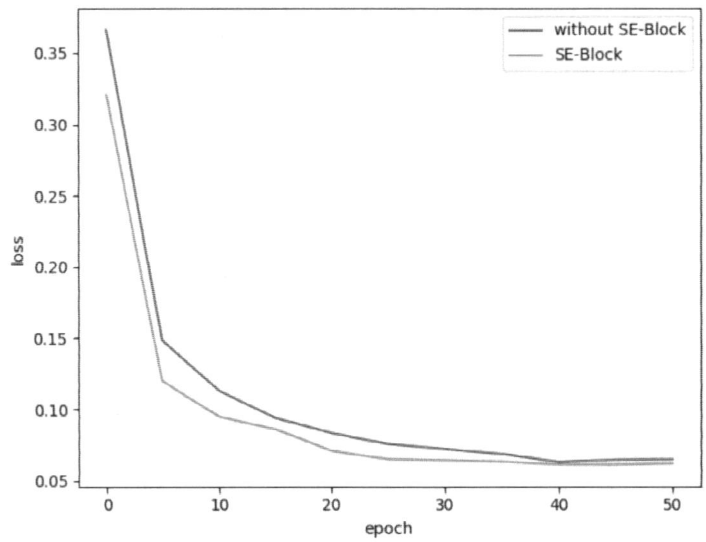

Figure 5. Comparison of Improved Deeplabv3+(SE-Block and without SE-Block)

5. Conclusion

In this paper, we have made a series of improvements based on the original Deeplabv3 + to improve the convergence speed of the training model and the accuracy of segmentation. In addition, a series of comparative experiments have been done in this article. By comparing the experimental results, it has been proved that the use of Xception as the backbone, Improved Deeplabv3+ and ASPP, as well as SE-block have the best segmentation effect on the human-parsing dataset.

Acknowledgment

This paper was financially supported by National Key R&D Program of China (2017YFB0304102) and Science and Technology Innovation Project of Shanghai (18511107400).

References

[1] Ma Z , Wang N , Gao X , et al. Semantic Segmentation Based Automatic Two-Tone Portrait Synthesis[J]. 2017.

[2] Aissa A A , Duchesne C , Rodrigue D . Polymer powders mixing part II: Multi-component mixing dynamics using RGB color analysis[J]. Chemical Engineering Science, 2010, 65(12):3729-3738.

[3] Long J ,Shelhamer E , Darrell T . Fully Convolutional Networks for Semantic Segmentation[J]. IEEE Transactions on Pattern Analysis & Machine Intelligence, 2014, 39(4):640-651.

[4] BadrinarayananV , Kendall A , Cipolla R . SegNet: A Deep Convolutional Encoder-Decoder Architecture for Image Segmentation[J]. 2015.

[5] RonnebergerO , Fischer P , Brox T . U-Net: Convolutional Networks for Biomedical Image Segmentation[J]. 2015.

[6] Chen L C , Papandreou G , Kokkinos I , et al. Semantic Image Segmentation with Deep Convolutional Nets and Fully Connected CRFs[J]. Computer Science, 2014(4):357-361.

[7] Chen L C , Papandreou G , Kokkinos I , et al. DeepLab: Semantic Image Segmentation with Deep Convolutional Nets, Atrous Convolution, and Fully Connected CRFs[J]. IEEE Transactions on Pattern Analysis & Machine Intelligence, 2016, 40(4):834-848.

[8] Chen L C , Papandreou G , Schroff F , et al. Rethinking Atrous Convolution for Semantic Image Segmentation[J]. 2017.

[9] He K , Zhang X , Ren S , et al. Spatial Pyramid Pooling in Deep Convolutional Networks for Visual Recognition[J]. IEEE Transactions on Pattern Analysis & Machine Intelligence, 2014, 37(9):1904-16.
[10] Chollet, François. Xception: Deep Learning with Depthwise Separable Convolutions[J]. 2016.
[11] Hu J , Shen L , Albanie S , et al. Squeeze-and-Excitation Networks[J]. IEEE Transactions on Pattern Analysis and Machine Intelligence, 2017.
[12] Chen L C ,ZhuY ,Papandreou G , et al. Encoder-Decoder with AtrousSeparableConvolution for Semantic Image Segmentation [J]. 2018.

IOP Conf. Series: Materials Science and Engineering **806** (2020) 012058 doi:10.1088/1757-899X/806/1/012058

Application of Activity-on-arrow Network in Product Development Management

Linsen Li[1*], Shanping Zhou[2], Ting Deng[3]

Academy of Art and Design, Anhui University of Technology, Ma'anshan, Anhui, 243032, China

*Corresponding author's e-mail: 769647428@qq.com

Abstract: Product development management is the core work of product preparation, which is restricted by many factors and requires regular adjustment of plans and real-time dynamic management. Taking the general process of product development management as an example, this paper adopts activity-on-arrow network in operational research to develop product management, and then draws and calculates the composition, key circuit and time parameters. Afterwards, the construction period is shorten and the consumption of cost is reduced through adopting the activity-on-arrow network in developing product management to achieve the design optimization and calculation of product development management, so as to save cost and improve economic efficiency of enterprises.

1. Introduction

Activity-on-arrow is also known as "arrow diagram method", which can make the time limit of each work well arranged, and make the beginning and ending time connect with each other in product development management, so that the product development organization and work preparation are more precise and accurate, and the execution time is more strictly planned. The most important goal of the diagram is to save labor, reduce product development cost, and shorten the process time, thereby improving work efficiency. The use of activity-on-arrow is conducive to ensuring that enterprise can profit and compensate the enterprise expense on time. It is beneficial to effectively coordinate enterprise resources, and enable the resources to be used at any time. Therefore, it helps to predict the level of funds and resources in different times, and is beneficial to meet strict deadlines. Therefore, activity-on-arrow plays an important role in product development management.

2. The composition of activity-on-arrow network

Activity-on-arrow network is mainly composed of arrows and nodes[1]. The nodes can be numbered by adding Arabic numerals in circles. One work should have only one arrow line and one pair of nodes. The number of arrow tail nodes is smaller than the number of the arrowhead nodes. The node numbers are arranged in order from small to large, which cannot be continuous, and also the same node is strictly prohibited to repeat. The arrow line indicates the event, which is the sign of the beginning or ending point of the job. The event does not consume resources, represented by one Arabic numeral in a circle. The node indicates the connection point between the production times. The left and right ends of the arrow line are numbered (i, j), in which, i indicates the beginning time, and j indicates the ending time. The upper part of the arrow shows work name, and the lower part indicates work duration. Addition, the arrow node refers to the end time j of the work, and the arrow tail node indicates beginning time i. there is only one start node or end node in activity-on-arrow network. It is strictly

IOP Conf. Series: Materials Science and Engineering **806** (2020) 012058 doi:10.1088/1757-899X/806/1/012058

forbidden to have line loop, arrows without arrow and line, two-way arrows, arrow lines without beginning or ending node. (Shown in Figure 1)

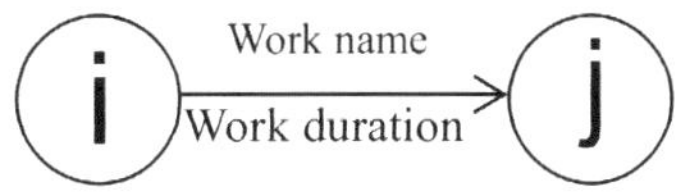

Figure 1. Nodes and work of activity-on-arrow network

3. Time code calculation of activity-on-arrow diagram

The activity-on-arrow network indicates the arrangement of scheduled projects and expenses and the relationship between each management links. Based on this, the system analyzes the time parameters required for each process, and infers the key lines according to the longest time required by the line, then finally uses the network map to optimize the process duration and cost. There are many planned tasks that are closely linked in one process duration, if any of the work processes delays upon the expected completion time, it eventually result in cost loss and reduction. How to save time and expense based on above phenomena is an issue to be studied in product development management.

The following is an in-depth study of how to use activity-on-arrow network to rationally arrange the logical relationship between work duration and cost in product development management. According to Table 1, the logic relationship of activity processes can be grasped at a glance from the logic between each process and required time of product project. Therefore, the initial activity-on-arrow network 2 of the activity can be conveniently and quickly drawn based on the theory.

Table 1. Logical relationship of activities

Activity	Activity Content	Close Front Time	Required Time（day）
A	Market research	/	4
B	Design concept	/	8
C	Product approval	A	2
D	Production preparation	A	2
E	Sample trial	B、C	2
G	Production and processing	B、C	2
H	Product evaluation	D、E	6
I	Customer trial	D、E	4
J	Mass production	H、G	6

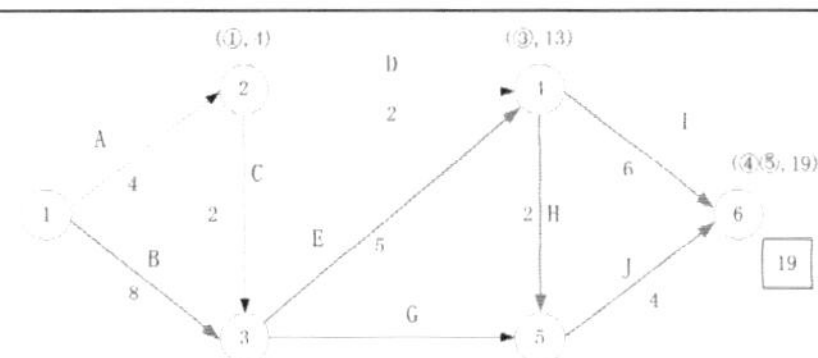

Figure 2. Initial activity-on-arrow network

3.1. Critical lines

The lines on the activity-on-arrow network refer to the lines from the starting node of the arrow in direction from left to right. In this process, the ending node passes through many intermediate nodes; the route trace is called the line. The sum of all work durations is the length of the line. Critical lines consist of critical works from beginning to the end, in another word, the largest sum of time used on all lines and with the longest total work duration are called critical lines. An activity may have more than one critical line, and there may be many lines. The critical lines are shown as double arrow lines and thick solid lines in activity-on-arrow network. The works on critical lines are called critical works[1]. The critical line is based on time or process parameters, calculated by various time parameters in computer network, and continuously optimize to obtain the shortest duration and minimum cost by time differences. According to Figure 2, logical process relationship concludes that there are two key lines with the longest total work durations: B-E-H-J and B-E-I. According to the estimated time, the two critical lines indicate that the maximum time required to complete the product development is 19 days.

3.2. Calculation of node time parameters

A node is one circle within a number with arrow head and tail of each arrow line. The circle is called node. The number inside is the node number. One node indicates the end of previous process and also the beginning to the next.

3.2.1. The earliest starting time of the node

The earliest time of the node is calculated by possible earliest time of the process, as at before time there might lack of start conditions, in another word, the earliest time of start node is 0, which shall be calculated from the beginning node of activity-on-arrow network, and calculated each node from left to right along the direction of the arrow line of the activity-on-arrow network. Its formula: ESj = max{ESi+tij}, ESj is the arrow node, j is the earliest end time; ESi is the arrow tail node, i is the earliest start time; tijis work time of the process i-j. The total duration is the earliest start time of the end node of activity-on-arrow network, which is the maximum sum of all the value of process durations from start to end node in all the lines of network diagram.

3.2.2. The latest ending time of the node

The latest ending time of the node refers to the time that the process of the node must be completed. If not, the next work will not start on time. The latest ending time of the node should be gradually calculated from the ending point against arrow line toward the start node. The end time of the end node is normally its earliest start time, which is the total duration. Its formula: LFi = max{LFj+tij}, where LFi is the latest start time of the tail node i; LFj is the latest end time of the arrow node j; tijis the work time of the process i-j.

3.2.3. Calculation of process time parameters

(1) The earliest start time of the process = the earliest start time of the previous node. Formula: ESij = ESi

(2) The earliest end time of the process = the earliest start time + the work time of the process. Formula: EFij = ESij+tij

(3) The latest end time of the process = the latest end time of the node afterwards. Formula: LFij = LFj

(4) The latest start time of the process = the latest end time of the process - the work time of the process. Formula: LSij = LFij-tij

(5) Time difference refers to the maneuver time available for an activity without affecting the completion time of the entire task.

(6) Node time difference = the latest end time of the node - the earliest start time, in which the node time difference=0 is the key node.

(7) The total time difference means free and maneuverable time between close front and after activity without affecting the total work duration. Formula: TFij = LSij-ESij = LFij-EFij

(8) The free time difference is free and maneuverable time when the work is started at the earliest time without affecting the work. Formula: EFij = ESj-ESi-tij

Table 2. Calculation of activity time parameter

Activity code	Earliest start time ESi	Earliest end time EFj	Latest start time LSi	Latest end time LFj	Total time difference TFij	Free time difference FFij
A	0	0+4=4	6-4=2	8-2=6	2-0=2	6-4=2
B	0	0+8=8	8-8=0	13-5=8	0-0=0	8-8=0
C	0+4=4	4+2=6	8-2=6	13-5=8	6-4=2	8-6=2
D	0+4=4	4+2=6	13-2=11	19-6=13	11-4=7	13-6=7
E	0+8=8	8+5=13	13-5=8	19-6=13	8-8=O	13-13=0
G	0+8=8	8+6=14	15-6=9	19-4=15	9-8=1	15-14=1
I	8+5=13	13+6=19	19-6=13	15+4=19	13-13=0	19-19=0
J	13+2=15	15+4=19	19-4=15	15+4=19	15-15=0	19-19=0

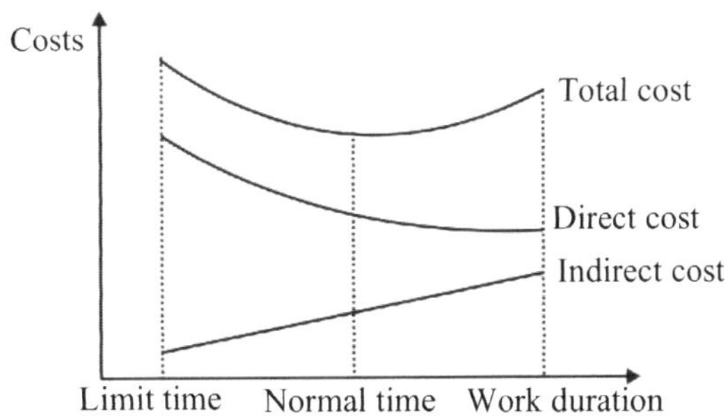

Figure 3. Work duration—cost optimization plan

4. Production time of product Development- cost optimization

The so-called optimization is the process of seeking an optimal network plan by continuously adjusting according to a certain target under the established constraints[1]. The ideal measure includes various factors such as expense, production time and resources. How to revile the three optimization of expense, production time and resources, which means the optimization of balancing the expense with limited production time; the optimization of shortest production time with limited expense; the optimization of the best production time and lowest cost, and optimization of the production time and cost discussed in this paper.

4.1. Production time-cost optimization plan

The production time-cost optimization plan mainly study the shortest production time, the low cost, and settle the problem of the shortest production time and lowest cost under the condition of guaranteed established completion period or minimum cost and under limited fees. Total cost = direct cost + indirect cost. Direct costs include direct surcharges, tools and equipment, worker wages, etc. Indirect costs include the salaries and administration expenses. As is shown in Figure 3, direct costs increase as the production time decreases, and indirect costs decrease as the total production time decreases.

4.2. Production time—cost optimization process

According to the initial activity-on-arrow network of Figure 2 drawn by the product development management process, the normal time in Figure 4 refers to the numbers outside the parentheses below the arrow line; the numbers in parentheses are the minimum duration of the work; and the numbers in parentheses are the direct fee required for the work to be completed in the shortest duration. The known indirect cost rate = 0.8 million Yuan / day.

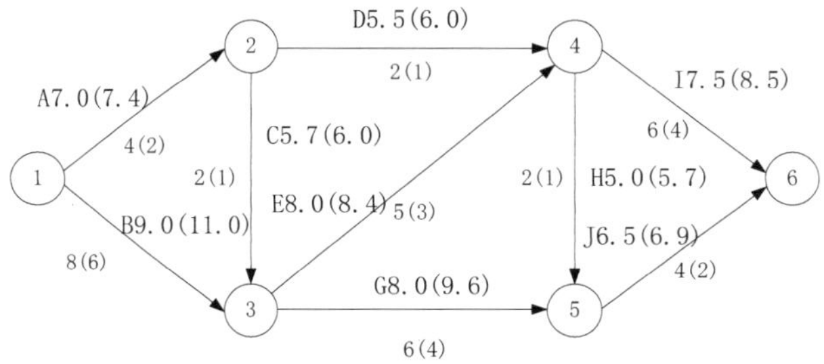

Figure 4. The initial plan of activity-on-arrow network

(1) As is shown in Figure 5, the production time calculated = 19 days, there are two key lines: ①-③-④-⑥ and ①-③-④-⑤-⑥.

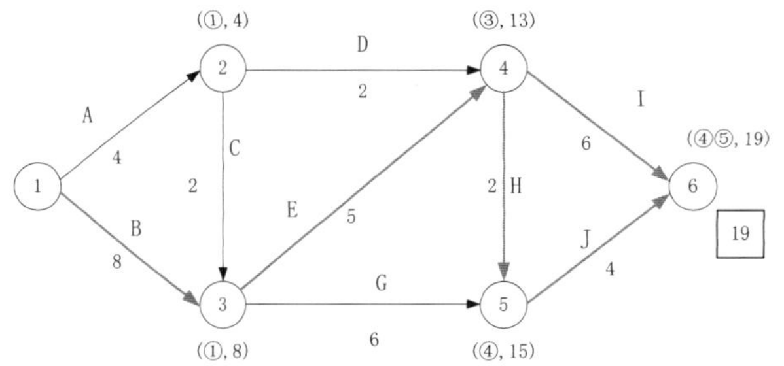

Figure 5. Key lines in the initial activity-on-arrow network

（2）The direct cost rate for each job caculation

△C1-2=（7.4-7.0）/（4-2）=0.2 million Yuan / day

△C1-3=（11.0-9.0）/（8-6）=1.0 million Yuan / day

△C1-2=（7.4-7.0）/（4-2）=0.2 million Yuan / day

△C2-3=0.3 million Yuan / day

△C2-4=0.5 million Yuan / day

△C3-4=0.2 million Yuan / day

△C3-5=0.8 million Yuan / day

△C4-5=0.7 million Yuan / day

△C4-6=0.5 million Yuan / day

△C5-6=0.2 million Yuan / day

(3) The total activity cost Calculation

①direct fee sum = Cd = 7.0 + 9.0 + 5.7 + 5.5 + 8.0 + 8.0 + 5.0 + 7.5 + 6.5 = 62.2 million Yuan;

②The sum of indirect costs = Ci = 0.8 × 19 = 15.2 million Yuan;

③total activity cost = Ct = Cd + Ci = 62.2 + 15.2 = 77.4 million Yuan.

(4) Production time-cost optimization requires constant compression of critical work

1) First compression

As is seen in Figure 5, there are two key lines in the activity-on-arrow network. To shorten the total duration of the two key lines at the same time, there are four compression schemes:

Compress B, the direct cost rate is B = 1.0 million Yuan/ day;

Compress E, direct expense rate = 0.2 million Yuan / day;

Compress H and I in one time, combined direct cost rate = 0.7 + 0.5 = 1.2 million Yuan / day;

Compress I and J in one time, combined direct cost rate = 0.5 + 0.2 = 0.7 million Yuan / day;

According to above compression scheme, the direct cost rate of E is the smallest, so E is selected as the compression object. The direct cost rate E = 0.2 million Yuan / day, less than the indirect expense rate E = 0.8 million Yuan / day, compression E can reduce the total cost. Compress E's duration time to minimum 3 days, recalculate the production time and critical lines shown in Figure 6. At this point, the key work E is compressed into non-critical work; therefore, 4-day time becomes key work. The first compression is shown in Figure 7. The number in parentheses above the arrow line in the figure is the direct expense rate.

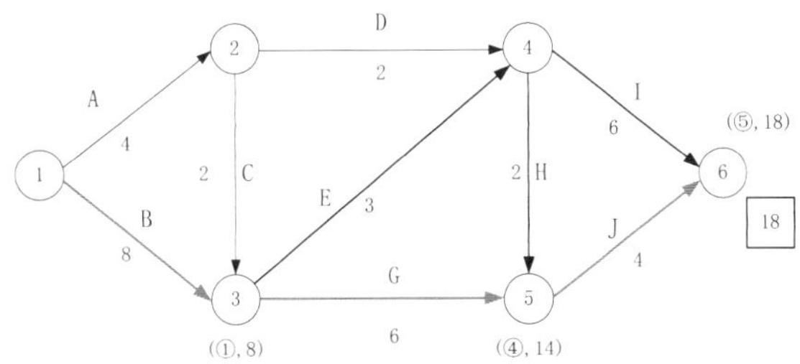

Figure 6. Work E's compressed shortest time of critical line

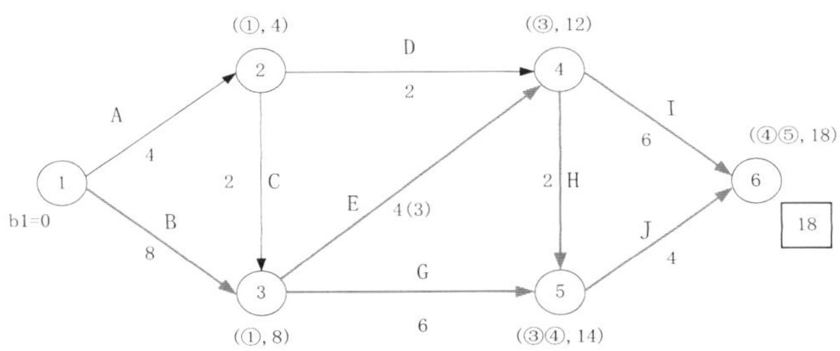

Figure 7. Network plan after first compression

2) Second compression

As is seen from Figure 7, there are three key lines in the dual-code network scheme: ①-③-④-⑥,①-③-④-⑤-⑥ and ①-③-⑤-⑥. To simultaneously reduce the total duration of three critical lines, there are five compression schemes:

①Compress B, the direct cost rate is 1 million Yuan / day;

②Compress E and G in one time, the combined direct cost rate is 0.2 + 0.8 = 1.0 million Yuan/ day;

③Compress E and J in one time, combined direct cost rate = 0.2+0.2 = 0.4 million Yuan/ day;

④Compress G, H and J in one time, combined direct cost rate = 0.8 + 0.7 + 0.5 = 2 million Yuan / day;

⑤Compress I and J in one time, combined direct expense rate = 0.5 + 0.2 = 0.7 million Yuan/ day.

According to the above compression scheme, since the combined direct cost rate of work E and work J is the smallest, E and J should be selected as compression targets. The combined direct cost rate of work E and work J is 0.4 million Yuan/day, which is less than the indirect cost rate of 0.8 million Yuan/day, indicating that compress work E and work J in one time can reduce the total project expense. Since the compression time of work E last only one day, the duration of work J can also be compressed for one day. After the duration of work E and J is compressed simultaneously for one day, the work duration and critical line are re-determined. Meanwhile, the key lines are changed from three before compression to two: ①-③-④-⑥ and ①-③-⑤-⑥. The original key work H was passively turned into non-critical works without compression. The network plan after the second compression is shown in Figure 8. At this point, the critical work E has the shortest duration and can no longer be compressed, so its direct cost rate becomes infinity.

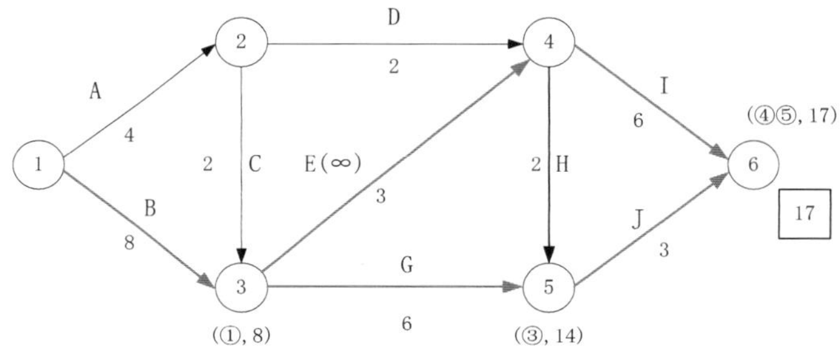

Figure 8. Network plan after the second compression

3) Third compression

As is shown in Figure 8, work E cannot be compressed. In order to shorten the total duration of the two key lines ①-③-④-⑥ and ①-③-⑤-⑥, there are only three compression schemes:

Compress work B, the direct expense rate is 1 million Yuan / day;

Compress G and I in one time, combined direct cost rate is 0.8 + 0.5 = 1.3 million Yuan / day;

Compress I and J in one time, combined direct cost rate: 0.5 + 0.2 = 0.7 million Yuan/ day.

In the scheme above, since the combined direct cost rate of I and J is the smallest, I and J are selected as the compression objects. The combined direct expense rate of work I and J is 0.7 million Yuan, which is less than the indirect expense rate of 0.8 million Yuan, indicating that the total cost of work I and J is reduced when compressed in one time. Work J can only be compressed for one day, and I can only compress it for one-day afterword. When I and J are compressed one day in one day, the duration and critical lines are recalculated. At this time, the key lines are still two: ①-③-④-⑥ and ①-③-⑤-⑥. The network diagram after the third compression is shown in Figure 9. At this time, the critical work has the shortest duration and cannot be compressed any more, so the direct cost rate becomes infinity.

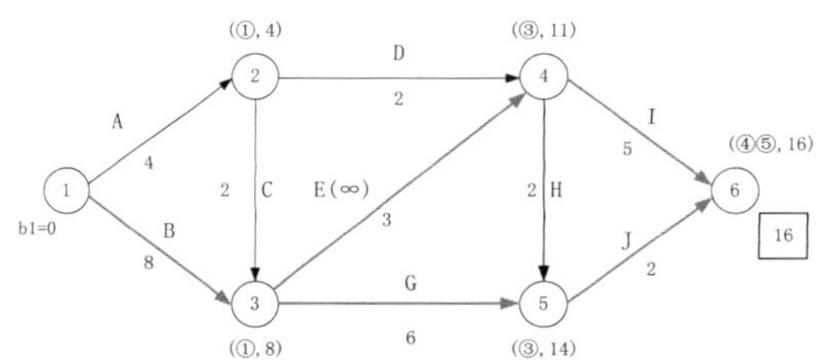

Figure 9. Network plan after the third compression

4) Fourth compression

As seen from the figure, E can no longer be compressed, meanwhile, shorten the total duration of the critical lines ①-③-④-⑥ and ①-③-⑤-⑥, there are two options:

① compression B, direct cost rate = 1.0 million Yuan / day;

② compression work G and I in one time, combined direct cost rate = 0.8 + 0.5 = 1.3 million Yuan. In the scheme above, the direct cost rate of B is the smallest, and B is selected as compression object. B's direct cost rate = 1.0 million Yuan / day, greater than the indirect rate 8 million Yuan / day, B will increase the total cost. Therefore, there is no need to compress B, and it is known from the optimization scheme that the optimized network plan is as shown in Figure 10. The number in parentheses above the arrow line is the direct costs of the work.

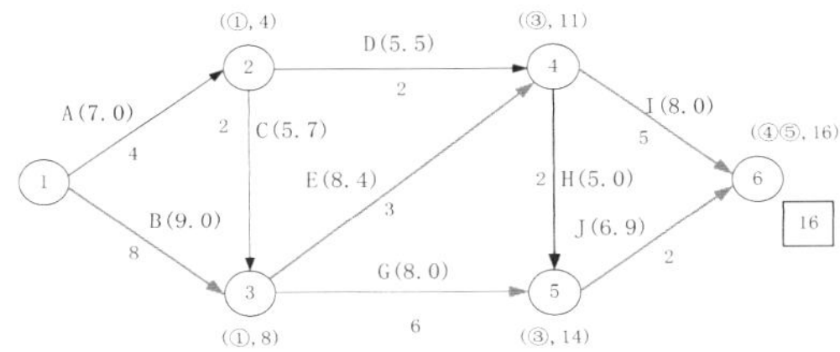

Figure 10. the optimized cost network plan

(5) Calculation of the total optimized activity costs

①The sum of direct cost=Cd0=7.0+9.0+5.7+5.5+8.4+8.0+5.0+8.0+6.9= 63.5 million Yuan

②The sum of indirect cost= Ci0=0.8×16=12.8 million Yuan

③The total activity cost= Ct0 = Cd0 + Ci0 = 63.5+12.8=76.3 million Yuan

Table 3. production time-cost optimization plan statistics

Compression time	Compressed Work code	Compressed work name	Direct cost rate (Million Yuan/day)	Rate difference (Million Yuan/day)	Shorten time (day)	Cost added value (Million Yuan)	Total production time (day)	Total cost (Million Yuan)
0	—	—	—	—	—	—	19	77.4
1	3-4	E	0.2	-0.6	1	-0.6	18	76.8
2	3-4 5-6	E、J	0.4	-0.4	1	-0.4	17	76.4
3	4-6 5-6	I、J	0.7	-0.1	1	-0.1	16	76.3
4	1-3	B	1.0	+0.2	—	—	—	—

According to the table analysis above, after several optimization schemes are designed and calculated, the total duration of the activity is reduced by 3 days, and the total cost is saved by 1.1 million Yuan, thereby reducing the cost of product development management and further improving the economic benefits of the enterprise.

5. Conclusions

Through mapping out the logical relationship of activity-on-arrow network and the calculation of time parameters, the key lines are obtained, and the best solution to increase economic benefits for the enterprise can be achieved through continuously network plan to optimize the production time and cost of product development. Taking full use of activity-on-arrow network in product development management can effectively save the production time and cost, and it has high application value in product management, in the meantime, to achieve good results in time and schedule planning. Therefore, the developed new product needs to seize the market opportunity to quickly win the purchasing power of consumers. At this moment, it is significantly necessary to shorten the production time and accelerate product development.

Reference

[1] Wang Dongsheng, Pan Feng. (2009)Construction Engineering Construction Management. China Mining University Press, Xuzhou.pp. 52-78
[2] Chi Renyong. (2014)Project Management. Tsinghua University Press,Beijing.
[3] Sun Jun.(2015) Project Planning and Control. Publishing House of Electronics Industry, Beijing.
[4] Cui Bin. (2016)Production Operation Management . People's University of China Press,Beijing.

Construction of Digital Library Precision Service System under Big Data Environment

Luyang Meng, Bin Zhang, Di Wang, Jun Fan

Library, University of Jinan, Jinan, Shandong Province, 250022, China

Abstract. With the promotion of Internet communication technology and information technology, university digital libraries are also in the process of continuous reform and development. With the passage of time, the information resources of digital libraries accumulate more and more, coupled with the continuous update of computer technology and social networks, users will generate more and more information resources. The rapid development of communication and Internet technology in the era of big data brings opportunities to digital library's accurate service, and at the same time, it also makes it face huge challenges. The library should rely on the massive book resources platform, combine the big data processing technology with the digital library service concept, and actively explore the data acquisition, sorting and storage technology, so as to provide more effective and humanized services for users. This paper introduces the goal of precision service of Digital Library Based on big data, expounds the characteristics of precision service of Digital Library Based on big data, and puts forward the construction method of precision service system of Digital Library under big data environment.

1. Introduction

With the rapid development of information technology, communication technology and network technology in the era of big data, digital libraries are facing unprecedented challenges and difficulties. How to use big data technology in digital libraries to benefit and avoid harm, give full play to the advantages of data, avoid user diversion caused by big data networking, and provide more satisfactory services for users are issues that librarians must seriously consider under the background of big data era [1]. With the passage of time, the information resources of the digital library accumulate more and more, coupled with the continuous update of computer technology and social networks, the information resources generated by users will also increase [2]. In the era of big data, the rapid development of communication technology and Internet technology has provided a better development prospect for digital libraries, and has also brought certain challenges to the development of digital libraries [3]. The complexity, variety and rapid development of internal information in the library make big data unable to meet its development needs and information cannot be fully utilized, which hinders the progress of digital library [4]. Libraries should rely on massive book resource platforms, closely integrate big data processing technology with digital library service concepts, and actively explore data acquisition, sorting, and storage technologies to provide users with more effective and user-friendly services [5].

In the era of planned economy, as a non-profit organization, college libraries have no competitive pressure, waiting for readers to arrive. Libraries are facing fast data growth and diverse data types, while the development and utilization of big data are very limited, and information resources are not effectively used, which has hindered the development of digital libraries to some extent [6]. In the era of big data, the rapid development of communication and Internet technologies has brought opportunities to the precise services of digital libraries, but also faced huge challenges [7]. With the development of information technology, many similar institutions of university digital libraries have appeared, such as information consulting service agencies and data analysis agencies. Libraries, as providers of traditional book information resources, should also keep up with the times and try to obtain more accurate and effective data information [8]. Digital library precision services are built on the basis of big data, which will greatly improve user satisfaction with digital library precision services. If the digital library of a university wants to maintain its own uniqueness all the time, it needs to conduct precise marketing, narrow the relationship between users and libraries, and increase readers' loyalty to the library [9]. This article analyzes the connotation of big data and digital library, and based on the background of the era of big data, puts forward some digital library precise service strategies to improve the service level of digital libraries.

2. Digital Library Service in Big Data Environment

2.1 Set up a correct concept

The digital library service under the big data environment must first establish the correct service concept, which is the premise for the digital library to carry out the big data service. The proposal of big data knowledge service model provides new solutions, ideas and approaches to solve the integration problem of big data and digital libraries, especially the development and distribution of knowledge service products in universities and research libraries with abundant resources and large service demand. The digital library service under the big data environment has the characteristics of digitalization, convenience and high efficiency. The information ecology emphasizes that the contents of each part of the information construction should not be seen in isolation, but should be studied skillfully from the perspective of coordinated development, so as to better meet the needs of users and provide reference for university library website staff. In the information age, creating a document integration process has become a new concept. It is a process of integration of work links, work posts and work steps after division of labor. Fig. 1 shows the structure of digital resources construction under the information ecology.

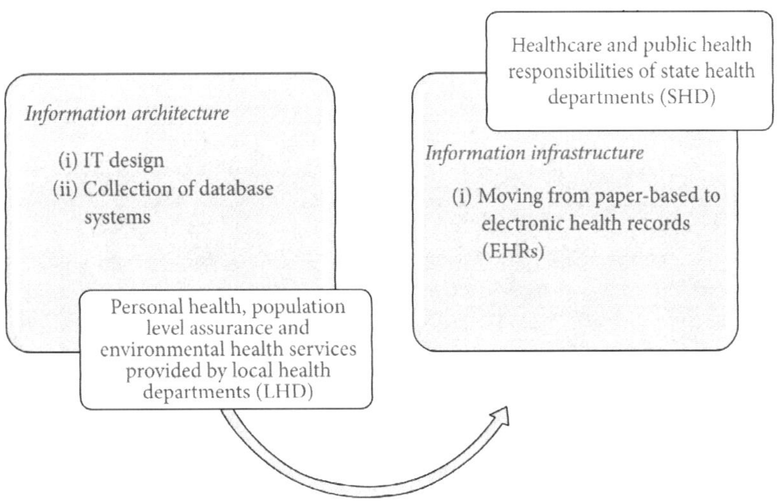

Fig. 1 Digital resource construction structure under information ecology

Digital library organizes and manages document information resources and reader behavior information by means of computer technology and communication technology through digital information resource system constructed by information technology. According to the survey results of readers' needs, digital libraries mainly use relevant cloud computing technologies to provide effective digital knowledge services for readers. Whether it is a digital library or a digital library based on big data, its foothold is in service, and the ultimate goal of service is to meet the information needs of users [10]. In the construction of service resources, it is necessary to set up detailed and standardized development plans, allocate professional personnel and equipment for resource construction, and rationalize the organization and classification of big data resources. Under the big data environment, the foundation of digital library's accurate service is the analysis of big data, and the difficulty of big data processing lies in the complexity of data structure. In library management service, the application of data mining technology can well grasp readers' preferences and behaviors, thus promoting the integration of digital resources, the adjustment of collection resources and the improvement of librarians' service level.

2.2 Building the service model of Digital Library

Building a service model of digital library based on big data architecture is the foundation for digital library to carry out relevant services. With the wide application of data mining technology and the increasing specialization of readers' information needs, digital libraries should deeply analyze readers' preferences and characteristics of information needs to help readers find the required data information from the huge data information. As the main form of library development in the future, intelligent library has intelligent requirements in its service mode and mode. Libraries should adhere to the principles of intellectual property rights and legal security, pay attention to the protection of readers' privacy and the convenience of transmission, and provide users with diversified digital service experiences through diversified service methods and channels. In big data, the proportion of semi-structured information and unstructured data is increasing. It is necessary to analyze these data in order to truly realize personalized and accurate service for users. The data of the digital library is a very valuable resource. These data with the characteristics of standardization, structure and integrity should be fully utilized for information mining and knowledge discovery through data mining technology. In the specific service process, we should establish the overall concept, fully understand and understand the system service elements. The promotion of information technology is the main factor for the rapid development of intelligent services. Libraries should make good use of information technology and continuously develop new intelligent services, which will provide an important guarantee for the sustainable development of digital libraries.

The purpose of the informatization construction experiment based on information ecology is to study the influence of different vector dimensions on the judgment results of emotional orientation. Table 1 shows the accuracy results of data sets in different feature dimensions. The change trend of accuracy rate under different feature dimensions is shown in Fig. 2.

Table 1 Information experiment results based on information ecology

Feature dimension	20	40	60	80	100
Accuracy (%)	79.18	75.31	78.59	82.17	71.76

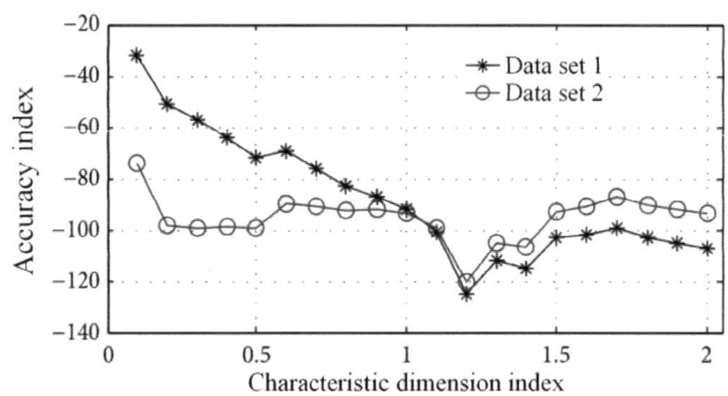

Fig. 2 Trends in the construction of informatization construction under different characteristic dimensions

Digital information consumers are users or readers of digital libraries. Their greatest concern is whether they can receive and effectively screen out useful digital information resources and make use of them. The integration of digital information resources is to organize relatively scattered resources and mutually independent information services together according to certain purposes and corresponding methods for the convenience of users. At the data level, the goal of information resource planning of university libraries is to improve the data sharing ability of libraries by formulating departmental data element standards, information classification standards and user view standards. From information promotion to knowledge, it is mainly to establish a connection between information and users' needs, explain information and give meaning to information. Based on the information construction of information ecology, information construction theory is applied to adjust the balance between users and environment, between users and content, and between content and environment. Integrating traditional libraries, digital libraries and other digital information resources can not only avoid repeated development, but also provide users with convenient and efficient information services. Information demanders are no longer satisfied with obtaining only documentary clues or original documents from the library, but require the library to provide them with excellent information that can solve practical problems.

3. Accurate Service System of Digital Library under Big Data Environment

When integrating library resources under the background of big data, the actual needs of users and subject development should be considered in advance, and the quality should be ensured while the quantity meets the needs. In order to apply the relevant technologies of big data to the service of digital libraries, first of all, it is necessary to define the big data architecture of digital libraries, which is also the combination of big data and digital libraries. The digital library should pay attention to the quantity construction of the collection resources and ensure the quality construction at the same time. Through the correlation analysis of the literature information consulted by readers, the collection resources should be rationally allocated. The digital library can recombine the current audio and video resources, self-built database resources, etc. to establish an information resource bank, which is convenient for readers to consult and use [11]. The digital library combines the readers' information needs with the daily services of the digital library closely, excavates the readers' professional needs and potential needs, and provides the readers with accurate service contents. Reasonable allocation of various collection resources can optimize collection resources, save construction funds and improve service efficiency. Digital libraries should work out targeted service modes after mastering users' personal hobbies, so as to replace the traditional passive mode with the modern active mode.

As a place for information gathering and knowledge service, digital library needs to use data mining technology to process and analyze readers' preference and behavior information. The behavior of users in the library will leave many traces. Therefore, the staff should sort out the information and summarize the user's hobbies so as to provide more accurate services. The use of big data technology by digital libraries must first be based on relevant data resources. All kinds of network information resources, database resources and reader data are "big data" available to the library [12]. These heterogeneous resources are stored in different nodes in a distributed manner, with various types and lack of in-depth mining, integration and analysis. The integration of resources in the museum plays an important role in improving the service level and maximizing the economic benefits. The precise service of digital library is embodied in grouping users according to their different ages or hobbies, determining their unique attributes, and thus providing targeted services for users. The rich diversity of digital libraries is reflected in the appearance of audio and other resource forms, and the establishment of a resource library provides convenience for users to find. The core layer of digital library's big data architecture is big data collection, storage, processing and application. Relevant data is the basic condition and user satisfaction is the ultimate goal.

4. Summary

The digital library service under the big data environment must first establish a correct concept based on big data and have a comprehensive understanding of the digital library service mode under the big data environment. In the era of big data, the demand for readers in digital libraries is showing a new trend. It is becoming more and more important to provide accurate service for readers with high quality. Under the background of big data, the demand of users is obviously increasing, and the demand for services is getting higher and higher. Therefore, if the digital library wants to get better development, it must formulate accurate service mode and provide users with targeted humanized service experience. The rapid development of modern information technology and network technology has created more convenient technical means for readers to provide accurate services, and service models and content are changing. The big data environment provides a new digital platform for the development of digital library services. At the same time that digital libraries use massive data resources, they should pay attention to selecting and identifying effective information data, and fully understand the important role of information security technology. The rapid development of information technology and network technology has provided very convenient conditions for readers of digital libraries, and the service model and content of digital libraries have also continuously changed.

References

[1] Li Xingkun. Knowledge Service to Big Data: Library Service Model Innovation in the Era of Big Data. Science and Fortune, 2014 (9): 79-80.
[2] Xu Rong. Library Service Strategies in the Era of Big Data. Jin Media, 2016, 24 (4): 25-26.
[3] Sang Ping. Innovation of Library Service System under the Background of Big Data. Network Security Technology and Application, 2015 (1): 162-163.
[4] Chen Guolan. How to Use Big Data to Build a New Library Knowledge Service System. Modern Information, 2014, 34 (9): 149-152.
[5] Xu Biwen. On the Development and Innovation of Library Services in the Era of Big Data. Library Theory and Practice, 2014 (12): 19-20.
[6] Zhao Jie. Probe into Library Services in the Era of Big Data. Foreign Electronic Measurement Technology, 2014, 33 (9): 72-74.
[7] Chen Xia, Yang Wanqin. Innovation of Library Service Model in the Big Data Era. Computer CD-ROM Software and Application, 2014 (18): 112-113.
[8] Yuan Jun. Application Research of User Portrait in University Library under Big Data Environment. Library Research and Work, 2019 (6): 22-26.
[9] Zhou Shiping. Analysis of the Construction of Library Service System Based on Big Data. Science and Technology Innovation, 2015 (20): 157-157.
[10] Huang Wenbo. Research on Reading Promotion in Public Libraries in the Era of Big Data. Divine China, 2016 (2): 191-192.
[11] Wang Chuan. Innovative Services of Library Digital Resources in the Era of Big Data. Silicon Valley, 2015 (1): 207-208.
[12] Jinxia Xia, Ma Yumeng. Thoughts on data service capabilities of open information resources in the era of big data. Digital Library Forum, 2016 (8): 54-59.

IOP Conf. Series: Materials Science and Engineering **806** (2020) 012060 doi:10.1088/1757-899X/806/1/012060

Computer Communication Technology and Electronic Information in Artificial Intelligence Application

Yi Yang[1], Ying Feng Fu[2]

[1] SiChuan Aerospace Vocational College

[2] Sichuan Academy of Aerospace Technology

Abstract: In the context of stable social and economic development, the core position of mankind has gradually emerged. People's behavior is controlled by their consciousness, which in turn creates a good living environment. With the development of science and technology, manpower has been largely liberated. Social productivity is constantly growing, and socio-economic order is constantly being reformed. To take full advantage of technological innovation, artificial intelligence is the most advanced technology. Domestic big data technology is rapidly innovating, and computer communication technology is increasingly used in the field of artificial intelligence. Communication and electronic information technology are important supports for the spread of artificial intelligence and can process various data in a timely manner. Process the complex data information and upload the processing results to the artificial intelligence platform. Build different data matrix models, analyze data parameters, and finally apply them to the field of artificial intelligence. The organic integration of computer communication technology and electronic information technology to successfully complete various tasks. It fully demonstrates the close connection between information technology and artificial intelligence, and provides theoretical support for the subsequent work. This article focuses on the application of computer communication technology and electronic information in artificial intelligence, and puts forward comments and suggestions based on actual conditions.

1. Introduction

Artificial intelligence is an emerging technology, starting from Internet technology and big data technology. With the support of new technologies such as integrated mobile Internet, cloud computing and other disciplines, the way in which the human brain handles problems is to make decisions about the problems. Discuss the data processing and decision-making methods based on big data technology, and apply it to the development of various industries. Further enhance intelligence to achieve innovative development of the industry. Artificial intelligence can not only strengthen the management level of the enterprise, but also save production and operation costs, thereby effectively improving economic benefits. In addition, the application of information technology to artificial intelligence can effectively promote the development of intelligence. Embedding information technology into artificial intelligence can effectively improve the accuracy and effectiveness of artificial intelligence. Artificial intelligence technology can be used in the information industry to achieve mutual promotion.

2. Artificial Intelligence

2.1. The Concept of Artificial Intelligence

Artificial intelligence can be subdivided into three levels: ①Simulation ability. After collecting information and data, artificial intelligence imitates human thinking activities. For example, machine learning, human-computer interaction, and so on, as close as possible to human thinking activities based on logical rules. ② Simulate brain structure. The neural cell theory is adopted to realize the simulation of human neural networks. Ensure that computer algorithms are similar to human brain structures, such as perceptron models. ③ Have the ability to imitate human behavior. Artificial intelligence realizes the purpose of self-learning and evolution by sensing the surrounding environment. This shows the adaptive characteristics of artificial intelligence to the environment, such as biological intelligence algorithms. The three levels are the summary of the development of artificial intelligence. First, it has a brain structure, and then continuously adapts to the surrounding environment by imitating human thinking. This completes self-evolution and better reflects the development process of artificial intelligence.

Artificial intelligence itself has very obvious characteristics, which can be briefly summarized as follows: ① Ability to perceive computing across media. Artificial intelligence must be able to collect information from different channels and platforms, and lay a solid foundation for the follow-up work. ② Big data processing technology. The core content of artificial intelligence processing capabilities is computer communication technology and electronic information, which must show more powerful data storage, computing, and processing capabilities. Cloud computing technology and 5G network technology are used to process data to ensure work efficiency. ③ Autonomous learning ability. Artificial intelligence needs to have the ability to learn and think independently, and understand and solve problems according to certain logical rules. Artificial intelligence continuously learns and adapts to rules, thereby gaining more powerful processing capabilities. ④ Bionic ability. Through the use of artificial intelligence and sensing devices, human-machine synchronization based on virtual reality can be achieved. ⑤ Language ability. The language of artificial intelligence is not just a simple dialogue. It is necessary to ensure that artificial intelligence can understand the core content, thereby strengthening the attributes of communication.

2.2. Artificial Intelligence Promotes Social Development

The use of artificial intelligence technology has made many contributions to social and economic development. First, the application of artificial intelligence technology can effectively reduce the actual work intensity. In the use of industrial production and social services, scientific evasion can lead to various mistakes. The service quality has been greatly improved, and the information processing speed has also been improved. Secondly, artificial intelligence is more efficient in controlling and processing information than traditional management models. Artificial intelligence performs services according to a set program, and can adapt to changes in the external environment, thereby achieving reasonable adjustments. Artificial intelligence can also judge the customer's satisfaction level by analyzing the emotions of the service objects. Corresponding feedback links have been added to effectively improve the overall service quality. Artificial intelligence can provide personalized service content and provide users with a comfortable application experience. Artificial intelligence not only has great value in social services and industrial production, but also a strong support for scientific research. Using artificial intelligence for calculations can save a lot of time. We can further increase the scientific research conversion rate and make more contributions to social and economic development. The development and application of artificial intelligence is an inevitable trend of social development and needs more attention.

3. Computer Communication Technology and Electronics

Computer communication technology has a crucial impact on the development of artificial intelligence. It takes full advantage of the multilayer network structure and scientifically processes linear and nonlinear information. The computer communication technology is widely used in the development of artificial intelligence technology, the most common speech recognition and response links. Relying on computer fusion technology to identify and process massive amounts of data information and responsible for data, thereby ensuring computing speed and performance. The development of artificial intelligence is built on the basis of computer communication, and it is necessary to actively use the data marking and processing capabilities of computer communication technology. Based on neural network learning and data propagation training, artificial intelligence data processing models are formulated to better highlight the most basic learning capabilities of artificial intelligence. Electronic information technology can be subdivided into two levels, including electronic science and information technology. Both are based on computer hardware research. In other words, both electronic science and information technology require the cooperation of computer hardware to achieve data transmission and processing.

With the continuous development of modern electronic information technology toward the intelligent and cloud computing level, the electronic information data processing center can also complete the collection and processing of information in the cloud, and then realize the transmission of cloud computing results supported by computer communication technology. Ensure the enhancement of artificial intelligence in deep learning and capabilities, and integrate it into various industries, laying a solid foundation for social and economic development. You can use electronic technology data information processing technology to mine the most valuable computing information in the complex data, which belongs to artificial intelligence data mining. In the actual use of computer communication technology and electronic information technology, deep learning should be performed on data information. Further improve the recognition of the computer's own information, and then better use it in artificial intelligence information processing.

4. Features and Development of Artificial Intelligence

Machine calculations similar to "artificial intelligence" use regular calculation codes for analysis and actively complete the computer's simulation thinking process. In the actual operation process, the verification and inference of mathematical theorems are mainly completed by means of the computer's symbol verification. The proof of machine theorem includes the steps of language understanding, program verification, problem solving, etc. The computer must complete the proof process of the steps according to accurate program settings. With the rapid development of information technology, machine learning has become an independent discipline. Search engine technology provides new ideas for the development of artificial intelligence. Machine learning is subdivided into teaching learning, mechanical learning, inductive learning, and analog learning. As the core part of artificial intelligence, inductive learning has become a hot topic of discussion. With the improvement of the theoretical system and practical exploration, the statistics and algorithm learning of artificial intelligence have gradually become the mainstream direction. The innovation of artificial intelligence computing methods has greatly improved the computing efficiency of artificial intelligence. The application of artificial intelligence data computing and deep learning systems is to continuously add learning information content through the network. Artificial intelligence simulates the brain's thinking activities, and gradually realizes intelligent data processing and information service content.

5. Application of Computer Communication Technology and Electronic Information Technology in Artificial Intelligence

Artificial intelligence uses intelligent methods to simulate human thinking and has gradually become an important part of computer science. In the development stage of artificial intelligence, data analysis has important value. The development of computer communication technology and electronic information based on data analysis has made certain achievements. Actively exploring the application of computer communication technology and electronic information in the field of artificial intelligence. Artificial intelligence technology is mainly based on the deep learning technology of neural network, which integrates communication technology and neural network organically. After the data information is combined with the labeling, mathematical statistical analysis is performed on the data.

5.1. Application of Computer Communication Technology and Electronic Information in Intelligent Driving

Driving is the most important part of traffic safety. The use of new technologies to improve driving safety has gradually become a hot topic. Computer technology and social development interact with each other and provide technical support for intelligent driving. The main purpose is to use computer communication technology and electronic information for driving, such as intelligent transportation systems and fully automated unmanned driving systems. Intelligent transportation systems need to adopt a variety of technical measures to collect and analyze information technology, and then achieve the purpose of coordinating road travel and commanding traffic. At the level of actual technological development, more attention must be paid to vehicle navigation and autonomous driving. Fully automated unmanned driving is the use of intelligent means to drive. The operation of the vehicle will be controlled by the dispatching center to reasonably respond to various road conditions. In the actual application stage, the information and data show the characteristics of sharing, achieving a high degree of unity of the electromechanical system. Fully autonomous driving is the development direction of car driving, and it is of great value for solving urban congestion and environmental pollution. The full potential of intelligent driving applications has been brought into full play, and the use of computer communication technology and electronic information has been continuously deepened. Promote a higher level of intelligent driving and further improve living standards.

5.2. Application of Computer Communication Technology and Electronic Information in Network Management

The purpose of network intelligent management is to comprehensively analyze massive data, and then obtain convenient information retrieval and consulting channels. Communication technology and electronic information play an important role in network management. Combining it with artificial intelligence can make data searching more convenient and intelligent, and network intelligent management has developed into a new system.

Communication technology and electronic information can detect network intrusions and make them the first line of defense for network security. Perform security analysis on data information, and find feedback in time when virus data is found. We need to build a good network security environment and create a better application experience for users. The use of communication technology and information security in smart firewalls is also very common. Smart firewalls have the ability to block viruses, and identification technology is even more important. The recorded virus data can be identified in time, and statistics, decision making, memory, and probability are used to reduce the amount of calculation. Network security has increased significantly, providing people with a more secure network environment. The defensiveness of the network is significantly enhanced, and the degree of intelligentization of network security is also continuously improved.

In addition, the use of communication technology and electronic information can complete the detection of spam. At this time, artificial intelligence recognition is very important, and data analysis is used to improve the success rate of spam interception. The actual application situation is fed back to the user, thereby ensuring secure acceptance of the mail. The full use of communication technology and electronic information in network intelligent security management has added new impetus to the entire industry. It is necessary to continue to deepen research and promote the application level of computer communication technology and electronic information at the network management level.

6. Conclusion:

To sum up, this article is mainly aimed at computer communication technology and electronic information in the application of artificial intelligence. Clarify the meaning of artificial intelligence, and carefully analyze the impact of artificial intelligence on social and economic development. We need to discuss the application of computer communication technology and electronic information, and provide theoretical support for the subsequent work. In fact, computer communication technology and electronic information have made some achievements in artificial intelligence driving and network security management. Fully automated unmanned driving is the use of intelligent means to drive. The operation of the vehicle will be controlled by the dispatching center to reasonably respond to various road conditions. Network intelligent management has developed into a new system that can make data search more convenient and intelligent.

References:

[1] Zhou Yanfang. Application of artificial intelligence in computer network technology [J]. Science and Technology Wind, 2017 (15): 130.[2]Sun Xiaoxia . Research on the Application of Artificial Intelligence in Computer Network Technology [J]. Network Security Technology and Application, 2018 (03): 99.

[2] Lu Huize, Dai Na. Application of computer network technology in hospital information construction [J]. Value Engineering, 2017, 36 (13): 71-72.

[3] Teng Sikai. Application and Analysis of Computer Communication Technology in the Field of Artificial Intelligence——Taking Intelligent Driving as an Example [J]. New Business Weekly, 2018, (21): 193.

[4] Yuan Hengke. Analysis of the application of traditional mathematical operations in the field of artificial intelligence in the future computer science and technology [J]. Silk Road Vision, 2018, (24): 139.

[5] Li Yuling. Application of artificial intelligence in the field of computer vision and network [J]. Computer Programming Skills and Maintenance, 2018, (8): 156-157, 161.

[6] Cui Tianming. Analysis on the Practical Application of Computer Communication Technology and Electronic Information in the Field of Artificial Intelligence [J]. Computer and Telecommunications, 2018, (6): 75-77,80.

[7] Wang Yuanzhe. Artificial Intelligence and Its Application in Computer Communication [J]. ChinaScience and Technology Investment, 2018, (4): 286.

Research on Japanese Teaching Analysis System Based on Artificial Intelligence

Yao Jing

Dalian Neusoft Univeristy of Information Zip code: 116023

Abstract: As one of the current popular technology, artificial intelligence has an important impact on education. Therefore, teachers need to strengthen the use of artificial intelligence equipment in teaching so that it can play its due role in education. Japanese talents are the key required by the current country. Teachers need to analyze their Japanese mastery skills in teaching to grasp the actual learning situation of students. At this time, teachers can use artificial intelligence Japanese teaching analysis system. This article mainly researches the Japanese teaching analysis system based on artificial intelligence, hoping to provide help for Japanese teaching.

1. Introduction

Although some colleges and universities have begun to use the Japanese teaching analysis system based on artificial intelligence, there are still situations in the application of the analysis system. Teachers do not understand the significance of the application and can not really play the role of the analysis system. It can affect the role of the artificial intelligence Japanese teaching analysis system. Therefore, every university should timely study the Japanese teaching analysis system, and constantly tap the teaching auxiliary function.

2. Artificial intelligence(AI)

Artificial intelligence can be said to be a branch of computer science, which can simulate the way that the human brain analyzes and judges data to process related data, and then draw corresponding conclusions. AI has two cores, one is artificial, and the other is intelligence. The so-called artificial means to manually set the relevant procedures to ensure that the smart device can perform prescribed by the person. The intelligence is the device that can perform information, symbol and problem solving. Now that AI is used more and more frequently in education, teachers need to strengthen research on AI equipment to improve students' learning abilities.

3. Function of Japanese teaching analysis system based on artificial intelligence

3.1 Representation and access of knowledge

The analysis system of AI Japanese teaching takes knowledge as the object, computer hardware and software, computer science and technology as the foundation, and philosophy and psychology as the guidance. Relevant knowledge is directly input into the database, the corresponding knowledge conclusion is found from the database, and then output to the big screen. The following figure is the structure diagram of artificial intelligence teaching analysis system.

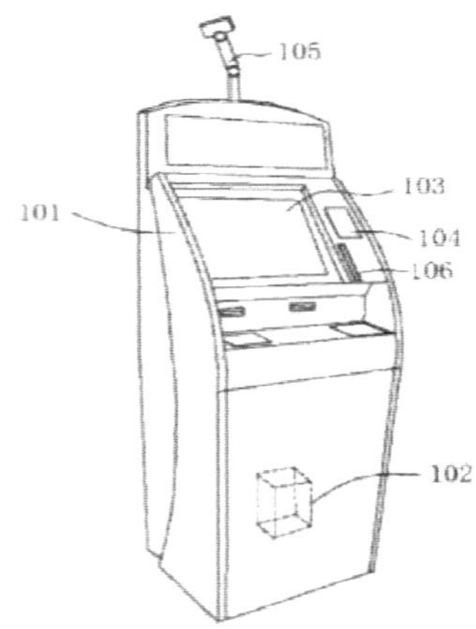

Figure. 1 Structure diagram of artificial intelligence teaching analysis system

3.2 Symbolic operation

Symbolic operation is also an important function of the Japanese teaching analysis system of AI. When teachers put students' scores in the AI system for each semester, the system will call the corresponding formula and symbols to process the data, thus reaching the corresponding conclusion[1]. The current AI analysis system for Japanese teaching can be used for numerical calculation, symbol calculation and function drawing. As shown in the figure, the intelligent host structure diagram of the teaching analysis system based on artificial intelligence is shown.

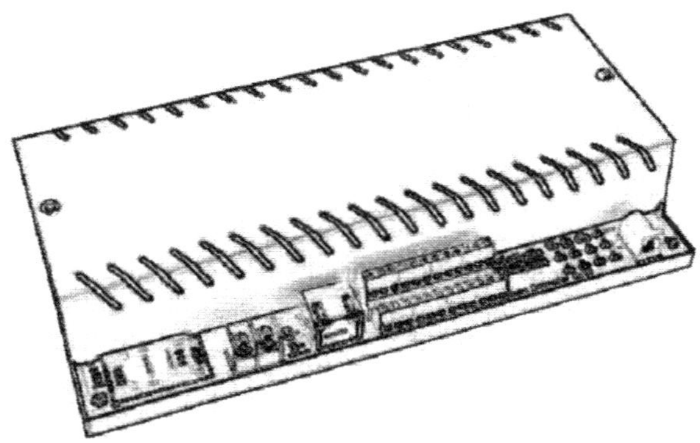

Figure. 2 Schematic diagram of intelligent host of AI teaching analysis system

3.3 Automatic analysis of students' mistakes

One of the main functions of the AI analysis system is to analyze the learning status of students, and to make post-study suggestions based on the current learning status of students. For example, when the AI analysis system changes the student's performance and the type of knowledge errors in the test paper, it may conclude that the students' basic knowledge is not solid. When it draws such a conclusion, the system will help the students to formulate the later learning plans and goals, thereby effectively improving the students' Japanese learning ability.

4. Measures to effectively use artificial intelligence Japanese teaching analysis system

4.1 The AI teaching analysis system is used to analyze academic achievements

As an important part of the current teaching system of language in colleges and universities, Japanese has a good support for the talents needed by the country. With the application of AI teaching equipment in teaching, the quality has been further improved. However, in terms of the current application of the AI Japanese teaching analysis system, most teachers still have more concerns about the performance, and they can not really use it in their daily teaching, which limits the role of the AI Japanese teaching analysis system. Therefore, teachers should use the system of AI to analyze students' achievements, and draw the problems in test papers.

For example, at the end of each test, the teacher can input the students' test paper into the artificial intelligence Japanese teaching analysis system, and use this system to draw the conclusion that what are the missing items of the students' whole test paper, what are the missing items of simple questions, and what are the missing items of students' grammatical errors, and then the teacher can input the test paper into the artificial intelligence Japanese teaching analysis system. Based on the analysis, it can make the next teaching plan. In addition, when teachers use the AI Japanese teaching analysis system to analyze students' scores, they can also use the comments function of the AI analysis system to comment on each students' problems when completing the whole set of test papers. Through the analysis of the students' achievements by the artificial intelligence teaching evaluation and analysis system, not only can the students be more clear about the problems existing in their learning process, but also can help the teachers to define the next step of teaching objectives, and make the later teaching plan.

4.2 The AI teaching analysis system is used to analyze students' learning state

Students' learning status has an important impact on students' knowledge of Japanese. If students have a good learning status, students' Japanese learning efficiency will be higher. There will be a decline otherwise. In traditional teaching, teachers only evaluate students' learning status based on their test scores and daily situation. Although this evaluation method can objectively analyze students' learning status, they may have large conclusion errors. The AI Japanese teaching analysis system can analyze the students' daily performance, participation in classroom activities, test scores, and students' comprehensive questions and answers to obtain analytical data on student attitudes. Therefore, teachers need to strengthen the research on AI Japanese teaching analysis system , and further exert the analysis function on students' learning attitudes[2].

For example, in Japanese teaching, teachers only need to input students' daily performance, test scores, and participation in classroom activities into an artificial intelligence analysis system, and the analysis system can analyze students' learning status, and combine student applications in a large database. There is a learning status to score the current student learning status, and the teacher can determine students' current learning status by analyzing the scores given by the system. In addition, the artificial intelligence analysis system can also summarize students' learning status data in recent months to form a corresponding data line chart. Teachers only need to enter students' recent test scores and usual activities in the artificial intelligence analysis system. After getting the map of students' learning status changes in recent months, teachers can then talk to the students according to the students' learning status map to understand the reasons for the changes in their learning status and encourage students to work hard to overcome difficulties. By using the artificial intelligence English teaching analysis system to analyze the learning status of students, teachers can more accurately grasp the changes in student learning status, and then encourage and guide students based on the changes in student learning status, so as to improve student learning interest.

4.3 The teaching strategy is analyzed by AI teaching analysis system

Although various universities and colleges have formed their own unique teaching systems for Japanese teaching as a foreign language, Japanese has a great challenge to students' learning ability basis, which still needs the guidance of teachers' teaching strategies. From the perspective of current teaching strategies, some teachers fail to combine the actual situation of students in the process of making teaching strategies, which affects the improvement of students' Japanese learning performance. However, the AI teaching analysis system can analyze the rationality and feasibility of teachers' teaching strategies in combination with students' learning conditions and the changes of students' grades to ensure the applicability of teaching strategies to the maximum extent. Therefore, teachers need to use the AI teaching analysis system to analyze their own teaching strategies, timely find the shortcomings, and correct the problems.

For example, a teacher in the Japanese teaching could be applied to "situational teaching method", "multimedia teaching equipment" and other advanced teaching methods. When teachers use the teaching way to students before the ability performance analysis, the data and the use of the above teaching strategies for students to test the learning ability of students through comparison of results before and after the application of the teaching strategies. In addition, teachers also can take use of various teaching methods to the student to carry on the teaching of lesson plan written into AI analysis system of teaching, using the rich cases in the system to analyze the rationality of their teaching plans, and concluding their various problems existing in making teaching plans. It can correct the problem in production to ensure that the teaching strategies and plans can play a guiding role to students' knowledge of Japanese study.

4.4 AI analysis system is used to analyze the teaching progress

For example, teachers only need to input students' daily performance, test scores, and participation in classroom activities into an AI analysis system, and the system can analyze students learning status, and combine student applications in a large database. There is a learning status to score the current student learning status, and the teacher can determine students current learning status by analyzing the scores given by the system. In addition, it can also summarize students learning status data in recent months to form a corresponding data line chart. Teachers only need to enter students recent test scores and usual activities in the AI analysis system. After getting the map of students' learning status changes in recent months, teachers can then talk to the students according to the students' learning status map to understand the reasons for the changes in their learning status and encourage students to work hard to overcome difficulties. By using the AI English teaching analysis system to analyze the status of students, teachers can more accurately grasp the changes in student learning status, and then encourage and guide students based on the changes in student learning status to improve learning interest.

For example, when a teacher can input his own Japanese teaching situation into the system before teaching Japanese, the analysis system will give the teacher an average teaching progress based on the previous situation. Then in future Japanese teaching, teachers can enter the time spent in their own chapters into the AI analysis system after each chapter ends, and the analysis system will score the teacher's progress in conjunction with the chapter's teaching progress collected by themselves and evaluation. If the evaluation score given by AI is low, teachers should slow down the teaching progress and consolidate the students' knowledge of Japanese. Through this detailed AI analysis system, teachers' teaching progress can be guaranteed to meet the learning needs of most students, and the phenomenon that students cannot consolidate their knowledge due to too fast teaching progress can be avoided [3].

5. Conclusion:

The use of AI teaching analysis system can effectively stimulate students interest in learning Japanese to improve the students learning ability with important significance. But in the current research, part of the teachers still cannot effectively know the importance of teaching of AI analysis system in Japanese teaching, which limits its development of teaching effect. Therefore, in the later stage of Japanese teaching, teachers need to apply the AI to students' learning status, students' learning results, and teachers' teaching progress and other aspects.

References

[1] Li Qing, Liu Chao. 2019. Research on Two-way Management System of Classroom Teaching Process Based on AI [J]. Journal of Higher Education, (19)
[2] Liu Qingtang, He Haoyi, Wu Linjing. 2019. Analysis Method and Application of Classroom Teaching Behavior Based on AI [J]. China Audio Visual Education, (9)
[3] Chao Yiwen, Guo Wei. 2019. Application of AI system in Music Education and Teaching in Primary and Secondary Colleges [J]. Art Evaluation, (15)

IOP Conf. Series: Materials Science and Engineering **806** (2020) 012062 doi:10.1088/1757-899X/806/1/012062

Analysis on the Economic Development and Innovation Path of Tourism Cultural and Creative Industry Based on Big Data Statistics

Wu Ke

Zhejiang Noraml University 321000

Abstract: As Computer Internet improves by leaps and bounds, human beings have been in an era full of data and information. Various efficient and advanced management software has filled every aspect of human life, and the rapid growth of it is more and more suitable for people's travel and lifestyle. As an information-intensive industry, tourism cultural and creative industry is naturally integrated with information technology. Therefore, the use of big data on the Internet has triggered two revolutions as the core. All of this have profoundly changed the mode of running, management and operation, and play a vital role in the development of such enterprises.

1. Introduction:

The so-called big data statistical analysis refers to the process of analyzing a large amount of collected data with appropriate statistical analysis methods, extracting useful information, and conducting detailed research and generalization of the data. With the vigorous promotion of Internet computers in China, especially the rapid growth of mobile phones this year, people's consumption concepts and lifestyles have undergone tremendous changes. A lot of consumption can be completed with a little. Therefore, the marketing strategy of tourism cultural and creative enterprises is facing huge changes to adapt to the development of society. Marketing in the era of big data has rejected the previous "focus on sales", instead products for making full use of the disruptive changes brought by the Internet, improving services through some online technologies, user experience, looking for more effective marketing channels and innovation, as well as promoting the industry.

2. The concept and connotation of big data

The Internet and the information industry have grown by leaps ans downs in recent years, and "big data" has attracted attention in various industries. The concept of "big data" was originally proposed by the world-renowned company McKinsey in the report "Big Data: The Next Frontier of Innovation, Competition and Productivity". With the rapid growth of computers and the Internet technologies, the relationship between people and the Internet is becoming more and more inseparable. Traces of people browsing on the Internet have also been recorded, and the data has grown tremendously. These data provide information and resources for people to reshape the world with an important impact on businesses, individuals and even countries. How large is big data? A set of data called "One Day on the Internet" tells us that in one day, the entire generated by the Internet can be engraved with 168 million DVDs; there are 294 billion e-mails (equivalent to two years of paper letters in the United States); 2 million community posts were sent (equivalent to the amount of text in 770 in Time Magazine); 378,000 mobile phones were sold, which is higher than the global number of babies born daily at 371,000 ... It has become a resource for governments in various countries to compete for development. In 2012, US President Barack Obama announced an investment of 200 million US dollars to promote big data and related industries. And many domestic and foreign companies are also aware of the impact of big data and began to collect and analyze data, therefore, the era of big data has come.

3. Economy path of tourism industry under big data statistical analysis

3.1 The impact of big data on the tourism industry

In the era of such rapid development of big data, tourism welcomes opportunities first, and opportunities and challenges coexist. The development of traditional tourism is gradually replaced by the Internet-based. Online travel service agents are the first to face the difficulties. With the help of big data, online travel service agents first provide customers with a series of services such as hotels that are closer or cheaper than their tourist attractions, or hotels that better match their travel destinations. All of this makes tourism in a dominant position. Big data brings new changes to the tourism industry. For example, personalized recommendation of tourism products achieve the scenic spot, unmanned shopping experience, using the operator's data to achieve tourist flow analysis, using face technology recognition and management evading tickets, using crawler technology to achieve the scenic spot's word of mouth monitoring, using block chain technology to improve the tourism experience.

3.2 Tourists' new demand for tourism in the era of big data

China's tourism industry started relatively late, but it has developed rapidly. In recent years, the rapid development of information technology makes the tourism industry catch up. With the popularity of smart phones, people are more and more dependent on mobile app, such as Microblog, Wechat, Twitter and other applications. Almost everyone will run similar software, and directly query and pay tourism products online through these software. Ctrip's wireless client has been downloaded more than 5 million times. The smart client represented by Ctrip and Qunar has occupied a certain market share and deeply loved by consumers. According to the statistics of the "Forecasting and Investment Strategic Planning Analysis Report of China's Tourism Industry" released by the Institute of foresight industry, the total revenue of China's tourism industry in 2012 was only 2.58 trillion yuan, and the total revenue of China's tourism industry in 2013 increased to 4 trillion yuan. Then, following the guidance of the national leading department, the tourism industry will carry out the cultural tourism integration industry, integrate the cultural connotation of the scenic spots, expand the development space of quality tourism industry with the combination of cultural content, and promote the development of tourism economy with the supply side reform. In 2017, China's tourism industry continued to grow at a high speed, exceeding 5.4 trillion yuan, up 15.14% year-on-year. In 2018, the domestic tourism market continued to grow again, and the total income of the industry in the whole year reached 5.97 trillion yuan, up 10.56% year-on-year.

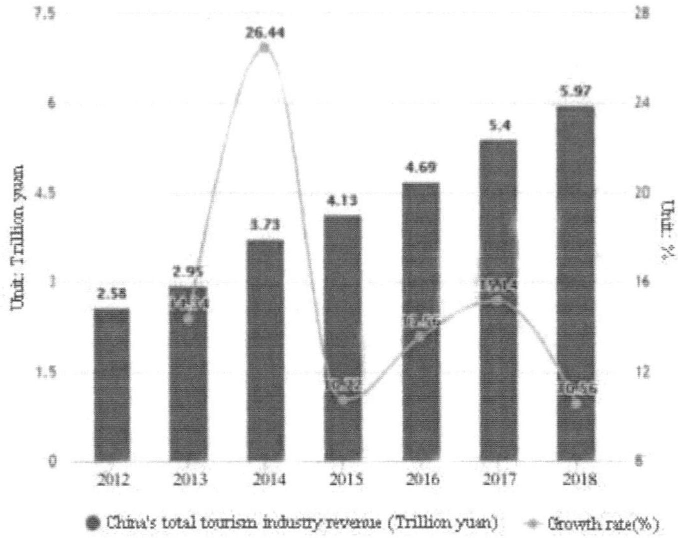

Figure 1 Statistics and growth of total revenue of Chin's tourism industry in 2012-2018

While people are enjoying the changes brought by the big data of information technology, the demand for tourism are also changing accordingly. More personalization are showed, specifically in the following aspects: first, with the development of the Internet as well as the popularity, people can access the Internet at any time, so tourists' travel needs are more independent. They no longer rely solely on travel and other agencies. Secondly, they are more independent to choose travel destinations. Problems encountered during travel will be solved as long as they search the Internet, which makes tourists gain more dependence. Taking tourism in Wuzhen from the city of Jiaxing as an example, a tourist returned from Wuzhen can write down what she saw and heard in there and the travel strategy, and put it on the Internet. If you plan to go to Wuzhen, you can search for these information strategies from the Internet and learn the experience. This can make your travel route more convenient. For example, the following itinerary can be used.

Day 1: Home-Hangzhou East Railway Station-Wuzhen

Day 2: Wuzhen-Xitang Ancient Town-Hangzhou

Day 3: Hangzhou-Qiandao Lake-Hangzhou

Day 4: Hangzhou-Hengdian-Hangzhou

Day 5: Hangzhou-West Lake-Pleasant Return

This is the convenience that big data brings to tourists, and a new travel option for tourists. Second, the interaction between tourists is more frequent. Nowadays, the development of big data on the Internet has enabled them to share their post-tourism experiences in online communities such as Weibo and Zhihu. For those who want to travel to Wuzhen, they can use this information to comprehensively consider itinerary routes, hotel reservations, special food and snack options, so that tourists can spend more reasonably and travel. Third, tourists more pursue spiritual tourism, not only superficial information synthesis, but also the best experience of traveling at the spiritual level.

3.3 New path analysis of tourism under big data

With the increasing development and popularization of the Internet, the traditional marketing mode should be changed. In the past, marketing was only focused on sales when the product was sold. It is more important to focus on a series of changes brought by comprehensive information compared to current sales. For example, people's experience in Wuzhen can be published on the platform by tourists. It may affect the demand of other tourists in the later period, so the marketing means must be consistent with the demand of tourists. Second, pay attention to the service and follow-up service in the process of tourism. People focus on the services obtained in tourism and the follow-up service.

Tourism is a way to relax, and people hope to experience another lifestyle. It is important to highlight the special services and emotional needs of tourists.

4. Big data statistics analysis of innovation path of industrial economy

4.1 The demand for cultural and creative products under big data

With the rapid development of Internet big data, people's demand for cultural and creative products is also increasing. Taking Wuzhen as an example, it has always been a place where scholars gather together and Mao Dun, the giant talent grew up and died. The background implication in the works written by Mao Dun, people want to experience this mood in person, so the experience of the former residence of Mao Dun in Wuzhen has been told. Combined with the local characteristics of water town and the mention of ancient buildings in film and TV plays and novels, the shooting location and other places can also be the reason for people to want to visit Wuzhen. Combined with this kind of cultural and creative products derived from local customs, such as Wujin with inch brocade and gold, Wuzhen Lake with exquisite craftsmanship, and Wuzhen cloth shoes adapted to the climate of Jiangnan water town, they will naturally become the best choice for souvenirs and gifts. Such cultural and creative products can better meet the needs of tourists' spiritual experience, and developers can also predict more needs based on tourists' recent needs.

4.2 Analysis on innovation path of cultural and creative enterprises under big data

With the rapid growth of the Internet, people are increasingly relying on the convenience brought by the Internet. In this context, cultural and creative enterprises should think about how to create products to meet the needs.

First, use the trend of Internet big data to re-plan the construction mode of cultural and creative enterprises. In recent years, with the changes in the market, many Internet cafes have targeted the industry of cultural and creative enterprises, so these changes have led people to believe that the operating mode will have great changes. Film, television, literature, and drama have made people see the integration of the Internet with these traditional companies, and their market share has increased year by year. Then, the joint cooperation between some traditional cultural and creative enterprises and Internet companies is an inevitable mode, so cultural and creative enterprises must also change the original traditional single enterprise mode.

Second, make use of big data analysis to make corresponding creative planning early publicity to guide consumption. Operators not only use the traces left by users browsing the Internet to track their consumption preference, but also integrate this to predict their consumption behavior in this period or in the next step. For example, "Sanzhenzhai" the favorite foods brand of tourists', bookmarks for fans of water villages, etc., may be more well-known through online marketing, which will attract more consumers. As more users have entered the Internet in recent years, the demand of cultural and creative products have become large. For example, the number of users of online has gradually increased, and the number of users of online games has also increased year by year. In addition, the Internet big data can track the personalized pursuit according to their browsing traces, and carry out targeted promotion, such as soft text marketing, game mall props to further enhance their consumer demand.

Third, use Internet big data to integrate the marketing mode of cultural and creative products. The bundled marketing of the Internet and traditional cultural and creative industries has become a trend. Cultural and creative enterprises should combine the advantages to develop business cooperation, establish network platforms, online and offline simultaneous marketing, two-dimensional and multi-dimensional cooperation. In addition, enterprises should keep pace with the times, formulate various marketing strategies, and improve the existing promotion mode. For example, Tao bao and Dou yin not only enhance the users' live experience of various cultural and creative products, but also expand customer group rapidly.

Fourth, it should be noted that the everyone can access the Internet and the information is completely transparent. Therefore, protecting the user's information security is an issue that must be considered. The originality of cultural and creative products must be guaranteed, the intellectual property rights of the original authors must be firmly maintained, and piracy must be resolutely resisted.

5. Conclusions:

The rapid growth of the Internet has made huge changes in many industry modes, especially tourism, cultural and creative industries, which have paid more attention to the integration of information technology. Therefore, it is more important to integrate the Internet' s big data according to the existing mode to drive the entire industry.

References:

[1] Chen Gaohuan. 2014. Analysis of the Impact of Big Data Application on China's Cultural Creative Industry [J]. Divine Land.
[2] Sun Yurong. 2014. Path Choice of Intellectual Property Protection of Cultural and Creative Industries in the Big Data Era [J]. Journal of Beijing Union University (Humanities and Social Sciences).
[3] Xiao Yanfei, Wang Yunyun, Wan Zijie. 2014. Influential Factors and Empirical Research on the Development of China's Cultural and Creative Industries [J]. Science and Technology Management Research.

AUTHOR INDEX

Institute of Physics
Dirac House, Temple Back
Bristol BS1 6BE UK

ISSN: 1757-8981
ISBN 978-1-7138-3510-3